PHARMACOGENOMICS OF HUMAN DRUG TRANSPORTERS

PHARMACOGENOMICS OF HUMAN DRUG TRANSPORTERS
Clinical Impacts

Edited by

TOSHIHISA ISHIKAWA
RIKEN Omics Science Center
Yokohama, Japan

RICHARD B. KIM
University of Western Ontario
London, Canada

JÖRG KÖNIG
Friedrich-Alexander-University
Erlangen-Nurenberg, Germany

WILEY

A JOHN WILEY & SONS, INC., PUBLICATION

Published by John Wiley & Sons, Inc., Hoboken, New Jersey.
Published simultaneously in Canada.

Library of Congress Cataloging-in-Publication Data

Pharmacogenomics of human drug transporters : clinical impacts / edited by Toshihisa Ishikawa, Richard B. Kim, Jörg König.
p. ; cm.
Includes bibliographical references and index.
ISBN 978-0-470-92794-6
I. Ishikawa, Toshihisa. II. Kim, Richard B. III. König, Jörg, 1965–
[DNLM: 1. Biological Transport–physiology. 2. Membrane Transport Proteins–physiology. 3. Drug Interactions. 4. Pharmacogenetics–methods. QU 120]
615.7–dc23

2013001312

10 9 8 7 6 5 4 3 2 1

CONTENTS

PREFACE vii

CONTRIBUTORS ix

CHAPTER 1 *INTRODUCTION TO PHARMACOGENOMICS OF DRUG TRANSPORTERS* 1

Marianne K. DeGorter and Richard B. Kim

CHAPTER 2 *ADME PHARMACOGENOMICS IN DRUG DEVELOPMENT* 13

Liangfu Chen and Joseph W. Polli

CHAPTER 3 *REGULATORY PERSPECTIVE ON PHARMACOGENOMICS OF DRUG-METABOLIZING ENZYMES AND TRANSPORTERS* 39

Lei Zhang, Gilbert J. Burckart, Lawrence J. Lesko and Shiew-Mei Huang

CHAPTER 4 *THE PHARMACOGENOMICS OF MEMBRANE TRANSPORTERS PROJECT* 73

Sook Wah Yee, Deanna L. Kroetz and Kathleen M. Giacomini

CHAPTER 5 *EMERGING NEW TECHNOLOGY OF SNP TYPING* 109

Toshihisa Ishikawa and Yoshihide Hayashizaki

CHAPTER 6 *OATP1A2, OAT1, AND OAT3* 125

Rommel G. Tirona

CHAPTER 7 *OATP1B1, OATP1B3, AND OATP2B1* 141

Jörg König and Martin F. Fromm

CHAPTER 8 *OCT (SLC22A) AND OCTN FAMILY* 171

Sophie L. Stocker, Arian Emami Riedmaier, Matthias Schwab and Kathleen M. Giacomini

CHAPTER 9 *MATE (SLC47) FAMILY* 209

Atsushi Yonezawa and Ken-ichi Inui

CHAPTER 10 *PEPT (SLC15A) FAMILY* 223

Tomoko Sugiura, Saki Umeda, Akira Tsuji and Yukio Kato

CHAPTER 11 *NUCLEOSIDE TRANSPORTERS (SLC28 AND SLC29) FAMILY* 243

Míriam Molina-Arcas and Marçal Pastor-Anglada

CHAPTER 12 *P-GLYCOPROTEIN (MDR1/ABCB1)* 271

Ingolf Cascorbi

CHAPTER 13 *BSEP (ABCB11)* 295

Bruno Stieger and Gerd A. Kullak-Ublick

CHAPTER 14 *BCRP (ABCG2)* 311

Agnes Basseville, Susan E. Bates, William D. Figg and Alex Sparreboom

CHAPTER 15 *MRP2 (ABCC2) AND MRP3 (ABCC3)* 345

Anne T. Nies

CHAPTER 16 *MRP4 (ABCC4)* 365

Satish B. Cheepala, Mugdha Sukthankar and John D. Schuetz

CHAPTER 17 *MRP8 (ABCC11)* 387

Yu Toyoda and Toshihisa Ishikawa

CHAPTER 18 *FUTURE PERSPECTIVES* 401

Toshihisa Ishikawa and Joseph Ware

INDEX 417

PREFACE

Interindividual variability in drug response and the often unpredictable adverse drug reactions that result from such variability are critical issues during drug development as well as in clinical pharmacotherapy. Accumulating evidence strongly suggests that genetic polymorphisms in drug-metabolizing enzymes, transporters, receptors, and other drug targets can often be the molecular and mechanistic basis for the observed interindividual differences, efficacy, as well as toxicity to a number of clinically important medications. During the past two decades, the role of carrier-mediated transport in determining the pharmacokinetics of drugs has become increasingly evident with the discovery of genetic variants that affect expression, localization, and/or function of a given drug transporter. For example, it is well documented that genetic polymorphisms in the *SLCO1B1* gene encoding the hepatic uptake transporter OATP1B1 may cause statin-induced myopathy.

Drug transporters are expressed at numerous epithelial barriers, such as intestinal epithelial cells, hepatocytes, renal tubular cells, and the blood–brain barrier, and aid in the vectorial movement of substrate drugs. There has been remarkable progress in the field of pharmacogenomics and there is far greater understanding of the inherited nature of differences that underlie pharmacokinetic and pharmacodynamic effects. Accordingly, the progress in this area has aided drug discovery and provides a stronger scientific basis for optimizing drug therapy tailored to a patient's genetic makeup. Therefore, pharmacogenomics of drug transporters will lead to a better understanding and prediction of drug interactions and adverse drug reactions.

Common genetic variations, frequently referred to as single-nucleotide polymorphisms (SNPs), can in some cases result in a significant variation in response to drugs; validation of clinically important genetic polymorphisms and new technologies capable of rapidly detecting clinically important variants are key issues for advancing the emerging field of personalized medicine. Pharmacogenomics data, when linked to robust *in vitro* and *in vivo* data, can serve as a predictive biomarker that enhance our understanding of the mechanistic basis of the observed variability in drug response, and thereby improve safety and efficacy of drug therapy for individual patients. The pharmaceutical regulatory authorities are now requesting that drug companies integrate genetic information into their clinical trial design. Therefore, a book such as this, covering new and emerging aspects of drug transporter pharmacogenomics has the potential to serve as a valuable reference and guide for researchers in industry and academia.

Accordingly, for this book we have focused on the following areas of interest in relation to drug transporter pharmacogenomics: (1) provide an overview on pharmacogenomics of drug transporters; (2) introduce emerging new technologies for rapid detection of genetic polymorphisms; (3) review the clinical aspect of genetic polymorphisms in major drug transporter genes; and (4) discuss future research directions of drug transporter pharmacogenomics. This book consists of 18 chapters, authored by world-leading scientists who describe the latest knowledge and new developments in the field. The breadth of information provided in this book may be of particular interest to pharmaceutical scientists, biochemists, pharmacologists, genetics and genomics researchers, as well as clinician scientists.

As editors, we are grateful to all of the authors who have contributed to this book for their time and dedication. In addition, we thank Jonathan Rose and Amanda Amanullah for their constant support and valuable advice from the planning to publication of this book.

Toshihisa Ishikawa
Richard B. Kim
Jörg König

CONTRIBUTORS

Agnes Basseville Medical Oncology Branch, Center for Cancer Research, National Cancer Institute, NIH, Bethesda, MD

Susan E. Bates Medical Oncology Branch, Center for Cancer Research, National Cancer Institute, NIH, Bethesda, MD

Gilbert J. Burckart Office of Clinical Pharmacology, Office of Translational Sciences, Center for Drug Evaluation and Research, Food and Drug Administration, Silver Spring, MD

Ingolf Cascorbi Institute of Experimental and Clinical Pharmacology, University Hospital Schleswig-Holstein, Kiel, Germany

Satish B. Cheepala Department of Pharmaceutical Sciences, St. Jude Children's Research Hospital, Memphis, TN

Liangfu Chen Preclinical Drug Metabolism and Pharmacokinetics, GlaxoSmithKline, Inc., King of Prussia, PA

Marianne K. DeGorter Division of Clinical Pharmacology, Department of Medicine, Department of Physiology and Pharmacology, University of Western Ontario, London, Canada

Arian Emami Riedmaier Dr. Margarete Fischer-Bosch-Institute of Clinical Pharmacology, Stuttgart, Germany

William D. Figg Clinical Pharmacology Program, Center for Cancer Research, National Cancer Institute, NIH, Bethesda, MD

Martin F. Fromm Institute of Experimental and Clinical Pharmacology and Toxicology, Clinical Pharmacology and Clinical Toxicology, Friedrich-Alexander-Universität Erlangen-Nürnberg, Erlangen, Germany

Kathleen M. Giacomini Department of Bioengineering and Therapeutic Sciences, University of California, San Francisco, CA

Yoshihide Hayashizaki Omics Science Center, RIKEN Yokohama Institute, Yokohama, Japan

Shiew-Mei Huang Office of Clinical Pharmacology, Office of Translational Sciences, Center for Drug Evaluation and Research, Food and Drug Administration, Silver Spring, MD

Ken-ichi Inui Department of Pharmacy, Kyoto University Hospital, Faculty of Medicine, Kyoto University, Kyoto Pharmaceutical University, Yamashina-ku, Kyoto, Japan

Toshihisa Ishikawa Omics Science Center, RIKEN Yokohama Institute, Yokohama, Japan

Yukio Kato Faculty of Pharmacy, Kanazawa University, Kakuma-machi, Kanazawa, Japan

Richard B. Kim Division of Clinical Pharmacology, Department of Medicine, Department of Physiology and Pharmacology, University of Western Ontario, London, Canada

Jörg König Institute of Experimental and Clinical Pharmacology and Toxicology, Clinical Pharmacology and Clinical Toxicology, Friedrich-Alexander-Universität Erlangen-Nürnberg, Erlangen, Germany

Deanna L. Kroetz Department of Bioengineering and Therapeutic Sciences, University of California, San Francisco, CA

Gerd A. Kullak-Ublick Department of Clinical Pharmacology and Toxicology, University Hospital, Zurich, Switzerland

Lawrence J. Lesko Office of Clinical Pharmacology, Office of Translational Sciences, Center for Drug Evaluation and Research, Food and Drug Administration, Center for Pharmacometrics and Systems Pharmacology, University of Florida at Lake Nona, Orlando, FL

Míriam Molina-Arcas Cancer Research UK London Research Institute, London, United Kingdom

Anne T. Nies Dr. Margarete Fischer-Bosch Institute of Clinical Pharmacology, Stuttgart, Germany, and University of Tübingen, Germany

Marçal Pastor-Anglada Department of Biochemistry and Molecular Biology, Institute of Biomedicine (IBUB) and CIBER EHD, University of Barcelona, Barcelona, Spain

Joseph W. Polli Preclinical Drug Metabolism and Pharmacokinetics, GlaxoSmithKline, Inc., Research Triangle Park, NC

John D. Schuetz Department of Pharmaceutical Sciences, St. Jude Children's Research Hospital, Memphis, TN

Matthias Schwab Dr. Margarete Fischer-Bosch-Institute of Clinical Pharmacology, Stuttgart, Germany and Department of Clinical Pharmacology, Institute of Experimental and Clinical Pharmacology and Toxicology, University Hospital, Tuebingen, Germany

Alex Sparreboom Department of Pharmaceutical Sciences, St Jude Children's Research Hospital, Memphis, TN

Bruno Stieger Department of Clinical Pharmacology and Toxicology, University Hospital, Zurich, Switzerland

Sophie L. Stocker Department of Bioengineering and Therapeutic Sciences, University of California, San Francisco, CA

Tomoko Sugiura Faculty of Pharmacy, Kanazawa University, Kakuma-machi, Kanazawa, Japan

Mugdha Sukthankar Department of Pharmaceutical Sciences, St. Jude Children's Research Hospital, Memphis, TN

Rommel G. Tirona Department of Physiology and Pharmacology, Department of Medicine, Schulich School of Medicine and Dentistry, The University of Western Ontario, Ontario, Canada

Yu Toyoda Graduate School of Bioscience and Biotechnology, Tokyo Institute of Technology, Yokohama, Japan

Akira Tsuji Faculty of Pharmacy, Kanazawa University, Kakuma-machi, Kanazawa, Japan

Saki Umeda Faculty of Pharmacy, Kanazawa University, Kakuma-machi, Kanazawa, Japan

Joseph Ware Clinical Pharmacology, Genentech, San Francisco, CA

Atsushi Yonezawa Department of Pharmacy, Kyoto University Hospital, Faculty of Medicine, Kyoto University, Kyoto, Japan

Sook Wah Yee Department of Bioengineering and Therapeutic Sciences, University of California, San Francisco, CA

Lei Zhang Office of Clinical Pharmacology, Office of Translational Sciences, Center for Drug Evaluation and Research, Food and Drug Administration, Silver Spring, MD

CHAPTER 1

INTRODUCTION TO PHARMACOGENOMICS OF DRUG TRANSPORTERS

Marianne K. DeGorter
Richard B. Kim

1.1 INTRODUCTION

Understanding the molecular mechanisms and clinical relevance of interindividual variability in drug response remains an important challenge. Pharmacogenomics, the study of genetic variation in the genes that influence drug effect, can provide insight into interindividual variability and a more accurate prediction of drug response than may be obtained by relying solely on a patient's clinical information. The goal of drug transporter pharmacogenomics is to understand the impact of genetic variation on the function of transporters that interact with medications. For many drugs in clinical use, transporters are important determinants of absorption, tissue accumulation, and elimination from the body, and thereby transporters significantly influence drug efficacy and toxicity. Adverse drug reactions can result from toxicity associated with high drug concentrations and lack of efficacy can result from subtherapeutic drug exposure. By understanding the genetic basis for drug transporter activity, it will be possible to enhance a predictive approach to individualization of drug therapy.

The purpose of this book is to highlight the advances in transporter pharmacogenomics that have been made since polymorphisms in drug transporter genes were first described in the late 1990s (Mickley et al., 1998; Hoffmeyer et al., 2000; Kim et al., 2001). As we enter the genomic era of medicine, pharmacogenomics will inform prescribing practices to maximize drug efficacy while minimizing risk for toxicity. Given the importance of transporters to the absorption, distribution, and elimination of many drugs, there is no doubt that transporter pharmacogenomics will make significant contributions to this aim.

Pharmacogenomics of Human Drug Transporters: Clinical Impacts, First Edition.
Edited by Toshihisa Ishikawa, Richard B. Kim and Jörg König

1.2 OVERVIEW OF DRUG TRANSPORTERS

Membrane transporters have diverse and important roles in maintaining cellular homeostasis by the uptake and efflux of endogenous compounds to regulate solute and fluid balance, facilitate hormone signaling, and extrude potential toxins. Drug transport proteins are a functional subset of membrane transporters that also interact with drugs and their metabolites. Compounds that are most likely to rely on carrier mechanisms are polar and bulky, and less likely to pass through cell membranes by simple diffusion. Transporter substrates include numerous drugs, their hydroxylated metabolites, and the glutathione-, sulfate-, or glucuronide-conjugated products of Phase II metabolism. Transporters that are expressed in the epithelia of intestine, liver, and kidney are of particular importance for vectorial or directional movement of drugs, resulting in efficient and rapid drug absorption, distribution, metabolism, and elimination. Moreover, expression of drug transporters on the basolateral versus apical domain of polarized epithelial cells in organs such as the intestine and liver may also be critical for a drug to enter the tissue and interact with its target (Ho and Kim, 2005; Giacomini et al., 2010).

Membrane transporters are comprised of multiple transmembrane domains (TMDs) that form a pore in the membrane through which the substrates pass. These domains are joined by alternating intracellular and extracellular loops which, together with TMDs, facilitate substrate recognition, binding, and translocation. The functional mechanism and conformational changes required for transport are not completely understood, and remain an active area of investigation (Kerr et al., 2010). Of particular interest to transporter pharmacogenomics is the ability to predict the functional effect of novel mutations that are discovered in individual genomes.

Drug transporters belong to two major classes, the solute carrier (SLC) superfamily and ATP-binding cassette (ABC) superfamily. In the human genome, there are 350 transporters in the SLC superfamily and 48 ABC transporters; these transporters are divided into subfamilies based on sequence homology (Giacomini et al., 2010). ABC transporters are distinguished by the presence of an intracellular nucleotide-binding domain that catalyzes the hydrolysis of ATP to generate the energy required to transport substrates against their concentration gradient (Schinkel and Jonker, 2003). In contrast, SLC transporters utilize facilitated diffusion, ion coupling, or ion exchange to translocate their substrates. In some cases, transport relies on an ion gradient that is actively maintained by ABC transporters (Hediger et al., 2004).

Transporter function may be influenced by multiple factors, and interindividual variability in transporter function is now recognized as a major source of variability in drug disposition and response. We know that drug transporters can be inhibited by numerous compounds, typically by competition for recognition and binding, resulting in unexpected pharmacokinetics of substrate drugs, and drug–drug interactions. Genetic variants may also affect transporter function, and, in recent years, the discovery of genetic variation in drug transporters has opened up an area of research in transporter pharmacogenomics (Giacomini et al., 2010; Sissung et al., 2010a).

1.3 OVERVIEW OF PHARMACOGENOMICS

The study of inherited differences in drug response dates back to observations made in the 1950s; in the late 1980s, molecular advances provided a mechanistic explanation for these findings (Evans and Relling, 1999; Weinshilboum and Wang, 2006). Many early achievements in pharmacogenetics were in cytochrome P450 (CYP) drug-metabolizing enzyme research and the effect of genetic variation in these enzymes on metabolite concentrations. Pharmacogenomics studies have benefited from having well-defined phenotypes: a pharmacokinetic measure such as the plasma or urine concentration of a drug or its metabolite, or a measure of drug response, such as a change in blood pressure or heart rate. For monogenic traits, this approach has led to new insights into our understanding of the factors underlying drug disposition and response, and provided a solid foundation to study traits that are influenced by multiple genes and other clinical factors. Today, pharmacogenomics encompasses a broad spectrum of genes involved in metabolism as well as transport, and in drug targets and related pathways (Sim and Ingelman-Sundberg, 2011).

Genetic variants include single-nucleotide polymorphisms (SNPs), which are typically present in less than 1% of the population, while more rare variants are considered to be genetic mutations. SNPs in the coding regions of proteins may be classified as synonymous or nonsynonymous, depending on whether the amino acid sequence is altered in the variant allele. SNPs may also come in the form of small insertions or deletions, which result in frameshift of amino acid sequence or premature truncation of the protein, and likely a nonfunctional product (Urban et al., 2006). Duplication or deletion of larger regions of genomic sequence (>50 bp) are classified as copy number or structural variants (Alkan et al., 2011). A classic example of copy-number variation comes from the field of pharmacogenomics: CYP2D6 is commonly duplicated or deleted, resulting in profound differences in the rate of metabolism of its substrate drugs in individuals with these alleles (Zanger et al., 2004). There is a growing appreciation for the importance of structural differences as a source of variation in the human genome, and further study of this variation, as it relates to transporter genes, is expected (Alkan et al., 2011).

Pharmacogenomic information may be used to predict treatment outcomes and choose the best drug and its optimal dose. Pharmacogenomics may also be used to predict a patient's risk for an adverse drug reaction, including drug–drug interactions that may be more severe due to the genetics of the proteins involved. At the time of writing, the US Food and Drug Administration (FDA) listed nearly 80 drugs for which pharmacogenomic biomarkers in over 30 genes were included in some part of the label recommendations. To date, the FDA has focused on drug-metabolizing enzymes and target proteins; however, transporter genes are expected to be added in the future, following the work of the International Transporter Consortium, sponsored by the FDA's Critical Path Initiative (Giacomini et al., 2010).

1.4 PHARMACOGENOMICS OF DRUG TRANSPORTERS

Transporter polymorphisms may increase or reduce an individual's overall exposure to a substrate, depending on the tissue expression and localization of the transporter. For example, reduced function of an uptake transporter on the luminal membrane of the intestine would result in reduced systemic exposure of its substrate, whereas reduced function of an uptake transporter on the basolateral membrane of the liver or kidney may result in increased systemic exposure if the drug in question relies on these organs for its elimination. On the other hand, reduced function of an ABC efflux transporter present on the luminal membrane of the intestine will result in increased plasma concentration of the substrate drug, as less drug is returned to the intestinal lumen by the transporter. In some cases, the precise *in vivo* contribution of a transporter may be difficult to define, particularly if the transporter is present in multiple tissues, or has overlapping function with transporters of similar expression patterns. The extent of phenotypic variation observed will depend on how much the substrate relies on the single transporter in question, and the extent of genetic variation present in the other transporters, metabolizing enzymes and targets that interact with the drug.

To date, the best studied transporter polymorphisms have been those in the coding regions of transporter genes. Some variants cause reduced trafficking of the transporter to the cell membrane, resulting from incorrect folding or an inability to interact with molecular chaperones, and other variants may affect substrate recognition or binding. Certain amino acid changes, particularly in substrate binding regions, have been shown to alter transport in a substrate-specific fashion, making it difficult to fully predict the effect of a polymorphism on transport of a particular compound without testing that compound directly. Although numerous polymorphisms in transporter genes have been identified, not all polymorphisms appear to affect transporter function. One method to test the function of a SNP is to express its protein product and measure its transport function *in vitro*. Of the 88 protein-altering variants studied in 11 SLC transporters, 14% had decreased or total loss of functional activity in *in vitro* assays (Urban et al., 2006). This is likely an underestimation, due to the possibility of substrate-specific differences in effect.

Analysis of large numbers of SNPs in the coding regions of transporters demonstrated that genetic diversity is significantly higher in loop domains compared to TMDs, suggesting that there is selective pressure against amino acid changes in these regions (Leabman et al., 2003). Polymorphisms may also occur in intronic regions, affecting splicing, or in promoter and enhancer regions, affecting RNA expression. Analysis of proximal promoter region variation showed that SLC transporter promoters are more likely to contain variants than ABC transporter promoters, and highly active promoters are more likely to contain variants than less active ones (Hesselson et al., 2009). Genetic diversity in transporter genes also appears to be related to ethnicity. In a study of 680 SNPs identified from samples representing five ethnic populations, only 83 SNPs were present in all the five populations (Leabman et al., 2003). Thus, differences in transporter polymorphism frequency may account for some variability in drug response observed across ethnicities.

The pharmacogenomics of SLC transporters of particular importance to drug transport are described in the following chapters: organic anion transporting polypeptides (OATPs/SLCO; Chapters 6 and 7), organic anion transporters (OATs/SLC22A; Chapter 6), organic cation transporters (OCTs/SLC22A; Chapter 8), organic cation and carnitine transporters (OCTN/SLC22A; Chapter 8), multidrug and toxin extrusion transporters (MATEs/SLC47; Chapter 9), peptide transporters (PEPTs/SLC15A; Chapter 10), and nucleoside transporters (NTs/SLC28 and SLC29; Chapter 11). The pharmacogenomics of ABC transporters important to drug transport are covered in subsequent chapters: P-glycoprotein (ABCB1; Chapter 12), bile salt export pump (BSEP/ABCB11; Chapter 13), breast cancer resistance protein (BCRP/ABCG2; Chapter 14), multidrug resistance-associated proteins MRP2 (ABCC2; Chapter 15), MRP3 (ABCC3; Chapter 15), MRP4 (ABCC4; Chapter 16), and MRP8 (ABCC11; Chapter 17). Significant advances in transporter biology and pharmacogenomics of transporters have been made through the contributions of individual labs as well as large multi-investigator projects such as the Pharmacogenomics of Membrane Transporters project, funded by the National Institutes of Health as part of the Pharmacogenomics Research Network, and described in Chapter 4 (Kroetz et al., 2010).

1.5 TECHNIQUES TO STUDY DRUG TRANSPORTER FUNCTION

The application of advances in molecular biology techniques to the study of transporters over the last 20 years has made a significant contribution to our understanding of transporter biology. *In vitro*, transporter activity is often characterized in primary cells and in expression systems, including transiently and stably transfected cultured human cell lines, inside-out membrane vesicles, and insect cells. One challenge to studying transporters *in vivo* is the overlapping substrate specificity and tissue distribution of many transporters, which can lead to difficulties in the precise identification of the transporter(s) responsible for a particular effect. Knockout mouse models of transporters have proven to be useful to delineate the contribution of certain transporters to drug disposition (DeGorter and Kim, 2011). Knockout mice exist for many of the SLC and ABC transporters, and double and triple ABC transporter knockout models have been used to characterize the contribution of multiple transporters with overlapping substrate specificities (Keppler, 2011). It is important to bear in mind that there are species-related differences in transporter expression and substrate specificity that may make it difficult to interpret and extrapolate the results obtained in mice to the human situation. The relative contribution of a given transporter *in vivo* has also been examined by drug-specific pharmacokinetic and pharmacodynamic studies in individuals with and without polymorphisms in the transporter gene of interest.

In the last decade, the field of genomics has developed rapidly, with the sequencing of the human genome (International Human Genome Sequencing Consortium,

2001; Venter et al., 2001) and subsequent efforts to determine haplotype structure by the HapMap project (The International HapMap Consortium, 2007), and sequence variation by the 1000 Genomes project (1000 Genomes Project Consortium, 2010). Genome-wide association studies (GWAS) incorporating clinical and genetic data have been widely used to identify genetic variants that predict risk for disease and also to assess drug response or toxicity. For pharmacogenomics studies, GWAS offer to identify candidate genes unrelated to our current knowledge of drug mechanism (Motsinger-Reif et al., 2010).

Methods for detecting transporter polymorphisms and predicting the functional consequences of unique polymorphisms in real time will be required to use pharmacogenomics in the clinical setting. To address this need, genotyping platforms for a focused set of important pharmacogenetic genes are being developed for clinical use (Sissung et al., 2010b). QSAR and molecular dynamics simulations are in silico approaches that are active areas of research aimed at addressing this challenge of SNP prediction (Ishikawa et al., 2010); see Chapters 5 and 18 for emerging technologies with applications to transporter pharmacogenomics.

1.6 TRANSPORTER PHARMACOGENOMICS IN DRUG DISCOVERY AND DEVELOPMENT

An understanding of transporter pharmacogenomics is important for the design and development of new drugs that are safe and effective. Transporters interacting with drug candidates may be identified during the preclinical stage of drug development, taking into consideration the limitations inherent to extrapolating *in vitro* and animal data to predict human response. For this reason, pharmacogenomic studies in later phases of drug development and postmarketing surveillance are crucial to identify potential transporter-mediated drug interactions, and individuals with transporter polymorphisms who may require dose adjustment or an alternative compound (Stingl Kirchheiner and Brockmoller, 2011). The International Transporters Consortium is a group of academic, industry, and regulatory leaders formed to create guidelines for the systematic inclusion of transporter studies in the drug development and approval process (Giacomini et al., 2010). Transporter pharmacogenomics and the role of diagnostic tests to support the clinical use of pharmacogenomics is discussed in Chapter 2, and a regulatory perspective on the contribution of drug transporters to the drug development process is provided in Chapter 3.

1.7 CLINICAL IMPLICATIONS OF TRANSPORTER PHARMACOGENOMICS

As our understanding of transporter pharmacogenomics matures, and pharmacogenomics technologies are more widely adopted in the clinic, transporter genomics could be used to select an appropriate dose, or the best medication from a particular class of compounds, and identify those individuals who may be at increased risk for an adverse drug reaction. Transporters that affect drug response are numerous and diverse in their effect; key examples from the SLC and ABC superfamilies are summarized in Tables 1.1 and 1.2, respectively.

P-glycoprotein is an example of an efflux transporter that can significantly limit the accumulation of its substrates in certain tissues. The expression of P-glycoprotein at the blood–brain barrier prevents the CNS accumulation of drugs such as protease inhibitors, and its overexpression in cancer cells is associated with a multidrug-resistant phenotype (Cascorbi, 2011). Genetic variants in the cation transporter OCT1 (SLC22A1) have been associated with reduced efficacy of metformin, an antidiabetic drug that targets the liver as its site of action (Shu et al., 2007). The OATP1B1 (SLCO1B1) polymorphism c.521T>C has been associated with increased risk for statin-induced muscle toxicity (Link et al., 2008) and genotyping patients for this variant has been proposed to identify those at greater risk for side effects (Niemi, 2010).

Transporter pharmacogenomics have not yet been widely used in a clinical setting. Moving forward, studies are needed to show that the risk–benefit ratio of a drug is improved by pharmacogenomic testing, and some efforts are being made to determine the key components to be included in pharmacoeconomic evaluations of pharmacogenomic tests (Beaulieu et al., 2010). As sequencing becomes more cost-efficient, the possibility of sequencing relevant genes or even genomes in a clinical setting poses a new challenge of interpreting pharmacogenomic information on an individual level (Ashley et al., 2010).

Finally, it is important to bear in mind that many factors contribute to variability in drug responsiveness, including renal and hepatic functions, underlying disease processes, and drug interactions. At the end of the day, a patient's actual drug-response phenotype, in terms of efficacy and toxicity, is the key clinically relevant endpoint, and pharmacogenomics should be integrated with other parameters such as drug levels, biomarkers, and measures of drug response in order to provide truly personalized medicine.

1.8 CONCLUSION

Genetic variation in transporters contributes significantly to observed interindividual variability in drug response. In future, systematic inclusion of drug transporter studies that include genetic variation, whether affecting transporter function or expression, will be essential to the development of drugs that are safe and effective. There is little doubt that drug transporter pharmacogenomics is expanding rapidly and new insights will continue to inform improved drug prescribing and thereby enhance the delivery of optimal medical care.

ACKNOWLEDGMENTS

This work was supported by grants from the Canadian Institutes of Health Research (MOP-89753) (RBK). RBK is the Medical Research Chair of Pharmacogenomics at the University of Western Ontario. MKD is the recipient of a Vanier Canada Graduate Scholarship from the Canadian Institutes of Health Research.

TABLE 1.1 Drug Transporters of the Solute Carrier Superfamily

Transporter (gene)	Tissue(s) of predominant expression in humans	Key drug substrates	Key inhibitors	SNPs associated with drug response	References
OCT1 (SLC22A1)	Hepatocyte (basolateral)	Metformin, oxaliplatin	Quinine	Multiple associated with metformin response	Nies et al. (2011)
OCT2 (SLC22A2)	Renal proximal tubule (basolateral)	Metformin, oxaliplatin	Cimetidine	None to date	Nies et al. (2011)
MATE1 (SLC47A1)	Hepatocyte (canalicular membrane); renal proximal tubule (luminal)	Cimetidine, metformin, procainamide	Cimetidine, pyrimethamine	Possibly rs2289669	Nies et al. (2011)
OAT1 (SLC22A6)	Renal proximal tubule (basolateral)	Acyclovir	Probenecid, NSAIDs	None to date	Burckhardt and Burckhardt (2011)
OAT3 (SLC22A8)	Renal proximal tubule (basolateral)	NSAIDs, furosemide	Probenecid, NSAIDs	None to date	Burckhardt and Burckhardt, 2011
OATP1B1 (SLCO1B1)	Hepatocyte (basolateral)	Statins, repaglinide	Rifampicin, gemfibrozil cyclosporine	c.521T>C (rs4149056)	Niemi et al. (2011)
OATP1B3 (SLCO1B3)	Hepatocyte (basolateral)	Statins, taxanes	Rifampicin, cyclosporine	Possibly c.334T>G (rs4149117)	Kalliokoski and Niemi (2009)
OATP2B1 (SLCO2B1)	Hepatocyte (basolateral); enterocyte (luminal)	Statins, fexofenadine	Cyclosporine	Possibly c.935G>A (rs12422149)	Kalliokoski and Niemi (2009)

TABLE 1.2 Drug Transporters of the ATP-binding Cassette Superfamily

Transporter (gene)	Tissue(s) of predominant expression in humans	Key drug substrates	Key inhibitors	SNPs associated with drug response	References
P-gp (ABCB1)	Hepatocyte (canalicular); enterocyte (luminal); blood–brain barrier	HIV protease inhibitors, antineoplastics	Cyclosporine, verapamil	Possibly c.3435T>C (rs1045642)	Cascorbi (2011)
BSEP (ABCB11)	Hepatocyte (canalicular)	Pravastatin	Cyclosporine, glibenclamide	Possibly p.V444A (rs2287622)	Klaassen and Aleksunes (2010); Stieger (2011)
MRP2 (ABCC2)	Hepatocyte (canalicular)	B-lactam antibiotics, methotrexate, multiple Phase II conjugates	Cyclosporine	None to date	Klaassen and Aleksunes (2010)
MRP4 (ABCC4)	Hepatocyte (basolateral); renal proximal tubule (luminal)	Nucleoside-based antivirals, methotrexate, topotecan	Dipyridamole, losartan	Possibly p.E757K (rs3765534)	Russel et al. (2008)
BCRP (ABCG2)	Hepatocyte (canalicular); enterocyte (luminal); blood–brain barrier	Statins, antineoplastics	Dipyridamole, cyclosporine	c.421C>A (rs2231142)	Schwabedissen and Kroemer (2011)

REFERENCES

1000 Genomes Project Consortium (2010) A map of human genome variation from population-scale sequencing. *Nature*, 467, 1061–1073.

Alkan C, Coe BP, Eichler EE (2011) Genome structural variation discovery and genotyping. *Nat Rev Genet*, 12, 363–376.

Ashley EA, Butte AJ, Wheeler MT, Chen R, Klein TE, Dewey FE, Dudley JT, Ormond KE, Pavlovic A, Morgan AA, et al. (2010) Clinical assessment incorporating a personal genome. *Lancet*, 375, 1525–1535.

Beaulieu M, de Denus S, Lachaine J (2010) Systematic review of pharmacoeconomic studies of pharmacogenomic tests. *Pharmacogenomics*, 11, 1573–1590.

Burckhardt G, Burckhardt BC (2011) In vitro and in vivo evidence of the importance of organic anion transporters (OATs) in drug therapy. In: *Handbook of Experimental Pharmacology*, Vol. 201. Heidelberg, Germany: Springer. pp 29–104.

Cascorbi I (2011) P-glycoprotein: tissue distribution, substrates, and functional consequences of genetic variations. In: *Handbook of Experimental Pharmacology*, Vol. 201. Heidelberg, Germany: Springer. pp 261–283.

DeGorter MK, Kim RB (2011) Use of transgenic and knockout mouse models to assess solute carrier transporter function. *Clin Pharmacol Ther*, 89, 612–616.

Evans WE, Relling MV (1999) Pharmacogenomics: translating functional genomics into rational therapeutics. *Science*, 286, 487–491.

Giacomini KM, Huang SM, Tweedie DJ, Benet LZ, Brouwer KL, Chu X, Dahlin A, Evers R, Fischer V, Hillgren KM, et al. (2010) Membrane transporters in drug development. *Nat Rev Drug Discov*, 9, 215–236.

Hediger MA, Romero MF, Peng JB, Rolfs A, Takanaga H, Bruford EA (2004) The ABCs of solute carriers: physiological, pathological and therapeutic implications of human membrane transport proteins. *Pflugers Arch*, 447, 465–468.

Hesselson SE, Matsson P, Shima JE, Fukushima H, Yee SW, Kobayashi Y, Gow JM, Ha C, Ma B, Poon A, et al. (2009) Genetic variation in the proximal promoter of ABC and SLC superfamilies: liver and kidney specific expression and promoter activity predict variation. *PLoS One*, 4, e6942.

Ho RH, Kim RB (2005) Transporters and drug therapy: implications for drug disposition and disease. *Clin Pharmacol Ther*, 78, 260–277.

Hoffmeyer S, Burk O, von Richter O, Arnold HP, Brockmoller J, Johne A, Cascorbi I, Gerloff T, Roots I, Eichelbaum M, Brinkmann U (2000) Functional polymorphisms of the human multidrug-resistance gene: multiple sequence variations and correlation of one allele with P-glycoprotein expression and activity in vivo. *Proc Natl Acad Sci U S A*, 97, 3473–3478.

International Human Genome Sequencing Consortium (2001) Initial sequencing and analysis of the human genome. *Nature*, 409, 860–921.

Ishikawa T, Sakurai A, Hirano H, Lezhava A, Sakurai M, Hayashizaki Y (2010) Emerging new technologies in pharmacogenomics: rapid SNP detection, molecular dynamic simulation, and QSAR analysis methods to validate clinically important genetic variants of human ABC transporter ABCB1 (P-gp/MDR1). *Pharmacol Ther*, 126, 69–81.

Kalliokoski A, Niemi M (2009) Impact of OATP transporters on pharmacokinetics. *Br J Pharmacol*, 158, 693–705.

Keppler D (2011) Multidrug resistance proteins (MRPs, ABCCs): importance for pathophysiology and drug therapy. In: *Handbook of Experimental Pharmacology*, Vol. 201. Heidelberg, Germany: Springer. pp 299–323.

Kerr ID, Jones PM, George AM (2010) Multidrug efflux pumps: the structures of prokaryotic ATP-binding cassette transporter efflux pumps and implications for our understanding of eukaryotic P-glycoproteins and homologues. *FEBS J*, 277, 550–563.

Kim RB, Leake BF, Choo EF, Dresser GK, Kubba SV, Schwarz UI, Taylor A, Xie HG, McKinsey J, Zhou S, Lan LB, Schuetz JD, Schuetz EG, Wilkinson GR (2001) Identification of functionally variant MDR1 alleles among European Americans and African Americans. *Clin Pharmacol Ther*, 70, 189–199.

Klaassen CD, Aleksunes LM (2010) Xenobiotic, bile acid, and cholesterol transporters: function and regulation. *Pharmacol Rev*, 62, 1–96.

Kroetz DL, Yee SW, Giacomini KM (2010) The pharmacogenomics of membrane transporters project: research at the interface of genomics and transporter pharmacology. *Clin Pharmacol Ther*, 87, 109–116.

Leabman MK, Huang CC, DeYoung J, Carlson EJ, Taylor TR, de la Cruz M, Johns SJ, Stryke D, Kawamoto M, Urban TJ, Kroetz DL, Ferrin TE, Clark AG, Risch N, Herskowitz I, Giacomini KM (2003) Natural variation in human membrane transporter genes reveals evolutionary and functional constraints. *Proc Natl Acad Sci U S A*, 100, 5896–5901.

Link E, Parish S, Armitage J, Bowman L, Heath S, Matsuda F, Gut I, Lathrop M, Collins R (2008) SLCO1B1 variants and statin-induced myopathy–a genomewide study. *N Engl J Med*, 359, 789–799.

Mickley LA, Lee JS, Weng Z, Zhan Z, Alvarez M, Wilson W, Bates SE, Fojo T (1998) Genetic polymorphism in MDR-1: a tool for examining allelic expression in normal cells, unselected and drug-selected cell lines, and human tumors. *Blood*, 91, 1749–1756.

Motsinger-Reif AA, Jorgenson E, Relling MV, Kroetz DL, Weinshilboum R, Cox NJ, Roden DM (2010) Genome-wide association studies in pharmacogenomics: successes and lessons. *Pharmacogenet Genomics*. [Epub ahead of print]

Niemi M (2010) Transporter pharmacogenetics and statin toxicity. *Clin Pharmacol Ther*, 87, 130–133.

Niemi M, Pasanen MK, Neuvonen PJ (2011) Organic anion transporting polypeptide 1B1: a genetically polymorphic transporter of major importance for hepatic drug uptake. *Pharmacol Rev*, 63, 157–181.

Nies AT, Koepsell H, Damme K, Schwab M (2011) Organic cation transporters (OCTs, MATEs), in vitro and in vivo evidence for the importance in drug therapy. In: *Handbook of Experimental Pharmacology*, Vol. 201. Heidelberg, Germany: Springer. pp 105–167.

Russel FG, Koenderink JB, Masereeuw R (2008) Multidrug resistance protein 4 (MRP4/ABCC4): a versatile efflux transporter for drugs and signalling molecules. *Trends Pharmacol Sci*, 29, 200–207.

Schinkel AH, Jonker JW (2003) Mammalian drug efflux transporters of the ATP binding cassette (ABC) family: an overview. *Adv Drug Deliv Rev*, 55, 3–29.

Schwabedissen HE, Kroemer HK (2011) In vitro and in vivo evidence for the importance of breast cancer resistance protein transporters (BCRP/MXR/ABCP/ABCG2). In: *Handbook of Experimental Pharmacology*, Vol. 201. Heidelberg, Germany: Springer. pp 325–371.

Shu Y, Sheardown SA, Brown C, Owen RP, Zhang S, Castro RA, Ianculescu AG, Yue L, Lo JC, Burchard EG, Brett CM, Giacomini KM (2007) Effect of genetic variation in the organic cation transporter 1 (OCT1) on metformin action. *J Clin Invest*, 117, 1422–1431.

Sim SC, Ingelman-Sundberg M (2011) Pharmacogenomic biomarkers: new tools in current and future drug therapy. *Trends Pharmacol Sci*, 32, 72–81.

Sissung TM, Baum CE, Kirkland CT, Gao R, Gardner ER, Figg WD (2010a) Pharmacogenetics of membrane transporters: an update on current approaches. *Mol Biotechnol*, 44, 152–167.

Sissung TM, English BC, Venzon D, Figg WD, Deeken JF (2010b) Clinical pharmacology and pharmacogenetics in a genomics era: the DMET platform. *Pharmacogenomics*, 11, 89–103.

Stieger B (2011) The role of the sodium-taurocholate cotransporting polypeptide (NTCP) and of the bile salt export pump (BSEP) in physiology and pathophysiology of bile formation. In: *Handbook of Experimental Pharmacology*, Vol. 201. Heidelberg, Germany: Springer. pp 205–259.

Stingl Kirchheiner JC, Brockmoller J (2011) Why, when, and how should pharmacogenetics be applied in clinical studies?: current and future approaches to study designs. *Clin Pharmacol Ther*, 89, 198–209.

The International HapMap Consortium (2007) A second generation human haplotype map of over 3.1 million SNPs. *Nature*, 449, 851–861.

Urban TJ, Sebro R, Hurowitz EH, Leabman MK, Badagnani I, Lagpacan LL, Risch N, Giacomini KM (2006) Functional genomics of membrane transporters in human populations. *Genome Res*, 16, 223–230.

Venter JC, Adams MD, Myers EW, Li PW, Mural RJ, Sutton GG, Smith HO, Yandell M, Evans CA, Holt RA, et al. (2001) The sequence of the human genome. *Science*, 291, 1304–1351.

Weinshilboum RM, Wang L (2006) Pharmacogenetics and pharmacogenomics: development, science, and translation. *Annu Rev Genomics Hum Genet*, 7, 223–245.

Zanger UM, Raimundo S, Eichelbaum M (2004) Cytochrome P450 2D6: overview and update on pharmacology, genetics, biochemistry. *Naunyn Schmiedebergs Arch Pharmacol*, 369, 23–37.

CHAPTER 2

ADME PHARMACOGENOMICS IN DRUG DEVELOPMENT

Liangfu Chen
Joseph W. Polli

2.1 INTRODUCTION

It is well recognized that drugs can exhibit wide interpatient variability in their efficacy and toxicity. For many drugs, these interindividual differences are due in part to polymorphisms in genes encoding drug-metabolizing enzymes (DMEs), drug transporters, and/or drug targets (e.g., receptors, enzymes) (Evans and Johnson, 2001). Pharmacogenetics and pharmacogenomics (PGx) involve the study of the role of inheritance in individual variation in drug response, yielding phenotypes that can vary from the intended therapeutic response in one patient to little or no response in another patient. As well, pharmacogenomics can also be a determinant in adverse drug response such as hepatotoxicity (Shah, 2006). Therefore, there has been high interest in pharmacogenomics to be a key part of the strategy to optimize drug therapy based on a patient's genetic background. The discipline of pharmacogenetics evolved from the convergence of advances in molecular pharmacology and genomics. Originally, pharmacogenetic studies focused mainly on monogenic traits, often in drug-metabolism enzymes as observed for cytochrome P450 (CYP) 2D6 (Kimura et al., 1989; Johansson et al., 1993). Despite more than 50 years of research, most of the progress in pharmacogenetics that is applicable to drug development has been made in the past 20 years. Contemporary pharmacogenomic studies increasingly involve entire "pathways" such as panels of absorption, distribution, metabolism, and excretion (ADME) enzymes that encode proteins which influence pharmacokinetics, processes that determine the concentration of a drug reaching the site of action, and pharmacodynamics, the response by the drug target itself. For example, genome-wide association studies (GWAS) and candidate gene approach have led to identification of drug transporters as a new functional determinant pathway involved in the pharmacokinetics, pharmacodynamics, and toxicity of several drugs (Ayrton and Morgan, 2008; Giacomini et al., 2010). This integrated approach has enabled pharmacogenomics to have more "translational" use in the clinical setting. Because

Pharmacogenomics of Human Drug Transporters: Clinical Impacts, First Edition.
Edited by Toshihisa Ishikawa, Richard B. Kim and Jörg König

of the increased value that pharmacogenomics has demonstrated, its application has been extensively incorporated into drug development and governmental regulation over the past decade. This chapter provides an overview of the ADME-related pharmacogenomics literature and offers insights for the potential impact of this field on drug development and clinical practices to achieve safe and effective use of drugs, with emphasis on drug transporters. The discussion focuses on four areas: (1) Current ADME pharmacogenomic practices in drug development and clinical practices; (2) current pharmacogenomic approaches and methodologies; (3) recent advances in pharmacogenomics of membrane transporters; and (4) ADME pharmacogenomics: challenges and opportunities. Although pharmacogenetics focuses on the effect of a single gene on drug response and pharmacogenomics deals with the effects of multiple genes on drug response, both the terms are used interchangeably in this chapter to simplify the discussion.

2.2 CURRENT ADME PHARMACOGENOMIC PRACTICES IN DRUG DEVELOPMENT AND CLINICAL PRACTICES

From the drug development perspective, the subject of pharmacogenomics may be defined by three general areas: (1) Genetic polymorphisms in systems involved in the ADME of a drug, leading to variations in drug levels; (2) variations in proteins influencing the drug action pathway, from receptors to elements in signal transduction pathways; and (3) variations contributing to the development of subtypes of patients suffering from a given disease (Raaijmakers et al., 2010). PGx research on the ADME properties of drugs has begun to have impact for both drug development and how an agent is prescribed.

2.2.1 Pharmacogenomic Study Conduct in Drug Development

The Pharmaceutical Research and Manufacturers of America (PhRMA) conducted a survey of major 14 pharmaceutical companies on their PGx practices and applications during 2003–2005, and published a white paper to provide a cross-industry perspective on the practices and utility of ADME PGx (Williams et al., 2008). Below are selected highlights from the white paper.

A majority of companies find that there is utility at times to collect PGx samples in clinical studies, as polymorphisms in ADME genes can have consequences for safety and efficacy (Williams et al., 2008). For example, collection of DNA samples with informed consent is standard industry practice for clinical studies where intensive pharmacokinetic data are collected and typically is done in phase I multiple dose or drug interaction studies. This is due to the belief that observed PK anomalies might be explained through genetic testing and the knowledge of the drug's metabolism and disposition along with the well-characterized pharmacokinetic profile. In contrast, the collection of DNA samples is not consistent in later development (phase II onward) or in studies of special populations. Even when DNA samples are collected, the majority of the time the samples are not used, typically because there is not a need in that particular study or at that moment in development. However, ADME-related genotyping for subject selection (inclusion criterion) or screening (exclusion criterion) is used and is often based on genes categorized as known valid biomarkers by the US Food and Drug Administration (FDA) (http://www.fda.gov/Drugs/ScienceResearch/ResearchAreas/Pharmacogenetics/ucm083378.htm) (Table 2.1). This indicates that early consideration of ADME PGx as an important element of drug development is becoming more frequent (Williams et al., 2008). Within the ADME genes, six polymorphic DMEs (i.e., CYP2D6, CYP2C9, CYP2C19, CYP3A5, CYP3A4, and UGT1A1) and three drug transporters (i.e., multidrug resistance 1 (MDR1, *ABCB1*), breast cancer resistance protein (BCRP, *ABCG2*), and organic anion-transporting polypeptide 1B1 (OATP1B1, SLCO1B1) are the genes most often considered when a PGx study is being applied in clinical development. Genotyping of other ADME genes, in particular transporters, is expanding across the pharmaceutical industry (Zhang et al., 2008; Sissung et al., 2010; Yee et al., 2010).

2.2.2 Regulatory Perspective and Available PGx Tests in Clinical Practices

The rapid development in our understanding of the genetic basis behind interindividual differences in drug response has been the result of two interwoven processes: human genome sequencing and the development of new technologies enabling automated and efficient genetic testing (Gervasini et al., 2010). One manner by which pharmacogenetic knowledge can be translated into routine clinical practice is by the establishment of guidelines and support from regulatory agencies. Pharmacogenomics is one of the fields that the FDA has had a leadership position to influence the safety and efficacy of new products by translating the research on genetic variability into regulatory actions such as drug labels (Huang et al., 2006). Ultimately, the agency expects to have the knowledge assimilated into standards of care that can be used to "individualize" drug therapy (Huang et al., 2006).

To provide guidance to industry about what type of pharmacogenomic information the Agency expects to receive, a final "FDA Guidance for Industry: Pharmacogenomic Data Submissions" has been published (http://www.fda.gov/cder/guidance/6400fnl.pdf), together with two companion documents and a newly created website for Genomics at the FDA (http://www.fda.gov/drugs/scienceresearch/researchareas/pharmacogenetics/default.htm). The guidance is intended to clarify what type of genomic information needs to be submitted to the Agency and when, and it offers a new submission path called "Voluntary Genomic Data Submission (VGDS)" to encourage sponsors who are using pharmacogenomics in exploratory research to submit such information for early discussion with the FDA, but without regulatory implications. In addition, various guidance documents on the development of pharmacogenomic testing have been published (Goodsaid and Frueh, 2006; Amur et al., 2008; Frueh et al., 2008).

TABLE 2.1 List of Clinically Valid Pharmacogenetic Biomarkers in FDA-Approved Drug Labels

Pharmacogenetic marker	Representative drug	Therapeutic area
c-KIT expression	Imatinib	Oncology
CCR5 expression	Maraviroc	Antiviral
Chromasome 5q	Lenalidomide	Hematology
CYP2C9 variants	Warfarin, celecoxib	Hematology and analgesics
CYP2C19 variants	Clopidogrel, diazepam, nelfinavir	Cardiovascular, psychiatry, and antiviral
CYP2D6 variants	Atomoxetine, codeine, fluoxetine	Psychiatry and analgesics
DPD deficiency	Capecitabine, fluorouracil	Oncology, dermatology, and dental
EGFR expression	Erlotinib, gefitinib, panitumumab	Oncology
Estrogen receptor	Tamoxifen	Oncology
G6PDH deficiency	Chloroquine, dapsone, rasburicase	Antiinfectives, dermatology, dental, and oncology
HER2/NEU overexpression	Lapatinib, trastuzumab	Oncology
HLA-B*1502[a]	Carbamazepine	Neurology
HLA-B*5701	Abacavir	Antiviral
KRAS	Cetuximab, panitumumab	Oncology
NAGS; CPS; ASS; OTC; ASL; ARG	Phenylbutyrate, valproic acid	Gastroenterology and psychiatry
NAT1/2 variants	Isoniazid, rifampin	Anti-infectives
Ph1 chromosome	Busulfan, dasatinib, imatinib	Oncology
PML/RARα translocation	Arsenic trioxide, tretinoin	Oncology, dermatology, and dental
TPMT variants	Azathioprine, 6-MP, thioguanine	Rheumatology and oncology
UGT1A1 variants	Irinotecan, Nilotinib	Oncology
VKORC1 variants	Warfarin	Hematology

Source: US Food and Drug Administration (2011).

6-MP, 6-mercaptopurine; c-KIT, C-kit receptor, a cytokine receptor; CCR5, chemokine receptor type 5; DPD, dihydropyrimidine dehydrogenase; EGFR, epidermal growth factor receptor; G6PDH, glucose-6-phosphate dehydrogenase; HER, human epidermal growth factor receptor; HLA, human leukocyte antigen; KRAS, V-Ki-ras2 Kirsten rat sarcoma viral oncogene homolog; NAGS; CPS; ASS; OTC; ASL; ARG, enzymes participating in urea cycle; NAT, N-acetyltransferase; OCD, obsessive–compulsive disorders; Ph1, Philadelphia; PML/RAR, retinoic acid receptor; TPMT, thiopurine S-methyltransferase; UGT1A1, uridine diphosphate glucuronosyltransferase 1A1; VKORC1, vitamin K epoxide reductase complex subunit 1.

The pharmacogenetic revolution also prompted the European Medicines Agency (EMEA) to establish in 2001 (formalized in 2005) the Pharmacogenetics (since 2008, Pharmacogenomics) Working Party (PGWP) group. It is a permanent and multidisciplinary core group of up to 14 experts that provide recommendations to the EMEA's Committee for Medicinal Products for Human

Use (CHMP) on all matters relating directly or indirectly to pharmacogenomics (Gervasini et al., 2010). Further details of the composition of PGWP group and scheduled meetings are available at website (http://www.ema.europa.eu/docs/en_GB/document_library/Report/2010/07/WC500094119.pdf).

Today, only a few genetic tests are common in clinical practice despite pharmacogenetic information being contained in more than 200 drug labels in the United States (Flockhart et al., 2009); these include a wide range of drugs across many therapeutic areas such as anticancer, anti-HIV, antifungal, antiepileptics, antipsychotics, and lipid lowering. In many cases, the identified drug labels provide pharmacogenomic information without recommending a specific action, and only a few labels recommend or require biomarker testing as a basis for reaching a therapeutic decision. While the list includes a number of DMEs known to carry genetic variants (Table 2.1), no membrane transporter gene is currently listed as a pharmacogenomic biomarker.

2.3 CURRENT PHARMACOGENOMICS APPROACHES AND METHODOLOGIES

While many methodologies have been employed to characterize the pharmacogenetics/pharmacogenomics of various agents, studies are typically designed in three different ways: candidate gene approach, GWAS, and pathway-based approach (Roden et al., 2006; Wu et al., 2008; Shin et al., 2009; Sissung et al., 2010). Each type of technique is useful in certain contexts, although each is also limited in certain ways. A brief overview is provided below. Detailed descriptions of the technologies in sequencing, SNP detection, and genotyping can be found in few recent reviews (Morozova and Marra, 2008; Ragoussis, 2009; Ishikawa et al., 2010).

2.3.1 Candidate Gene Approach

Most pharmacogenetic studies have employed the "candidate gene approach" to detect associations between known SNPs and clinical or pharmacological end points. The candidate gene approach tests whether a particular allele or a set of alleles is more frequent in patients who have a better (or worse) drug response (Kwon and Goate, 2000). Most often, genes are selected based on their known physiological or pharmacological effect on disease, drug response or knowledge of the metabolism, transporter, and disposition of the drug. Thus, prior knowledge about the function of a gene is essential for selecting a gene to study. If there is a known genetic polymorphism that affects the function of a protein, that polymorphism is often selected for the study. For example, OCT1 alleles R61C, G401S, 420del, and G465R were chosen to study their effect on pharmacokinetics and efficacy of antidiabetic drug metformin (Table 2.2; Shu et al., 2008; Tzvetkov et al., 2009) because these polymorphisms had previously been shown to have reduced uptake function of OCT1 (Shu et al., 2007). Conversely, the evolving knowledge of functional effect of genetic polymorphisms is prompted by the discovery of a new genetic variant, often in a particular ethnic group. The functional characterization of three new nonsynonymous OCT variants (i.e., Q97K, P117L, and R206C), which were identified from the 1000

TABLE 2.2 Examples of Clinically Relevant Genetic Polymorphisms of Membrane Transporters Influencing Drug Disposition and Response

Genetic polymorphism of membrane transporter genes	Clinical effect	Affected drug (reference)
OATP *OATP1B1* (*SLCO1B1*5*, Val174Ala, c.521T>C) Higher frequency in American natives and Europe, lower in sub-Sahara Africa and Oceania.	Reduced uptake function. Increases in drug plasma levels, and individual susceptibility to drug-induced adverse events. Reduced therapeutic effect	Atorvastatin (Pasanen et al., 2007) Fexofenadine (Niemi et al., 2005b) Fluvastatin no effect (Niemi et al., 2006) Irinotecan (Xiang et al., 2006) Pitavastatin (Chung et al., 2005; Ieiri et al., 2007; Deng et al., 2008) Pravastatin (Niemi et al., 2005c, 2006; Ho et al., 2007; Zhang et al., 2007a); Deng et al., 2008) Repaglinide (Niemi et al., 2005a; Kalliokoski et al., 2008a, 2008b) Rosuvastatin (Lee et al., 2005; Pasanen et al., 2007; Choi et al., 2008), Simvastatin acid (Pasanen et al., 2006; Link et al., 2008) Endobiotics: Bile acids (Xiang et al., 2009) Bilirubin (Zhang et al., 2007b; Johnson et al., 2009)
BCRP *ABCG2* (rs2231142, Gln141Lys, c.421C>A) Higher frequency in Asians, lower in sub-Sahara Africa and African Americans	Reduced efflux function. Increases in drug plasma levels and incidence of drug-induced toxicity	Atorvastatin (Keskitalo et al., 2009c) Diflomotecan (Sparreboom et al., 2004) Fluvastatin (Keskitalo et al., 2009b) Gefitinib (Cusatis et al., 2006) Pitavastatin no effect (Ieiri et al., 2007) Pravastatin no effect (Ho et al., 2007; Keskitablo et al., 2009b) Rosuvastatin (Zhang et al., 2006; Keskitalo et al., 2009c; Tomlinson et al., 2010) Simvastatin lactone (Keskitalo et al., 2009b) Sulfasalazine (Urquhart et al., 2008; Yamasaki et al., 2008; Adkison et al., 2010) Topotecan (Sparreboom et al., 2005)

Pgp *ABCB1* (c.3435C>T, 2677G/T/A)	Various associations with therapeutic and adverse response to anticancer, antiviral, and antiepileptic drugs	Atorvastatin (Keskitalo et al., 2008) Carboplatin (Hamidovic et al., 2010) Digoxin (Johne et al., 2002) Docetaxel (Sissung et al., 2008) Fluvastatin, Pravastatin, Lovastatin, and Rosuvastatin, no effect (Keskitalo et al., 2009a) Paclitaxel (Sissung et al., 2006; Chang et al., 2010; Hamidovic et al., 2010) Simvastatin acid (Keskitalo et al., 2008) Various antiepileptic drugs (Sisodiya and Goldstein, 2007; Seo et al., 2008; Kwan et al., 2009; Haerian et al., 2010; Nurmohamed et al., 2010)—Controversial findings on the relevance
MRP2 *ABCC2* (-24C>T, c.1249G>T, c.3972C>T, c.4544C>T)	Various associations with therapeutic and adverse response to anticancer, antiviral, and antiepileptic drugs	Docetaxel (Kiyotani et al., 2008) Irinotecan (de Jong et al., 2007) Mycophenolic acid (Naesens et al., 2006; Yang et al., 2009) Tenofovir (Izzedine et al., 2006; Rodriguez-Novoa et al., 2009) Various antiepileptic drugs (Ufer et al., 2009) Endobiotics: Bilirubin glucuronide—Dubin–Johnson Syndrome (Tsujii et al., 1999) Tamoxifen (Kiyotani et al., 2010)

(continued)

TABLE 2.2 ***(Continued)***

Genetic polymorphism of membrane transporter genes	Clinical effect	Affected drug (reference)
OCTs *OCT1* (R61C, P160L, G401S, 420del, and G465R) *OCT2* (A270S)	Reduced uptake function. Association with effects on drug plasma levels, renal drug clearance, adverse response, and pharmacological effects	Cisplatin (Filipski et al., 2009) Imatinib (White et al., 2007, 2010a, 2010b; Engler et al., 2010) Metformin (Shu et al., 2007, 2008; Tzvetkov et al., 2009)
5-HTT The serotonin transporter gene, *SLC6A4* (5-*HTTLPR*), a 5-HTT-linked promoter region (5-HTTLPR) insertion/deletion polymorphism	Various associations with diseases and behavioral illnesses, and response to drug therapy related to depression	Citalopram (Eichhammer et al., 2003; Smith et al., 2004; Capozzo et al., 2009) Fluoxetine (Serretti et al., 2004) Paroxetine (Serretti et al., 2004) Respiridone (Dolzan et al., 2008)

Genomes Project in Chinese and Japanese populations was one of these examples (Chen et al., 2009).

If there are many SNPs in a gene of interest, it is often not feasible to genotype all of them. It is a common practice to set the minor allele frequency (MAF) to more than 5% to select SNPs, because SNPs with a frequency of less than 5% usually do not provide enough power to the study and may be of limited clinical relevance (Shin et al., 2009). The advantage of the candidate gene approach is that it is less expensive and requires a smaller sample size than GWAS (Zhu and Zhao, 2007). A major disadvantage of the candidate gene approach is that it requires prior knowledge of the function of the gene regarding the drug response. If information on the function of the gene is limited, the selection of the gene is difficult to justify.

2.3.2 Genome-wide Association Studies

The GWAS approach is useful to determine the most significant SNPs associated with a phenotype among a high-density set of polymorphisms (Sissung et al., 2010). The GWAS surveys the common genetic variations for a role in disease or drug response by genotyping large sets of SNPs across the genome (Wang et al., 2005; Shin et al., 2009). The human genome is estimated to have about 12 million common SNPs. There are many small regions (10–100 kb) in the genome where SNPs are in linkage disequilibrium and form two to four common haplotypes (Shin et al., 2009). Thus, an SNP in a region can be selected to represent its genetic variation. In other words, a tag SNP can be used as a proxy for the other SNPs in the region. This enables the genetic variations across the genome to be surveyed by genotyping tag SNPs. The number of SNPs used in a GWAS ranges from 100,000 to 1,000,000, with a higher number generally providing better coverage of the variations in the genome (Pe'er et al., 2006; Shin et al., 2009). Most GWAS have been conducted as a case-control, cohort, or family study (Link et al., 2008; Sandhu et al., 2008). The strength of an association is usually assessed using an odds ratio (OR) or a relative risk and p value. As with the candidate gene association study, the findings should be replicated in multiple independent populations (Kathiresan et al., 2004); Shin et al., 2009). GWAS can be used to identify new biomarkers that could explain the underlying mechanisms of adverse drug reactions. For example, this approach has led to the discovery that the common variants identified in SLCO1B1 were strongly associated with an increased risk of simvastatin-induced myopathy (Table 2.2; Link et al., 2008), and as one of the top loci associated with serum bilirubin levels (Table 2.2; Johnson et al., 2009). Also, GWAS is increasingly recognized as a useful tool to identify disease-associated genes, and has been utilized in the identification of MDR1 genetic variants in association with susceptibility to ulcerative colitis and Crohn's disease (Ho et al., 2006; Krupovcs et al., 2009).

The requirement for a large clinical sample size and the high cost of whole-genome SNP panels for GWAS compared with the candidate gene approach have been the limiting factors in using GWAS (Tilson and Ro, 2006; Shin et al., 2009). The coverage of genetic variations also differs among various commercial SNP panels (Hirschhorn and Daly, 2005; Shin et al., 2009). For example, rare SNPs and copy-number variations may not be included in a certain set of SNPs in a GWAS. Despite the above-mentioned limitations, the GWAS holds great potential for contributing to the understanding of complex disease development and identifying the factors that affect variable drug responses.

2.3.3 Pathway-Based Approach

Pathway-based approach utilizes foreknowledge of both the genetic variants, genes, and the pathways that these genes participate in. It is particularly useful in identifying and characterizing pharmacogenetics end points given that studies are conducted to test the interaction between genes, rather than assuming that each SNP confers a monogenic trait (Wu et al., 2008). However, the incorporation of interaction testing requires the utilization of machine learning techniques, and these techniques can often be complex and require larger sample sizes than candidate gene approaches. Moreover, validation of gene–gene interactions is often difficult because a fundamental understanding of the biology of the interactions is required, yet the current knowledge base is often incomplete (Wu et al., 2008). Examples of pathway-based approaches include the study of multiple genotypes of *CYP3A4, CYP3A5, SLCO1B3, ABCB1,* and *ABCC2* in the docetaxel metabolism and elimination pathway (Baker et al., 2009), and the study of polymorphisms of gamma-aminobutyric acid (GABA) transporter 3 (*SLC6A11*), together with seven other genes belonging to the GABA receptor signaling pathway, proposed to be involved in genetic susceptibility to treatment-resistant tardive dyskinesia (Inada et al., 2008).

The recently developed Affymetrix Drug Metabolizing Enzymes and Transporters (DMET) genotyping platform is essentially designed for scaled-up pathway-based pharmacogenetics studies. It offers the ability to scan 1936 variants in 225 genes related to drug metabolism and disposition (Deeken, 2009; Sissung et al., 2010). Utility of the platform has yielded several previously unknown associations between polymorphisms and therapy with widely used drugs, for example, docetaxel, warfarin, and clopidogrel (Caldwell et al., 2008; Baker et al., 2009; Mega et al., 2009; Deeken et al., 2010). The DMET platform represents an exploratory, pathway-based approach that scans the genome for SNPs and haplotypes in ADME genes that may correlate with interindividual variation in drug response. Like the genome-wide approach, it offers a comprehensive analysis of the genome, but lessens the possibility of type I error associated with GWAS. It is similar to the pathway-based approach for testing certain hypotheses, yet less likely to overlook an important variant (Sissung et al., 2010).

2.4 RECENT ADVANCES IN PHARMACOGENOMICS OF MEMBRANE TRANSPORTERS

During the last decade, a great focus has been given to the impact of genetic variation in membrane transporters on the pharmacokinetics and toxicity of numerous therapeutic drugs. While the majority of transporter-related pharmacogenomic research has been in regard to genes encoding the outward-directed ATP-binding cassette (ABC) transporters, such as *ABCB1* (P-glycoprotein), *ABCG2* (BCRP), and *ABCC2*

(MRP2), more studies have been conducted in recent years evaluating genes encoding solute carriers (SLC) that mediate the cellular uptake of drugs, such as SLCO1B1 (OATP1B1) and *SLC22A1* (OCT1). The distribution of ABC and SLC transporters in tissues key to pharmacokinetics, such as intestine (absorption), blood–brain barrier (distribution), liver (metabolism and biliary clearance), and kidneys (excretion), strongly suggests that genetic variation associated with changes in protein expression or function of these transporters may have a substantial impact on systemic drug exposure and toxicity (Ayrton and Morgan, 2008; Giacomini et al., 2010). In addition, there is increasing evidence that genetic variants of some transporters are positively associated with, if not solely responsible for, interindividual differences in drug efficacy and toxicity. Table 2.2 lists out the examples of genetic polymorphisms of critical ABC and SLC transporters and their contribution to interindividual variability in pharmacokinetics, therapeutic, and toxicity response of substrate drugs. Their genotypes ranged from nonsynonymous single-nucleotide polymorphisms of many transporters to insertion/deletion polymorphisms in the promoter region of the serotonin transporter gene (*SLC6A4*). Their relevance in drug development and clinical practice is highlighted below. Detailed overviews on these and other transporters can be found in other chapters of this book.

2.4.1 Effect on Drug Level

Since a majority of membrane transporters are highly expressed in tissues primarily responsible for drug absorption and clearance, and they transport drugs by active uptake or efflux (Ayrton and Morgan, 2008; Giacomini et al., 2010), it is not surprising that a functional genetic variant can lead to an unusual increase, or in some cases decrease, in systemic drug levels (Yee et al., 2010). The understanding of this causal relationship has been greatly enhanced with recent advancement in genotyping technology, in concert with carefully designed clinical studies. The effect of OATP1B1 (SLCO1B1*5, c.521T>C) and BCRP (ABCG2, c.421C>A) variants on statin exposure is probably one of the examples with a most prominent change (Table 2.2; Zhang et al., 2006; Ieiri et al., 2007; Keskitalo et al., 2009a; Niemi, 2010; Rodrigues, 2010).

The knowledge of the effect of transporter genetic polymorphism on drug levels will significantly benefit drug development and patient care because of the following reasons: (1) It may explain interindividual and interethnic variability of pharmacokinetics; (2) it may provide guidance for dose adjustment for a pertinent patient population for an optimal therapy and minimal side effects; (3) it would enhance mechanistic understanding of the drug exposure which may be influenced by multiple factors, including drug transporters and drug metabolism; (4) it may provide an explanation for drug-induced toxicity or adverse events; (5) it would allow informed decision making on the potential target-efficacy populations, and thus a more focused clinical study plan early in drug development; and (6) it may serve as a surrogate for maximum pharmacokinetic drug interaction potential if the genetic variant results in the loss of function of the studied transporter. This could be of particular value given that a specific inhibitor is still lacking for any transporter.

2.4.2 Association with Drug-induced Toxicity

The increase in drug blood level due to transporter polymorphism has been shown to link to drug-induced toxicity. For example, cholesterol-lowering drug statins are substrates of OATP1B1 (SLCO1B1) (Fujino et al., 2005; Kameyama et al., 2005; Kopplow et al., 2005; Ho et al., 2006; Pasanen et al., 2006; Kitamura et al., 2008). The area under the plasma concentration–time curve (AUC) of active simvastatin acid has been shown to be approximately 2.2-fold greater in individuals with the c.521CC genotype than in those with the c.521TT genotype (Pasanen et al., 2006). The SLCO1B1 c.521T>C SNP is strongly associated with simvastatin-induced myopathy in a GWAS, where approximately 300,000 genome markers were determined in 85 patients who had developed myopathy while receiving 80 mg simvastatin daily and in 90 controls without myopathy (Link et al., 2008). More than 60% of the myopathy cases could be attributed to the c.521T>C SNP, with an OR of 4.5 per copy of the c.521C allele. Furthermore, this association was seen to be replicated in another study of 20,000 patients receiving 40 mg simvastatin daily, yielding a relative risk of 2.6 per copy of the c.521C allele (Link et al., 2008). Genotyping these variants may help to achieve the benefits of statin therapy more safely and effectively. Of note, this information has not been translated to the drug label as of finishing of this manuscript.

Chemotherapeutic agents have been effective in anticancer treatment of a wide range of tumor types. However, unpredictable interindividual variation in severe adverse reactions remains an important determinant for limiting their use. By conducting genotype and phenotype association studies directly in cancer patients, genetic variants of a number of membrane transporters (in particular ABC efflux transporters) have been identified to contribute to the toxicities (Table 2.2). For example, development of leucopenia/neutropenia is common in some cancer patients who received docetaxel. To identify possible genetic factor(s) of the toxicity, Kiyotani et al. (2008) genotyped 113 patients for a total of 79 SNPs in seven genes possibly involved in the metabolism/transport of the drug. There is a significant association of rs12762549 in ABCC2 (MRP2) and rs11045585 in SLCO1B3 (OATP1B3) with docetaxel-induced leucopenia/neutropenia (Kiyotani et al., 2008). It is worth noting that both the variants were tag SNPs in the noncoding region, while three nonsynonymous SNPs found in these genes: One in ABCC2 (rs2273697, I417V) and two in SLCO1B3 (rs4149117, S112A and rs7311358, M233I) showed no association with the toxicity. It was suggested that these tag SNPs may influence the expression levels of these genes directly, or in concert with other related genetic variations, which affect the risk of leucopenia/neutropenia.

2.4.3 Impact on Drug Efficacy

Statins are widely used as prophylactic treatment for lowering serum cholesterol levels and reducing cardiovascular risk. Since the therapeutic effect of statins is dependent on inhibiting HMG-CoA reductase in hepatocytes, any decrease in the uptake of these drugs into the liver could reduce their efficacy, in addition to a potential increase in systemic drug level and the risk of statin-induced myopathy as described above

(Table 2.2; Link et al., 2008; Neuvonen, 2010). Indeed, the SLCO1B1*5, c.521T>C SNP was associated with a slight reduction in the cholesterol-lowering efficacy of simvastatin, whereas the c.388A>G SNP was associated with a slightly enhanced efficacy (Link et al., 2008). These findings are consistent with the hypothesis that the c.521T>C SNP is associated with a reduction in hepatic uptake, whereas the SLCO1B1*1B haplotype, an enhancement in uptake. In another study, the ability of pravastatin to reduce plasma cholesterol and lathosterol is significantly impaired in individuals carrying the SLCO1B1*17 haplotype, although this was accompanied by a significant inverse relationship with pravastatin AUC values (Niemi et al., 2005c). Similarly, the c.521T>C SNP SLCO1B1 was found to significantly attenuate the total cholesterol-lowering efficacy of pravastatin in Chinese patients with coronary heart disease (Zhang et al., 2007b). Furthermore, an impaired efficacy of pravastatin, simvastatin, and atorvastatin on cholesterol reduction was found in Japanese patients expressing variants in the 521 T>C SNP (Tachibana-Iimori et al., 2004).

Conversely, the BCRP (*ABCG2*) c.421A variant was found to be significantly associated with greater reduction in the low-density lipoprotein cholesterol, a key therapeutic effect of rosuvastatin, in a gene dose-dependent manner (Tomlinson et al., 2010). This is consistent with the data from pharmacokinetics studies in Chinese and Caucasian subjects that showed approximately twice the level of systemic exposure to rosuvastatin in subjects with at least one c.421A allele as compared with those with the c.421CC genotype (Zhang et al., 2006; Keskitalo et al., 2009b). The c.421A allele results in lower expression levels of the *ABCG2* efflux transporter protein and a reduced ability to export substrate, leading to an increase in drug accumulation both in hepatocytes and in the systemic circulation (Kondo et al., 2004). Rosuvastatin is one of the few drugs for which regulatory authorities, including the FDA, have recommended starting with lower doses (5 mg instead of 10 mg) in Asian patients (US FDA, 2005). This recommendation was based on the data that systemic exposure to rosuvastatin was approximately twice as high in Asians as in non-Asians (Grundy, 2005). A later study by Lee et al. (2005) to compare rosuvastatin systemic exposure of Chinese, Malay, and Asian-Indian with that of white subjects found that SLCO1B1 genotypes did not account for the observed pharmacokinetic differences between Asians and white subjects, indicating that OATP1B1 may have a minor role in rosuvastatin disposition in Asians. The differences are unlikely due to ethnicity-related variations in DMEs, as metabolism (primarily CYP2C9) plays a minor role in rosuvastatin clearance with 90% of orally administered rosuvastatin recovered unchanged in feces (Carswell et al., 2002; Martin et al., 2003), Rosuvastatin, on the other hand, is a substrate for a number of drug transporters that influence its disposition, including OATPs, NTCP, Pgp, MRP2 and BCRP (Tirona, 2005; Ho et al., 2006; Kitamura et al., 2008). The higher frequency of the c.421A allele in Chinese and Japanese populations (∼35%) as compared with Caucasians (14%) suggests that this polymorphism likely contributes to the interethnic variation in the pharmacokinetics and potentially the treatment outcomes of rosuvastatin.

The human serotonin transporter (5-HTT) is expressed in brain and blood cells and implicated in mood and anxiety regulation. It is where antidepressant and anti-anxiety drugs initially act in the brain. A 5-HTT-linked promoter region (5-HTTLPR) insertion/deletion polymorphism with long (l) and short (s) forms affects the transporter expression and function. The s variant reduces 5-HTT gene transcription resulting in reduced transporter levels and 5-HT uptake (Greenberg et al., 1999). Selective serotonin reuptake inhibitors (SSRIs) have been widely used in treatment of depression and bipolar disorder (Table 2.2). Patients who are carriers of the short variant of 5-HTTLPR were found to have a poor response to SSRIs, citalopram, fluvoxamine, or paroxetine (Serretti et al., 2004; Capozzo et al., 2009). Interestingly, the long variant of 5-HTTLPR also showed a significantly better mood amelioration after total sleep deprivation than those who were heterozygotic and homozygotic for the short variant.

2.4.4 Underlining Disease Mechanism

The clearance of bilirubin consists of two sequential steps: Glucuronidation of unconjugated bilirubin catalyzed by UDP-glucuronosyl transferase 1A1 (UGT1A1) and biliary secretion of conjugated bilirubin via multidrug resistance protein 2 (MRP2), expressed on the canalicular membrane of hepatocytes. Impairment of either process can lead to hyperbilirubinemia. Gilbert's syndrome (GS) is recessively inherited, unconjugated hyperbilirubinemia caused by reduced UGT1A1 activity associated mostly with homozygous A(TA) 7 TAA variant of the TATAA-box in the *UGT1A1* gene promoter, named *UGT1A1*28* Allele (Bosma, 2003). GS is the most frequent hereditary hyperbilirubinemia affecting 5–10% of the white population. Dubin–Johnson syndrome (DJS) is a rare, benign, predominantly conjugated hyperbilirubinemia, characterized by impaired secretion of anionic conjugates from hepatocytes into bile. DJS is associated with over a dozen various genetic variants of MRP2 gene (ABCC2), which result in partial or complete loss of the transporter activity (Cebecauerova et al., 2005). Furthermore, recent studies showed that interindividual variability in the serum bilirubin level is also affected by genetic polymorphism in the SLCO1B1 gene, encoding hepatic uptake transporter OATP1B1 (Zhang et al., 2007b; Johnson et al., 2009). In particular, the SLCO1B1*15 carriers have higher baseline serum levels of total, unconjugated, and direct bilirubin compared with subjects with the SLCO1B1*1a/*1a and SLCO1B1*1b/*1b genotypes. In addition, SLCO1B1*15/*15 homozygotes are more susceptible to hyperbilirubinaemia in the presence of OATP1B1 inhibitor rifampicin, even with a low dose (Zhang et al., 2007b).

Therefore, it seems there are multiple endogenous mechanisms which could contribute to hyperbilirubinemia. This is of particular interest as drug-induced hyperbilirubinemia is not an uncommon event during drug development. When it occurs, one would need to consider collectively the genetic predispositions as well as the potential perturbation of bilirubin disposition mechanism by the investigative new drug. For example, anticancer drug pazopanib induced hyperbilirubinemia and was shown to be associated with Gilbert's syndrome UGT1A1 polymorphism (Xu et al., 2010). However, approximately 16% of the patients who had isolated hyperbilirubinemia had the UGT1A1 TA6/TA6 genotype. As pazopanib is a potent inhibitor of OATP1B1 as well as UGT1A1 *in vitro*, the authors suggested the possibility that the drug-induced hyperbilirubinemia may be the result of inhibition of both

OATP1B1 and UGT1A1 activity combined with genetic defects of the UGT1A1 (Xu et al., 2010).

2.5 ADME PHARMACOGENOMICS: CHALLENGES AND OPPORTUNITIES

As discussed above, many impactful progresses have been made over the last decades in our understanding of pharmacogenomic effect of DMEs and transporters on a drug therapy, as well as their applications in drug discovery and clinical practices. However, significant challenges remain to be overcome if the goal of genomic science is to achieve its full potential as a major medical application, for example, personalized medication, an area currently viewed by some as having disappointing progress, yet carrying many promises and expectations from both the industry and regulatory bodies. On the other hand, the rapid advances in pharmacogenomics are anticipated to offer exciting opportunities in drug development and clinical therapies.

2.5.1 Establish Genotype and Phenotype Correlations

Correlating an individual's genotype with the *in vivo* activity of a given DME or transporter can be difficult. While in the majority of cases there is a good correlation between genotype and phenotype, there are examples where genotype and phenotype are not correlated, often due to factors that can affect the regulation or activity of the polymorphic gene, or other nongenetic influences on phenotype. Such factors may include ethnicity, disease, age, diet, use of alcohol, and coadministered drugs (Huang and Temple, 2008). In these instances, correlations that exist between genotype and phenotype that exist in the "normal" state can be lost when induction or downregulation of an enzyme and/or transporter occurs. Further, there are examples where genotypic differences (e.g., different alleles) exist but do not impact the phenotype. Such associations may not be sufficient for applications such as diagnostics, but may still be useful for other purposes such as informing on mechanism of drug metabolism and disposition (Williams et al., 2008). The rapid progresses in the development and applications of PGx technologies and methodologies as discussed earlier, in concert with sophisticated functional analysis of the identified genetic variants should vastly advance the knowledge of correlations between genotype (a static phenomenon, and often single factor) and phenotype (a dynamic phenomenon, and often multiple factors), as well as their relevance to drug therapy.

2.5.2 Improve Understanding Transporter Role in Drug Disposition

There are a few unique features of membrane transporters, as compared to DME, which pose challenges when it comes to understanding their role in drug disposition and hence the drug response. These unique features include function in a membrane-bound environment; have a broad overlapping substrate specificity; lack specific inhibitors; express in multiple tissues with various protein and functional levels; and can have differential effect on drug level in systemic versus local tissue exposure (Ayrton and Morgan, 2008; Giacomini et al., 2010). For example, although drug response is often driven by the systemic drug level, changes in transporter-mediated local drug distribution into the target organ can have a profound effect. This is exemplified by the differential influence of the genetic polymorphisms of OATP1B1 and BCRP on statin efficacy as discussed above (Kondo et al., 2004; Link et al., 2008; Tomlinson et al., 2010).

As one consequence of the aforementioned unique features, the lack of, and hence the effort to identify a sensitive probe drug or specific inhibitor for drug transporters still remains a key challenge. For example, sulfasalazine has recently been evaluated in three separate studies as a BCRP probe substrate and the influence of C421A polymorphism (Urquhart et al., 2008; Yamasaki et al., 2008; Adkison et al., 2010). Although the C421A polymorphism was found to associate with significant increases in sulfasalazine AUC in two studies (Urquhart et al., 2008; Yamasaki et al., 2008), no significant pharmacokinetic differences were observed in the last study between the genetic variants, nor with coadministrated BCRP inhibitor pentoprazole (Adkison et al., 2010). Nonetheless, all the three studies had observed high intersubject variability in the salfasalazine pharmacokinetics. This suggests that the majority of the pharmacokinetic variability cannot be attributed to differences in BCRP-mediated efflux, rather interindividual differences in gastrointestinal physiology, absorption, and other factors, such as the extensive bacteria-mediated metabolism of sulfasalazine in the gut may be major contributors (Adkison et al., 2010). Further comparisons of the studies suggested vast differences in dose selection and formulation use, indicating an apparent dose dependence in the sensitivity of sulfasalazine. The authors concluded that sulfasalazine is a less-than-optimal probe drug for assessing BCRP activity in clinical phenotyping or drug interaction studies, and further research into BCRP probe substrates is warranted (Adkison et al., 2010).

In the next decade, continued advances in the field of membrane transporter pharmacogenomics are anticipated. Ultimately, it is anticipated that a detailed functional and mechanistic understanding of the variants in coding and noncoding regions of transporter genes that contribute to variation in drug response will be revealed (Yee et al., 2010). This will enhance not only our knowledge on the effect of genetic polymorphisms on the function of a particular transporter but also understanding of the overall contribution of the transporter in the ADME of a drug. The established knowledge may build toward a transporter-based biomarker for pharmacogenomic testing, potentially for guiding personalized medication.

2.5.3 Adopt a Holistic and Integrated Approach

Current pharmacogenomic studies have largely focused on the association of single genetic variants with drug response. It is anticipated that gene–gene interactions will be identified and will reveal epistatic interactions among transporter genes, or transporter genes and genes encoding enzymes or other proteins that interact with transporters (Yee et al., 2010). The aforementioned evolving knowledge on multiple contributors of enzymes and transporters to drug-induced hyperbilirubinemia is one of these examples. Collectively, such interactions may contribute to the vast amount of variation in drug response that remains unexplained. Furthermore, as factors other than genetic can also contribute to interindividual variation in drug response, pharmacogenomic test and knowledge will need to couple with molecular diagnostic and drug monitoring for optimal individualization of drug therapy (Gervasini et al., 2010). Finally, the integration of pharmacogenomics and systems biology can help elucidate the mechanisms of diseases and drug actions at various levels and connect information between different levels. For example, altered genetic structure may cause malfunctions at the molecular level, which would influence the downstream interactions, pathways, and networks at the cellular level. Such changes may then lead to tissue or organ disorders that are disease phenotypes reflected as symptoms of the whole body. The understanding of such an interwoven network may be the ultimate key to accurately identify drug targets and to avoid adverse reactions (Yan, 2003).

In 2004, the FDA introduced the Critical Path Initiative with the intent of modernizing drug development by incorporating recent scientific advances, such as genomics and advanced imaging technologies, into the process. An important part of the initiative is the use of public–private partnerships and consortia to accomplish the needed research (Woodcock and Woosley, 2008). In addition, the NIH Pharmacogenetics Research Network (PGRN) is also a collaborative group of investigators with a wide range of research interests, but all attempting to correlate drug response with genetic variation. Several research groups concentrate on drugs used to treat specific medical disorders (asthma, depression, cardiovascular disease, addiction of nicotine, and cancer), whereas others are focused on specific groups of proteins that interact with drugs (membrane transporters and phase II DMEs). The diverse scientific information is stored and annotated in a publicly accessible knowledge base, the Pharmacogenetics and Pharmacogenomics Knowledge base (PharmGKB) (Giacomini et al., 2007). With the significant challenges still ahead but great promises to be gained, one can expect more collaboration and integration efforts in pharmacogenomics across the pharmaceutical industries, academic labs, and regulatory bodies. A process map of pharmacogenomic application in drug development is proposed in Figure 2.1.

2.5.4 Educate Clinical Practitioners

It is important to note that despite significant scientific progress, a critical challenge in bringing pharmacogenomics "from the bench to the bedside" is educating many different health care professionals about the logistics and benefits of using genetic and genomic information to individualize drug therapy (Shin et al., 2009). This has not reached a level of critical mass yet by which translation of this knowledge can be measured by its use in the clinic. This is not unexpected given the relatively short time for pharmacogenomics. Consequently, significantly more effort is needed to not only ensure good science but also to invest in educational programs that inform physicians, pharmacists, clinical chemists, laboratory directors, third party providers, and patients about the potential of this exciting field to improve public health (Huang et al., 2006).

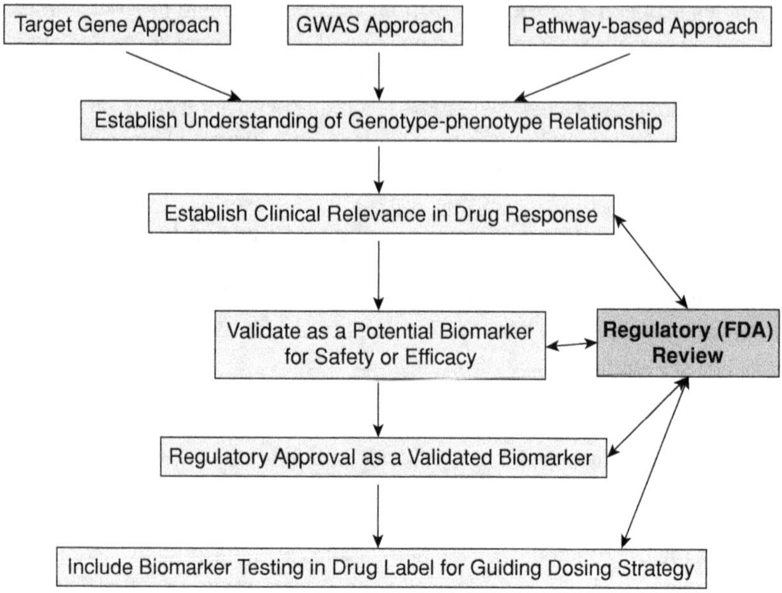

Figure 2.1 Process map of pharmacogenomic application in drug development.

2.6 CONCLUSION

In a review in Science more than a decade ago, Evans and Relling thoroughly described the possibility to translate functional genomics into rational therapeutics (Evans and Relling, 1999). They brought together, at that time, the available well-documented knowledge on a number of adverse events and/or therapeutic failures that can be causally related to genetic variations. It was anticipated that, in the last decade, the Human Genome Project, coupled with DNA array technology, high-throughput screening systems, and advanced bioinformatics, would permit rapid elucidation of complex genetic components of human health and disease. It was also anticipated that such pharmacogenomic studies should also permit the development of therapeutic agents targeted for specific, but genetically identifiable, subgroups of the population (Evans and Relling, 1999). The pharmaceutical industry, in consortia with scientists from academia and regulatory bodies, has gained considerable experience in recent years in understanding how ADME pharmacogenomics, in particular genetic variation in membrane transporters, can be applied to drug discovery and clinical development. Because transporters, like DMEs, are involved in the pharmacokinetics and pharmacodynamics of most clinically used drugs, it is anticipated that the field of pharmacogenomics of membrane transporters will contribute enormously to understanding the determinants of variation in drug response in the next decade. In return, the knowledge of genotype–phenotype relationships will enable engaging right patient populations in drug development such that the PK–PD relationship can be defined early, and safety risks identified and managed prospectively. As our understanding of the underlying science evolves, and the clinical practice of enzyme and transporter genotyping becomes more commonplace, more prescription labels will contain recommendations, or even requirements, to target dosing according to genotype. This evolving practice should help to minimize drug-related adverse events and maximize drug efficacy through prospective understanding and management of the ADME in individual patient.

REFERENCES

Adkison KK, Vaidya SS, Lee DY, Koo SH, Li L, Mehta AA, Gross AS, Polli JW, Humphreys JE, Lou Y, Lee EJ (2010) Oral sulfasalazine as a clinical BCRP probe substrate: pharmacokinetic effects of genetic variation (C421A) and pantoprazole coadministration. *J Pharm Sci*, 99, 1046–1062.

Amur S, Frueh FW, Lesko LJ, Huang SM (2008) Integration and use of biomarkers in drug development, regulation and clinical practice: a US regulatory perspective. *Biomark Med*, 2, 305–311.

Ayrton A, Morgan P (2008) Role of transport proteins in drug discovery and development: a pharmaceutical perspective. *Xenobiotica*, 38, 676–708.

Baker SD, Verweij J, Cusatis GA, van Schaik RH, Marsh S, Orwick SJ, Franke RM, Hu S, Schuetz EG, Lamba V, Messersmith WA, Wolff AC, Carducci MA, Sparreboom A (2009) Pharmacogenetic pathway analysis of docetaxel elimination. *Clin Pharmacol Ther*, 85, 155–163.

Bosma PJ (2003) Inherited disorders of bilirubin metabolism. *J Hepatol*, 38, 107–117.

Caldwell MD, Awad T, Johnson JA, Gage BF, Falkowski M, Gardina P, Hubbard J, Turpaz Y, Langaee TY, Eby C, King CR, Brower A, Schmelzer JR, Glurich I, Vidaillet HJ, Yale SH, Qi ZK, Berg RL, Burmester JK (2008) CYP4F2 genetic variant alters required warfarin dose. *Blood*, 111, 4106–4112.

Capozzo MA, Schillani G, Aguglia E, De VM, Grassi L, Conte MA, Giraldi T (2009) Serotonin transporter 5-HTTLPR polymorphism and response to citalopram in terminally ill cancer patients: report of twenty-one cases. *Tumori*, 95, 479–483.

Carswell CI, Plosker GL, Jarvis B (2002) Rosuvastatin. *Drugs*, 62, 2075–2085.

Cebecauerova D, Jirasek T, Budisova L, Mandys V, Volf V, Novotna Z, Subhanova I, Hrebicek M, Elleder M, Jirsa M (2005) Dual hereditary jaundice: simultaneous occurrence of mutations causing Gilbert's and Dubin–Johnson syndrome. *Gastroenterology*, 129, 315–320.

Chang H, Rha SY, Jeung HC, Im CK, Noh SH, Kim JJ, Chung HC (2010) Association of the ABCB1 3435C>T polymorphism and treatment outcomes in advanced gastric cancer patients treated with paclitaxel based chemotherapy. *Oncol Rep*, 23, 271–278.

Chen Y, Li S, Brown C, Cheatham S, Castro RA, Leabman MK, Urban TJ, Chen L, Yee SW, Choi JH, Huang Y, Brett CM, Burchard EG, Giacomini KM (2009) Effect of genetic variation in the organic cation transporter 2 on the renal elimination of metformin. *Pharmacogenet Genomics*, 19, 497–504.

Choi JH, Lee MG, Cho JY, Lee JE, Kim KH, Park K (2008) Influence of OATP1B1 genotype on the pharmacokinetics of rosuvastatin in Koreans. *Clin Pharmacol Ther*, 83, 251–257.

Chung JY, Cho JY, Yu KS, Kim JR, Oh DS, Jung HR, Lim KS, Moon KH, Shin SG, Jang IJ (2005) Effect of OATP1B1 (SLCO1B1) variant alleles on the pharmacokinetics of pitavastatin in healthy volunteers. *Clin Pharmacol Ther*, 78, 342–350.

Cusatis G, Gregorc V, Li J, Spreafico A, Ingersoll RG, Verweij J, Ludovini V, Villa E, Hidalgo M, Sparreboom A, Baker SD (2006) Pharmacogenetics of ABCG2 and adverse reactions to gefitinib. *J Natl Cancer Inst*, 98, 1739–1742.

de Jong FA, Scott-Horton TJ, Kroetz DL, McLeod HL, Friberg LE, Mathijssen RH, Verweij J, Marsh S, Sparreboom A (2007) Irinotecan-induced diarrhea: functional significance of the polymorphic ABCC2 transporter protein. *Clin Pharmacol Ther*, 81, 42–49.

Deeken J (2009) The Affymetrix DMET platform and pharmacogenetics in drug development. *Curr Opin Mol Ther*, 11, 260–268.

Deeken JF, Cormier T, Price DK, Sissung TM, Steinberg SM, Tran K, Liewehr DJ, Dahut WL, Miao X, Figg WD (2010) A pharmacogenetic study of docetaxel and thalidomide in patients with castration-resistant prostate cancer using the DMET genotyping platform. *Pharmacogenomics J*, 10, 191–199.

Deng JW, Song IS, Shin HJ, Yeo CW, Cho DY, Shon JH, Shin JG (2008) The effect of SLCO1B1*15 on the disposition of pravastatin and pitavastatin is substrate dependent: the contribution of transporting activity changes by SLCO1B1*15. *Pharmacogenet Genomics*, 18, 424–433.

Dolzan V, Serretti A, Mandelli L, Koprivsek J, Kastelic M, Plesnicar BK (2008) Acute antipyschotic efficacy and side effects in schizophrenia: association with serotonin transporter promoter genotypes. *Prog Neuropsychopharmacol Biol Psychiatry*, 32, 1562–1566.

Eichhammer P, Langguth B, Wiegand R, Kharraz A, Frick U, Hajak G (2003) Allelic variation in the serotonin transporter promoter affects neuromodulatory effects of a selective serotonin transporter reuptake inhibitor (SSRI). *Psychopharmacology (Berl)*, 166, 294–297.

Engler JR, Frede A, Saunders VA, Zannettino AC, Hughes TP, White DL (2010) Chronic myeloid leukemia CD34 +cells have reduced uptake of imatinib due to low OCT1 activity. *Leukemia*, 24, 765–770.

Evans WE, Johnson JA (2001) Pharmacogenomics: the inherited basis for interindividual differences in drug response. *Annu Rev Genomics Hum Genet*, 2, 9–39.

Evans WE, Relling MV (1999) Pharmacogenomics: translating functional genomics into rational therapeutics. *Science*, 286, 487–491.

Filipski KK, Mathijssen RH, Mikkelsen TS, Schinkel AH, Sparreboom A (2009) Contribution of organic cation transporter 2 (OCT2) to cisplatin-induced nephrotoxicity. *Clin Pharmacol Ther*, 86, 396–402.

Flockhart DA, Skaar T, Berlin DS, Klein TE, Nguyen AT (2009) Clinically available pharmacogenomics tests. *Clin Pharmacol Ther*, 86, 109–113.

Frueh FW, Amur S, Mummaneni P, Epstein RS, Aubert RE, DeLuca TM, Verbrugge RR, Burckart GJ, Lesko LJ (2008) Pharmacogenomic biomarker information in drug labels approved by the United States food and drug administration: prevalence of related drug use. *Pharmacotherapy*, 28, 992–998.

Fujino H, Saito T, Ogawa S, Kojima J (2005) Transporter-mediated influx and efflux mechanisms of pitavastatin, a new inhibitor of HMG-CoA reductase. *J Pharm Pharmacol*, 57, 1305–1311.

Gervasini G, Benitez J, Carrillo JA (2010) Pharmacogenetic testing and therapeutic drug monitoring are complementary tools for optimal individualization of drug therapy. *Eur J Clin Pharmacol*, 66, 755–774.

Giacomini KM, Brett CM, Altman RB, Benowitz NL, Dolan ME, Flockhart DA, Johnson JA, Hayes DF, Klein T, Krauss RM, et al. (2007) The pharmacogenetics research network: from SNP discovery to clinical drug response. *Clin Pharmacol Ther*, 81, 328–345.

Giacomini KM, Huang SM, Tweedie DJ, Benet LZ, Brouwer KL, Chu X, Dahlin A, Evers R, Fischer V, Hillgren KM, et al. (2010) Membrane transporters in drug development. *Nat Rev Drug Discov*, 9, 215–236.

Goodsaid F, Frueh F (2006) Process map proposal for the validation of genomic biomarkers. *Pharmacogenomics*, 7, 773–782.

Greenberg BD, Tolliver TJ, Huang SJ, Li Q, Bengel D, Murphy DL (1999) Genetic variation in the serotonin transporter promoter region affects serotonin uptake in human blood platelets. *Am J Med Genet*, 88, 83–87.

Grundy SM (2005) The issue of statin safety: where do we stand? *Circulation*, 111, 3016–3019.

Haerian BS, Roslan H, Raymond AA, Tan CT, Lim KS, Zulkifli SZ, Mohamed EH, Tan HJ, Mohamed Z (2010) ABCB1 C3435T polymorphism and the risk of resistance to antiepileptic drugs in epilepsy: a systematic review and meta-analysis. *Seizure*, 19, 339–346.

Hamidovic A, Hahn K, Kolesar J (2010) Clinical significance of ABCB1 genotyping in oncology. *J Oncol Pharm Pract*, 16, 39–44.

Hirschhorn JN, Daly MJ (2005) Genome-wide association studies for common diseases and complex traits. *Nat Rev Genet*, 6, 95–108.

Ho RH, Choi L, Lee W, Mayo G, Schwarz UI, Tirona RG, Bailey DG, Michael SC, Kim RB (2007) Effect of drug transporter genotypes on pravastatin disposition in European and African American participants. *Pharmacogenet Genomics*, 17, 647–656.

Ho RH, Tirona RG, Leake BF, Glaeser H, Lee W, Lemke CJ, Wang Y, Kim RB (2006) Drug and bile acid transporters in rosuvastatin hepatic uptake: function, expression, and pharmacogenetics. *Gastroenterology*, 130, 1793–1806.

Huang SM, Goodsaid F, Rahman A, Frueh F, Lesko LJ (2006) Application of pharmacogenomics in clinical pharmacology. *Toxicol Mech Methods*, 16, 89–99.

Huang SM, Temple R (2008) Is this the drug or dose for you? Impact and consideration of ethnic factors in global drug development, regulatory review, and clinical practice. *Clin Pharmacol Ther*, 84, 287–294.

Ieiri I, Suwannakul S, Maeda K, Uchimaru H, Hashimoto K, Kimura M, Fujino H, Hirano M, Kusuhara H, Irie S, Higuchi S, Sugiyama Y (2007) SLCO1B1 (OATP1B1, an uptake transporter) and ABCG2 (BCRP, an efflux transporter) variant alleles and pharmacokinetics of pitavastatin in healthy volunteers. *Clin Pharmacol Ther*, 82, 541–547.

Inada T, Koga M, Ishiguro H, Horiuchi Y, Syu A, Yoshio T, Takahashi N, Ozaki N, Arinami T (2008) Pathway-based association analysis of genome-wide screening data suggest that genes associated with the gamma-aminobutyric acid receptor signaling pathway are involved in neuroleptic-induced, treatment-resistant tardive dyskinesia. *Pharmacogenet Genomics*, 18, 317–323.

Ishikawa T, Sakurai A, Hirano H, Lezhava A, Sakurai M, Hayashizaki Y (2010) Emerging new technologies in Pharmacogenomics: rapid SNP detection, molecular dynamic simulation, and QSAR analysis methods to validate clinically important genetic variants of human ABC Transporter ABCB1 (P-gp/MDR1). *Pharmacol Ther*, 126, 69–81.

Izzedine H, Hulot JS, Villard E, Goyenvalle C, Dominguez S, Ghosn J, Valantin MA, Lechat P, Deray AG (2006) Association between ABCC2 gene haplotypes and tenofovir-induced proximal tubulopathy. *J Infect Dis*, 194, 1481–1491.

Johansson I, Lundqvist E, Bertilsson L, Dahl ML, Sjoqvist F, Ingelman-Sundberg M (1993) Inherited amplification of an active gene in the cytochrome P450 CYP2D locus as a cause of ultrarapid metabolism of debrisoquine. *Proc Natl Acad Sci U S A*, 90, 11825–11829.

Johne A, Kopke K, Gerloff T, Mai I, Rietbrock S, Meisel C, Hoffmeyer S, Kerb R, Fromm MF, Brinkmann U, Eichelbaum M, Brockmoller J, Cascorbi I, Roots I (2002) Modulation of steady-state kinetics of digoxin by haplotypes of the P-glycoprotein MDR1 gene. *Clin Pharmacol Ther*, 72, 584–594.

Johnson AD, Kavousi M, Smith AV, Chen MH, Dehghan A, Aspelund T, Lin JP, van Duijn CM, Harris TB, Cupples LA, Uitterlinden AG, Launer L, Hofman A, Rivadeneira F, Stricker B, Yang Q, O'Donnell CJ, Gudnason V, Witteman JC (2009) Genome-wide association meta-analysis for total serum bilirubin levels. *Hum Mol Genet*, 18, 2700–2710.

Kalliokoski A, Backman JT, Neuvonen PJ, Niemi M (2008a) Effects of the SLCO1B1*1B haplotype on the pharmacokinetics and pharmacodynamics of repaglinide and nateglinide. *Pharmacogenet Genomics*, 18, 937–942.

Kalliokoski A, Neuvonen M, Neuvonen PJ, Niemi M (2008b) The effect of SLCO1B1 polymorphism on repaglinide pharmacokinetics persists over a wide dose range. *Br J Clin Pharmacol*, 66, 818–825.

Kameyama Y, Yamashita K, Kobayashi K, Hosokawa M, Chiba K (2005) Functional characterization of SLCO1B1 (OATP-C) variants, SLCO1B1*5, SLCO1B1*15 and SLCO1B1*15+C1007G, by using transient expression systems of HeLa and HEK293 cells. *Pharmacogenet Genomics*, 15, 513–522.

Kathiresan S, Newton-Cheh C, Gerszten RE (2004) On the interpretation of genetic association studies. *Eur Heart J*, 25, 1378–1381.

Keskitalo JE, Kurkinen KJ, Neuvoneni PJ, Niemi M (2008) ABCB1 haplotypes differentially affect the pharmacokinetics of the acid and lactone forms of simvastatin and atorvastatin. *Clin Pharmacol Ther*, 84, 457–461.

Keskitalo JE, Kurkinen KJ, Neuvonen M, Backman JT, Neuvonen PJ, Niemi M (2009a) No significant effect of ABCB1 haplotypes on the pharmacokinetics of fluvastatin, pravastatin, lovastatin, and rosuvastatin. *Br J Clin Pharmacol*, 68, 207–213.

Keskitalo JE, Pasanen MK, Neuvonen PJ, Niemi M (2009b) Different effects of the ABCG2 c.421C>A SNP on the pharmacokinetics of fluvastatin, pravastatin and simvastatin. *Pharmacogenomics*, 10, 1617–1624.

Keskitalo JE, Zolk O, Fromm MF, Kurkinen KJ, Neuvonen PJ, Niemi M (2009c) ABCG2 polymorphism markedly affects the pharmacokinetics of atorvastatin and rosuvastatin. *Clin Pharmacol Ther*, 86, 197–203.

Kimura S, Umeno M, Skoda RC, Meyer UA, Gonzalez FJ (1989) The human debrisoquine 4-hydroxylase (CYP2D) locus: sequence and identification of the polymorphic CYP2D6 gene, a related gene, and a pseudogene. *Am J Hum Genet*, 45, 889–904.

Kitamura S, Maeda K, Wang Y, Sugiyama Y (2008) Involvement of multiple transporters in the hepatobiliary transport of rosuvastatin. *Drug Metab Dispos*, 36, 2014–2023.

Kiyotani K, Mushiroda T, Imamura CK, Hosono N, Tsunoda T, Kubo M, Tanigawara Y, Flockhart DA, Desta Z, Skaar TC, et al. (2010) Significant effect of polymorphisms in CYP2D6 and ABCC2 on clinical outcomes of adjuvant tamoxifen therapy for breast cancer patients. *J Clin Oncol*, 28, 1287–1293.

Kiyotani K, Mushiroda T, Kubo M, Zembutsu H, Sugiyama Y, Nakamura Y (2008) Association of genetic polymorphisms in SLCO1B3 and ABCC2 with docetaxel-induced leukopenia. *Cancer Sci*, 99, 967–972.

Kondo C, Suzuki H, Itoda M, Ozawa S, Sawada J, Kobayashi D, Ieiri I, Mine K, Ohtsubo K, Sugiyama Y (2004) Functional analysis of SNPs variants of BCRP/ABCG2. *Pharm Res*, 21, 1895–1903.

Kopplow K, Letschert K, Konig J, Walter B, Keppler D (2005) Human hepatobiliary transport of organic anions analyzed by quadruple-transfected cells. *Mol Pharmacol*, 68, 1031–1038.

Krupoves A, Seidman EG, Mack D, Israel D, Morgan K, Lambrette P, Costea I, Deslandres C, Grimard G, Law L, Levy E, Amre DK (2009) Associations between ABCB1/MDR1 gene polymorphisms and Crohn's disease: a gene-wide study in a pediatric population. *Inflamm Bowel Dis*, 15, 900–908.

Kwan P, Wong V, Ng PW, Lui CH, Sin NC, Poon WS, Ng HK, Wong KS, Baum L (2009) Gene-wide tagging study of association between ABCB1 polymorphisms and multidrug resistance in epilepsy in Han Chinese. *Pharmacogenomics*, 10, 723–732.

Kwon JM, Goate AM (2000) The candidate gene approach. *Alcohol Res Health*, 24, 164–168.

Lee E, Ryan S, Birmingham B, Zalikowski J, March R, Ambrose H, Moore R, Lee C, Chen Y, Schneck D (2005) Rosuvastatin pharmacokinetics and pharmacogenetics in white and Asian subjects residing in the same environment. *Clin Pharmacol Ther*, 78, 330–341.

Link E, Parish S, Armitage J, Bowman L, Heath S, Matsuda F, Gut I, Lathrop M, Collins R (2008) SLCO1B1 variants and statin-induced myopathy—a genomewide study. *N Engl J Med*, 359, 789–799.

Martin PD, Warwick MJ, Dane AL, Hill SJ, Giles PB, Phillips PJ, Lenz E (2003) Metabolism, excretion, and pharmacokinetics of rosuvastatin in healthy adult male volunteers. *Clin Ther*, 25, 2822–2835.

Mega JL, Close SL, Wiviott SD, Shen L, Hockett RD, Brandt JT, Walker JR, Antman EM, Macias W, Braunwald E, Sabatine MS (2009) Cytochrome p-450 polymorphisms and response to clopidogrel. *N Engl J Med*, 360, 354–362.

Morozova O, Marra MA (2008) Applications of next-generation sequencing technologies in functional genomics. *Genomics*, 92, 255–264.

Naesens M, Kuypers DR, Verbeke K, Vanrenterghem Y (2006) Multidrug resistance protein 2 genetic polymorphisms influence mycophenolic acid exposure in renal allograft recipients. *Transplantation*, 82, 1074–1084.

Neuvonen PJ (2010) Drug interactions with HMG-CoA reductase inhibitors (statins): the importance of CYP enzymes, transporters and pharmacogenetics. *Curr Opin Investig Drugs*, 11, 323–332.

Niemi M (2010) Transporter pharmacogenetics and statin toxicity. *Clin Pharmacol Ther*, 87, 130–133.

Niemi M, Backman JT, Kajosaari LI, Leathart JB, Neuvonen M, Daly AK, Eichelbaum M, Kivisto KT, Neuvonen PJ (2005a) Polymorphic organic anion transporting polypeptide 1B1 is a major determinant of repaglinide pharmacokinetics. *Clin Pharmacol Ther*, 77, 468–478.

Niemi M, Kivisto KT, Hofmann U, Schwab M, Eichelbaum M, Fromm MF (2005b) Fexofenadine pharmacokinetics are associated with a polymorphism of the SLCO1B1 gene (encoding OATP1B1). *Br J Clin Pharmacol*, 59, 602–604.

Niemi M, Neuvonen PJ, Hofmann U, Backman JT, Schwab M, Lutjohann D, von BK, Eichelbaum M, Kivisto KT (2005c) Acute effects of pravastatin on cholesterol synthesis are associated with SLCO1B1 (encoding OATP1B1) haplotype *17. *Pharmacogenet Genomics*, 15, 303–309.

Niemi M, Pasanen MK, Neuvonen PJ (2006) SLCO1B1 polymorphism and sex affect the pharmacokinetics of pravastatin but not fluvastatin. *Clin Pharmacol Ther*, 80, 356–366.

Nurmohamed L, Garcia-Bournissen F, Buono RJ, Shannon MW, Finkelstein Y (2010) Predisposition to epilepsy—does the ABCB1 gene play a role? *Epilepsia*, 51, 1882–1885.

Pasanen MK, Fredrikson H, Neuvonen PJ, Niemi M (2007) Different effects of SLCO1B1 polymorphism on the pharmacokinetics of atorvastatin and rosuvastatin. *Clin Pharmacol Ther*, 82, 726–733.

Pasanen MK, Neuvonen M, Neuvonen PJ, Niemi M (2006) SLCO1B1 polymorphism markedly affects the pharmacokinetics of simvastatin acid. *Pharmacogenet Genomics*, 16, 873–879.

Pe'er I, de Bakker PI, Maller J, Yelensky R, Altshuler D, Daly MJ (2006) Evaluating and improving power in whole-genome association studies using fixed marker sets. *Nat Genet*, 38, 663–667.

Raaijmakers JA, Koster ES, Maitland-van der Zee AH (2010) Pharmacogenetics and the pharmaceutical industry. *Curr Pharm Des*, 16, 238–244.

Ragoussis J (2009) Genotyping technologies for genetic research. *Annu Rev Genomics Hum Genet*, 10, 117–133.

Roden DM, Altman RB, Benowitz NL, Flockhart DA, Giacomini KM, Johnson JA, Krauss RM, McLeod HL, Ratain MJ, Relling MV, Ring HZ, Shuldiner AR, Weinshilboum RM, Weiss ST (2006) Pharmacogenomics: challenges and opportunities. *Ann Intern Med*, 145, 749–757.

Rodrigues AC (2010) Efflux and uptake transporters as determinants of statin response. *Expert Opin Drug Metab Toxicol*, 6, 621–632.

Rodriguez-Novoa S, Labarga P, Soriano V, Egan D, Albalater M, Morello J, Cuenca L, Gonzalez-Pardo G, Khoo S, Back D, Owen A (2009) Predictors of kidney tubular dysfunction in HIV-infected patients treated with tenofovir: a pharmacogenetic study. *Clin Infect Dis*, 48, e108–e116.

Sandhu MS, Waterworth DM, Debenham SL, Wheeler E, Papadakis K, Zhao JH, Song K, Yuan X, Johnson T, Ashford S, et al. (2008) LDL-cholesterol concentrations: a genome-wide association study. *Lancet*, 371, 483–491.

Seo T, Ishitsu T, Oniki K, Abe T, Shuto T, Nakagawa K (2008) ABCC2 haplotype is not associated with drug-resistant epilepsy. *J Pharm Pharmacol*, 60, 631–635.

Serretti A, Cusin C, Rossini D, Artioli P, Dotoli D, Zanardi R (2004) Further evidence of a combined effect of SERTPR and TPH on SSRIs response in mood disorders. *Am J Med Genet B Neuropsychiatr Genet*, 129B, 36–40.

Shah RR (2006) Can pharmacogenetics help rescue drugs withdrawn from the market? *Pharmacogenomics*, 7, 889–908.

Shin J, Kayser SR, Langaee TY (2009) Pharmacogenetics: from discovery to patient care. *Am J Health Syst Pharm*, 66, 625–637.

Shu Y, Brown C, Castro RA, Shi RJ, Lin ET, Owen RP, Sheardown SA, Yue L, Burchard EG, Brett CM, Giacomini KM (2008) Effect of genetic variation in the organic cation transporter 1, OCT1, on metformin pharmacokinetics. *Clin Pharmacol Ther*, 83, 273–280.

Shu Y, Sheardown SA, Brown C, Owen RP, Zhang S, Castro RA, Ianculescu AG, Yue L, Lo JC, Burchard EG, Brett CM, Giacomini KM (2007) Effect of genetic variation in the organic cation transporter 1 (OCT1) on metformin action. *J Clin Invest*, 117, 1422–1431.

Sisodiya SM, Goldstein DB (2007) Drug resistance in epilepsy: more twists in the tale. *Epilepsia*, 48, 2369–2370.

Sissung TM, Baum CE, Deeken J, Price DK, ragon-Ching J, Steinberg SM, Dahut W, Sparreboom A, Figg WD (2008) ABCB1 genetic variation influences the toxicity and clinical outcome of patients with androgen-independent prostate cancer treated with docetaxel. *Clin Cancer Res*, 14, 4543–4549.

Sissung TM, English BC, Venzon D, Figg WD, Deeken JF (2010) Clinical pharmacology and pharmacogenetics in a genomics era: the DMET platform. *Pharmacogenomics*, 11(1), 89–103.

Sissung TM, Mross K, Steinberg SM, Behringer D, Figg WD, Sparreboom A, Mielke S (2006) Association of ABCB1 genotypes with paclitaxel-mediated peripheral neuropathy and neutropenia. *Eur J Cancer*, 42, 2893–2896.

Smith GS, Lotrich FE, Malhotra AK, Lee AT, Ma Y, Kramer E, Gregersen PK, Eidelberg D, Pollock BG (2004) Effects of serotonin transporter promoter polymorphisms on serotonin function. *Neuropsychopharmacology*, 29, 2226–2234.

Sparreboom A, Gelderblom H, Marsh S, Ahluwalia R, Obach R, Principe P, Twelves C, Verweij J, McLeod HL (2004) Diflomotecan pharmacokinetics in relation to ABCG2 421C>A genotype. *Clin Pharmacol Ther*, 76, 38–44.

Sparreboom A, Loos WJ, Burger H, Sissung TM, Verweij J, Figg WD, Nooter K, Gelderblom H (2005) Effect of ABCG2 genotype on the oral bioavailability of topotecan. *Cancer Biol Ther*, 4, 650–658.

Tachibana-Iimori R, Tabara Y, Kusuhara H, Kohara K, Kawamoto R, Nakura J, Tokunaga K, Kondo I, Sugiyama Y, Miki T (2004) Effect of genetic polymorphism of OATP-C (SLCO1B1) on lipid-lowering response to HMG-CoA reductase inhibitors. *Drug Metab Pharmacokinet*, 19, 375–380.

Tilson MD, III, Ro CY (2006) The candidate gene approach to susceptibility for abdominal aortic aneurysm: TIMP1, HLA-DR-15, ferritin light chain, and collagen XI-Alpha-1. *Ann N Y Acad Sci*, 1085, 282–290.

Tirona RG (2005) Ethnic differences in statin disposition. *Clin Pharmacol Ther*, 78, 311–316.

Tohkin M, Ishiguro A, Kaniwa N, Saito Y, Kurose K, Hasegawa R (2010) Prediction of severe adverse drug reactions using pharmacogenetic biomarkers. *Drug Metab Pharmacokinet*, 25, 122–133.

Tomlinson B, Hu M, Lee VW, Lui SS, Chu TT, Poon EW, Ko GT, Baum L, Tam LS, Li EK (2010) ABCG2 polymorphism is associated with the low-density lipoprotein cholesterol response to rosuvastatin. *Clin Pharmacol Ther*, 87, 558–562.

Tsujii H, Konig J, Rost D, Stockel B, Leuschner U, Keppler D (1999) Exon-intron organization of the human multidrug-resistance protein 2 (MRP2) gene mutated in Dubin–Johnson syndrome. *Gastroenterology*, 117, 653–660.

Tzvetkov MV, Vormfelde SV, Balen D, Meineke I, Schmidt T, Sehrt D, Sabolic I, Koepsell H, Brockmoller J (2009) The effects of genetic polymorphisms in the organic cation transporters OCT1, OCT2, and OCT3 on the renal clearance of metformin. *Clin Pharmacol Ther*, 86, 299–306.

Ufer M, Mosyagin I, Muhle H, Jacobsen T, Haenisch S, Hasler R, Faltraco F, Remmler C, von SS, Kroemer HK, Runge U, Boor R, Stephani U, Cascorbi I (2009) Nonresponse to antiepileptic pharmacotherapy is associated with the ABCC2 -24C>T polymorphism in young and adult patients with epilepsy. *Pharmacogenet Genomics*, 19, 353–362.

Urquhart BL, Ware JA, Tirona RG, Ho RH, Leake BF, Schwarz UI, Zaher H, Palandra J, Gregor JC, Dresser GK, Kim RB (2008) Breast cancer resistance protein (ABCG2) and drug disposition: intestinal expression, polymorphisms and sulfasalazine as an in vivo probe. *Pharmacogenet Genomics*, 18, 439–448.

US FDA (2005) Crestor label. Available at: http://www.accessdata.fda.gov/drugsatfda_docs/label/2005/21366slr005lbl.pdf. Accessed on 25 January 2011.

US Food and Drug Administration (2011). Table of valid genomic biomarkers in the context of approved drug labels. Available at: http://www.fda.gov/Drugs/ScienceResearch/ResearchAreas/Pharmacogenetics/ucm083378.htm. Accessed on 24 January 2011.

Wang WY, Barratt BJ, Clayton DG, Todd JA (2005) Genome-wide association studies: theoretical and practical concerns. *Nat Rev Genet*, 6, 109–118.

White DL, Dang P, Engler J, Frede A, Zrim S, Osborn M, Saunders VA, Manley PW, Hughes TP (2010a) Functional activity of the OCT-1 protein is predictive of long-term outcome in patients with chronic-phase chronic myeloid leukemia treated with imatinib. *J Clin Oncol*, 28, 2761–2767.

White DL, Saunders VA, Dang P, Engler J, Hughes TP (2010b) OCT-1 activity measurement provides a superior imatinib response predictor than screening for single-nucleotide polymorphisms of OCT1. *Leukemia*, 24, 1962–1965.

White DL, Saunders VA, Dang P, Engler J, Venables A, Zrim S, Zannettino A, Lynch K, Manley PW, Hughes T (2007) Most CML patients who have a suboptimal response to imatinib have low OCT-1 activity: higher doses of imatinib may overcome the negative impact of low OCT-1 activity. *Blood*, 110, 4064–4072.

Williams JA, Andersson T, Andersson TB, Blanchard R, Behm MO, Cohen N, Edeki T, Franc M, Hillgren KM, Johnson KJ, et al. (2008) PhRMA white paper on ADME pharmacogenomics. *J Clin Pharmacol*, 48, 849–889.

Woodcock J, Woosley R (2008) The FDA critical path initiative and its influence on new drug development. *Annu Rev Med*, 59, 1–12.

Wu X, Gu J, Spitz MR (2008) *Strategies to Identify Pharmacogenomic Biomarkers: Candidate Gene, Pathway-Based, and Genome-Wide Approaches*. 1st ed. New Jersey: Humana Press.

Xiang X, Han Y, Neuvonen M, Pasanen MK, Kalliokoski A, Backman JT, Laitila J, Neuvonen PJ, Niemi M (2009) Effect of SLCO1B1 polymorphism on the plasma concentrations of bile acids and bile acid synthesis marker in humans. *Pharmacogenet Genomics*, 19, 447–457.

Xiang X, Jada SR, Li HH, Fan L, Tham LS, Wong CI, Lee SC, Lim R, Zhou QY, Goh BC, Tan EH, Chowbay B (2006) Pharmacogenetics of SLCO1B1 gene and the impact of *1b and *15 haplotypes on irinotecan disposition in Asian cancer patients. *Pharmacogenet Genomics*, 16, 683–691.

Xu CF, Reck BH, Xue Z, Huang L, Baker KL, Chen M, Chen EP, Ellens HE, Mooser VE, Cardon LR, Spraggs CF, Pandite L (2010) Pazopanib-induced hyperbilirubinemia is associated with Gilbert's syndrome UGT1A1 polymorphism. *Br J Cancer*, 102, 1371–1377.

Yamasaki Y, Ieiri I, Kusuhara H, Sasaki T, Kimura M, Tabuchi H, Ando Y, Irie S, Ware J, Nakai Y, Higuchi S, Sugiyama Y (2008) Pharmacogenetic characterization of sulfasalazine disposition based on NAT2 and ABCG2 (BCRP) gene polymorphisms in humans. *Clin Pharmacol Ther*, 84, 95–103.

Yan Q (2003) Bioinformatics and data integration in membrane transporter studies. *Methods Mol Biol*, 227, 37–60.

Yang JW, Lee PH, Hutchinson IV, Pravica V, Shah T, Min DI (2009) Genetic polymorphisms of MRP2 and UGT2B7 and gastrointestinal symptoms in renal transplant recipients taking mycophenolic acid. *Ther Drug Monit*, 31(5), 542–548.

Yee SW, Chen L, Giacomini KM (2010) Pharmacogenomics of membrane transporters: past, present and future. *Pharmacogenomics*, 11, 475–479.

Zhang W, Chen BL, Ozdemir V, He YJ, Zhou G, Peng DD, Deng S, Xie QY, Xie W, Xu LY, Wang LC, Fan L, Wang A, Zhou HH (2007a) SLCO1B1 521T–>C functional genetic polymorphism and lipid-lowering efficacy of multiple-dose pravastatin in Chinese coronary heart disease patients. *Br J Clin Pharmacol*, 64, 346–352.

Zhang W, He YJ, Gan Z, Fan L, Li Q, Wang A, Liu ZQ, Deng S, Huang YF, Xu LY, Zhou HH (2007b) OATP1B1 polymorphism is a major determinant of serum bilirubin level but not associated with rifampicin-mediated bilirubin elevation. *Clin Exp Pharmacol Physiol*, 34, 1240–1244.

Zhang W, Yu BN, He YJ, Fan L, Li Q, Liu ZQ, Wang A, Liu YL, Tan ZR, Fen J, Huang YF, Zhou HH (2006) Role of BCRP 421C>A polymorphism on rosuvastatin pharmacokinetics in healthy Chinese males. *Clin Chim Acta*, 373, 99–103.

Zhang L, Zhang YD, Strong JM, Reynolds KS, Huang SM (2008) A regulatory viewpoint on transporter-based drug interactions. *Xenobiotica*, 38, 709–724.

Zhu M, Zhao S (2007) Candidate gene identification approach: progress and challenges. *Int J Biol Sci*, 3, 420–427.

CHAPTER 3

REGULATORY PERSPECTIVE ON PHARMACOGENOMICS OF DRUG-METABOLIZING ENZYMES AND TRANSPORTERS

Lei Zhang
Gilbert J. Burckart
Lawrence J. Lesko
Shiew-Mei Huang

3.1 INTRODUCTION

Various intrinsic (e.g., age, race, gender, organ dysfunction, genetic) and extrinsic (e.g., drug interactions, food, smoking) factors can affect a drug's exposure and/or response (Huang and Temple, 2008). Understanding the effect of these factors on a drug's pharmacokinetics (PK) or pharmacodynamics (PD) can help explain the sources of variability in drug exposure and/or response, which can potentially lead to altered safety and efficacy of drug therapy for individual subgroups of patients. Genetic polymorphism or heterogeneity is one of the important intrinsic factors that can potentially affect all aspects of disease management and its medical treatment including the incidence of disease, the prognosis of disease progression or recurrence, the drug or drug class most likely to result in optimal benefit, the ideal therapeutic dose, the nature and extent of beneficial responses to treatment, and the likelihood of drug toxicity. The genetic factors include (1) genes relevant to the drug's PK (absorption, distribution, metabolism (including formation of active metabolites), and excretion (ADME)), (2) genes that code for intended or unintended drug targets and other pathways related to the drug's pharmacology, (3) genes not directly related to a drug's pharmacology that can predispose to toxicities such as immune reactions, and (4) genes that influence disease susceptibility or progression. All of these genetic factors can affect the benefit/risk profile of a drug product in a given disease situation.

Pharmacogenomic (PGx) research on genes that impact the ADME properties and pharmacodynamic or toxicity properties of drugs has impacted both drug

Pharmacogenomics of Human Drug Transporters: Clinical Impacts, First Edition.
Edited by Toshihisa Ishikawa, Richard B. Kim and Jörg König

development and utilization (Lesko and Zineh, 2010;Williams et al., 2008). The FDA has encouraged the sponsor to collect pharmacogenomic information early in the drug development process, explore their potential effect on drug exposure and response, and integrate genetic information into clinical trial design where appropriate through publications, public workshops, advisory committee meetings (Lesko and Woodcock, 2002; Lesko and Woodcock, 2004; Surh et al., 2010; Zineh and Huang, 2011), and, more recently, published a guidance on "Clinical Pharmacogenomics: Premarketing Evaluation in Early-Phase Clinical Studies and Recommendations for Labeling" (http://www.fda.gov/Drugs/GuidanceComplianceRegulatoryInformation/Guidances/ucm064982.htm).

This chapter describes the pharmacogenomics of drug-metabolizing enzymes (DMEs), transporters, and receptors, and their impact on drug PK, PD, efficacy, or safety. Several regulatory review examples are provided to demonstrate how the PGx information was incorporated into the labeling to improve the safe and effective use of various therapies. Highlights of the recently published draft on clinical pharmacogenomic guidance are also discussed. Finally, future perspectives are provided.

3.2 PHARMACOGENOMICS OF METABOLIZING ENZYMES

Contribution of genetics to variability in drug metabolism is well recognized. Pharmacogenetics of DMEs and drug transporters (DTPs) are known to affect drug exposure and thus may affect efficacy and safety. Four major phenotypes of cytochrome P450 (CYP) enzymes can be typically identified by phenotyping or genotyping: poor metabolizer (PM), intermediate metabolizer (IM), extensive metabolizer (EM), and ultrarapid metabolizer (UM). Per definition, in a PM phenotype, both alleles are null alleles, whereas individuals with only one null allele and one wild-type allele (heterozygotes) or individuals with two wild-type alleles (homozygotes) are called IM or EMs, respectively. For active drugs (where the administered drug has the desired pharmacological effect), PMs or IMs receiving the same dose would be predicted to accumulate higher drug concentrations than EMs due to decreased metabolism. The UMs, in contrast, are likely to have a less than expected drug concentrations. However, for prodrugs (where the administered drug has little or no pharmacological activity and is metabolized *in vivo* to form an active metabolite), the UMs are expected to have higher active metabolite exposure compared to other phenotypes. The DMEs, such as CYPs, are responsible for the majority of drug metabolism. Polymorphisms in CYP2C9, CYP2C19, and CYP2D6 have been shown to cause clinically significant differences in exposure of several drugs. Comparative PK in PM versus EM subjects of a polymorphic CYP enzyme can indicate the importance of this pathway in a drug's overall clearance when PK/PD relationships are known, and may indicate that important drug interaction by a strong inhibitor of this pathway can occur in clinical practice. Recently, the FDA has included genetic information related to CYP enzymes in the product labeling either at the time of drug approval or postmarketing to inform prescribers of the impact of these genetic factors on drug dosing alterations or increased safety risks. Pharmacogenomic information obtained from genomic investigations during the course of drug development and from post-marketing studies/trials have improved the dosing, effectiveness, and safety of drugs, as well as enhanced new product development and successful regulatory approvals (Lesko and Zineh, 2010; Surh et al., 2010; Zineh and Huang, 2011).

Three examples and their clinical relevance are provided below to illustrate the value of genomic information: CYP2C9 (and VKORC1) and warfarin; CYP2C19 and clopidogrel; and CYP2D6 and tetrabenazine. Although information from the warfarin and clopidogrel examples was obtained postmarketing rather than premarketing, it supports the potential values of having such data earlier when it could be linked to the outcomes in the clinical trials supporting safety and effectiveness.

3.2.1 CYP2C9 and Warfarin

There are more than 35 variants of CYP2C9 (http://www.cypalleles.ki.se/cyp2c9.htm). Two variants of CYP2C9 (CYP2C9*2 and *3) are known to be associated with reduced activity of the enzyme. The frequency of the CYP2C9 polymorphisms exhibits large interethnic variability (Williams et al., 2008; Xie et al., 2002). The frequencies of CYP2C9*2 and CYP2C9*3 in Caucasians are approximately 11% and 7%, respectively. In African Americans, the frequencies of CYP2C9*2 and CYP2C9*3 variants are less common, showing a frequency of 0.5–1%, respectively. The CYP2C9*2 variant is absent in East Asian populations (Korean, Japanese, and Chinese), but the frequency of the CYP2C9*3 variant is 1.1% in Korean, 2.2% in Japanese, and 3.3% in Chinese populations. Other CYP2C9 alleles associated with reduced enzymatic activity occur at lower frequencies, including *5, *6, and *11 alleles in populations of African ancestry and *5, *9, and *11 alleles in Caucasians.

The CYP2C9*2 and CYP2C9*3 proteins generally exhibit a reduced catalytic activity with increased K_m values and/or decreased V_{max} values, resulting in decreased apparent intrinsic clearance (V_{max}/K_m) for various substrates (Yamazaki et al., 1998). There are numerous studies indicating the clinical importance of CYP2C9 polymorphisms, and genotyping for CYP2C9*2 and CYP2C9*3 predicts slow metabolism of drugs cleared mainly by this enzyme, for example, warfarin and tolbutamide (Xie et al., 2002). The FDA has discussed the impact of CYP2C9 polymorphisms on dosing in the warfarin label (www.accessdata.fda.gov/drugsatfda_docs/label/2010/009218s108lbl.pdf; Kim et al., 2009).

Warfarin, an oral anticoagulant, is a vitamin K antagonist. Treatment with warfarin is complicated because of its narrow therapeutic range and complex dose–response relationship (Daly and King, 2003; Kim et al., 2009). The FDA's Adverse Event Reporting System indicated that warfarin is among the top 10 drugs with the largest number of serious adverse event reports submitted during the 1990 and 2000 decade (Wysowski et al., 2007). Major bleeding frequencies have been reported to be as high as 10–16%, although lower estimates have been reported in other studies. It is also essential to achieve adequate anticoagulation to prevent thromboembolic events that warfarin is intended to prevent. Relatively large variations in warfarin dose and resulting response were influenced by genes that affect warfarin PK and PD, as described below.Warfarin is a racemate. The S-enantiomer, which is primarily responsible for warfarin's pharmacologic effect, is mainly metabolized by

CYP2C9. Patients who are homozygous or heterozygous of CYP2C9 variant alleles cannot metabolize warfarin at the same rate as the wild-type CYP2C9 (*1/*1). For example, the clearance of S-warfarin was found to be threefold lower in CYP2C9*2 homozygous carriers and tenfold lower in CYP2C9*3 homozygous carriers (Scordo et al., 2002). Reduced clearance in patients with CYP2C9 variants is associated with higher exposures which lead to the need for reduced warfarin maintenance doses, longer time to achieve a stable dose, and a higher risk of serious bleeds especially during the induction period. Correlation between average warfarin dose requirements and CYP2C9 (and VKORC1) genotype in patient with various ethnicity/race has been established (Sanderson et al., 2005; Scordo et al., 2002), which further highlights the importance of understanding the heterogeneity and the impact of CYP2C9 polymorphism in clinical practice. In August 2007, the FDA included the impact of CYP2C9 polymorphisms on dosing in the warfarin label to provide general information about genetic testing and to encourage healthcare providers to use this information for initial dosing to reduce the risk of both bleeding and undertreatment (www.accessdata.fda.gov/drugsatfda_docs/label/2010/009218s108lbl.pdf). For example, a pharmacogenetic algorithm was shown to predict the therapeutic steady-state warfarin dose more accurately than the one using clinical factors and international normalized ratio (INR) response alone (Lenzini et al., 2010).

In addition to CYP2C9 polymorphisms, variations in the vitamin K epoxide reductase (VKOR), which is the target for vitamin K antagonists, were shown to have an important impact on warfarin response. Polymorphisms in *VKORC1* affect an individual's response to warfarin. The major polymorphism, mutation in the *VKORC1* promoter region (-1639G>A), for example, decreases the expression of the gene and increases the responsiveness to warfarin. Patients with these polymorphisms (either heterozygous -1639G/A genotype or -1639A/A homozygous genotype) will generally require lower doses than patients with the -1639G/G genotype. There is a relatively high frequency of the -1639A/A genotype in different racial groups, ranging from approximately 3% and 20% in African Americans and Caucasians, respectively to approximately 80–90% in Asians (Marsh et al., 2006; Owen et al., 2010; Yuan et al., 2005). On the other hand, the prevalence of -1639G/G genotype is approximately 58%, 39%, and 1.3% in African Americans, Caucasians, and Asians, respectively (Yang et al., 2010).

Polymorphisms of *CYP2C9* and *VKORC1* genes alone consistently account for approximately 30% of the variability in warfarin dose requirement (Gage and Lesko, 2008). Common genetic variants in the CYP2C9 and VKORC1 enzymes, along with known nongenetic factors (e.g., age, sex), account for approximately 50% of warfarin dose variability (Johnson et al., 2011). Studies have shown that patients carrying variant *CYP2C9* and/or *VKORC1* genotypes had a higher chance of major hemorrhage during warfarin therapy as a result of overdosing, and that genotypes play an important role in determining a more customized dose of warfarin compared to a one-size-fits-all dose (Caraco et al., 2008; Gage et al., 2008; Kim et al., 2009). Pharmacogenetics-based dose adjustments are one tool to individualize drug treatment according to the genetic factors. In January 2010, the warfarin labeling was updated by the FDA to include a dosing table to be considered for initial dosing based specifically on both *CYP2C9* and *VKORC1* genotypes (www.accessdata.

fda.gov/drugsatfda_docs/label/2010/009218s108lbl.pdf) (see below). In the **Dosage and Administration** Section of the warfarin label, it states that:

> The patient's CYP2C9 and VKORC1 genotype information, when available, can assist in selection of the starting dose. Table 5 describes the range of stable maintenance doses observed in multiple patients having different combinations of CYP2C9 and VKORC1 gene variants. Consider these ranges in choosing the initial dose.
>
> In all patients, subsequent dosage adjustments must be made based on the results of PT/INR determinations.

TABLE 5 Range of Expected Therapeutic Warfarin Doses Based on CYP2C9 and VKORC1 Genotypes†

	CYP2C9					
VKORC1	*1/*1	*1/*2	*1/*3	*2/*2	*2/*3	*3/*3
GG	5–7 mg	5–7 mg	3–4 mg	3–4 mg	3–4 mg	0.5–2 mg
AG	5–7 mg	3–4 mg	3–4 mg	3–4 mg	0.5–2 mg	0.5–2 mg
AA	3–4 mg	3–4 mg	0.5–2 mg	0.5–2 mg	0.5–2 mg	0.5–2 mg

†Ranges are derived from multiple published clinical studies. Other clinical factors (e.g., age, race, body weight, sex, concomitant medications, comorbidities) are generally accounted for along with genotype in the ranges expressed in the Table. VKORC1 −1639 G->A (rs9923231) variant is used in this table. Other co-inherited VKORC1 variants may also be important determinants of warfarin dose. Patients with CYP2C9 *1/*3, *2/*2, *2/*3 and *3/*3 may require more prolonged time (>2–4 wks) to achieve maximum INR effect for a given dosage regimen.

3.2.2 CYP2C19 and Clopidogrel

More than 25 alleles for CYP2C19 have been identified (http://www.cypalleles.ki.se/cyp2c19.htm). Two variants, CYP2C19*2 and CYP2C19*3, which are nonfunctional alleles, account for the majority of Caucasian (85%) and Asian (99%) PMs. Some of the less frequent variants associated with PM status include but not limited to CYP2C19*4, *5, *6, *7, and *8 (http://www.cypalleles.ki.se/cyp2c19.htm). Individuals with PM status possess two nonfunctional alleles. The PM frequency varies between different populations, and there are more PMs among Asians (~15%) than among Caucasians and Africans/blacks (~3%) (Desta et al., 2002). On the other hand, the frequency of ultrarapid metabolizers of CYP2C19 (CYP2C19*17) is about 18%, 4%, and 1.3% in Swedish, Chinese, and Japanese populations, respectively (Sim et al., 2006).

Clopidogrel is an oral, thienopyridine class of antiplatelet agent. Clopidogrel is a prodrug with no inherent antiplatelet activity, but about 15% of the dose is metabolized to an active metabolite in a two-step process involving multiple CYP enzymes, one of which is the polymorphic CYP2C19 (http://www.accessdata.fda.gov/drugsatfda_docs/label/2011/020839s051lbl.pdf). If a prodrug (inactive) is converted to the active metabolite(s) by a polymorphic enzyme, the therapeutic effect may be absent in a subgroup of the population. Clopidogrel represents a case where a polymorphic enzyme plays an important role in the formation of its active metabolite that affects the drug's PD.

The PM status of CYP2C19 is associated with a diminished clinical response to clopidogrel. Namely, data from several PK/PD studies demonstrated that carriers of nonfunctional allele of CYP2C19 showed reduced exposure to the active metabolite and less inhibition of platelet aggregation, which is considered an intermediate biomarker for efficacy (Mega et al., 2009). These observations have been independently replicated in multiple clinical pharmacology studies. The relationship between the CYP2C19 genotype and PK/PD was further extended to clinical outcomes in several population or clinical trial-based cohort studies (Mega et al., 2010). For example, in TRITON-TIMI 38 study, carriers of nonfunctional CYP2C19 alleles had a 1.5–3-fold higher rate of death, nonfatal myocardial infarction, or nonfatal stroke as compared to noncarriers (no nonfunctional allele) following percutaneous coronary intervention (PCI). Additionally, the rate of stent thrombosis over the same time period was approximately three times higher in carriers of CYP2C19*2 or CYP2C19*3 (Mega et al., 2010; Mega et al., 2011).

Between May 2009 and December 2011, several sections of the clopidogrel label were updated to include PGx information related to the diminished antiplatelet responses and the increased risk of cardiovascular events in patients with reduced CYP2C19 function (Tables 3.1 and 3.2).

The CYP2C19 information was added to the box warning (see below), where "consider alternative treatment" was one of the recommendations in patients identified as CYP2C19 PMs, in addition to Dosage and Administration (see below) and other sections of the label (Table 3.2).

WARNING: DIMINISHED EFFECTIVENESS IN POOR METABOLIZERS

The effectiveness of Plavix is dependent on its activation to an active metabolite by the cytochrome P450 (CYP) system, principally CYP2C19 (see Warnings and Precautions (5.1)). Plavix at recommended doses forms less of that metabolite and has a smaller effect on platelet function in patients who are CYP2C19 poor metabolizers. Poor metabolizers with acute coronary syndrome or undergoing percutaneous coronary intervention treated with Plavix at recommended doses exhibit higher cardiovascular event rates than do patients with normal CYP2C19 function. Tests are available to identify a patient's CYP2C19 genotype; these tests can be used as an aid in determining therapeutic strategy (see Clinical Pharmacology (12.5)). Consider alternative treatment or treatment strategies in patients identified as CYP2C19 poor metabolizers (see Dosage and Administration (2.3)).

DOSAGE AND ADMINISTRATION

2.3 CYP2C19 Poor Metabolizers

CYP2C19 poor metabolizer status is associated with diminished antiplatelet response to clopidogrel. Although a higher dose regimen in poor metabolizers increases antiplatelet response (see Clinical Pharmacology (12.5)), an appropriate dose regimen for this patient population has not been established.

TABLE 3.1 Ethnic Differences in Allele Frequencies for Selected CYP Enzymes and Transporters

Gene	Affected drugs	Variant allele	Alteration in allele function	Allele frequencies (%)				
				White	Black	Asian	Chinese	Japanese
CYP2A6	Nicotine	*1	Normal	46	54		34	
		*4	Deletion	0; 1.2	0.6; 1.9	10.8[a]	6.7	19; 24.2
		*7	Reduced	0; 0.3	0	9.8[a]	3.1	9.8; 6.3
		*9	Reduced	8.0; 7.1	8.5; 7.1	19.3[a]	15.6	19.0; 20.3
		*10	Reduced or virtually absent	0	0		0.4	1.6
		*17	Reduced	0	10.5	0		0
CYP2B6	Cyclophosphamide, Efavirenz, Nevirapin, Bupropion, Methadone	*1	Normal	50.7	44.3	30		68.0
		*4	Increased	6	2	4		
		*5	No change	3; 12	5; 3	3		
		*6	Reduced	28; 25.6	34; 32.8	23	16.2	18
		*7	Reduced	3	1	0		
		*9	Reduced	1	1	0		
		*11	Reduced	1	0	0		
		*15	Unknown	0.4	0	0		
		*16	Reduced		7			
		*17	Unknown	0	6; 7.1	0		
		*18	Reduced	0	9; 2.9	0		
		*22	Unknown	3.3	1.1	0		

(continued)

TABLE 3.1 (*Continued*)

Gene	Affected drugs	Variant allele	Alteration in allele function	Allele frequencies (%)				
				White	Black	Asian	Chinese	Japanese
CYP2C8	Repaglinide	*1	Normal	87; 77.5	80		100	
		*2	Reduced	0.4	18			0
		*3	Reduced	13–15	2			0
		*4	Reduced	7.5				0
CYP2C9	Warfarin, Celecoxib	*1	Normal	80	>90	>90	96.3	98.9
		*2	Reduced	10–13	3	Absent or rare		none
		*3	More reduced	5.6–8	1			3.5
CYP2C19	Omeprazole, Pantoprazole, Esomeprazole, Rabeprazole, Voriconazole, Clopidogrel, Prasugrel	*1	Normal	86; 85.3	75; 81.4; 82.3	62	54.1; 64.7	67
		*2	Nonfunctional	13.6–15	17		30	34.5[b]
		*3	Nonfunctional	<1	<1		5	9[b]
CYP2D6[c]	Atomoxetine, Venlafaxine, Risperidone, Tamoxifen, Timlol, Fluoxetine, Tolterodine, Terbinafine, Tramadol, Clozapine, Aripiprazole, Metoprolol, Propranol, Carvedilol, Terbinafine, Propafenone, Thioridazine, Protriptyline, Tetrabenazine, Codeine	EM	Normal	71; 70; 73.4	37; 48	51.5	45	55.2
		PM	Nonfunctional	7.7	1.9–7.3	0–4.8		
		IM	Decreased	1–2		51		
		UM	Increased	4.3	4.9	0.9		

CYP3A5 (WE = white European; WC = white Canadian; AA = African American; Z = Zimbabwean)	Tacrolimus	*1A	Normal	8–15; 5–7 WE	45; 40 AA	23–40; 25	22	23
		*1B	None (controversial)	0 WC; 0.5–3.0 WE	0AA; 0Z			
		*1C	Unknown	4.6 WC; 3.0 WE	7.0 AA; 0 Z	0		
		*2	Unknown	0.7WC; 2.0 WE	0 AA; 0 Z			
		*3	No function	92.9 WC; 70 WE	27.0–50.0 AA; 77.6 Z	75.0		
		*5	Unknown	0 WC	0Z	0.9		
		*6	Decreased function	0 WC; 0 WE	13.0 AA; 22.0 Z	0		
		*7	No function	0	10.0 AA; 10.0 Z	0		
ABCB1 (P-gp)[d]	Digoxin	Wild-type (C1236T) C allele	Normal	34	6		20.0	45.5
		*1	Unknown	15	15	15		
		*13	Unknown	34	5	34–37		
		*26	Unknown	10	9	5		
		*21	Unknown	3	8			
		*11	Unknown	1	2	23		
		*14	Unknown	10	2	2		

(continued)

TABLE 3.1 (*Continued*)

Gene	Affected drugs	Variant allele	Alteration in allele function	Allele frequencies (%)				
				White	Black	Asian	Chinese	Japanese
ABCG2 (BCRP)	Topotecan, Irinotecan, Mitoxantrone, Rosuvastatin, Methotrexate	Wild type	Normal	81; 94.4	93.7			82.4
		34G>A (V12M)	Reduced	2	4	45[e]; 19.8[f]	20	15
		421C>A (Q141K)	Reduced	14, 10	0–5	15[e]; 27.8[f]	35	35
SCLCO1B1 (OATP1B1)	Pravastatin, Rosuvastatin, Simvastatin	*1a	Normal	32.5				35
		*1b (388A>G)	Increased (possibly protein expression)	30; 30–51; 38	74; 75; 77	57–88	59.9	46.9; 53.7; 63–67
		*2	Unknown	2	0			0
		*4	Unknown	16	2			0
		*5 (521T>C)	Decreased	14; 2.7% (Finnish)	2			0.7
		*6	Unknown	2	0			
		*7	Unknown	1	0			
		*8	Unknown	1	0			
		*9	Unknown	0	9			
		*10	Unknown	2	0			
		*11	Unknown	2	34			

	*15 (both 388A>G and 521T>C)	Decreased	2.4 (Finnish)		14	3.7; 10.3
	*16	Unknown	7.9 (Finnish)			3.7
	*17 (388A>G and 521T>C, and -11187G>C)	Decreased	6.9 (Finnish)			13.3

Source: Modified from Yasuda et al. (2008).

Note: Data from different references are separated by semicolons. Blank cells indicate that data are not found in the literature.

[a] for CYP2A6, Asian refers to Korean.

[b] Native Japanese.

[c] for CYP2D6,

PM: denotes individuals with 2 of the following non-functional alleles: *3, *4, *5, *6, *7, *8, *14, *18, *21, and *44.

IM: denotes individuals with one non-functional alleles and one functional allele or two reduced function alleles: *10 and *17.

UM: denotes individuals with two or more copies of the functional alleles.

[d] Mexican Americans had allele frequencies of 34% for *26 and 20% for *11; Asian refers to Asian-Americans.

[e] South East Asia (non-Chinese, non-Japanese).

[f] Korean.

TABLE 3.2 Examples of Inclusion of Pharmacogenomics Information in the FDA Drug Labeling

Drug name	Genetic biomarkers	Specific issues	Clinical impact	Labeling sections	Date of last label update
Warfarin	*CYP2C9, VKORC1*	Over-anticoagulation	Efficacy, safety	Clinical pharmacology (metabolism and pharmacogenomics subsections), precautions, dosage and administration	Oct 2011
Clopidogrel	*CYP2C19*	Reduced efficacy	Efficacy, safety	Highlights, boxed warning, dosage and administration, warnings and precautions, clinical pharmacology (pharmacogenomic subsection)	Dec 2011
Tetrabenazine	*CYP2D6*	Increased risk of depression, suicidal, QT prolongation	Safety	Highlights, dosage and administration, warnings and precautions (laboratory tests subsection), use in specific populations, clinical pharmacology	Aug 2012
Carbamazepine	*HLA-B**1501*	SJS–TEN	Safety	Boxed warning, warnings, precautions (laboratory tests subsection)	Dec 2012
Abacavir	*HLA-B**5701*	Hypersensitivity reaction (HSR)	Safety	Boxed warning, highlights, contraindications, warnings and precautions, patient counseling information	May 2012
Panitumumab	*KRAS*	Absence of benefit	Efficacy, safety	Highlights, indication and usage, clinical pharmacology, clinical studies	Aug 2012
Irinotecan	*UGT1A1*	Neutropenia	Safety	Highlights, dose and administration, warnings and precautions, clinical pharmacology	July 2012
Mercaptopurine	*TPMT*	Myelosuppression	Safety	Clinical pharmacology, warnings, precautions (laboratory testing and drug–drug interaction subsections), adverse reactions, dosage and administration	May 2011

Source: http://www.fda.gov/Drugs/ScienceResearch/ResearchAreas/Pharmacogenetics/ucm083378.htm and Drugs@FDA.

Some practitioners have developed an algorithm to indicate when alternate treatments or treatment strategies (including dose increases) are recommended for patients with variant CYP2C19 genotypes (Scott et al., 2011). One algorithm included recommendation for ultrarapid metabolizers of CYP2C19*17 (http://www.stsiweb.org/index.php/news_events/detail/vanderbilt_now_also_routinely_gene_testing_for_clopidogrel_metabolizer_stat).

3.2.3 CYP2D6 and Tetrabenazine

Polymorphism of CYP2D6 is one of the well-studied CYPs in terms of gene polymorphisms. More than 100 variants have been identified including a number of nonfunctional alleles (due to deletion, insertion, splicing defects, frame shifts, or premature stop codons) (e.g., *CYP2D6*3, *4, *5, *6, *7, *8, *14, *18, *21,* and **44*), reduced functional alleles (e.g., *CYP2D6*10* and **17*), and gene amplification leading to ultrarapid metabolism (e.g., multiple copies of CYP2D6*2 genotype) (www.imm.ki.se/cypalleles/). Based on the genotype, individuals could be UMs (with two or more copies of the functional alleles), EMs with normal activity, IMs with reduced activity (individuals with one nonfunctional alleles and one functional allele or two reduced functional alleles), or PMs with little or no activity (with two nonfunctional alleles). Approximately 10% of Caucasians appear to have the nonfunctional *CYP2D6*4* allele, whereas approximately 50% of Asians possess the reduced function *CYP2D6*10* allele (Droll et al., 1998). Both Africans and African Americans have reduced function alleles representing 35% of allele variation, mainly *CYP2D6*17* (Bradford, 2002). The prevalence of CYP2D6 PMs is approximately 6–10% in Caucasians, but is lower in Asians (2%). Reports on the prevalence of PMs in African populations differ widely, with estimates varying in the range of 0–19% (Bernard et al., 2006).

Tetrabenazine is a vesicular monoamine transporter 2 (VMAT) inhibitor approved by the FDA in 2008 for the treatment of chorea associated with Huntington's disease (HD), a rare disease. Tetrabenazine is mainly metabolized by carbonyl reductase to the active metabolites α-dihydrotetrabenazine (HTBZ); and β-HTBZ. α-HTBZ and β-HTBZ are metabolized primarily by polymorphic CYP2D6 (http://www.accessdata.fda.gov/drugsatfda_docs/nda/2008/021894s000TOC.cfm; http://www.accessdata.fda.gov/drugsatfda_docs/label/2011/021894s004lbl.pdf). Although the PK of tetrabenazine and its metabolites have not been studied in CYP2D6 PM subjects, it is likely that the exposure to α-HTBZ and β-HTBZ would be increased in CYP2D6 PM subjects based on observed increased exposure in subjects taking strong CYP2D6 inhibitors (three- and ninefold, respectively). Using modeling and simulation, the relative exposure in PMs versus EMs was estimated based on the information from drug interaction studies, and various genotype-appropriate doses were recommended. Because there is an increased risk of drug-associated depression, suicidality, and QTc prolongation with increased dose of tetrabenazine and anticipated increased exposure in CYP2D6 PMs compared to CYP2D6 EMs, the FDA recommends genotyping the patients for CYP2D6 prior to administering a higher dose (i.e., >50 mg/day) for tetrabenazine; the maximum single dose for CYP2D6 PMs is 25 mg and the maximum daily dose is 50 mg (i.e., 25 mg twice daily) (http://www.accessdata.fda.gov/drugsatfda_docs/nda/2008/021894s000TOC.cfm; http://www.accessdata.fda.gov/drugsatfda_docs/label/2011/021894s004lbl.pdf).

The following recommendations are in the tetrabenazine label.

DOSAGE AND ADMINISTRATION

Dosing Recommendations Above 50 mg per day

Patients who require doses of XENAZINE greater than 50 mg/day should be first tested and genotyped to determine if they are PM or EMs by their ability to express the DME, CYP2D6. The dose of XENAZINE should then be individualized accordingly to their status as PMs or EMs (see Warnings and Precautions (5.2, 5.4), Use in Specific Populations (8.8), and Clinical Pharmacology (12.3)).

Extensive and Intermediate CYP2D6 Metabolizers

Genotyped patients who are identified as EMs or IMs of CYP2D6, who need doses of XENAZINE above 50 mg/day, should be titrated up slowly at weekly intervals by 12.5 mg daily to allow the identification of a tolerated dose that reduces chorea.

Poor CYP2D6 Metabolizers

In PMs, the initial dose and titration is similar to EMs except that the recommended maximum single dose is 25 mg, and the recommended daily dose should not exceed a maximum of 50 mg (see Warnings and Precautions (5.2), Use in Specific Populations (8.8), and Clinical Pharmacology (12.3)).

USE IN SPECIFIC POPULATIONS

8.8 Use in Poor or Extensive CYP2D6 Metabolizers

Patients who require doses of XENAZINE greater than 50 mg/ day should be first tested and genotyped to determine if they are PMs or EMs by their ability to express the DME, CYP2D6. The dose of XENAZINE should then be individualized accordingly to their status as either PMs or EMs (see Dosage and Administration (2.2), Warnings and Precautions (5.2, 5.4), and Clinical Pharmacology (12.3)).

Poor Metabolizers

Poor CYP2D6 metabolizers (PMs) will have substantially higher levels of exposure to the primary metabolites (about threefold for α-HTBZ and ninefold for β-HTBZ) compared to EMs. The dosage should, therefore, be adjusted according to a patient's CYP2D6 metabolizer status by limiting a single dose to a maximum of 25 mg and the recommended daily dose not to exceed a maximum of 50 mg/day in patients who are CYP2D6 PMs (see Dosage and Administration (2.2), Warnings and Precautions (5.2, 5.4), and Clinical Pharmacology (12.3)).

Extensive / Intermediate Metabolizers

In EMs or IMs, the dosage of XENAZINE can be titrated to a maximum single dose of 37.5 mg and a recommended maximum daily dose of 100 mg (see Dosage and Administration (2.2), Drug Interaction (7.1), and Clinical Pharmacology (12.3)).

Tetrabenazine represents a case where drug product labeling has included a recommendation of genotyping before administering a specific dose (i.e., >50 mg/day). Conceptually, in this case, an individual who is a CYP2D6 EM can be converted to a *de facto* CYP2D6 PM by concomitant administration of a strong CYP2D6 inhibitor. Drug interaction data with a strong CYP2D6 inhibitor were used to project the increased exposure in CYP2D6 PMs for labeling dose recommendations.

3.3 PHARMACOGENOMICS OF TRANSPORTERS

Membrane transporters are the major determinants of the absorption, distribution, and elimination, and thereby the systemic and tissue-specific drug levels of commonly used drugs (Giacomini et al., 2010). The genetic polymorphisms of DTPs can therefore contribute to the interindividual variability of PK, PD, effectiveness, or toxicity in a way similar to that of the CYP enzymes.

In the past decade, the field of membrane transporter pharmacogenomics has undergone enormous growth. Genetic variants in many transporters belonging to the two major superfamilies of transporters, ATP-binding cassette (ABC) and solute carrier (SLC), have been identified and functionally characterized (Fahrmayr et al., 2010; Maeda and Sugiyama, 2008; Yee et al., 2010). Some of these transporters include the ABC transporters, such as P-glycoprotein (P-gp or MDR1), breast cancer resistance protein (BCRP), and multidrug resistance-associated protein 2 (MRP2), and SLC transporters such as organic cation transporters 1 and 2 (OCT1 and OCT2), organic anion transporters 1 and 3 (OAT1 and OAT3), organic anion transporting polypeptides (OATP1B1, OATP1B3, OATP2B1, and OATP1A2), and multidrug and toxin extrusion transporters 1 and 2 (MATE1 and MATE2K). The review of the National Human Genome Research Institute (NHGRI) reveals catalogs of numerous genome-wide association studies (GWAS) that describe significant associations of single-nucleotide polymorphisms (SNPs) within or in close proximity to membrane transporter genes with phenotypes related to drug concentrations in plasma, drug efficacy and drug toxicity. Results from clinical studies have suggested that the genetic polymorphisms of DTPs (e.g., OATP1B1, MDR1, BCRP, and OCT1) influence the PK and related pharmacological and toxicological effects of drugs. The functional regulation of a transporter located in drug clearance organs such as liver and kidney should affect the drug concentration in the blood circulation, whereas a transporter located in a drug target can alter the concentration at the target site and affect the resulting response, without concurrent changes in plasma concentrations.

Increasing numbers of GWAS have included the assessment of gene expression levels in various human tissues such as lymphoblastoid cells, liver, adipose tissue, and cortical cells. GWAS have identified important quantitative trait loci that inform clinically relevant phenotypes (Schadt et al., 2008). With continuing advances in sequencing technologies and large-scale human variation studies, functional variants in noncoding regions of transporters will be discovered, and large clinical studies will result in the identification of variants in multiple genes including transporter genes, which may contribute to variation in clinical drug response.

Although various polymorphisms of transporters have been identified and their clinical significance has been evaluated, the information has not been included in the FDA product labeling to date. The limited availability of genetic testing for transporter polymorphism may have posed a challenge in the utilization of transporter polymorphism information in guiding clinical use of therapeutics in patients. Examples of polymorphisms of ABC and SLC transporters and their potential impact on drug exposure and response are discussed below. A more detailed discussion of polymorphisms of various DTPs is provided in this book (Chapters 6–17). As screening of patients for drug-related metabolic enzyme and transporter polymorphisms becomes routine, the information regarding the magnitude of the effect (toxicity, drug–drug interaction, or therapeutic effect) of the transporter polymorphism will become more important.

3.3.1 ABC Transporters

3.3.1.1 P-glycoprotein (ABCB1) P-glycoprotein (P-gp) encoded in the *ABCB1* gene is the most extensively studied membrane transporter. The substrates for P-gp include a wide variety of lipophilic molecules. Early studies in the 1980s established P-gp's association with drug resistance in cancer chemotherapy. Since that time, studies in cancer resistance now include many of the ABC and SLC transporters (Huang, 2007).

The expression of P-gp is concentrated in several areas of the body where it serves as a protective mechanism against xenobiotics. At the level of the central nervous system, P-gp is highly expressed in the endothelial cells of blood vessels to form the "blood–brain barrier." In the intestine, P-gp acts to limit the absorption of P-gp substrates. In the liver and kidney, P-gp plays a role as an excretory mechanism. Even in T-lymphocytes, P-gp has readily measurable activity.

Therefore, inhibition or induction of the function of P-gp can have physiological and pharmacological consequences. The pharmacological effects of altering P-gp function then could range from changes in drug absorption and bioavailability to changes in drug distribution to the central nervous system and other organs, to changes in drug elimination. For this reason, drugs are now routinely tested during development as to whether they are substrates or inhibitors of P-gp. Although this is a sensible approach, considerable variability can be encountered in the clinical testing of a drug's ability to interact with P-gp function.

The question then arises as to whether part of this variability can be explained by genetic differences in the expression of P-gp. In the *ABCB1* gene encoding P-gp, more than 100 SNPs have been identified (http://pharmacogenetics.ucsf.edu/ and http://www.pharmgkb.org/gene/PA267?tabType=tabVip). Through extensive studies, two primary SNPs are found to be associated with changes in P-gp expression and/or function. In the *ABCB1* gene, a nonsynonymous G–T or A transition at position 2677 within exon 21 (rs2032582) is associated with altered function, as is the synonymous C–T transition at position 3435 within exon 26 (rs1045642). The explanations for the effect of the *ABCB1* C3435T SNP on P-gp function include linkage with the G2677T SNP, decreased mRNA stability, and change in substrate specificity. Most studies have examined associations with the homozygous *ABCB1* 2677 TT and 3435 TT genotypes, who are considered to be low expressors of P-gp function, or their corresponding haplotypes. Examples where these *ABCB1* genotypes or haplotypes have been associated with lower P-gp function include higher digoxin exposure and higher dose-adjusted tacrolimus blood concentrations. Most of the previous studies of associations between drug effect or drug concentration and *ABCB1* genotype have demonstrated only a modest effect difference between genotypes and haplotypes. For example, in organ transplantation, a marginally significant effect of *ABCB1* genotype has been demonstrated on steroid dependency, acute rejection, and tacrolimus dose requirements (Wang et al., 2006; Zheng et al., 2004; Zheng et al., 2005).

One study from 2003 investigated the effects of *ABCB1* genotypes on the interaction between loperimide, a P-gp substrate, and quinidine, a strong P-gp inhibitor, in healthy subjects (Skarke et al., 2003). Loperimide is an opioid, so it provides the opportunity to examine both the plasma concentration changes and the changes in effect through a biomarker, the change in pupillary diameter. In the 20 healthy subjects in this study, the co-administration of quinidine with loperimide produced both higher plasma concentrations of loperimide and significant changes in restriction in pupil size. When examining the *ABCB1* genotypes and haplotypes, the investigators found that the *ABCB1* 3435 TT genotype was associated with the largest percent decrease in pupil size when both loperimide and quinidine were administered. The haplotype that included 3435 TT also had the largest increase in loperimide plasma concentrations when both of the drugs were administered. The differences between *ABCB1* genotypes and haplotypes were not dramatically different.

In a study of the interaction between tacrolimus and fluconazole in transplant patients (Kuypers et al., 2008), the investigators genotyped subjects for both *CYP3A* and *ABCB1* polymorphisms. They examined the tacrolimus trough blood concentrations in relation to drug doses at steady state, and found that changes in exposure associated with the administration of fluconazole were not associated with *ABCB1* genotypes. However, they did find that *CYP3A5**1* expressors had a reduced susceptibility to the inhibitory effects of fluconazole on tacrolimus metabolism. Although this study suggests that CYP3A genetics may be more important than *ABCB1* genetics, many drugs are substrates for both CYP3A and P-gp so that polymorphisms in both the enzyme and the transporter may determine the final impact of an interaction.

The interaction between P-gp and CYP3A extends to the genetic level, and is an important consideration for the large number of drugs that are substrates of both. In a 2006 study using human liver cells and intestinal cells, Lamba et al, 2006 found that the *ABCB1* 2677T allele was associated with enhanced constitutive expression of CYP3A4 in human liver and intestine, as compared to the *ABCB1* 2677G allele. Their results suggest that this is accomplished through the interaction between the *ABCB1* 2677 SNP and the PXR 6-base pair deletion that influences CYP3A4 expression. Therefore, examination of *ABCB1* genetics alone in the context of use of substrates for both P-gp and CYP3A may not be appropriate, as was observed in the tacrolimus–fluconazole example above.

Although the role of *ABCB1* in the disposition of drugs has become clearer in the past decade, the impact of polymorphisms on its function remains to be clarified. For drug development purposes, testing for interactions between a new drug and P-gp remains important, especially for the agents in which penetration into the central nervous system is crucial to efficacy or toxicity. Many FDA-approved drug labels include information on whether drug is a substrate, inhibitor, non-substrate, or non-inhibitor for P-gp (Agarwal et al., 2011, 2012; Huang et al., 2010) (Table 3.3). Currently, however, there are no recommendations for testing for *ABCB1* variants related to dosage or drug interactions, and such tests for clinical use are not widely available.

3.3.1.2 Breast Cancer Resistance Protein (ABCG2) BCRP encoded by *ABCG2* gene is a half transporter, and probably functions as a homodimer. It was first identified by Doyle et al. (1998) in a human breast cancer cell line selected for doxorubicin resistance in the presence of verapamil. This BCRP is now recognized as an important contributor to the blood–brain barrier function, in the placenta as a fetoprotective mechanism, in cancer cells as a mechanism of resistance, and in the gastrointestinal tract to alter the absorption of drugs such as the cholesterol-lowering statins. Similar to P-gp, BCRP has a wide variety of substrates and inhibitors. Although there is considerable overlap between BCRP and P-gp substrates or inhibitors, the quantitative structure-activity relationship (QSAR) analysis has been used to identify key structural elements for a molecule to modulate BCRP or P-gp and thus differentiate substrates or inhibitors of BCRP and P-gp (Saito et al., 2006; Ishikawa et al., 2010).

Although more than 40 variants of the *ABCG2* gene are known, a nonsynonymous SNP 421C>A in exon 5 that results in a glutamine to lysine amino acid change at position 141 (Q141K) has been studied most extensively (Hu et al., 2011). This variant has been associated with markedly decreased levels of BCRP protein expression and also low levels of drug resistance. The *ABCG2* c.421C>A SNP has a frequency of approximately 10–15% in Caucasians, 25–35% in Asians, and 0–5% in sub-Saharan Africans and African Americans (Ishikawa et al., 2005; Keskitalo et al., 2009b; Robey et al., 2009; Nakagawa et al., 2011).

Association studies have linked the low-activity variant of Q141K to gefitinib-induced diarrhea (Cusatis et al., 2006). The reduced protein levels and altered ATPase activity of the *ABCG2* C421C>A variant might affect the oral absorption and/or elimination pathways of gefitinib and thereby increase the steady-state gefitinib plasma concentrations leading to diarrhea. The PK of a number of other substrates of BCRP has also been linked to this genetic variant, but findings are inconsistent for the agents known to be substrates of BCRP. Most statin lipid-lowering agents are substrates of BCRP (Keskitalo et al., 2009a, 2009b; Niemi, 2010), and rosuvastatin appears to be the most important substrate for BCRP with an impact of the Q141K polymorphism. The area under the curve (AUC) of the plasma concentration–time graph for rosuvastatin was 144% greater, for inactive simvastatin lactone 111% greater, and for atorvastatin and fluvastatin 72% greater in the individuals with the ABCG2 c.421A/A genotype than in those with the c.421C/C genotype (Keskitalo et al., 2009a, 2009b).

TABLE 3.3 Transporters in the labeling

Transporter	Drug names[a]
ABC Transporters	
P-gp	Aliskiren, ambrisentan, (aprepitant), bendamustine, ***boceprevir, cabazitaxel,*** *clarithromycin*, colchicine, *conivaptan,* ***corizotinib, cyclosporine,*** dabigatran extile, (dalfampridine), (desvenlafaxine), *dronedarone*, (eltrombopag), *etravirine*, *eribulin*, ***everolimus***, (ezogabine), fexofenadine, fidaxomicin, (fosaprepitant), (gabapentin), (ixabepilone), ***lapatinib***, lenalidomide, ***maraviroc***, ***nilotinib***, *paliperidone*, pazopanib, (pralatrexate), posaconazole, (prasugrel), ((propafenone)), propranolol, ***ranolazine***, rivaroxaban, saxagliptin, silodosin, sirolimus, sitagliptin, (tetrabenazine), ***ticagrelor, tipranavir***[b], ***tolvaptan***, topotecan, ***vemurafenib***, (vorinostat)
BCRP	Bendamustine, *cabazitaxel*, ***eltrombopag***[c], (fingolimod), ***lapatinib***, pazopanib, pralatrexate, ***rivaroxaban***, topotecan
MRP1	(Cabazitaxel), (lenalidomide)
MRP2	(Cabazitaxel), (fingolimod), (indacaterol), (lenalidomide), mycophenolate, ***pralatrexate***, valsartan
SLC Transporters	
OATP1B1	Atorvastatin, ambrisentan, [*cyclospoprine*], *dornedorone, eltrombopag*[c], (fingolimod), *lapatinib*, (lenalidomide), *panzopanib*, (pralatrexate), *telithromycin*, valsartan
OAT	(Dronedarone) (OAT3), (Lenalidomide) (OAT1, OAT3), pemetrexed (OAT3), (pralatrexate) (OAT1, OAT3), sitagliptin (OAT3)
OCT	Gabapentin (OCT2), (linagliptin), metformin, pramipexole, (saxagliptin), (sitagliptin), (vandetanib) (OCT2), varenicline (OCT2)
MATE	(indacaterol) (MATE1, MATE2K)

Source: Updated from Huang et al. (2010) and Agarwal et al. (2012).

P-gp, P-glycoprotein; BCRP, breast cancer resistance protein; MRP, multidrug resistance associated protein; OATP, organic anion transporting polypeptide; OAT, organic anion transporter; OCT, organic cation transporter.

[a]This is not an extensive list: Data based on a preliminary survey of electronic PDR and Drugs@FDA (http://www.accessdata.fda.gov/scripts/cder/drugsatfda/).

Note that the table contains substrates unless they are italicized (inhibitors) and bolded (both substrates and inhibitors).

[b]Tipranavir is also a P-gp inducer.

[c]In eltrombopag labeling, the following were mentioned as OATP1B1 or BCRP substrates: Benzylpenicillin, atorvastatin, fluvastatin, pravastatin, rosuvastatin, methotrexate, nateglinide, repaglinide, rifampin, and doxorubicin.

(): mentioned in the labeling as "not a substrate" or "not an inhibitor."

(()): stated in the labeling as "not studied."

[]: stated as an inhibitor in the labeling for other drugs.

These increases in plasma concentration are most likely due to increased bioavailability of the orally administered drug, as a consequence of decreased intestinal efflux of these statins by the ABCG2 c.421A/A genotype. These findings could translate into an increased risk of muscle toxicity in individuals with the c.421A/A genotype who use rosuvastatin or other stain drugs that are BCRP substrates. Two recent studies were published on the effect of the *ABCG2* polymorphism on the lipid-lowering effect of rosuvastatin in Caucasian and Chinese patients (Bailey et al., 2010; Tomlinson et al., 2010). The *ABCG2* 421C>A variant was associated with greater reductions in low-density lipoprotein (LDL)-C in Chinese and Caucasian patients in a gene dose-dependent manner. The difference in LDL-C response to rosuvastatin between the two homozygous groups was equivalent to more than doubling the dose of rosuvastatin. The clinical impact of the *ABCG2* polymorphism in the context of statin therapy requires further confirmation. A recent GWAS study in 6,989 patients with European ancestry taking 20 mg rosuvastatin identified an intronic *ABCG2* SNP rs219936 associated with the LDL-cholesterol-lowering effects of rosuvastatin, at a genome wide significance. The rs219936 SNP is in strong linkage disequilibrium with c.421C>A (10). Three other genes, i.e., *LPA*, *APOE*, and *PSCK*, were also found to be associated with differential effects on LDL-C reduction compared to that in patients on placebo (Chasman D I et al., 2012). This study further supports the importance of *ABCG2* SNPs, especially c.421C>A, in rosuvastatin response. Besides the statin drugs, the tyrosine kinase inhibitors are substrates for BCRP, and concern about a role for BCRP in cancer resistance to tyrosine kinase inhibitors has been expressed (Lemos et al., 2008).

In FDA-approved drug labels, BCRP has been mentioned in several labels for agents that are substrates or inhibitors of BCRP (Table 3.3) (Agarwal et al., 2012). For example, the cabazitaxel label mentions that the drug inhibits BCRP, but only in concentrations 38 times that observed in humans (http://www.accessdata.fda.gov/drugsatfda_docs/label/2010/201023lbl.pdf). The labels of two of the tyrosine kinase inhibitors, lapatinib and pazopanib, mention that these agents are substrates for BCRP and P-gp (http://www.accessdata.fda.gov/drugsatfda_docs/label/2011/022059s011s012lbl.pdf; http://www.accessdata.fda.gov/drugsatfda_docs/label/2011/022465s003s005lbl.pdf) (Table 3.3). To date no mention of *ABCG2* polymorphism is made in any FDA-approved label. With an augment of documentation of clinical pharmacogenetic effect related to BCRP, and increased availability of the genetic test for *ABCG2,* clinical relevant recommendation based on *ABCG2* polymorphism may start to be utilized in clinical practice and drug label, similar to CYP polymorphisms.

3.3.1.3 Multidrug Resistance Protein 2 (ABCC2) The MRP2, encoded by the *ABCC2* gene, is an extensively studied transporter in the biliary elimination for endogenous and xenobiotic substrates. MRP2 is important for the enterohepatic recirculation of drugs such as pravastatin and mycophenolic acid (MPA). MRP2 is also in the gut where it has an impact on the absorption of drugs, and in the renal proximal tubule where it may play a role in drug elimination.

The genetic influence on the expression and function of MRP2 is well known due to an inborn genetic mutation named Dubin–Johnson syndrome. Bilirubin is one of the endogenous substrates for MRP2. Dubin–Johnson syndrome produces conjugated hyperbilirubinemia secondary to a mutation in MRP2, which plays a role in the secretion of conjugated bilirubin to bile. Suzuki and Sugiyama (2002) have reviewed the initial understanding of the polymorphic nature of the ABC gene (*ABCC2*) encoding for MRP2. The effects of these polymorphisms have been investigated for several drugs.

Mycophenolic acid (MPA) is a substrate for MRP2 and has been extensively studied in relation to MRP2 function and *ABCC2* polymorphisms. MRP2 inhibition by cyclosporine explained the original observation in transplant patients of differing plasma concentrations of MPA in patients receiving cyclosporine versus tacrolimus. A promoter polymorphism (*ABCC2* rs717620, previously called the -24C>T polymorphism) has been associated with adverse gastrointestinal effects in transplant patients receiving MPA (Ohmann et al., 2010), leading to drug discontinuation with potentially poorer outcomes. Although the FDA-approved labels for some MPA products, for example, CellCept, indicate that MPA is a substrate for MRP2 (http://www.accessdata.fda.gov/drugsatfda_docs/label/2009/050722s024,050723s023,050758s022,050759s028lbl.pdf) (Table 3.3), they currently do not carry any warning about *ABCC2* genotypes and adverse effects.

Inhibition of MRP2 can cause drug interactions. For example, indomethacin is an inhibitor of MRP2, and significantly interacts with sulfasalazine, a drug which exerts its action locally within the gastrointestinal tract. By giving both drugs together, sulfasalazine gets absorbed, and therefore may produce both a decreased local gastrointestinal therapeutic effect and the potential for systemic toxicity due to sulfasalazine absorption (Dahan and Amidon, 2010). The magnitude of this drug–drug interaction could be modified by MRP2 genotypes, but has yet to be examined.

3.3.2 SLC Transporters

3.3.2.1 Organic Anion Transporting Polypeptide 1B1 (SLCO1B1)

OATP1B1 (encoded by *SLCO1B1* gene) is a major uptake transporter in the liver. It is located at the basolateral (sinusoidal) membrane of human hepatocytes and mediates the hepatic uptake of its substrates from portal blood into hepatocytes. In addition to OATP1B1, two other OATP transporters—OATP1B3 and OATP2B1—are also expressed on the hepatocyte basolateral membrane in quantities roughly similar to that of OATP1B1. OATP1A2 and OATP2B1 are the major OATP transporters in other tissues such as intestine. These SLC transporters may affect the absorption of various drugs (Tamai, 2011). The effects of polymorphisms in OATP1A2 and OATP2B1 on drug absorption have been described in Chapters 6 and 7. This section focuses on the role of *SLCO1B1* polymorphism on the systemic exposure, safety, and/or efficacy of the cholesterol-lowering statin drugs.

OATP1B1 is polymorphic. More than 40 mutations in *SLCO1B1* gene have been identified (Fahrmayr et al., 2010; Maeda and Sugiyama, 2008; Niemi, 2010). Two common SNP variants of the *SLCO1B1* gene—c.388A>G (p.Asn130Asp; rs2306283) and c.521T>C (p.Val174Ala; rs4149056)—have been shown to affect the transport function of OATP1B1. The effects, however, depend on their combination in individual haplotypes (Table 1). When the c.388A>G exists without the c.521T>C SNP, the haplotype is termed *SLCO1B1*1b*; it is usually associated with increased activity of OATP1B1 and therefore with lower plasma concentrations of OATP1B1 substrates (Maeda et al., 2006; Mwinyi et al., 2004). The c.521T>C SNP reduces the transport activity of OATP1B1 and increases the plasma concentrations of OATP1B1 substrates, both in the *SLCO1B1*5* (c.521T>C alone) haplotype and in the *SLCO1B1*15* (c.521T>C along with c.388A>G) haplotype (Niemi et al., 2004; Nishizato et al., 2003). The low-activity *SLCO1B1*5* and *SLCO1B1**15 haplotypes (i.e., c.521T>C SNP) have a combined allele frequency of approximately 15–20%

in Caucasians, 10–15% in Asians, and 2% in sub-Saharan Africans and African Americans (Pasanen et al., 2008). The *SLCO1B1**1b* haplotype has a frequency of approximately 25–30% in Caucasians, 40% in South/Central Asians, 60% in East Asians, and 80% in sub-Saharan Africans and African Americans (Pasanen et al., 2008).

Most currently marketed statin drugs are substrates of OATP1B1 (Niemi, 2010). The liver is the major site of action for statins. The wide interindividual variability in response to statin therapy (lipid lowering and adverse drug reactions) may be in part due to polymorphisms of OATP1B1, which affect the PK of statin drugs. The effect is greatest with simvastatin; the AUC of active simvastatin acid has been shown to be 221% greater in individuals with the c.521C/C genotype than in those with the c.521T/T genotype (Pasanen et al., 2006). A recent GWAS showed that the *SLCO1B1* c.521T>C SNP, the same SNP with a documented influence on simvastatin PK, is strongly associated with simvastatin-induced myopathy (SEARCH Collaborative Group et al., 2008). Approximately 300,000 genome markers were determined in 85 patients who had developed myopathy and in 90 patients without myopathy after receiving 80 mg daily doses of simvastatin for an average of 6 years. A noncoding SNP in the *SLCO1B1* gene, which is in nearly complete linkage disequilibrium with the c.521T>C SNP, was associated with myopathy at a genome-wide significance level. More than 60% of the myopathy cases could be attributed to the c.521T>C SNP, with an odd ratio of 4.5 per copy of the c.521C allele. Among participants taking simvastatin 80 mg/day, C/C homozygotes had a cumulative risk for myopathy of 18%. This risk is in contrast to those with the C/T genotype (3%) or T/T homozygotes (0.6%). The association was replicated in a study of 20,000 patients receiving 40 mg of simvastatin daily, which yielded a relative risk of 2.6 per copy of the c.521C allele. Moreover, the c.521T>C SNP was associated with a slight reduction in the cholesterol-lowering efficacy of simvastatin, whereas the c.388A>G SNP was associated with a slightly enhanced efficacy. These findings are consistent with the hypothesis that the c.521T>C SNP is associated with a reduction in hepatic uptake and that the *1b haplotype is associated with an enhancement in the uptake of simvastatin acid. Although other statins have not been studied in a GWAS, the *SLCO1B1* c.521T>C SNP may probably be associated with an increased risk of myopathy for other statins as well. However, the risk attributable to the c.521T>C SNP would probably depend on the specific statin and the relative contribution of OATP1B1 in the overall clearance of the statins. Recently, the Clinical Pharmacogenomics Implementation Consortium proposed a dosing algorithm for simvastatin based on *SLCO1B1* polymorphism (Wilke et al., 2012).

Although OATP1B1-related drug interactions have been included in the drug label (e.g., the labeling of eltrombopag described its interactions with rosuvastatin) (http://www.accessdata.fda.gov/drugsatfda_docs/label/2011/022291s006lbl.pdf) (Table 3.3), to date OATP1B1-related polymorphisms have yet to be included in the FDA approved statin drug labels. With an augment of documentation of clinical pharmacogenetic effect related to OATP1B1, and increased availability of the genetic test for *SLCO1B1*, clinical relevant recommendation based on *SLCO1B1* polymorphism will start to be utilized in clinical practice and drug label, similar to CYP polymorphisms.

3.3.2.2 Organic Cation Transporters (SLC22A) and Multidrug and Toxin Extrusion Transporters (SLC47A) Various OCTs are expressed in excretory organs such as liver and kidney. The OCT1 encoded by *SLC22A1* gene is mainly expressed in the liver (basolateral side of hepatocytes), and the OCT2 (encoded by *SLC22A2*) is mainly expressed in the kidney (basolateral side). The MATE1 (*SLC47A1*) is expressed in both kidney and liver at the apical side of the cell membrane, whereas MATE2-K (*SLC47A2*) is mainly expressed in the kidney. Recently, genetic variations of these transporters and their potential impact on metformin PK or PD have been studied.

Many cationic drugs including metformin are substrates of these OCTs and MATEs. Metformin is a drug widely used for the treatment of type 2 diabetes mellitus. Metformin is not metabolized by hepatic enzymes, but is excreted unchanged by the kidneys. DTPs play a major role in the distribution of metformin to tissues and elimination of metformin through renal excretion. Several transporters have been shown to be involved in the metformin transport in humans including OCT1, OCT2, MATE1, and MATE2-K.

The pharmacological basis for the glucose-lowering effect of metformin is not completely understood, although inhibition of hepatic gluconeogenesis has a key role. The uptake of metformin in the hepatocytes by OCT1 is an essential step for the inhibition of hepatic gluconeogenesis and the glucose-lowering effect of metformin. Genetic variation in the *SLC22A1* gene, which encodes variants (R61C, G401S, M420del, G465R) of the OCT1 transporter, is associated with differences in metformin blood levels and glucose levels after an oral glucose tolerance test in healthy volunteers (Shu et al., 2007; Shu et al., 2008). Genetic variation in *SLC22A*, which encodes variants (T199I, T201M, and A270S) of the OCT2 transporter, is associated with differences in metformin blood levels and renal excretion (Song et al., 2008).

Recently, two additional genetic polymorphisms in OCT1 (*SLC22A1* rs622342 A>C) and in MATE1 (*SLC47A1* rs2289669 G>A) were identified to be associated with the glucose-lowering effect of metformin (Becker et al., 2009a, 2009b). In metformin users, the presence of *SLC22A1* rs622342 minor C alleles (OCT1) can result in low levels of the OCT1 protein, which results in the decreased transport of metformin into the hepatocytes, and is associated with a reduced glucose-lowering effect (HbA1c 0.28; 95% confidence interval (CI): 0.09–0.47, $p = .005$). In contrast, the presence of *SLC47A1* rs2289669 minor A alleles (MATE1), which can result in low levels of MATE1 protein, produces a reduced efflux out of the hepatocytes, higher intracellular metformin levels, and is associated with an increased glucose-lowering effect (HbA1c 0.30; 95% CI: 0.51–0.10; $p = .005$). Interestingly, both polymorphisms are situated in an intron, suggesting that polymorphisms in linkage disequilibrium may be responsible for the effect.

A recent study demonstrated an interaction between these two SNPs (Becker et al., 2010). In patients homozygous for a normal functioning OCT1 influx transporter, genetic variations in the gene encoding for the MATE1 efflux transporter did not affect the glucose-lowering effect of metformin. In contrast, in patients homozygous for a reduced functioning OCT1 uptake transporter, the genetic variation in the gene encoding for the MATE1 efflux transporter has a substantial impact on the glucose-lowering effect of metformin. The glucose-lowering effect of metformin was most impaired in the patients with the rs622342 C/C genotype, which has a reduced functioning OCT1 uptake transporter, and with the rs2289669 G/G genotype, which has normal functioning MATE1 efflux transporter. Patients heterozygous for one of these genes had glucose-lowering effects somewhere in between. Besides the opposite effects between OCT1 (uptake) and MATE1 (efflux) in hepatocytes, MATE1 also impacts the renal clearance of metformin. The effect of the rs2289889 polymorphism also resulted in reduced renal excretion and increased metformin plasma levels.

More recently, a common gain-of-function 5′-UTR variant of *SLC47A2* g.−130G>A was found to be significantly associated with a poor glycemic response to metformin in newly diagnosed type 2 diabetic patients (Choi et al., 2011).

Besides OCT and MATE genes, variations in the drug target of metformin, adenosine monophosphate-activated protein kinase (AMPK), and genes in AMPK signaling may be associated with the glycemic response to metformin. The risk of treatment failure or adverse drug reactions in patients receiving metformin therapy is most likely not monogenic but based on the combinations of variants in multiple genes.

The recent advances in the science and technology of genome-wide analysis and continued efforts in the evaluation of transporters in drug disposition will improve our understanding of the role of transporters in drug safety and effectiveness.

3.4 PHARMACOGENOMICS OF OTHER GENES

Besides ADME genes, genetic variations in genes that code for intended or unintended drug targets and other pathways of a drug's pharmacology, or genes that can predispose individual patients to toxicities such as immune reactions have demonstrated importance in the safe and effective use of various therapeutic agents. A few recent US FDA labeling examples include warfarin and *VKORC1* (see Section 2.1 on warfarin), panitumumab (and cetuximab) and *KRAS*, carbamazepine and *HLA-B**1501*, and abacavir and *HLA-B**5701* (Table 3.2).

3.4.1 Panitumumab (and Cetuximab) and *KRAS*

Panitumumab, a fully human monoclonal antibody, is an epidermal growth factor receptor (EGFR) antagonist indicated as a single agent for the treatment of metastatic colorectal carcinoma (mCRC) with disease progression (http://www.accessdata.fda.gov/drugsatfda_docs/label/2009/125147s080lbl.pdf). KRAS is a G-protein with an intrinsic GTPase, which plays a key role in the mitogen activating phosphatase (MAP) kinase pathway. *KRAS* mutations in codon 12 or 13 induce a defective GTPase activity with an increased population of active KRAS protein that activates the RAS/MAP kinase pathway EGFR-independently and make EGFR blockage to be ineffective (Garcia-Saenz et al., 2009). This mutation exists in approximately 40% of colorectal tumors. Retrospective analyses across seven randomized clinical trials suggest that panitumumab is not effective for the treatment of patients with mCRC containing *KRAS* mutations (http://www.accessdata.fda.gov/

xdrugsatfda_docs/label/2009/125147s080lbl.pdf). For example, in one phase III study in patients with mCRC, the response rate to panitumumab was 17% in *KRAS* wild-type tumors versus 0% in patients with *KRAS* mutation. Additionally, the progression-free survival (PFS) was prolonged from 7.3 to 12.3 weeks in patients with no *KRAS* mutation (Amado et al., 2008). Therefore, the Vectibix (panitumumab) labeling states that "use of Vectibix is not recommended for the treatment of colorectal cancer with these mutations" (http://www.accessdata.fda.gov/drugsatfda_docs/label/2009/125147s080lbl.pdf).

Similar analyses of clinical trial data on cetuximab have resulted in similar conclusion that

> *"retrospective subset analyses of metastatic or advanced colorectal cancer trials have not shown a treatment benefit for Erbitux in patients whose tumors had KRAS mutations in codon 12 or 13. Use of Erbitux is not recommended for the treatment of colorectal cancer with these mutations" (http://www.accessdata.fda.gov/drugsatfda_docs/label/2011/125084s153lbl.pdf).*

3.4.2 Carbamazepine and HLA

Carbamazepine, an antiepileptic drug, is known to cause serious cutaneous adverse drug reactions including Stevens–Johnson syndrome (SJS) and toxic epidermal necrosis (TEN). Recently, carbamazepine-induced SJS–TEN was found to be strongly associated with *HLA-B*1502* in Han Chinese. Chung et al. have shown that 44 out of 44 (100%) of the carbamazepine-induced SJS–TEN patients were positive for the allele, whereas only 3% of the tolerant controls and 8.6% of the healthy controls carried the allele (Chung et al., 2004; Hung et al., 2006). A strong association between *HLA-B*1502* and carbamazepine-induced SJS–TEN has also been found in Asian populations other than the Han Chinese, including Malay, Thai, and South Asian Indians (Lonjou et al., 2006; Mehta et al., 2009; Tassaneeyakul et al., 2010). Across Asian populations, notable variations exist in the prevalence of *HLA-B*1502*. More than 15% of the population is reported positive in Hong Kong, Thailand, Malaysia, and parts of the Philippines, compared to about 10% in Taiwan and 4% in North China. South Asians, including Indians, appear to have intermediate prevalence of *HLA-B*1502*, averaging 2–4% (range 0–6%) (Mehta et al., 2009). However, *HLA-B*1502* is present in less than 1% of the population in Japan and Korea and is largely absent in individuals not of Asian origin (e.g., Caucasians, African Americans, Hispanics, and Native Americans) (http://www.allelefrequencies.net/; http://www.accessdata.fda.gov/drugsatfda_docs/label/2011/016608s100s102,018281s049s050,018927s041s042,020234s031s033lbl.pdf).

Because of serious, life-threatening nature of cutaneous adverse event caused by carbamazepine, the US FDA updated the carbamazepine label in 2008 to include a boxed warning (see text in the box below) (http://www.accessdata.fda.gov/drugsatfda_docs/label/2011/016608s100s102,018281s049s050,018927s041s042,020234s031s033lbl.pdf; Ferrell and McLeod, 2008). The label recommends prospectively screening patients for *HLA-B*1502* before treating with carbamazepine, especially in the patients with ancestry in genetically at-risk populations (e.g., Han Chinese). A recent report from Chen et al. reported how prospective *HLA-B*1502* typing can prevent carbamazepine-induced SJS–TEN in Han Chinese populations (Chen et al., 2011).

Blackbox warning:

SERIOUS DERMATOLOGIC REACTIONS AND HLA-B*1502 ALLELE

Serious and sometimes fatal dermatological reactions, including TEN and SJS, have been reported during the treatment with TEGRETOL. These reactions are estimated to occur in 1–6 per 10,000 new users in countries with mainly Caucasian populations, but the risk in some Asian countries is estimated to be about 10 times higher. Studies in patients of Chinese ancestry have found a strong association between the risk of developing SJS–TEN and the presence of HLA-B*1502, an inherited allelic variant of the HLA-B gene. HLA-B*1502 is found almost exclusively in patients with ancestry across broad areas of Asia. Patients with ancestry in genetically at-risk populations should be screened for the presence of HLA-B*1502 before initiating treatment with TEGRETOL. Patients testing positive for the allele should not be treated with TEGRETOL unless the benefit clearly outweighs the risk (see **Warnings** and **Precautions, Laboratory Tests**).

Recently, GWAS studies have also identified an association of a different allele, *HLA-A*3101*, with carbamazepine-induced adverse drug reactions in Northern European (prevalence 2–5% in Northern European) (McCormack et al., 2011) and in Japanese (prevalence 9.1% in Japanese) (Ozeki et al., 2011), indicating that different genetic markers may be responsible for the adverse cutaneous reactions among different ethnic groups.

3.5 FDA CLINICAL PHARMACOGENOMIC GUIDANCE

In February 2011, the FDA published a draft Clinical Pharmacogenomic Guidance and in January 2013 published the final Clinical Pharmacogenomics Guidance entitled "*Clinical Pharmacogenomics: Premarketing Evaluation in Early-Phase Clinical Studies and Recommendations for Labeling*" (http://www.fda.gov/Drugs/GuidanceComplianceRegulatoryInformation/Guidances/ucm064982.htm). The goal of this guidance was to recommend to the pharmaceutical industry and other investigators engaged in new drug development in evaluating how variations in the human genome, specifically DNA sequence variants, could affect a drug's PK, PD, efficacy, and/or safety. The guidance provides recommendations on when genomic information should be considered to address questions arising during drug development, and during regulatory review. Although the application of pharmacogenomic approaches during drug development is an evolving process that begins with discovery and continues through confirmation of clinical efficacy and safety outcomes, it is the focus of this guidance to provide advice on general approaches in incorporating PGx in early drug development phases, principles of study design, data collection, and data analysis.

DNA collection is fundamental in identifying sources of variability, understanding the impact of genetic variants on PK or PD, and explaining the potential causes of adverse events observed in phase 3 trials. The value of PGx was discussed in the context of recent regulatory examples. Although some examples included PGx data obtained postmarketing, they support the potential values of having such data early when it could be linked to the findings in the clinical trials supporting safety and effectiveness. The values of collecting such data early range from explaining the variability of PK, supporting the importance of a particular metabolizing enzyme or transporter for a drug, estimating drug interactions, to optimizing dosing selection for phase 3 trials to enhance efficacy and reduce adverse events. The guidance provides a framework on when and how to identify genes that may be important for a drug's PK, PD, efficacy, or safety along the drug development process based on drug characteristics and information collected through early exploratory phase 1 and phase 2 studies to inform later confirmatory studies. General considerations on study design and data analysis were also discussed. Mathematical simulations using population-based, physiological PK models (i.e., physiologically based pharmacokinetic (PBPK) models) that simultaneously integrate various patient-intrinsic and -extrinsic factors can provide an understanding of the potential complex changes in exposure–response relationships in patients where multiple covariates are present. Some applications of these models, including the design of clinical trials to evaluate the effects of DME polymorphisms on PK and PD, are increasingly being included in regulatory submissions (Huang and Rowland, 2012; Zhao et al., 2011). Finally, recommendations of incorporation of PGx into labeling were also briefly discussed.

3.6 FUTURE PERSPECTIVES

Membrane transporters have been increasingly recognized as contributing to variability in drug exposure and response and are important to evaluate during drug development and regulatory review (Giacomini et al., 2010; Huang and Woodcock, 2010; Zhang et al., 2011). It is therefore important to determine the contribution to altered drug response due to varied transporter activities because of either concomitantly administered drugs that are modulators of DTPs or individual subjects' transporter genotype.

The FDA recommends the collection of PGx information early in drug development (the 2013 FDA Pharmacogenomics Guidance) (http://www.fda.gov/Drugs/GuidanceComplianceRegulatoryInformation/Guidances/ucm064982.htm). Identification of the DMEs and transporters in the elimination of the new molecular entities will help determine if polymorphic DMEs and/or transporter(s) that are polymorphic are involved. Exploratory studies on the genes/proteins involved in the pharmacological pathways can also yield valuable information. Novel technologies such as candidate gene studies and GWAS may be used for hypothesis generation. The amount of evidence to demonstrate the clinical utility (risk/benefit) of genetic tests and the process to incorporate the genomic information during drug development have been discussed (Lesko et al., 2010; Zineh and Huang, 2011).

REFERENCES

Agarwal S, Chinn L, Zhang L (2012) An overview of transporter information in package inserts of recently approved new molecular entities. *Pharm Res.* 2012 Nov 28. [Epub ahead of print]

Agarwal S, Zhang L, Huang S-M (2011) Relevance and value of *in vitro* P-gp inhibition data for new molecular entities. *Clin Pharmacol Ther*, 89, S40.

Amado RG, Wolf M, Peeters M, Van CE, Siena S, Freeman DJ, Juan T, Sikorski R, Suggs S, Radinsky R et al. (2008) Wild-type KRAS is required for panitumumab efficacy in patients with metastatic colorectal cancer. *J Clin Oncol*, 26, 1626–1634.

Bailey KM, Romaine SP, Jackson BM, Farrin AJ, Efthymiou M, Barth JH, Copeland J, McCormack T, Whitehead A, Flather MD et al. (2010) Hepatic metabolism and transporter gene variants enhance response to rosuvastatin in patients with acute myocardial infarction: the GEOSTAT-1 Study. *Circ Cardiovasc Genet*, 3, 276–285.

Becker ML, Visser LE, van Schaik RH, Hofman A, Uitterlinden AG, Stricker BH (2009a) Genetic variation in the multidrug and toxin extrusion 1 transporter protein influences the glucose-lowering effect of metformin in patients with diabetes: a preliminary study. *Diabetes*, 58, 745–749.

Becker ML, Visser LE, van Schaik RH, Hofman A, Uitterlinden AG, Stricker BH (2009b) Genetic variation in the organic cation transporter 1 is associated with metformin response in patients with diabetes mellitus. *Pharmacogenomics J*, 9, 242–247.

Becker ML, Visser LE, van Schaik RH, Hofman A, Uitterlinden AG, Stricker BH (2010) Interaction between polymorphisms in the OCT1 and MATE1 transporter and metformin response. *Pharmacogenet Genomics*, 20, 38–44.

Bernard S, Neville KA, Nguyen AT, Flockhart DA (2006) Interethnic differences in genetic polymorphisms of CYP2D6 in the US population: clinical implications. *Oncologist*, 11 (2), 126–135.

Bradford LD (2002) CYP2D6 allele frequency in European Caucasians, Asians, Africans and their descendants. *Pharmacogenomics*, 3, 229–243.

Caraco Y, Blotnick S, Muszkat M (2008) CYP2C9 genotype-guided warfarin prescribing enhances the efficacy and safety of anticoagulation: a prospective randomized controlled study. *Clin Pharmacol Ther*, 83, 460–470.

Chasman DI, Giulianini F, MacFadyen J, Barratt BJ, Nyberg F, Ridker PM (2012) Genetic determinants of statin-induced low-density lipoprotein cholesterol reduction: the Justification for the Use of Statins in Prevention: an Intervention Trial Evaluating Rosuvastatin (JUPITER) trial. *Circ Cardiovasc Genet*, 5, 257–264.

Chen P, Lin JJ, Lu CS, Ong CT, Hsieh PF, Yang CC, Tai CT, Wu SL, Lu CH, Hsu YC, et al. (2011). Carbamazepine-induced toxic effects and HLA-B*1502 screening in Taiwan. *New Eng J Med*, 364, 1126–1133.

Choi JH, Yee SW, Ramirez AH, Morrissey KM, Jang GH, Joski PJ, Mefford JA, Hesselson SE, Schlessinger A, Jenkins G, et al. (2011) A common 5′-UTR variant in MATE2-K is associated with poor response to metformin. *Clin Pharmacol Ther*, 90, 674–684.

Chung WH, Hung SI, Hong HS, Hsih MS, Yang LC, Ho HC, Wu, JY, Chen YT (2004) Medical genetics: a marker for Stevens–Johnson syndrome. *Nature*, 428, 486.

Cusatis G, Gregorc V, Li J, Spreafico A, Ingersoll RG, Verweij J, Ludovini V, Villa E, Hidalgo M, Sparreboom A, Baker SD (2006) Pharmacogenetics of ABCG2 and adverse reactions to gefitinib. *J Natl Cancer Inst*, 98, 1739–1742.

Dahan A, Amidon, GL (2010) MRP2 mediated drug-drug interaction: indomethacin increases sulfasalazine absorption in the small intestine, potentially decreasing its colonic targeting *Int J Pharm*, 386 (1–2), 216–220. Available from: PM:19944137.

Daly AK, King BP (2003) Pharmacogenetics of oral anticoagulants. *Pharmacogenetics*, 13, 247–252.

Desta Z, Zhao X, Shin JG, Flockhart DA (2002) Clinical significance of the cytochrome P450 2C19 genetic polymorphism. *Clin Pharmacokinet*, 41, 913–958.

Doyle LA, Yang W, Abruzzo LV, Krogmann T, Gao Y, Rishi AK, Ross DD (1998) A multidrug resistance transporter from human MCF-7 breast cancer cells. *Proc Natl Acad Sci USA*, 95, 15665–15670.

Droll K, Bruce-Mensah K, Otton SV, Gaedigk A, Sellers EM, Tyndale RF (1998) Comparison of three CYP2D6 probe substrates and genotype in Ghanaians, Chinese, and Caucasians. *Pharmacogenetics*, 8, 325–333.

Fahrmayr C, Fromm MF, König J (2010) Hepatic OATP and OCT uptake transporters: their role for drug–drug interactions and pharmacogenetic aspects. *Drug Meta Rev*, 42, 380–401.

Ferrell PBJr., McLeod HL (2008) Carbamazepine, HLA-B*1502 and risk of Stevens–Johnson syndrome and toxic epidermal necrolysis: US FDA recommendations. *Pharmacogenomics*, 9, 1543–1546.

Gage BF, Eby C, Johnson JA, Deych E, Rieder MJ, Ridker PM, Milligan PE, Grice G, Lenzini P, Rettie AE, et al. (2008) Use of pharmacogenetic and clinical factors to predict the therapeutic dose of warfarin. *Clin Pharmacol Ther*, 84, 326–331.

Gage BF, Lesko, LJ (2008) Pharmacogenetics of warfarin: regulatory, scientific, and clinical issues. *J Thromb Thrombolysis*, 25, 45–51.

Garcia-Saenz JA, Sastre J, Diaz-Rubio GE (2009) Biomarkers and anti-EGFR therapies for KRAS wild-type metastatic colorectal cancer. *Clin Transl Oncol*, 11, 737–747.

Giacomini KM, Huang SM, Tweedie DJ, Benet LZ, Brouwer KL, Chu X, Dahlin A, Evers R, Fischer V, Hillgren KM, et al. (2010) Membrane transporters in drug development. *Nat Rev Drug Discov*, 9(3), 215–236.

Hu, M, To, KK, Mak, VW, Tomlinson, B. (2011). The ABCG2 transporter and its relations with the pharmacokinetics, drug interaction and lipid-lowering effects of statins. *Expert Opin Drug Metab Toxicol*, 7, 49–62.

Huang SM, Rowland M (2012) The role of physiologically based pharmacokinetic modeling in regulatory review. *Clin Pharm Ther*, 91, 542–549.

Huang SM, Temple R (2008) Is this the drug or dose for you? Impact and consideration of ethnic factors in global drug development, regulatory review, and clinical practice. *Clin Pharmacol Ther*, 84, 287–294.

Huang SM, Woodcock J (2010) Transporters in drug development: advancing on the Critical Path. *Nat Rev Drug Discov*, 9, 175–176.

Huang SM, Zhang L, Giacomini KM (2010) The International Transporter Consortium: a collaborative group of scientists from academia, industry, and the FDA. *Clin Pharmacol Ther*, 87(1), 32–36.

Huang Y (2007) Pharmacogenetics/genomics of membrane transporters in cancer chemotherapy. *Cancer Metastasis Rev*, 26 (1), 183–201. Available from: PM:17323126.

Hung SI, Chung WH, Jee SH, Chen WC, Chang YT, Lee WR, Hu SL, Wu MT, Chen GS, Wong TW, et al. (2006) Genetic susceptibility to carbamazepine-induced cutaneous adverse drug reactions. *Pharmacogenet Genomics*, 16, 297–306.

Ishikawa T, Sakurai A, Hirano H, Lezhava A, Sakurai M, Hayashizaki Y (2010) Emerging new technologies in Pharmacogenomics: rapid SNP detection, molecular dynamic simulation, and QSAR analysis methods to validate clinically important genetic variants of human ABC Transporter ABCB1 (P-gp/MDR1). *Pharmacol Ther*, 126, 69–81.

Ishikawa T, Tamura A, Saito H, Wakabayashi K, Nakagawa H (2005) Pharmacogenomics of the human ABC transporter ABCG2: from functional evaluation to drug molecular design. *Naturwissenschaften*, 92, 451–463.

Johnson JA, Gong L, Whirl-Carrillo M, Gage BF, Scott SA, Stein CM, Anderson JL, Kimmel SE, Lee MT, Pirmohamed M, et al. (2011) Clinical Pharmacogenetics Implementation Consortium Guidelines for CYP2C9 and VKORC1 genotypes and warfarin dosing. *Clin Pharmacol Ther*, 90, 625–629.

Keskitalo JE, Pasanen MK, Neuvonen PJ, Niemi M (2009a) Different effects of the ABCG2 c421C>A SNP on the pharmacokinetics of fluvastatin, pravastatin and simvastatin. *Pharmacogenomics*, 10, 1617–1624.

Keskitalo JE, Zolk O, Fromm MF, Kurkinen KJ, Neuvonen PJ, Niemi M (2009b) ABCG2 polymorphism markedly affects the pharmacokinetics of atorvastatin and rosuvastatin. *Clin Pharmacol Ther*, 86, 197–203.

Kim MJ, Huang SM, Meyer UA, Rahman A, Lesko LJ (2009) A regulatory science perspective on warfarin therapy: a pharmacogenetic opportunity. *J Clin Pharmacol*, 49, 138–146.

Kuypers DR, de JH, Naesens M, Vanrenterghem Y (2008) Effects of CYP3A5 and MDR1 single nucleotide polymorphisms on drug interactions between tacrolimus and fluconazole in renal allograft recipients. *Pharmacogenet Genom*, 18, 861–868.

Lamba J, Strom S, Venkataramanan R, Thummel KE, Lin YS, Liu W, Cheng C, Lamba V, Watkins PB, Schuetz E (2006) MDR1 genotype is associated with hepatic cytochrome P450 3A4 basal and induction phenotype. *Clin Pharmacol Ther*, 79, 325–338.

Lemos C, Jansen G, Peters GJ (2008) Drug transporters: recent advances concerning BCRP and tyrosine kinase inhibitors. *Brit J cancer*, 98, 857–862.

Lenzini P, Wadelius M, Kimmel S, Anderson JL, Jorgensen AL, Pirmohamed M, Caldwell MD, Limdi N, Burmester JK, Dowd MB, et al. (2010) Integration of genetic, clinical, and INR data to refine warfarin dosing. *Clin Pharmacol Ther*, 87, 572–578.

Lesko LJ, Woodcock J (2002) Pharmacogenomic-guided drug development: regulatory perspective. *Pharmacogenomics J*, 2, 20–24.

Lesko LJ, Woodcock J (2004) Translation of pharmacogenomics and pharmacogenetics: a regulatory perspective. *Nat Rev Drug Discov*, 3, 763–769.

Lesko LJZineh I (2010) DNA, drugs, and chariots: on a decade of pharmacogenomics at the US FDA. *Pharmacogenomics*, 11, 507–512.

Lesko LJ, Zineh I, Huang SM (2010) What is clinical utility and why should we care? *Clin Pharmacol Ther*, 88, 729–733.

SEARCH Collaborative Group, Link E, Parish S, Armitage J, Bowman L, Heath S, Matsuda F, Gut I, Lathrop M, Collins R (2008) SLCO1B1 variants and statin-induced myopathy–a genomewide study. *New Engl J Med*, 359(8), 789–799.

Lonjou C, Thomas L, Borot N, Ledger N, de TC, LeLouet H, Graf E, Schumacher M, Hovnanian A, Mockenhaupt M, Roujeau JC (2006) A marker for Stevens–Johnson syndrome: ethnicity matters. *Pharmacogenomics J*, 6, 265–268.

Maeda K, Ieiri I, Yasuda K, Fujino A, Fujiwara H, Otsubo K, Hirano M, Watanabe T, Kitamura Y, Kusuhara H, Sugiyama Y (2006) Effects of organic anion transporting polypeptide 1B1 haplotype on pharmacokinetics of pravastatin, valsartan, and temocapril. *Clin Pharmacol Ther*, 79, 427–439.

Maeda K, Sugiyama Y (2008) Impact of genetic polymorphisms of transporters on the pharmacokinetic, pharmacodynamic and toxicological properties of anionic drugs. *Drug Metab Pharmacokinet*, 23, 223–235.

Marsh S, King CR, Porche-Sorbet RM, Scott-Horton TJ, Eby CS (2006) Population variation in VKORC1 haplotype structure. *J Thromb Haemost*, 4, 473–474.

McCormack M, Alfirevic A, Bourgeois S, Farrell JJ, Kasperaviciute D, Carrington M, Sills GJ, Marson T, Jia X, de Bakker PI, et al. (2011) HLA-A*3101 and carbamazepine-induced hypersensitivity reactions in Europeans. *New Engl J Med*, 364, 1134–1143.

Mega JL, Close SL, Wiviott SD, Shen L, Hockett RD, Brandt JT, Walker JR, Antman EM, Macias W, Braunwald E, Sabatine MS (2009) Cytochrome p-450 polymorphisms and response to clopidogrel. *New Engl J Med*, 360, 354–362.

Mega JL, Hochholzer W, Frelinger ALIII, Kluk MJ, Angiolillo DJ, Kereiakes DJ, Isserman S, Rogers WJ, Ruff CT, Contant C, et al. (2011) Dosing clopidogrel based on CYP2C19 genotype and the effect on platelet reactivity in patients with stable cardiovascular disease. *JAMA*, 306, 2221–2228.

Mega JL, Simon T, Collet JP, Anderson JL, Antman EM, Bliden K, Cannon CP, Danchin N, Giusti B, Gurbel P, et al. (2010) Reduced-function CYP2C19 genotype and risk of adverse clinical outcomes among patients treated with clopidogrel predominantly for PCI: a meta-analysis. *JAMA*, 304, 1821–1830.

Mehta TY, Prajapati LM, Mittal B, Joshi CG, Sheth JJ, Patel DB, Dave DM, Goyal RK (2009) Association of HLA-B*1502 allele and carbamazepine-induced Stevens–Johnson syndrome among Indians. *Indian J Dermatol Venereol Leprol*, 75, 579–582.

Mwinyi J, Johne A, Bauer S, Roots I, Gerloff T (2004) Evidence for inverse effects of OATP-C (SLC21A6) 5 and 1b haplotypes on pravastatin kinetics. *Clin Pharmacol Ther*, 75, 415–421.

Nakagawa H, Toyoda Y, Wakabayashi-Nakao K, Tamaki H, Osumi M, Ishikawa T (2011) Ubiquitin-mediated proteasomal degradation of ABC transporters: a new aspect of genetic polymorphisms and clinical impacts. *J Pharm Sci*, 100, 3602–3619.

Niemi M (2010) Transporter pharmacogenetics and statin toxicity. *Clin Pharmacol Ther*, 87, 130–133.

Niemi M, Schaeffeler E, Lang T, Fromm MF, Neuvonen M, Kyrklund C, Backman JT, Kerb R, Schwab M, Neuvonen PJ, et al. (2004) High plasma pravastatin concentrations are associated with single nucleotide polymorphisms and haplotypes of organic anion transporting polypeptide-C (OATP-C, SLCO1B1). *Pharmacogenetics*, 14, 429–440.

Nishizato Y, Ieiri I, Suzuki H, Kimura M, Kawabata K, Hirota T, Takane H, Irie S, Kusuhara H, Urasaki Y, et al. (2003) Polymorphisms of OATP-C (SLC21A6) and OAT3 (SLC22A8) genes: consequences for pravastatin pharmacokinetics. *Clin Pharmacol Ther*, 73, 554–565.

Ohmann EL, Burckart, GJ, Brooks, MM, Chen, Y, Pravica, V, Girnita, DM, Zeevi A, Webber SA (2010) Genetic polymorphisms influence mycophenolate mofetil-related adverse events in pediatric heart transplant patients. *J Heart Lung Transplant*, 29, 509–516.

Owen RP, Gong, L, Sagreiya, H, Klein, TE, Altman, RB (2010) VKORC1 pharmacogenomics summary. *Pharmacogenet Genom*, 20, 642–644.

Ozeki T, Mushiroda T, Yowang A, Takahashi A, Kubo M, Shirakata Y, Ikezawa Z, Iijima M, Shiohara T, Hashimoto K, et al. (2011) Genome-wide association study identifies HLA-A*3101 allele as a genetic risk factor for carbamazepine-induced cutaneous adverse drug reactions in Japanese population. *Hum Mol Genet*, 20, 1034–1041.

Pasanen MK, Neuvonen M, Neuvonen PJ, Niemi M (2006) SLCO1B1 polymorphism markedly affects the pharmacokinetics of simvastatin acid. *Pharmacogenet Genomics*, 16, 873–879.

Pasanen MK, Neuvonen PJ, Niemi M (2008) Global analysis of genetic variation in SLCO1B1. *Pharmacogenomics*, 9, 19–33.

Robey RW, To KK, Polgar O, Dohse M, Fetsch P, Dean M, Bates SE (2009) ABCG2: a perspective. *Adv Drug Deliv Rev*, 61, 3–13.

Saito H, Hirano H, Nakagawa H, Fukami T, Oosumi K, Murakami K, Kimura H, Kouchi T, Konomi M, Tao E, et al. (2006) A new strategy of high-speed screening and quantitative structure-activity relationship analysis to evaluate human ATP-binding cassette transporter ABCG2-drug interactions. *J Pharmacol Exp Ther*, 317(3), 1114–1124.

Sanderson, SEmery, JHiggins, J (2005) CYP2C9 gene variants, drug dose, and bleeding risk in warfarin-treated patients: a HuGEnet systematic review and meta-analysis. *Genet Med*, 7, 97–104.

Schadt EE, Molony C, Chudin E, Hao K, Yang X, Lum PY, Kasarskis A, Zhang B, Wang S, Suver C, et al. (2008) Mapping the genetic architecture of gene expression in human liver. *PLoS Biol*, 6(5), e107.

Scordo MG, Pengo V, Spina E, Dahl ML, Gusella M, Padrini R (2002) Influence of CYP2C9 and CYP2C19 genetic polymorphisms on warfarin maintenance dose and metabolic clearance. *Clin Pharmacol Ther*, 72, 702–710.

Scott SA, Sangkuhl K, Gardner EE, Stein CM, Hulot JS, Johnson JA, Roden DM, Klein TE, Shuldiner AR (2011) Clinical Pharmacogenetics Implementation Consortium guidelines for cytochrome P450-2C19 (CYP2C19) genotype and clopidogrel therapy. *Clin Pharmacol Ther*, 90, 328–332.

Shu Y, Brown C, Castro RA, Shi RJ, Lin ET, Owen RP, Sheardown SA, Yue L, Burchard EG, Brett CM, Giacomini KM (2008) Effect of genetic variation in the organic cation transporter 1, OCT1, on metformin pharmacokinetics. *Clin Pharmacol Ther*, 83, 273–280.

Shu Y, Sheardown SA, Brown C, Owen RP, Zhang S, Castro RA, Ianculescu AG, Yue L, Lo JC, Burchard EG, et al. (2007) Effect of genetic variation in the organic cation transporter 1 (OCT1) on metformin action. *J Clin Invest*, 117, 1422–1431.

Sim SC, Risinger C, Dahl ML, Aklillu E, Christensen M, Bertilsson L, Ingelman-Sundberg M (2006) A common novel CYP2C19 gene variant causes ultrarapid drug metabolism relevant for the drug response to proton pump inhibitors and antidepressants. *Clin Pharmacol Ther*, 79, 103–113.

Skarke C, Jarrar M, Schmidt H, Kauert G, Langer M, Geisslinger G, Lotsch J (2003) Effects of ABCB1 (multidrug resistance transporter) gene mutations on disposition and central nervous effects of loperamide in healthy volunteers. *Pharmacogenetics*, 13, 651–660.

Song IS, Shin HJ, Shim EJ, Jung IS, Kim WY, Shon JH, Shin JG (2008) Genetic variants of the organic cation transporter 2 influence the disposition of metformin. *Clin Pharmacol Ther*, 84, 559–562.

Surh LC, Pacanowski MA, Haga SB, Hobbs S, Lesko LJ, Gottlieb S, Papaluca-Amati M, Patterson SD, Hughes AR, Kim MJ, et al. (2010) Learning from product labels and label changes: how to build pharmacogenomics into drug-development programs. *Pharmacogenomics*, 11, 1637–1647.

Suzuki H, Sugiyama Y (2002) Single nucleotide polymorphisms in multidrug resistance associated protein 2 (MRP2/ABCC2): its impact on drug disposition. *Adv Drug Deliv Rev*, 54, 1311–1331.

Tamai I (2011) Oral drug delivery utilizing intestinal OATP transporters. *Adv Drug Deliv Rev*, 64, 508–514.

Tassaneeyakul W, Tiamkao S, Jantararoungtong T, Chen P, Lin SY, Chen WH, Konyoung P, Khunarkornsiri U, Auvichayapat N, Pavakul K, et al. (2010) Association between HLA-B*1502 and carbamazepine-induced severe cutaneous adverse drug reactions in a Thai population. *Epilepsia*, 51, 926–930.

Tomlinson B, Hu M, Lee VW, Lui SS, Chu TT, Poon EW, Ko GT, Baum L, Tam LS, Li EK (2010) ABCG2 polymorphism is associated with the low-density lipoprotein cholesterol response to rosuvastatin. *Clin Pharmacol Ther*, 87, 558–562.

Wang J, Zeevi A, McCurry K, Schuetz E, Zheng H, Iacono A, McDade K, Zaldonis D, Webber S, Watanabe RM, Burckart GJ (2006) Impact of ABCB1 (MDR1) haplotypes on tacrolimus dosing in adult lung transplant patients who are CYP3A5 *3/*3 non-expressors. *Transpl Immunol*, 15, 235–240.

Wilke RA, Ramsey LB, Johnson SG, Maxwell WD, McLeod HL, Voora D, Krauss RM, Roden DM, Feng Q, Cooper-Dehoff RM, Gong L, Klein TE, Wadelius M, Niemi M, Clinical Pharmacogenomics Implementation Consortium (CPIC) (2012) The clinical pharmacogenomics implementation consortium: CPIC guideline for SLCO1B1 and simvastatin-induced myopathy. *Clin Pharmaco. Ther*, 92, 112–117.

Williams JA, Andersson T, Andersson TB, Blanchard R, Behm MO, Cohen N, Edeki T, Franc M, Hillgren KM, Johnson KJ, et al. (2008) PhRMA white paper on ADME pharmacogenomics. *J Clin Pharmacol*, 48, 849–889.

Wysowski DK, Nourjah P, Swartz L (2007) Bleeding complications with warfarin use: a prevalent adverse effect resulting in regulatory action. *Arch Intern Med*, 167, 1414–1419.

Xie HG, Prasad HC, Kim RB, Stein CM (2002) CYP2C9 allelic variants: ethnic distribution and functional significance. *Adv Drug Deliv Rev*, 54, 1257–1270.

Yamazaki H, Inoue K, Chiba K, Ozawa N, Kawai T, Suzuki Y, Goldstein JA, Guengerich FP, Shimada T (1998) Comparative studies on the catalytic roles of cytochrome P450 2C9 and its Cys- and Leu-variants in the oxidation of warfarin, flurbiprofen, and diclofenac by human liver microsomes. *Biochem Pharmacol*, 56, 243–251.

Yang L, Ge W, Yu F, Zhu H (2010) Impact of VKORC1 gene polymorphism on interindividual and interethnic warfarin dosage requirement–a systematic review and meta analysis. *Thromb Res*, 125, e159–e166.

Yasuda SU, Zhang L, Huang SM (2008) The role of ethnicity in variability in response to drugs: focus on clinical pharmacology studies. *Clin Pharmacol Ther*, 84, 417–423.

Yee SW, Chen L, Giacomini KM (2010) Pharmacogenomics of membrane transporters: past, present and future. *Pharmacogenomics*, 11, 475–479.

Yuan HY, Chen JJ, Lee MT, Wung JC, Chen YF, Charng MJ, Lu MJ, Hung CR, Wei CY, Chen CH, et al. (2005) A novel functional VKORC1 promoter polymorphism is associated with inter-individual and inter-ethnic differences in warfarin sensitivity. *Hum Mol Genet*, 14, 1745–1751.

Zhang L, Huang SM, Lesko LJ (2011) Transporter-mediated drug-drug interactions. *Clin Pharmacol Ther*, 89, 481–484.

Zhao P, Zhang L, Grillo JA, Liu Q, Bullock JM, Moon YJ, Song P, Brar SS, Madabushi R, Wu TC, et al. (2011) Applications of physiologically based pharmacokinetic (PBPK) modeling and simulation during regulatory review. *Clin Pharmacol Ther*, 89, 259–267.

Zheng HX, Webber SA, Zeevi A, Schuetz E, Zhang J, Lamba J, Boyle GJ, Wilson JW, Burckart GJ (2004) The impact of pharmacogenomic factors on steroid dependency in pediatric heart transplant patients using logistic regression analysis. *Pediatr Transplant*, 8, 551–557.

Zheng HX, Zeevi A, McCurry K, Schuetz E, Webber S, Ristich J, Zhang J, Iacono A, Dauber J, McDade K, et al. (2005) The impact of pharmacogenomic factors on acute persistent rejection in adult lung transplant patients. *Transpl Immunol*, 14, 37–42

Zineh I, Huang SM (2011) Biomarkers in drug development and regulation: a paradigm for clinical implementation of personalized medicine. *Biomark Med*, 5, 705–713.

Zineh I, Lesko LJ (2009) Pharmacogenetics in medicine: barriers, critical factors and a framework for dialogue. *Per Med*, 6, 359–361.

FDA WEB SITES

2008: Tetrabenazine Clinical Pharmacology Review. http://www.accessdata.fda.gov/drugsatfda_docs/nda/2008/021894s000TOC.cfm. (Accessed December 18, 2011) Ref Type: Electronic Citation

2009: CellCept® (Mycophenolate Mofetil) Labeling. http://www.accessdata.fda.gov/drugsatfda_docs/label/2009/050722s024,050723s023,050758s022,050759s028lbl.pdf. (Accessed December 18, 2011) Ref Type: Electronic Citation

2010: COUMADIN® (Warfarin) Prescribing Information. www.accessdata.fda.gov/drugsatfda_docs/label/2010/009218s108lbl.pdf. (Accessed December 18, 2011) Ref Type: Electronic Citation

2011a: ERBITUX® (Cetuximab) Labeling. http://www.accessdata.fda.gov/drugsatfda_docs/label/2011/125084s153lbl.pdf. (Accessed December 18, 2011) Ref Type: Electronic citation

2011c: JEVTANA (Cabazitaxel) Labeling. http://www.accessdata.fda.gov/drugsatfda_docs/label/2010/201023lbl.pdf. (Accessed December 18, 2011) Ref Type: Electronic Citation

2011d: PLAVIX (Clopidogrel) Labeling. http://www.accessdata.fda.gov/drugsatfda_docs/label/2011/020839s051lbl.pdf. (Accessed December 18, 2011) Ref Type: Electronic Citation

2011e: TEGRETOL® (Carbamazepine) Labeling. http://www.accessdata.fda.gov/drugsatfda_docs/label/2011/016608s100s102,018281s049s050,018927s041s042,020234s031s033lbl.pdf. (Accessed December 18, 2011) Ref Type: Electronic Citation

2011f: TYKERB (Lapatinib) Labeling. http://www.accessdata.fda.gov/drugsatfda_docs/label/2011/022059s011s012lbl.pdf. (Accessed December 18, 2011) Ref Type: Electronic Citation

2011g: VECTIBIX® (Panitumumab) Labeling. http://www.accessdata.fda.gov/drugsatfda_docs/label/2009/125147s080lbl.pdf. (Accessed December 18, 2011) Ref Type: Electronic Citation

2011h: VOTRIENT (Pazopanib) Labeling. http://www.accessdata.fda.gov/drugsatfda_docs/label/2011/022465s003s005lbl.pdf. (Accessed December 18, 2011) Ref Type: Electronic Citation

2011i: XENAZINE (Tetrabenazine) Labeling. http://www.accessdata.fda.gov/drugsatfda_docs/label/2011/021894s004lbl.pdf. (Accessed December 18, 2011) Ref Type: Electronic Citation

2011j: PROMACTA (Eltrombopag) Labeling. http://www.accessdata.fda.gov/drugsatfda_docs/label/2011/022291s006lbl.pdf. (Accessed February 3, 2012) Ref Type: Electronic Citation

2013: Guidance for Industry: "Clinical Pharmacogenomics: Premarketing Evaluation in Early-Phase Clinical Studies and Recommendations for Labeling." http://www.fda.gov/Drugs/GuidanceCompliance RegulatoryInformation/Guidances/ucm064982.htm. (Accessed February 2, 2013) Ref Type: Electronic Citation

CHAPTER 4

THE PHARMACOGENOMICS OF MEMBRANE TRANSPORTERS PROJECT

Sook Wah Yee
Deanna L. Kroetz
Kathleen M. Giacomini

4.1 INTRODUCTION

The pharmacogenomics of membrane transporters (PMT) project is in its thirteenth year of NIH funding. First funded in 2000, PMT was competitively renewed in 2005, and again in 2010. The overall goal of PMT, a discovery, hypothesis-generating project, continues to be the identification of genetic variants in transporters that underlie variation in therapeutic and adverse drug response (Fig. 4.1). PMT focuses on transporters in the ATP-binding cassette (ABC) and solute carrier (SLC) superfamilies, with particular attention placed on 60 high-priority transporters (Fig. 4.2). The 60 transporters were selected primarily based on compelling *in vivo* evidence (e.g., studies in knockout mice or volunteers with genetic polymorphisms) that the transporter is involved in drug disposition and response. Since its inception, the project has focused on studies that address three aims: (a) to discover sequence variants in membrane transporters; (b) to characterize the function of transporter variants in cellular systems; and (c) to determine the importance of transporter variants to clinical drug response. Key findings under each of these aims will be discussed below with respect to particular transporters.

Because large-scale human variation projects are now underway, single nucleotide polymorphism (SNP) discovery studies have been curtailed in the recent renewal of PMT and replaced with SNP data mining from the 1000 Genomes Project, which centers on 13 populations from various regions of the world (1000 Genomes Project Consortium et al., 2010; Mills et al., 2011). Prior to the 1000 Genomes Project, PMT was one of the world's leading genetic variation discovery projects. Many SNPs in membrane transporter genes in both coding and noncoding regions were discovered by the project. Figure 4.3 shows the number of SNPs in various regions of 129

Pharmacogenomics of Human Drug Transporters: Clinical Impacts, First Edition.
Edited by Toshihisa Ishikawa, Richard B. Kim and Jörg König

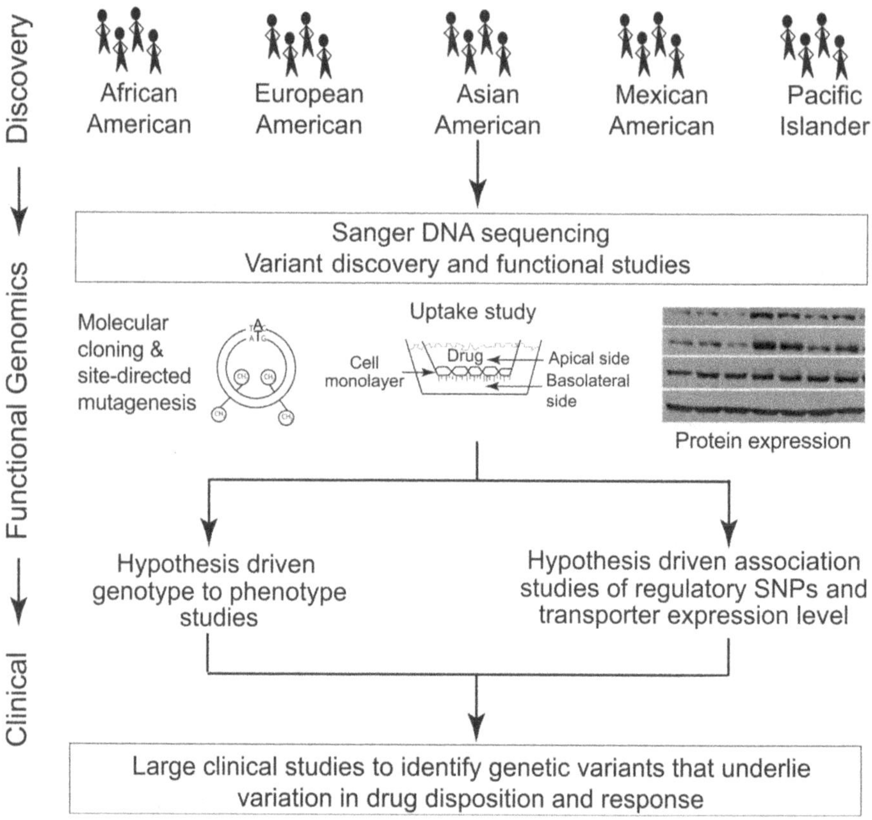

Figure 4.1 Pharmacogenomics of membrane transporters (PMT) project flow of studies. The project involved SNP discovery from major ethnic groups (SOPHIE cohort), functional studies of variants in coding and noncoding regions followed by investigations of functional variants in healthy volunteers (SOPHIE), tissue samples or large clinical studies. Examples of variants discovered from this project are summarized in Table 4.1.

transporter genes that was discovered by PMT along with the cumulative number of publications in which PMT SNPs have been cited (from 2001 through 2010). As shown, the discovery aspect of the project served to stimulate and support research in pharmacogenomics and led to a wealth of new information in the scientific literature. Further, using data generated in discovery studies, PMT investigators have performed large-scale functional genomic studies of membrane transporters (Table 4.1, Fig. 4.1).

A major focus of PMT is on ABC proteins, which are primary active transporters that rely on ATP hydrolysis to actively pump their substrates across biomembranes. Forty-nine genes for ABC proteins in the human genome are grouped into seven major families (*ABCA* to *ABCG*) (Borst and Elferink, 2002). ABC transporters differ from one another in terms of tissue and subcellular localization, as well as substrate specificities. Many of the transporters are responsible for efflux mechanisms of a variety of endogenous solutes such as cholesterol and bile salts (ABCA1, ABCB11, ABCG1, ABCG5, and ABCG8) (Brooks-Wilson et al., 1999; Berge et al., 2000;

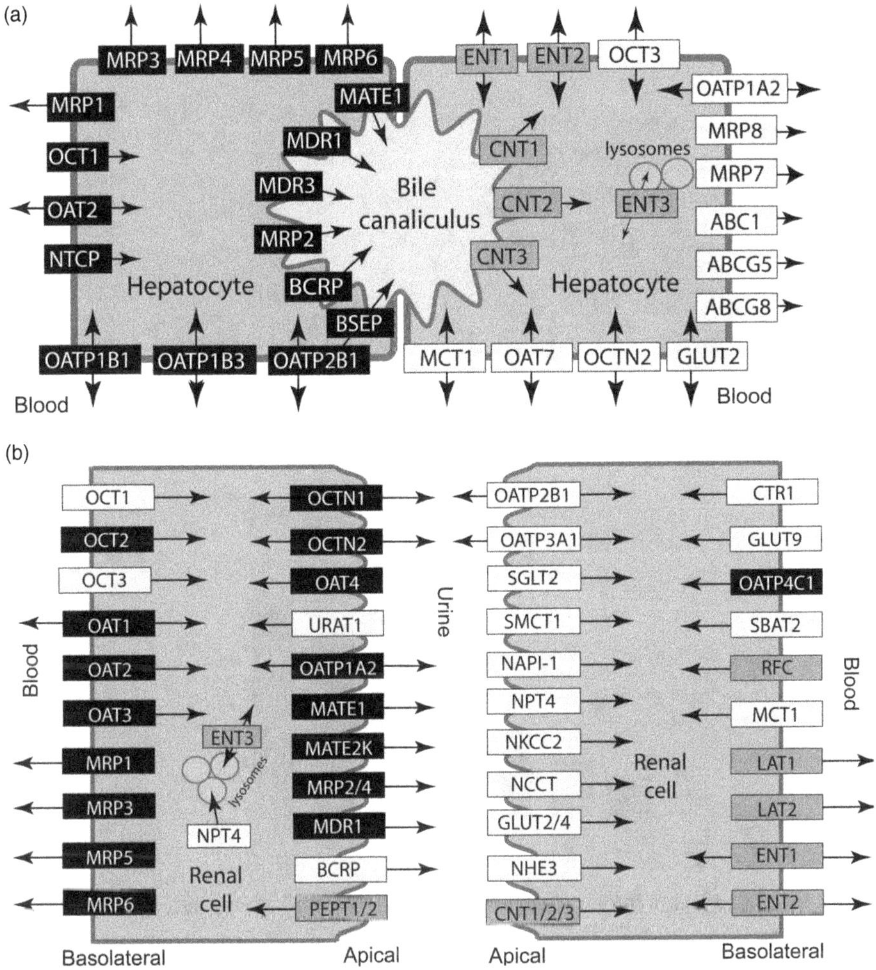

Figure 4.2 Membrane transporters that are the focus of the PMT project. Transporters on the basolateral and canalicular membranes of the hepatocyte (a) and on the apical and basolateral membranes of the renal epithelial cell (b) that play important roles in the uptake and efflux of drugs and other xenobiotics. The figure shows the key 60 membrane transporters in the kidney and/or liver which are the focus of the PMT project. Transporters that are well-studied, known to play many roles in the uptake or efflux of multiple clinical drugs and/or found at high expression levels are shaded dark. Transporters that recognize specific class of drugs are indicated by a medium shade. Many of the remaining transporters play specialized roles or may be drug targets (light shade). (*See insert for color representation of the figure.*)

Klucken et al., 2000; Lee et al., 2001), and long-chain fatty acids (ABCD1 and ABCD2) (Watkins et al., 1995; Lombard-Platet et al., 1996). Drug transporters, which are of particular interest to PMT, include transporters in the ABCB, ABCC, and ABCG families particularly, ABCB1, ABCB11, ABCG2, and several members of the ABCC family (Kruh and Belinsky, 2003; Schinkel and Jonker, 2003; Vlaming

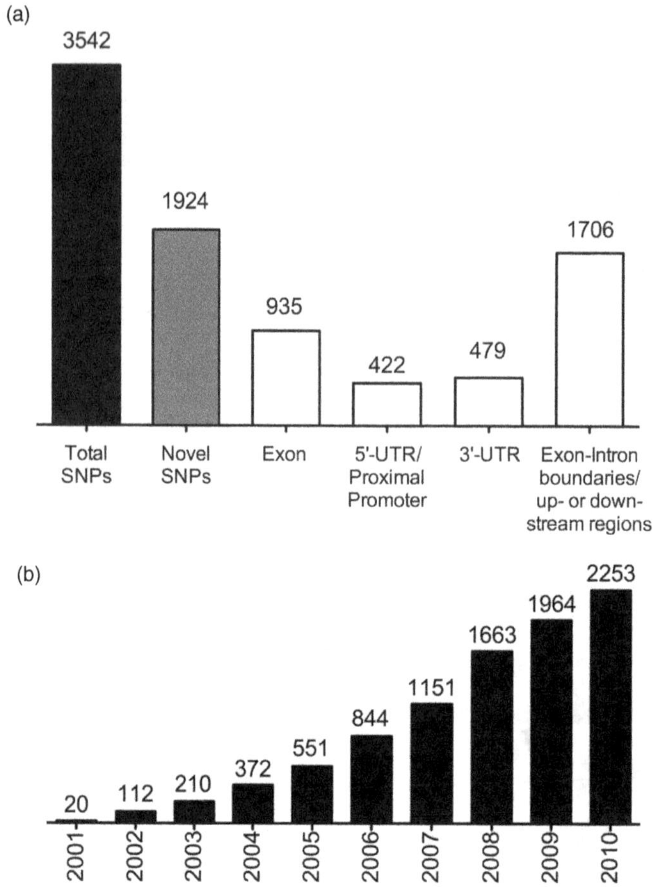

Figure 4.3 Measures of impact of the PMT project on the field of pharmacogenomics. (a) Total number of SNPs discovered by the PMT project since 2000. (b) Cumulative number of publications investigating the effect of genetic variants in SNPs in membrane transporter genes identified in PMT discovery studies from 2001 to 2010. Notes: **Total SNPs**: Total number of SNPs identified from sequencing exons, promoter regions and/or flanking intronic regions of 129 membrane transporter genes (from 2000 to 2010). **Novel SNPs**: SNPs that were exclusively discovered by PMT from the sequencing and had not been previously reported or deposited in dbSNP 130.

et al., 2009). Many of these transporters have distinct, but overlapping substrate specificities. For example, there is some overlapping specificity between substrates of ABCB1 (P-glycoprotein) and ABCG2 (BCRP). The SLC superfamily includes genes encoding facilitated transporters and ion-coupled, secondary-active transporters that reside in various cell membranes (Hediger et al., 2004). Fifty-one SLC families with a total of about 378 transporters have been identified in the human genome.The list of SLC tables could be accessed from http://www.bioparadigms.org/slc/menu.asp. PMT focuses on the ABC and SLC transporters known to play important pharmacological roles in pharmacokinetics and pharmacodynamics (Schloss and Williams, 1998; Hahn

TABLE 4.1 Functional Variants Discovered and Assessed by PMT Investigators

Variant (rsID)	AA (100)	CA (100)	AS (30)	ME (10)	Cell system	Substrates	Functional change compared to reference	Reference
SLC22A1 (OCT1) – Variants and functional studies								
R61C rs12208357	0	7.2	0	5.6	(i) Oocytes (ii) HEK293 (iii) HEK293 (iv) HEK293 (v) HEK-GFP tagged	(i) MPP+ (ii) Metformin (iii) Picoplatin (iv) ASP+	(i) ↓ uptake (ii) ↓ uptake; ↓ p-AMPK and p-ACC activities; ↓ protein expression on plasma membrane (iii) ↓ uptake and ↓ platinum-DNA adducts (iv) ↓ uptake of OCT1 reference; ↓ K_m, ↓ V_{max} and ↓ V_{max}/K_m (v) ↓ GFP staining on cell membrane and showed diffuse intracellular localization	(Shu et al., 2003, 2007; Ahlin et al., 2011; More et al., 2010)
F160L rs683369	0.5	6.5	1.7	5	(i) Oocytes (ii) HEK293 (iii) HEK293	(i) MPP+ (ii) Metformin (iii) Picoplatin	(i) uptake (ii) ←→ uptake (iii) ←→ uptake and ←→ platinum-DNA adducts	

(continued)

TABLE 4.1 (*Continued*)

SLC22A1 (OCT1) – Variants and functional studies							
G401S rs34130495	0.7	1.1	0	0	(i) Oocytes (ii) HEK293 (iii) HEK293	(i) MPP+ (ii) Metformin (iii) Picoplatin	(i) ↓ uptake (ii) ↓ uptake; ↓ p-AMPK and p-ACC activities. (iii) ↓ uptake and ↓ platinum-DNA adducts.
M420del rs35167514, rs34305973 and rs35191146	2.9	18.5	0	21.4	(i) Oocytes (ii) HEK293 (iii) HEK293 (iv) HEK293 (v) HEK-GFP tagged	(i) MPP+ (ii) Metformin (iii) Picoplatin (iv) ASP+	(i) ←→ uptake (ii) ↓ uptake; slight ↓ p-AMPK and p-ACC activities; ←→ K_m, ↓ V_{max} and ↓ V_{max}/K_m; ←→protein expression on plasma membrane (iii) ↓ uptake and ↓ platinum-DNA adducts (iv) ←→ uptake; ←→ K_m, ↓ V_{max} and ←→ V_{max}/K_m (v) clear membrane localization by GFP staining

G465R rs34059508	0	4	0	0	(i) Oocytes (ii) MDCK-GFP tagged (iii) HEK293 (iv) HEK293 (v) HEK293 (vi) HEK-GFP tagged	(i) MPP+ (ii) NA (iii) Metformin (iv) Picoplatin (v) ASP+	(i) ↓ uptake (ii) ↓ localization at basolateral surface (iii) ↓ uptake; ↓ p-AMPK and p-ACC activities; ↓ protein expression on plasma membrane (iv) ↓ uptake and ↓ platinum-DNA adducts (v) ↓ uptake (vi) subset of cells with GFP staining on plasma membrane.	

SLC22A2 (OCT2) – Variants and functional studies

Variant (rsID)	AA (100)	CA (100)	AS (30)	ME (10)	Cell system	Substrates	Functional change compared to reference	Reference
A270S rs316019	11	15.8	8.6	15	(i) Oocytes (ii) HEK293 (iii) HEK293 (iv) HEK293	(i) MPP+ (ii) Metformin (iii) Picoplatin	(i) ←→ uptake (ii) ↑ uptake (iii) ↑ uptake (iv) ↑ total protein expression	(Fujita et al., 2006; Chen et al., 2009a)

(*continued*)

TABLE 4.1 (*Continued*)

SLC22A3 (OCT3) – Variants and functional studies								
Variant (rsID)	AA (100)	CA (100)	AS (30)	PA (7)	Cell system	Substrates	Functional change compared to reference	Reference
T44M rs68187715	0.6	0.6	0	14.3	(i) HEK293	(i) Metformin (ii) MPP+ (iii) Histamine (iv) Tyramine (v) Serotonin (vi) Dopamine (vii) Norepinephrine (viii) Epinephrine	(i) ↑ uptake (160% of OCT3 reference); ↑ V_{max}; ←→ K_m (ii) ↑ uptake (iii) ↑ uptake (iv) – (viii) ←→ uptake	(Chen et al., 2010)
V423F Chr6:160778212 (build 36)	YRI (60)	CEU (60)	AS (60)	-				
	0	0	6.8	-				
	1000 Genome Data (Pilot 1). This SNP was removed from the 1000 Genome Data Release July 2010.							

SLC22A4 (OCTN1) – Variants and functional studies								
Variant (rsID)	AA (80)	CA (80)	AS (60)	ME (25)	Cell system	Substrates	Functional change compared to reference	Reference
L503F rs1050152	8.8	41.2	0	23	(i) HEK293 (ii) HEK293 (iii) HEK-GFP tagged	(i) Betaine (ii) ↑ TEA	(i) ↓ uptake (ii) ↑ TEA; ↓ K_m, ↑ V_{max} (iii) ←→ subcellular localization of OCTN1 variant-GFP protein	(Urban et al., 2007)

SLC22A5 (OCTN2) – Variants and functional studies

Variant (rsID)	AA (68)	CA (68)	AS (68)	ME (68)	Cell system	Substrates	Functional change compared to reference	Reference
-207G>C rs2631367	38.2	50	0	32.1	(i) Immortalized lymphoblasts	(i) L-carnitine	(i)↑ uptake (-207G)	(Urban et al., 2006a)

SLC22A6 (OAT1) – Variants and functional studies

Variant (rsID)	AA (80)	CA (80)	AS (60)	ME (50)	Cell system	Substrates	Functional change compared to reference	Reference
R454Q rs11568634	0.6	0	0	0	(i) Oocytes	(i) Adefovir (ii) Ochratoxin (iii) Methotrexate	(i)–(iii) ↓uptake	(Fujita et al., 2005)

SLC22A8 (OAT3) – Variants and functional studies

Variant (rsID)	AA (80)	CA (80)	AS (60)	ME (50)	Cell system	Substrates	Functional change compared to reference	Reference
I305F rs11568482	0	0	3.5	1.1	(i) HEK293 (ii) HEK293 (iii) HEK293-GFP tagged	(i) Estrone sulfate (ii) Cimetidine	(i) ↓uptake, ←→ V_{max}, ←→ K_m (ii) ←→uptake, ←→ V_{max}, ←→ K_m (iii) ←→ subcellular localization of OAT3 variant-GFP protein	(Erdman et al., 2006)

(*continued*)

TABLE 4.1 (Continued)

SLC22A11 (OAT4) – Variants and functional studies								
Variant (rsID)	AA (68)	CA (68)	AS (68)	ME (68)	Cell system	Substrates	Functional change compared to reference	Reference
L29P rs11231819	Presence of the variant in dbSNP but lacking frequency data.				(i) HEK293 (ii) HEK293 (iii) HEK293	(i) Esterone sulfate (ii) Ochratoxin A (iii) Uric acid	(i)–(iii) ↓uptake; ↓ plasma membrane protein levels	(Shima et al., 2010)
R48Stop rs35008345	0	2.3	0	0.8			(i)–(iii) ↓uptake; ↓ mRNA expression and plasma membrane protein levels	
H469R ss120239857	0	0	0	0.7			(i)–(iii) ↓uptake; ↓ plasma membrane protein levels	

SLC28A2 (CNT2) – Variants and functional studies								
Variant (rsID)	AA (100)	CA (100)	AS (30)	ME (10)	Cell system	Substrates	Functional change compared to reference	Reference
-146T>A rs2413775	60	72.5	26.7	40	(i) HCT116, HEPG2, ACHN cells (ii) Mouse	NA	(i) ↑ luciferase activity in cells (ii) higher luciferase assay in mouse liver	(Yee et al., 2009)
F355S rs17215633	3.5	0	0	0	(i) Oocytes	(i) Inosine (ii) Uridine	(i)–(ii) Altered substrate specificity	(Owen et al., 2005)

SLC47A1 (MATE1) – Variants and functional studies

Variant (rsID)	AA (68)	CA (68)	AS (68)	ME (68)	Cell system	Substrates	Functional change compared to reference	Reference
-66T>C rs2252281	44.5	32.1	23.1	28.9	(i) HCT116, ACHN, HepG2, HeLa cells (ii) Mouse (iii) Kidney tissue	NA	(i) ↓ luciferase activity in haplotype constructs with -66T>C variant allele (ii) ↓ luciferase activity in mouse liver (iii) ↓ mRNA level in kidney tissue	(Choi et al., 2009)
L125F rs77474263	0	0	0.7	5.1	(i) HEK293 (ii) HEK293 (iii) HEK293 (iv) HEK293 (v) HEK293-GFP tagged	(i) TEA (ii) Metformin (iii) Paraquat (iv) Oxaliplatin	(i) ↓ uptake (ii) ↓ uptake; ←→ V_{max}, K_m, V_{max}/K_m (iii) ↓ uptake; ←→ V_{max}, K_m, V_{max}/K_m (iv) ←→ uptake; similar IC_{50} (v) ←→ subcellular localization of MATE1 variant-GFP protein	(Chen et al., 2009b)
V338I rs35790011	5.1	0	0	0			(i) ↓ uptake (ii) ↓ uptake (iii) ←→ uptake (iv) ←→ uptake; similar IC_{50} (v) ←→ subcellular localization of MATE1 variant-GFP protein	

(continued)

TABLE 4.1 (Continued)

SLCO1A2 (OATP1A2) – Variants and functional studies								
Variant (rsID)	AA (80)	CA (80)	AS (60)	ME (50)	Cell system	Substrates	Functional change compared to reference	Reference
I13T rs10841795	2.5	16.3	0	0.05	(i) Oocytes (ii) Oocytes	(i) Esterone sulfate (ii) Methotrexate	(i) ↑ uptake (ii) ↑ uptake; ↑ $V_{\max}$, ←→ K_{m}	(Badagnani et al., 2006)
E172D rs11568563	0.6	1.9	0	2			(i)–(ii) ↓ uptake	
ABCB1 (P-gp) – Variants and functional studies								
Variant (rsID)	AA (100)	CA (100)	AS (30)	ME (10)	Cell system	Substrates	Functional change compared to reference	Reference
S1141T rs2229107	11	0	0	5	(i) Transformed yeast (ii) HEK293T	(i) Daunorubicin (ii) Doxorubicin (iii) Valinomycin (iv) Actinomycin D (v) BODIPY-FL-paclitaxel	(i) and (iv) ←→ resistance (ii) and (iii) ↑ resistance	(Jeong et al., 2007; Gow et al., 2008a)
A893S rs2032582	10	46.4	45	40			(ii) ↑ resistance (i), (iii)–(iv) ←→ resistance (v) ↑ intracellular BODIPY-FL-paclitaxel levels (decreased function)	
A893T rs2032582	0.5	3.6	6.7	0	(i) HEK293T	(i) Calcein-AM	(i) ↓ intracellular calcein levels (increased function)	

V1251I rs28364274	0	0	0	5	(i) HEK293T (ii) HEK293T	(i) Calcein-AM (ii) BODIPY-FL-paclitaxel	(i) ↓ intracellular calcein levels (increased function) (ii) ↑ intracellular BODIPY-FL-paclitaxel levels (decreased function)	
N21D/1236C>T/ A893S/3435C>T	2	7.5	1.6	0	(i) HEK293T	(i) BODIPY-FL-paclitaxel	↑intracellular BODIPY-FL-paclitaxel levels (decreased function), less sensitive to cyclosporine A inhibition	

AECB11 (BSEP) – Variants and functional studies

Variant (rsID)	AA (100)	CA (100)	AS (30)	ME (10)	Cell system	Substrates	Functional change compared to reference	Reference
I206V rs11568357	2.5	0	0	0	(i)–(ii) HeLa cells infected with vaccinia	(i) Taurocholate	(i) ↓ transport activity (ii) ←→ total and cell surface protein expression.	(Ho et al., 2010)
V444A rs2287622	53	57.1	66.7	50	Human liver tissue		Associated with significantly reduced hepatic BSEP mRNA levels in human liver samples.	
E1186K rs1521808	2.5	0	0	0	(i)–(ii) HeLa cells infected with vaccinia	(i) Taurocholate	(i) ↓ transport activity (ii) ↓ total and cell surface protein expression.	

ABCC4 (MRP4) – Variants and functional studies

Variant (rsID)	AA (80)	CA (80)	AS (60)	ME (50)	Cell system	Substrates	Functional change compared to reference	Reference
G187W rs11568658	0	2.5	10.8	13	(i)–(ii) HEK293T	(i) AZT (ii) PMEA	(i)–(ii) ↑ intracellular AZT and PMEA levels (decreased function)	(Abla et al., 2008; Kelly et al., 2010)

(*continued*)

TABLE 4.1 (*Continued*)

SLC22A1 (OCT1) – Genotype to phenotype studies		
OCT1-reference vs OCT1-variant (R61C, G401S, 420del, G465R)	Change in metformin kinetics and pharmacologic effects as measured by oral glucose tolerance tests (OGTT).	(Shu et al., 2007, 2008)
SLC22A2 (OCT2) – Genotype to phenotype studies		
A270S rs316019	Higher renal clearance and net tubular secretion of metformin in individual with the OCT2 variant.	(Chen et al., 2009a)
SLC22A4 (OCTN1) – Genotype to phenotype studies		
L503F rs1050152	Higher gabapentin exposure and lower renal clearance/tubular secretion in individuals carrying OCTN1 variant.	(Urban et al., 2008)
SLC22A5 (OCTN2) – Genotype to phenotype studies		
-207G>C rs2631367	No appreciable effect of this variant on carnitine disposition.	(Tahara et al., 2009)
SLC22A6 (OAT1) – Genotype to phenotype clinical studies		
R454Q rs11568634	No difference in renal clearance and secretory clearance of adefovir in family-based studies.	(Fujita et al., 2005)
SLCO1B3 (OAT1B3) – Mycophenolic acid pharmacokinetic in large clinical studies		
S112A and M233I rs4149117 and rs7311358	Mycophenolic acid (MPA) and MPA-7-*O*-glucuronide pharmacokinetics were significantly influenced by the OATP1B3 nonsynonymous variants in renal transplant patients.	(Picard et al., 2010)
SLCO1A2 (OAT1P1A2) – Methotrexate clearance in large clinical studies		
rs11045981 and rs4148981 (Haplotype *1 *3)	Significant association between the haplotype *1 *3 and increased methotrexate clearance.	(Kim et al., 2011)
ABCs and SLCOs – Irinotecan pharmacokinetic and neutropenia effects in large clinical studies		
Approximately 50% of the variation in ANC nadir in cancer patients is explained by rs3765129 (ABCC1, IVS11 -48C>T), rs2306283 (SLCO1B1, N130D) and UGT1A1*93; the AUC of irinotecan, SN-38, SN-38 glucuronide and APC are influenced by rs3740066 (ABCC2 3972C>T), rs2306283, rs35605 (ABCC1, 1684C>T), rs10276036 (ABCB1 IVS9-44A>G), and rs717620 (ABCC2, -24C>T). ABCC2*2 (-1019A>G) is associated with lower irinotecan clearance and with significant reduction of severe diarrhea in cancer patients.		(Innocenti et al., 2009)

The table summarizes the functional activities of the variants in different cellular systems to study the effects of the genetic variants on transport activity, protein expression, cellular localization, and/or expression levels. In some cases, the identification of functionally significant variants *in vitro* led to further studies in human volunteers or in clinical studies involving patients.

AA, African-Americans; CA, European-Americans; AS, Asian-Americans; ME, Mexican-Americans; PA, Pacific-Islander.

and Blakely, 2002), drug sensitivity/resistance (Germann, 1996; Gottesman et al., 1996; Lin and Yamazaki, 2003), and drug toxicity (SEARCH Collaborative Group et al., 2008).

In this chapter, we first describe three large functional genomic studies published by PMT investigators followed by representative in-depth studies characterizing the functional effects of genetic variants in specific transporters. Next, we describe clinical studies, most of which are focused on proof-of-concept genotype to phenotype studies in healthy volunteers. We end the chapter with a section on future directions of the project.

4.2 FUNCTIONAL GENOMIC STUDIES OF ABC AND SLC SUPERFAMILY TRANSPORTERS

Because the PMT project has generated large datasets of genetic variants (Fig. 4.3) in membrane transporters, and numerous studies focused on functional analyses of variants in cellular assays, it has accrued a wealth of information on membrane transporter polymorphisms and their functional characteristics (Table 4.1). Consequently, a major effort of PMT has been to mine this information and where possible to derive and publish general principles in the area of transporter genomics. Three such studies are described below, followed by studies of individual transporter genes and summaries of the functional SNPs discovered and functionally characterized by the PMT group (Table 4.1).

In 2006, Urban et al. published a meta-analysis of functional genomic studies carried out on 11 membrane transporters in heterologous cell systems (Urban et al., 2006b). The transporters were primarily in the SLC22, SLC28, and SLC29 families and included organic cation and anion transporters and concentrative and equilibrative nucleoside transporters. Fourteen percent of 87 nonsynonymous variants in the 11 transporters had virtually no activity or significantly reduced activity. The reduced function variants were found at lower allele frequencies than variants that retained function and were more likely to change a conserved amino acid residue. Interestingly, approximately 14% of variants for which more than one substrate was tested showed kinetic and specificity differences. Collectively, the data suggest that strong selective pressure in membrane transporters exists to retain function (i.e., reduced function variants are selected against) and that variants that alter substrate specificity are reasonably common. The analysis was important for our understanding of transporter polymorphisms. In particular, xenobiotic transporters may be thought to tolerate functional variants since the transporters are redundant and many are not essential for survival (based on studies in knockout mice). The data showed that functional variation was not tolerated even in transporters that seem to function primarily in the disposition of xenobiotics (e.g., organic cation transporters). Further, specificity variants may suggest that a variant will have distinct pharmacogenomic effects depending on the drug under study.

The vast majority of functional genomic studies of membrane transporters have focused on nonsynonymous variants. Though such variants may have large effects on function and therefore are important to analyze, noncoding region variants are more common and may have important effects on transcription rates and levels of mRNA transcripts. In particular, the proximal promoter region of genes is defined as the region immediately upstream and downstream of the transcription start site. This region is important in the initiation of gene transcription since it is the site of binding of many transcription factors and nucleotide polymerases responsible for DNA and RNA transcription. In a large-scale study, we discovered 579 genetic variants in the proximal promoter region of 107 membrane transporter genes in 272 DNA samples from ethnically diverse populations (Hesselson et al., 2009). Genetic variants were more frequent and more common in the proximal promoter region in comparison to nonsynonymous variants. Further, we observed that the proximal promoter of transporters primarily expressed in the liver had more genetic variation than those expressed primarily in the kidney. Finally, in reporter assays of 21 proximal promoters, we found that weak promoters, that is, ones that elicited low levels of luciferase activities in multiple cell lines, had reduced genetic variation in comparison to strong promoters, consistent with the idea that weak promoters are under more selective pressure. These findings provide important principles for understanding genetic variation in promoter regions of membrane transporters and their impact on function. Promoter variants could indeed contribute to variation in drug response because they are common and can modulate transcription rates.

Lymphoblastoid cell lines have been analyzed in several investigations to identify expression SNPs, expressed single nucleotide polymorphisms (eSNPs), which associate with variation in the expression levels of various genes. Because lymphoblastoid cell lines represent a powerful discovery resource with information on both expression levels of mRNA transcripts and genetic variation, we conducted extensive studies on publicly available data from lymphoblastoid cell lines derived from HapMap populations. Our goals were to discover gender and race differences in expression levels of ABC transporters and importantly, eSNPs that associated with expression levels of 49 ABC genes (Matsson et al., 2011). Our results showed that of the 49 ABC transporters, 19 showed significantly different expression levels among racial groups: African, Asian, and European (Matsson et al., 2011). Twenty-four eSNPs were discovered which significantly associated with the expression levels of ABC transporter genes. Several were experimentally validated in reporter assays. The discovery of eSNPs in ABC transporters is important for understanding genetic factors that contribute to differences in the expression levels of the transporters among people. Because ABC transporters are so critical in drug disposition and disease risk, these eSNPs may be important determinants of disease susceptibility and drug response. It will be essential to determine which eSNPs are associated with expression levels of ABC transporters in tissues of pharmacological importance such as the liver and kidney.

4.2.1 ABCB1 (MDR1/P-gp)

4.2.1.1 Description and Tissue Distribution *ABCB1*, located at chromosome 7q21.12, encodes P-glycoprotein, was the first cloned human ABC gene and is the most studied of the multidrug resistance transporters (Dean et al., 2001; Leslie et al., 2005). P-glycoprotein was first characterized as an efflux transporter overexpressed in cancer cells, which is considered an important mechanism for resistance to many cancer chemotherapy agents. It is apically expressed in the epithelial cells of the adrenal cortex, renal proximal tubules and mesangial cells, intestinal brush border membrane, and the canalicular membrane of the hepatocyte (Fig. 4.4). P-glycoprotein is also localized to the capillary endothelial cells of the blood–brain and blood–testis barriers as well as various leukocyte lineages. A wide range of hydrophobic cationic or neutral compounds are transported across membranes by P-glycoprotein. Substrates include anthracyclines, vinca alkaloids, paclitaxel, protease inhibitors, immunosuppressants, calcium channel blockers, opiates, digoxin, and numerous steroids.

4.2.1.2 Pharmacogenetics The functional consequences of genetic variants of *ABCB1* have been studied by PMT and others. The most frequently studied SNP is the triallelic variant at amino acid position 893, located in the sixth intracellular loop near the C-terminus (Ala893Ser/Thr; rs2032582), which is found at a minor allele frequency of 10% in African-Americans and 45–46% in Caucasians and Asians (Kroetz et al., 2003). The Ser893 P-gp variant was first shown to have enhanced efflux of the model substrate digoxin (Kim et al., 2001). However, subsequent studies using multiple substrates showed only modest if any changes in function for this polymorphism (Kimchi-Sarfaty et al., 2002; Kroetz et al., 2003; Morita et al., 2003; Gow et al., 2008a) (Table 4.1). Other P-gp variants, including Asn21Asp, Phe103Leu, and Ala998Thr, generally show no differences in function compared with reference P-gp (Kimchi-Sarfaty et al., 2002; Gow et al., 2008b).

A common synonymous variant of *ABCB1*, 3435C>T (rs1045642), has been associated with changes in mRNA expression and P-gp tertiary structure (Wang et al., 2005; Kimchi-Sarfaty et al., 2007). Allelic specific expression of the *ABCB1* 3435C allele in human livers was consistent with changes in mRNA stability, possibly caused by changes in mRNA secondary structure (Wang et al., 2005). The 3435C>T allele uses an infrequent codon that may alter the timing of cotranslational folding and results in changes in P-gp tertiary structure that affects inhibitor interactions with this variant transporter (Kimchi-Sarfaty et al., 2007). PMT efforts also discovered a number of SNPs in the 3′-UTR of ABCB1, and functional studies indicated that there was no effect of these variants on mRNA stability (Gow et al., 2008).

4.2.2 *ABCC2* (MRP2)

4.2.2.1 Description and Tissue Distribution Multidrug-resistance-associated protein 2 (MRP2) is encoded by ABCC2, located at chromosome 10q24, and is an important efflux transporter, particularly in the liver (Dean et al., 2001; Haimeur et al., 2004). MRP2 is highly expressed on the bile canalicular membrane and in the gut and kidney (Fig. 4.4). The function of MRP2 is to efflux anionic conjugates out of the cell, in most cases facilitating their elimination into bile or urine. The substrate specificity of MRP2 is for amphiphilic organic anions that are conjugated with glutathione, glucuronide or sulfate moieties. The heme metabolite bilirubin glucuronide is transported by MRP2 from within the hepatocyte into the bile canaliculi where it is eliminated. MRP2 has been shown to transport methotrexate, leukotriene C_4, 17β-estradiol glucuronide, and bile salt derivatives. Additional

Human tissue	[1]ABCB1	[1]ABCC2	[1]ABCC4	[1]ABCB11	[1]SLC22A1	[1]SLC22A2	[1]SLC22A3	[1]SLC22A4	[1]SLC22A5	[1]SLC22A6	[1]SLC22A8	[1]SLC22A11	[1]SLCO1A2	[1]SLC28A1	[1]SLC28A2	[1]SLC28A3	[1]SLC29A1	[1]SLC29A2	[2]SLC47A1
Adrenal gland	1			BLQ		BLQ					BLQ	BLQ							
Bone marrow				BLQ		BLQ				BLQ	BLQ	BLQ			BLQ	1			
Brain				BLQ								BLQ							
Colon										BLQ	BLQ	BLQ							
Heart				BLQ		BLQ					BLQ	BLQ			BLQ				
Kidney				BLQ		1		1		1	1	1		1					1
Liver		1		1	1							BLQ							
Lung				BLQ		BLQ					BLQ	BLQ							
Pancreas				BLQ		BLQ				BLQ	BLQ	BLQ			BLQ				NA
Peripheral leukocytes				BLQ		BLQ	BLQ			BLQ	BLQ	BLQ			BLQ				NA
Placenta				BLQ						BLQ	BLQ								
Prostate			1	BLQ		BLQ				BLQ	BLQ	BLQ							
Salivary gland				BLQ		BLQ					BLQ	BLQ			BLQ				
Skeletal muscle				BLQ		BLQ	1		1	BLQ	BLQ	BLQ			BLQ	BLQ	1	1	
Small intestine				BLQ		BLQ				BLQ	BLQ	BLQ			1				
Spinal cord				BLQ								BLQ	1						NA
Spleen				BLQ		BLQ				BLQ	BLQ	BLQ			BLQ				
Stomach				BLQ						BLQ	BLQ	BLQ							
Testis											BLQ	BLQ							
Thymus											BLQ	BLQ			BLQ				
Thyroid gland				BLQ		BLQ					BLQ	BLQ							
Trachea												BLQ							NA
Uterus				BLQ								BLQ							

Figure 4.4 mRNA expression levels of the 19 drug transporters described in this chapter. The mRNA expression levels are adapted from literature references [1]Nishimura and Naito, 2005 and [2]Masuda et al., 2006. Data from [1]Nishimura and Naito (2005) were expressed as the ratio of the target mRNA to PPIA housekeeping gene mRNA. Using this ratio, the relative value to the highest values among the various tissues (labeled as 1) were calculated and plotted as a color gradient in this figure. Data from [2]Masuda et al. (2006) were expressed as amol mRNA/microgram total RNA; using this value, the relative value to the highest values among the various tissues (labeled as 1) were calculated and plotted as a color gradient.

Ratio of the target mRNA to the highest expressed tissue within the target mRNA	Color gradient
>0.1-1.0	
>0.01-0.1	
>0.001-0.01	
>0.0001-0.001	
<0.0001	
Below the limit of quantitation	BLQ
Not Available	NA

Figure 4.4 *(continued)*

substrates for MRP2 are the chemotherapeutic agents vinblastine, methotrexate, and the glucuronide conjugate of SN-38, the active metabolite of irinotecan.

4.2.2.2 Pharmacogenetics A common nonsynonymous variant of *ABCC2* is a Val417Ile change (rs2273697) found at a frequency of 12–17% in major ethnic populations. Cellular studies showed that the transport of estradiol 17β-glucuronide, leukotriene C_4, and 2,4-dinitrophenol-S-glutathione was unaffected by the Ile417 variant (Hirouchi et al., 2004). Several nonsynonymous variants of ABCC2 are found at relatively high frequency in the African-American population, including Arg1181Leu (8.5%), Val1188Glu (7.5%), and Cys1515Tyr (19.6%) (see http://pharmacogenetics.ucsf.edu). Functional analysis of these variants is ongoing within the PMT group and will inform clinical studies in African-American populations.

4.2.3 *ABCC4* (MRP4)

4.2.3.1 Description and Tissue Distribution *ABCC4* encodes the multidrug resistance transporter MRP4. *ABCC4* is located on chromosome 13 at position q32.1 and is considered an important transporter for the efflux of nucleoside analogs. It has been implicated in the transport of antiviral agents, such as the nucleoside/nucleotide analogs azidothymidine, adefovir, tenofovir, lamivudine, and ganciclovir, anticancer drugs (methotrexate, 6-mercaptopurine, 6-thioguanine, camptothecins), as well as endogenous molecules such as prostaglandins, steroids, bile acids, cyclic nucleotides, and folate (Borst et al., 2007). MRP4 is ubiquitously expressed, with a high expression in the prostate, as well as in hematopoietic stem cells, blood cells, the kidney proximal tubules, liver, and brain (Fig. 4.3) (Haimeur et al., 2004; Borst et al., 2007). Interestingly, its localization in most tissues is apical but basolateral localization has been demonstrated in brain choroid plexus epithelial cells, in prostate, and in hepatocytes.

4.2.3.2 Pharmacogenetics Resequencing of the *ABCC4* coding region in 270 ethnically diverse samples identified 22 nonsynonymous variants, but only three of these had minor allele frequencies in at least one population of 5% or greater (Abla et al., 2008). Eleven nonsynonymous variants were functionally characterized *in vitro* using the antivirals azidothymidine and adefovir. The Gly187Trp (2.5, 11, and 13% minor allele frequency in European-, African-, and Mexican-Americans, respectively) and Gly487Glu (<1% minor allele frequency in Asian-Americans) showed significantly reduced transport of both substrates compared with the reference. Decreased transport by the Gly187Trp variant is attributed at least in part to decreased expression on the cell membrane (Abla et al., 2008). Additional studies demonstrated reduced transport of tenofovir by Gly487Glu and the rare African-American-specific Val1071Ile variant of MRP4 (Kelly et al., 2010).

4.2.4 *ABCB11* (BSEP)

4.2.4.1 Description and Tissue Distribution The bile salt export pump (BSEP) is encoded by *ABCB11* located on 2q24.3. BSEP is expressed on the bile canalicular membrane and plays an important role in bile formation (Stieger et al., 2007). In addition to high expression in the liver, BSEP is also found at significant levels in the testis (Fig. 4.3), although its role in this tissue is unclear. The only known substrates of BSEP are monovalent bile salts.

4.2.4.2 Pharmacogenetics Progressive familial intrahepatic cholestasis is associated with rare mutations in *ABCB11*, indicating a critical role for BSEP in maintaining bile flow (Stieger et al., 2007; Stieger, 2009). The PMT group resequenced the *ABCB11* gene to identify more common polymorphisms that might also be associated with the secretion of bile salts (Ho et al., 2010). Nine nonsynonymous variants were identified, most of which were rare and population specific. Exceptions were the Val444Ala (rs2287622) found at a minor allele frequency of >50% in all four populations studied and the Met677Val (rs11568364) found at 15, 5.5, 1.7, and 5% frequency in European-, African-, Chinese-, and Mexican-Americans, respectively. The bile acid uptake transporter sodium/taurocholate cotransporting polypeptide and BSEP were coexpressed in HeLa cells to investigate the effect of common and rare variants on bile salt transport (Ho et al., 2010). Several rare BSEP variant transporters, including Ile206Val, Gln558His, Asn591Ser, and Glu1186Lys had significantly reduced taurocholate transport compared to the reference BSEP (Table 4.1). The common Val444Ala and Met677Val variants had no effect on bile salt transport.

4.2.5 SLC22A1, SLC22A2, SLC22A3 (OCT1, OCT2, OCT3)

4.2.5.1 Description and Tissue Distribution Organic cation transporters, OCT1-3, play important roles in pharmacokinetics and pharmacodynamics. A thorough description of the transporters and their genetic polymorphisms is included in Chapter 8. Briefly, the genes encoding OCT1-3, SLC22A1-3, are highly homologous to one another and located on the long arm of chromosome 6 with SLC22A1 being 5′ of SLC22A2, which in turn is located on the 5′ end of SLC22A3 (Koehler et al., 1997; Grundemann et al., 1998; Grundemann and Schomig, 2000). The transporters are expressed in a number of tissues and have some overlapping tissue distribution (Fig. 4.4). OCT1 is expressed abundantly in the liver (Zhang et al., 1997), OCT2 is expressed abundantly in the kidney (Grundemann et al., 1997), and OCT3 is expressed in abundance in the prostate gland, the heart, and the skeletal muscle (Chen et al., 2010).

4.2.5.2 Genetic Polymorphisms Though many polymorphisms of the transporters have been identified and are discussed in great detail in Chapter 8, the most common nonsynonymous polymorphisms that have functional effects are described here.

OCT1 This polymorphism, Met420del, is a 3 bp deletion and carries with it three rs numbers: rs35167514, rs34305973, and rs35191146. The deletion results in an OCT1 protein devoid of a methionine at position 420 (Leabman et al., 2003). OCT1-Met420del appears to be an ancestral European allele with about a 20% allele frequency in individuals of European ancestries, and also present in African- and Mexican-American populations, which are admixed with Europeans (Shu et al., 2003). Functionally, the variant transports the model substrate, MPP^+, at a rate similar to the reference OCT1, but exhibits a reduced transport of the antidiabetic drug, metformin (Shu et al., 2007). This change in specificity is both interesting and challenging in pharmacogenomic studies because it suggests that variants such as OCT1-Met420del will be drug-specific in their clinical effects.

OCT2 The common variant, A270S (rs316019), is present at allele frequencies between 10% and 20% in ethnic and racial populations studied to date. Functionally, it seems to modulate the uptake of metformin. However, some data suggest that it increases (Chen et al., 2009a) and some that it decreases (Wang et al., 2008) metformin uptake. Clinical results are also controversial. That is, the variant has been associated with both an increased and a decreased renal clearance of metformin. However, these effects may be due to different ethnic populations under study and therefore different polymorphisms in linkage with the A270S (or 808C>T allele). Alternatively, the effect size is so small that different results may simply occur by chance.

OCT3 Coding variants in OCT3 have yet to be associated with clinical drug response and all nonsynonymous variants identified to date are present at low allele frequencies suggesting that the gene is under heavy selective pressure; however, functional genomic studies have suggested that several rare variants result in changes in the uptake of metformin (Chen et al., 2010).

4.2.6 *SLC22A4* and *SLC22A5* (OCTN1 and OCTN2)

4.2.6.1 Description and Tissue Distribution As with the OCTs, a thorough description of the novel organic cation transporters, OCTNs, can be found in Chapter 8. Briefly, the two transporters are located on chromosome 5, one on the Watson strand and the other on the Crick strand of the chromosome, suggesting that they share promoter elements in common. OCTN1 has a high expression level in red blood cells and is also expressed abundantly in the proximal tubule, skeletal muscle, and trachea and other tissues (Nishimura and Naito, 2005) (Fig. 4.4). OCTN2, also known as the carnitine transporter, is expressed abundantly in the kidney proximal tubule, the small intestine, the skeletal muscle, the prostate and the testis and at lower levels in the liver (Fig. 4.4). Though both transporters interact with various drugs and xenobiotics, each have naturally occurring ligands. OCTN2 transports carnitine, which is its major physiological role (Tamai et al., 1998). OCTN2 genetic mutations are associated with systemic carnitine deficiency (Filippo et al., 2011). OCTN1 transports ergothioneine and carnitine to a limited extent (Grundemann et al., 2005). Xenobiotic substrates of OCTN1 include quinidine, verapamil, and tetraethylammonium. For OCTN2, substrates also include quinidine and verapamil (Ohashi et al., 1999).

4.2.6.2 Genetic Polymorphisms Genetic polymorphisms and mutations in OCTN1 and OCTN2 have been associated with several diseases including systemic carnitine deficiency (OCTN2) (Ohashi et al., 2001) and Crohn's disease (OCTN1) (Peltekova et al., 2004) (see Chapter 8). OCTN1 has a common nonsynonymous polymorphism with a large functional effect (OCTN1-L503F) (see Chapter 8). This polymorphism has been associated with a greater uptake of TEA and a reduced uptake of betaine and carnitine (Urban et al., 2007). The allele appears to be a European allele and is present in European admixed populations (Peltekova et al., 2004; Urban et al., 2005). For OCTN2, nonsynonymous polymorphisms that have been characterized generally retain function if found at allele frequencies greater than 1%. The exception is OCTN2-P17L, found at 2% allele frequency in Asians. In comparison to the reference OCTN2, this variant has about 50% transport function of carnitine in cellular assays (Urban et al., 2006a).

4.2.7 *SLC47A1* (MATE1)

4.2.7.1 Description and Tissue Distribution Moriyama's group first cloned the cDNA of human multidrug and toxic compound extrusion 1 (hMATE1) from the human brain in 2005 (Otsuka et al., 2005). MATE1 is an electroneutral H^+/organic cation exchanger, encoded by the gene *SLC47A1* on chromosome 17p11.2. The transporter is responsible for the excretion of endogenous waste products and xenobiotics at the apical membrane of the renal tubule and canalicular membrane of hepatocytes (Otsuka et al., 2005; Moriyama et al., 2008). Northern blot analysis demonstrated that human gene *SLC47A1* is primarily expressed in the kidney, the liver, the skeletal muscle, the adrenal gland and is also ubiquitously expressed throughout the body (Otsuka et al., 2005; Masuda et al., 2006; Moriyama et al., 2008) (Fig. 4.4). Laboratory studies have demonstrated that upon expression of hMATE1 in HEK293 cells, the transporter is localized at the plasma membrane and mediates the H^+-coupled electroneutral exchange of (i) organic cation compounds, (e.g., tetraethylammonium (TEA) and 1-methyl-4-phenyl-pyridinium (MPP^+), metformin, paraquat, platinum compounds (Choi et al., 2009; Chen et al., 2007, 2009b) (ii) anionic compounds (e.g., estrone sulfate, acyclovir) and; (iii) zwitterionic compounds (e.g., cephalexin and cephradine) (Tanihara et al., 2007). These substrates are also transported by transporters in the *SLC22A* family, such as the OCTs, OCTNs and OATs (Koepsell et al., 2007; International Transporter Consortium et al., 2010; VanWert et al., 2010). Similarly, there is some overlap in the inhibitors of hMATE1 with inhibitors of OCTs. Several groups determined the transcellular translocation of organic cations via the basolateral transporters and apical transporters using double-transfected MDCK cells expressing human OCT1/MATE1 and OCT2/MATE1 as models of human hepatocytes and renal epithelial cells, respectively (Sato et al., 2008; Konig et al., 2010; Meyer zu Schwabedissen et al., 2010).

Recent *in vivo* studies demonstrated differences in drug distribution and potentiation of drug-induced toxicity between *Mate1(-/-)* and *Mate1(+/+)* mice. For example, higher plasma levels and reduced renal clearance of metformin (Tsuda et al., 2009) and cephalexin (Watanabe et al., 2010) were observed in *Mate1(-/-)*, in comparison to *Mate1(+/+)*. For cisplatin, higher plasma levels, greater renal accumulation, and greater nephrotoxicity were observed in *Mate1(-/-)* mice (Nakamura et al., 2010). These studies provide *in vivo* evidence for the clinical importance of hMATE1 in determining drug levels and associated toxicities.

4.2.7.2 Genetic Polymorphisms Several groups including PMT have characterized genetic variants in the promoter and coding regions of MATE1 in different ethnic populations (Chen et al., 2009b; Choi et al., 2009; Kajiwara et al., 2009; Meyer zu Schwabedissen et al., 2010). We and others identified two common nonsynonymous variants in *SLC47A1*: Leu125Phe (rs77474263) in Mexican-Americans at MAF of 5.1% and Val338Ile (rs35790011) in African-Americans at MAF of 5.1% (Chen et al., 2009b; Meyer zu Schwabedissen et al., 2010). In addition, a less common variant, Cys497Phe (rs35395280) was also found in African-Americans at a MAF of 2.4%. Our functional studies showed that both of the common variants, Leu125Phe (rs77474263) and Val338Ile (rs35790011), exhibited significant reduction (by 40–50%) in transport activities of the substrates TEA and metformin, but not of oxaliplatin. The rare variant Gly64Asp, which was also reported in Japanese populations (singleton) (Kajiwara et al., 2009), and the variant Val480Met exhibited >50% reduction of TEA, metformin, paraquat, and oxaliplatin uptake (Chen et al., 2009b). The greater reduction in transport activity caused by these rare variants may have been due to low surface protein expression as observed by cellular localization of green fluorescent protein (GFP)-tagged MATE1 (Chen et al., 2009b) and by biotinylation assays (Kajiwara et al., 2009).

Five polymorphic promoter variants have been identified in the basal promoter region of MATE1 (Choi et al., 2009) with three of the variants having MAF >5% in at least one ethnic group. Of the variants, the most interesting was the most common variant g.-66T>C which exhibited binding to AP-1 and AP-2rep transcription factors in electrophoretic mobility shift assays (EMSA) (Choi et al., 2009). The luciferase assay results in different cell lines and the EMSA assays suggested that the reduced transcriptional activity of g.-66T>C resulted from decreased binding potency of AP-1 and enhanced binding potency of the repressor, AP-2rep. This variant, g.-66T>C, was also associated with reduced MATE1 mRNA expression levels in human kidney samples (Choi et al., 2009).

4.2.8 *SLCO1A2* (OATP1A2)

4.2.8.1 Description and Tissue Distribution Human organic-anion transporting polypeptide transporter (OATP) 1A2, encoded by the gene *SLCO1A2* (solute carrier organic anion transporter family, member 1A2) on chromosome 12p12, is expressed in tissues such as the intestine (apical domain of intestinal enterocytes), kidney (apical domain of distal nephron), liver (cholangiocyte), placenta (villous syncytiotrophoblasts and cytotrophoblasts), and brain (endothelial capillary cells) (Lee et al., 2005; Glaeser et al., 2007; Bolger et al., 2009; Urquhart and Kim, 2009; Loubiere et al., 2010) (Fig. 4.4). OATP1A2 and other OATPs such as OATP1B1, OATP1B3, and OATP2B1, are important for facilitating sodium-independent cellular uptake of various drugs and endogenous compounds, including statins, protease inhibitors, methotrexate, bile acids, estrogen, and thyroid hormones and their conjugates (Badagnani et al., 2006; Kalliokoski and Niemi, 2009; Hartkoorn et al., 2010; Ianculescu et al., 2010; Rodrigues, 2010). A recent study using Oatp1a/1b-knockout mice demonstrated differences in the disposition of bilirubin, bile acids, methotrexate, and fexofenadine from wild-type mice. In particular, the knockout mice have decreased hepatic uptake of methotrexate and fexofenadine and consequently greater systemic exposure (van de Steeg et al., 2010).

4.2.8.2 Genetic Polymorphisms The functional effects of OATP1A2 coding region variants have been described in several studies including one from PMT (Lee et al., 2005; Badagnani et al., 2006). We identified 11 nonsynonymous variants in 4 major ethnic groups, namely the African- ($n = 80$ subjects), European- ($n = 80$ subjects), Asian- ($n = 60$ subjects), and Mexican-Americans ($n = 50$ subjects) (Badagnani et al., 2006). There are three common protein-altering variants in SLCO1A2, Ile13Thr (rs10841795), Glu172Asp (rs11568563), and Thr668Ser (rs11568557) and none of the variants are unique to a single ethnic group or have allele frequencies >2% in one of the ethnic groups. Functional studies showed that there were significant differences in the transport of estrone sulfate and/or methotrexate in cells expressing the two variants compared to OATP1A2 reference. The variant Ile13Thr showed approximately twofold increase in the transport of estrone sulfate and methotrexate whereas Glu172Asp exhibited approximately 40% reduction in the uptake of both substrates (Badagnani et al., 2006) (Table 4.1). Recently, PMT investigated the effect of genetic variants in the enhancer regions of membrane transporters highly expressed in the liver (Kim et al., 2011). Among the identified enhancer regions, the haplotype *3 (rs4148981) in the *SLCO1A2* intronic region (Chr12:21378175-21378474, hg18), was associated with lower luciferase activity in *in vivo* mouse tail-vein luciferase reporter assays ($p < 0.0001$) and with lower OATP1A2 mRNA levels ($p = 0.01$) compared with reference. Haplotype analysis showed a significant association between haplotype *1*3 with increased methotrexate clearance ($p = 0.03$) in 639 leukemic children treated with multiple courses of methotrexate 2–5 g/m^2 (Kim et al., 2011). This finding may have been due to the complete linkage disequilibrium of this variant rs11045981 in haplotype *1*3 with the OATP1A2 nonsynonymous variant I13T (rs10841795). This nonsynonymous variant resulted in higher methotrexate uptake compared to reference in oocytes (Badagnani et al., 2006).

4.2.9 *SLC28A* and *SLC29A* (CNT and ENT)

4.2.9.1 Description and Tissue Distribution Two major nucleoside transporter proteins, concentrative nucleoside transporters (CNT, SLC28) and equilibrative nucleoside transporters (ENT, SLC29), mediate the transport of naturally occurring nucleosides and a broad range of synthesized derivatives of nucleosides. Approximately 1 in 5 approved anticancer drugs are chemically related to nucleosides and are used in both solid tumors and in leukemia. Three isoforms of the CNT proteins namely, CNT1-3 (*SLC28A1-3*) and four ENT proteins, namely ENT1-4 (*SLC29A1-4*) have been cloned and characterized and are reviewed in recent articles (Borrelly et al., 1992; Pastor-Anglada et al., 2007; Zhou et al., 2007; Molina-Arcas et al., 2008; Molina-Arcas et al., 2009). CNTs are sodium-dependent transporters that mediate the unidirectional transport of their substrates, adenosine (CNT2, CNT3), guanosine (CNT2 and 3), uridine (CNT1-3), cytidine (CNT1 and 3) and thymidine (CNT1 and 3), with K_m values ranging from 5 to 100 μM, whereas, ENTs are bidirectional carriers that catalyze passive diffusion processes mediating the influx and efflux of substrates. In comparison to CNTs, ENTs have lower affinities for nucleosides (K_m 40–1000 μM). CNTs and ENTs are localized in apical and basolateral membranes of epithelial cells. This asymmetric distribution results in vectorial transport of nucleosides from extracellular to intracellular spaces in the intestinal tract and renal tubule. In hepatocytes, CNT2 is highly concentrated on the sinusoidal membrane, whereas CNT1 is a canalicular membrane protein (Pastor-Anglada et al., 2001). The importance of nucleosides and nucleobases in mammalian cell function have led to studies on the endogenous roles of CNTs and ENTs and their substrates in endothelial cells and skeletal muscle (Archer et al., 2004; Bone et al., 2010; Rose et al., 2010) using knockout mice or recombinantly expressed proteins. Chapter 11 reviews CNT and ENT families in more detail, covering information related to the transporters' roles in human diseases. The expression levels of the mRNA transcripts of CNTs and ENTs are shown in Figure 4.3. Briefly, ENTs are expressed ubiquitously, whereas CNTs have more tissue-specific localization patterns (Fig. 4.3).

4.2.9.2 Genetic Polymorphisms As part of the PMT project, DNA samples from the initial cohort available from the Coriell Institute of Medicine (PMT Set I cohort) and PMT-272 cohort were used to identify genetic variants in the coding regions (including 5'UTRs, 3'UTRs, and exons) or the basal promoter regions of CNT1-3 and ENT1-3. The nonsynonymous variants in these transporters were functionally characterized using *Xenopus laevis* oocytes (Gray et al., 2004; Badagnani et al., 2005; Owen et al., 2005; Owen et al., 2006), transfected mammalian cell lines (Gray et al., 2004) or transformed *Saccharomyces cerevisiae* (Osato et al., 2003). In addition, the promoter variants in CNT2 were also functionally characterized using *in vitro* and *in vivo* luciferase assays and EMSA (Yee et al., 2009). The SNPs that showed functional changes are summarized in Table 4.1 and briefly described below (Osato et al., 2003; Gray et al., 2004; Badagnani et al., 2005; Owen et al., 2005).

4.2.9.3 CNT1 Coding Region Variants CNT1 exhibits a high degree of nucleotide diversity. Of the 15 nonsynonymous SNPs identified in CNT1 in DNA samples from over 200 individuals from ethnically diverse U.S. populations, four common protein-altering variants (Leu140INS, Val189Ile, Gln237Lys, and Asp521Asn) were present at allele frequencies >20% in three of four major ethnic groups (African-, European-, and Asian-Americans). Two variants, Ser546Pro (singleton) and Val385DEL resulted in significant reduction in thymidine uptake and a common variant Val189Ile showed reduced affinity to gemcitabine compared to the reference CNT1 (Table 4.1). Overall, *SLC28A1* is a highly variable gene and the average nucleotide diversity (Π_T) of *SLC28A1* over the total population (11.8×10^{-4}) was higher than average values reported in the studies of Cargill et al. (5.05×10^{-4}) (Cargill et al., 1999) and Sachidanandam et al. (7.51×10^{-4}) (Sachidanandam et al., 2001). Recently, the common variant Val189Ile was found to be associated with a reduced bioavailability of mizoribine in renal transplant patients (Naito et al., 2010), perhaps reflecting a reduced affinity of mizoribine to Val189Ile.

4.2.9.4 CNT2 Coding Region Variants PMT identified six nonsynonymous variants and four synonymous variants in *SLC28A2* (Owen et al., 2005). Two common nonsynonymous variants (Pro22Leu and Ser75Arg) had allele frequencies >10% in all major ethnic groups (African-, European-, Asian-, and Mexican-Americans). Two variants, Phe355Ser (3.5%) and Leu462Phe (1%) were found only in the African-American sample. Among all the nonsynonymous variants, only CNT2-Phe355Ser showed altered specificity for naturally occurring nucleosides, which could have implications for nucleoside homeostasis (Owen et al., 2005, Table 4.1).

4.2.9.5 CNT2 Basal Promoter Region Variants Of the five variants identified in the proximal promoter region of CNT2, only one (-146T>A, rs2413775) was found at high allele frequencies >25% in all ethnic groups studied (Yee et al., 2009). Using luciferase assays and EMSA, the -146T>A showed an increase in promoter activity and a greater binding to the transcription factors HNF1α and β, compared to the reference allele -146T (Yee et al., 2009).

4.2.9.6 CNT3 Coding Region Variants PMT identified ten nonsynonymous and six synonymous SNPs in *SLC28A3* in about 240 DNA samples from ethnically diverse U.S. populations, of which 12 were novel at the time of discovery. Functional studies using oocytes showed that only CNT3-Gly367 (found only once in 60 Asian subjects) showed significant reduction in the uptake of purine and pyrimidine nucleosides. This complete loss of uptake activity was consistent with an alteration of an evolutionarily conserved site in the substrate recognition site of CNT3 (Badagnani et al., 2005). Recent pharmacogenomic studies of gemcitabine and cytarabine using lymphoblastoid cell line-based model systems have identified variants in CNT3 that showed significant association with the IC_{50} of gemcitabine and cytarabine in cytotoxicity assays (Li et al., 2009). These results may suggest that variants in other regions of CNT3 play larger roles in terms of function compared to nonsynonymous SNPs of CNT3. Alternatively, the oocyte expression system may not have recapitulated the transporter activity observed *in vivo*.

4.2.9.7 ENT1 Coding Region Variants Unlike coding region variants in the CNTs, the number of polymorphic variants in ENT1 was much lower. From the sequencing of 247 subjects, two novel nonsynonymous SNPs were identified. One variant (Ile216Thr) was found at a frequency of 0.5% in DNA samples from African-Americans, and at greater frequencies in samples from European-Americans (2.1%) and Mexican-Americans (5.6% in 10 subjects). Another variant, ENT1-Glu391Lys was found only in African-Americans with a MAF of 1%. Our functional analysis in *S. cerevisiae* revealed no differences in the kinetics of interactions of naturally occurring and synthetic nucleoside analogs with the two nonsynonymous variants (Osato et al., 2003). A recent study by Morello et al. (Morello et al., 2010) demonstrated that an intronic SNP in ENT1 (rs760370 A>G) influenced the rapid virological response to pegylated-interferon–ribavirin therapy in HIV patients with hepatitis C infection. This variant may modulate the intracellular levels of ribavirin in the liver.

4.2.9.8 ENT2 Coding Region Variants Among the five protein-altering variants identified in DNA samples from 247 individuals, only ENT2-del-845-846, which resulted in a frameshift mutation that prematurely truncated the protein, was associated with a significantly lower uptake of the tested substrates, inosine, uridine, hypoxanthine, fludarabine, and gemcitabine (Owen et al., 2006). Sequencing data from dbSNP132 and the 1000 Genomes project together with data from PMT suggest that the ENT2 coding region is highly conserved with few variants, all present at low allele frequencies.

4.3 CLINICAL STUDIES

A number of clinical studies have ensued from PMT discovery and functional genomic studies. That is, variants discovered and functionally characterized are tested clinically for effects on drug disposition and response (Fig. 4.1). For these studies, we use a cohort of healthy volunteers developed by PMT, termed Studies of Pharmacogenomics In Ethnic Populations (SOPHIE). SOPHIE volunteers have an age range of 18–40 years, and are from four ethnic groups (Fig. 4.4). Below, we briefly describe these important genotype to phenotype proof-of-concept studies and two clinical studies involving patients on drugs that are substrates of membrane transporters.

4.3.1 Metformin

Metformin studies are largely summarized in Chapter 8 and will only be briefly described. As an excellent substrate for organic cation transporters, OCT1-3 and multi-drug and toxin extrusion proteins (MATEs), metformin provides an ideal model drug to study the effects of genetic variants in these transporters on its disposition and clinical response. Moreover, the drug is among the top-prescribed drugs in the world with variable response; therefore, genetic variants that influence the pharmacokinetics and pharmacodynamics of metformin may be important in the optimum therapeutic use of the drug. PMT investigators have published three clinical studies on the effects of genetic variants in transporters on metformin disposition and response in SOPHIE volunteers (Shu et al., 2007, 2008; Chen et al., 2009a). Two of the studies focused on genetic variants in OCT1. In one study, the influence of OCT1 variants on the disposition of metformin was examined (Shu et al., 2008) and in the other study, the effect of the variants on the pharmacodynamics was probed in healthy volunteers (Shu et al., 2007). As described in Chapter 8, the studies showed that OCT1 variants played an important role in both the pharmacokinetics and pharmacodynamics of the drug. In particular, multiple genetic variants of OCT1 with reduced function including R61C, G401S, M420del, and G465R were associated with reduced response to metformin and altered pharmacokinetics in healthy volunteers. The third study demonstrated that the common variant of OCT2, A270S, was associated with a greater renal clearance of metformin in European-American populations (Chen et al., 2009a). Collectively, these exciting studies paved the way for further studies focused on genetic variants in transporters in patients with diabetes.

4.3.2 Gabapentin

PMT functional genomic studies demonstrated that OCTN1-L503F exhibited reduced uptake of the neuroleptic drug, gabapentin, in cellular assays (Urban et al., 2008). Because this variant is extraordinarily common in populations of European ancestries (e.g., allele frequencies of about 50%), clinical studies were performed (Urban et al., 2008). In brief, the pharmacokinetics of gabapentin were measured in SOPHIE volunteers who were homozygous for the L503 or the F503 allele of OCTN1. Gabapentin renal clearance was reduced in patients with the F503 allele, consistent with the reduced uptake in cells. In fact, the secretory clearance of gabapentin was abolished in individuals with the F503 allele compared to individuals homozygous for the reference allele (L503). Collectively the studies show an important role for OCTN1 in the renal elimination of gabapentin and that the variant modulates renal secretion. This study was among the first to focus on the effect of genetic variants in transporters on renal drug elimination.

Mycophenolic Acid

Mycophenolate mofetil is a prodrug of mycophenolic acid (MPA) and is a widely prescribed immunosuppressant for kidney transplantation. MPA undergoes enterohepatic recirculation to form the inactive metabolite MPA-7-*O*-glucuronide (MPAG). Studies have shown that the inhibition of MRP2 with cyclosporine results in the inhibition of MPA enterohepatic cycling and thus affects MPA levels. Picard et al. (Picard et al., 2010) determined the transport uptake of MPA and MPAG by OATPs. Their studies showed that the metabolite MPAG is a substrate of OATP1B3 but not of OATP1A2 and OATP1B1 in heterologous cell lines expressing the transporters. The common nonsynonymous variant in OATP1B3, c.334T>G (Ser112Ala, rs4149117) is found at allele frequencies of 46.2% (c.334G), 17.9% (c.334T), 30.1% (c.334T), and 16.7% (c.334T) in samples from African-, European-, Asian-, and Mexican-Americans in the SOPHIE cohort. c.334T>G is in complete linkage disequilibrium with a nonsynonymous variant c.699G>A (Met233Ile, rs7311358) not identified in the SOPHIE cohort. MPAG showed greater OATP1B3-mediated uptake (velocity) in cells stably transfected with the OATP1B3 haplotype T334-G699 compared to cells transfected with the haplotype G334-A699. These results were further supported by a clinical pharmacokinetic study which exhibited significant increases in the MPAG/MPA AUC ratio due to reduced enterohepatic recirculation in patients carrying the OATP1B3 334G-699A haplotype (Picard et al., 2010).

4.3.3 Irinotecan

Irinotecan is an important chemotherapeutic agent in the treatment of colorectal cancer and other solid tumors. The active metabolite of irinotecan, SN-38, is glucuronidated by uridine diphosphate glucuronosyltransferases (UGTs), and SNPs in *UGT1A1* have been associated with the rate of SN-38 glucuronidation and irinotecan toxicity (Iyer et al., 1999, 2002). More recently, irinotecan and SN-38 pharmacokinetics and toxicity have also been associated with ABC transporter polymorphisms. Both irinotecan and SN-38 are transported from hepatocytes into the bile by MRP2 (Ma and McLeod, 2003). A haplotype consisting of six SNPs in *ABCC2*, designated *ABCC2*2*, was significantly associated with lower irinotecan clearance and an increased risk of irinotecan-induced diarrhea in a cohort of Caucasian cancer patients, suggesting impaired MRP2 transport function and elevated irinotecan plasma levels leading to toxicity (de Jong et al., 2007). In a second analysis in 85 ethnically diverse advanced cancer patients a pathway-based approach was used to investigate the association of common genetic polymorphisms in 12 candidate metabolism and transport genes with irinotecan and SN-38 pharmacokinetics and their associated neutropenia (Innocenti et al., 2009). Irinotecan exposure was associated with the *ABCC2* -24C>T and *SLCO1B1**5 alleles while a synonymous variant in *ABCC1* and an intronic variant in *ABCB1* contributed to variation in SN-38 exposure. Polymorphisms in *UGT1A1*, *SLCO1B1*, and *ABCC1* were each associated with variation in ANC nadir (Innocenti et al., 2009) following irinotecan treatment. This study nicely illustrates the importance of considering drug response pathways in pharmacogenetic studies.

4.4 FUTURE DIRECTIONS

Going forward, PMT will employ RNAseq to quantitatively measure transcript levels of transporters and their common splice variants in liver, kidney, and other tissues. NextGen sequencing methods will be used to identify novel regulatory variants associated with expression levels or splicing. Current studies have largely focused on eSNPs associated with expression levels in lymphoblastoid cell lines. These studies have identified common eSNPs. In contrast, little is known about eSNPs that associate with transporter expression levels in tissues of pharmacologic importance such as liver and kidney. Further, less common eSNPs have not been identified. Using a combination of RNAseq and NextGen DNA sequencing, common and rare SNPs that associate with expression levels of transporters in liver and kidney samples from ethnically diverse U.S. populations will be identified. Another goal of our RNAseq studies will be to identify transporter containing co-expression networks. Such studies will provide important clues as to genes that are co-regulated with transporters in liver and kidney and which together with transporters may contribute to variation in drug disposition and response. In addition to regulatory variants, PMT will continue our computational and experimental studies characterizing coding region variants in transporters. Three types of clinical studies will be performed in the next few years: (a) hypothesis-driven clinical studies evaluating the role of functionally important genetic variants in drug disposition and response in our SOPHIE population; (b) a large genome-wide association and NextGen sequencing study of response to metformin; and (c) a number of large collaborative studies largely focused on genetic factors that contribute to efficacy and toxicity of anticancer drugs. Overall these studies will continue to contribute to understanding the role of genetic variants in membrane transporters in therapeutic and adverse drug response.

ACKNOWLEDGMENTS

PMT (Pharmacogenomics of Membrane Transporters) is supported by NIH/NIGMS Pharmacogenetics Research Network (PGRN; U01GM61390).

REFERENCES

1000 Genomes Project Consortium, Durbin RM, Abecasis GR, Altshuler DL, Auton A, Brooks LD, Durbin RM, Gibbs RA, Hurles ME, McVean GA (2010) A map of human genome variation from population-scale sequencing. *Nature*, 467, 1061–1073.

Abla N, Chinn LW, Nakamura T, Liu L, Huang CC, Johns SJ, Kawamoto M, Stryke D, Taylor TR, Ferrin TE, Giacomini KM, Kroetz DL (2008) The human multidrug resistance protein 4 (MRP4, ABCC4): functional analysis of a highly polymorphic gene. *J Pharmacol Exp Ther*, 325, 859–868.

Ahlin G, Chen L, Lazorova L, Chen Y, Ianculescu AG, Davis RL, Giacomini KM, Artursson P (2011) Genotype-dependent effects of inhibitors of the organic cation transporter, OCT1: predictions of metformin interactions. *Pharmacogenomics J*, 11, 400–411.

Archer RG, Pitelka V, Hammond JR (2004) Nucleoside transporter subtype expression and function in rat skeletal muscle microvascular endothelial cells. *Br J Pharmacol*, 143, 202–214.

Badagnani I, Castro RA, Taylor TR, Brett CM, Huang CC, Stryke D, Kawamoto M, Johns SJ, Ferrin TE, Carlson EJ, Burchard EG, Giacomini KM (2006) Interaction of methotrexate with organic-anion transporting polypeptide 1A2 and its genetic variants. *J Pharmacol Exp Ther*, 318, 521–529.

Badagnani I, Chan W, Castro RA, Brett CM, Huang CC, Stryke D, Kawamoto M, Johns SJ, Ferrin TE, Carlson EJ, Burchard EG, Giacomini KM (2005) Functional analysis of genetic variants in the human concentrative nucleoside transporter 3 (CNT3; SLC28A3). *Pharmacogenomics J*, 5, 157–165.

Berge KE, Tian H, Graf GA, Yu L, Grishin NV, Schultz J, Kwiterovich P, Shan B, Barnes R, Hobbs HH (2000) Accumulation of dietary cholesterol in sitosterolemia caused by mutations in adjacent ABC transporters. *Science*, 290, 1771–1775.

Bolger MB, Lukacova V, Woltosz WS (2009) Simulations of the nonlinear dose dependence for substrates of influx and efflux transporters in the human intestine. *AAPS Journal*, 11, 353–363.

Bone DB, Choi DS, Coe IR, Hammond JR (2010) Nucleoside/nucleobase transport and metabolism by microvascular endothelial cells isolated from ENT1-/- mice. *Am J Physiol Heart Circ Physiol*, 299, H847–H856.

Borrelly J, Martin F, Grosdidier G, Routiot T (1992) Value of limited resections in the surgical treatment of lung cancers. *Annales de Chirurgie*, 46, 766–769.

Borst P, de Wolf C, van de Wetering K (2007) Multidrug resistance-associated proteins 3, 4, and 5. *Pflugers Arch*, 453(5), 661–673.

Borst P, Elferink RO (2002) Mammalian ABC transporters in health and disease. *Ann Rev Biochem*, 71, 537–592.

Brooks-Wilson A, Marcil M, Clee SM, Zhang LH, Roomp K, van Dam M, Yu L, Brewer C, Collins JA, Molhuizen HO, Loubser O, et al. (1999) Mutations in ABC1 in Tangier disease and familial high-density lipoprotein deficiency. *Nature Genet*, 22, 336–345.

Cargill M, Altshuler D, Ireland J, Sklar P, Ardlie K, Patil N, Shaw N, Lane CR, Lim EP, Kalyanaraman N, et al. (1999) Characterization of single-nucleotide polymorphisms in coding regions of human genes. *Nature Genet*, 22, 231–238.

Chen L, Pawlikowski B, Schlessinger A, More SS, Stryke D, Johns SJ, Portman MA, Chen E, Ferrin TE, Sali A, Giacomini KM (2010) Role of organic cation transporter 3 (SLC22A3) and its missense variants in the pharmacologic action of metformin. *Pharmacogenet Genomics*, 20, 687–699.

Chen Y, Li S, Brown C, Cheatham S, Castro RA, Leabman MK, Urban TJ, Chen L, Yee SW, Choi JH, Huang Y, Brett CM, Burchard EG, Giacomini KM (2009a) Effect of genetic variation in the organic cation transporter 2 on the renal elimination of metformin. *Pharmacogenet Genomics*, 19, 497–504.

Chen Y, Teranishi K, Li S, Yee SW, Hesselson S, Stryke D, Johns SJ, Ferrin TE, Kwok P, Giacomini KM (2009b) Genetic variants in multidrug and toxic compound extrusion-1, hMATE1, alter transport function. *Pharmacogenomics J*, 9, 127–136.

Chen Y, Zhang S, Sorani M, Giacomini KM (2007) Transport of paraquat by human organic cation transporters and multidrug and toxic compound extrusion family". *J Pharmacol Exp Ther*, 322, 695–700.

Choi JH, Yee SW, Kim MJ, Nguyen L, Lee JH, Kang JO, Hesselson S, Castro RA, Stryke D, Johns SJ, et al. (2009) Identification and characterization of novel polymorphisms in the basal promoter of the human transporter, MATE1. *Pharmacogenet Genomics*, 19, 770–780.

de Jong FA, Scott-Horton TJ, Kroetz DL, McLeod HL, Friberg LE, Mathijssen RH, Verweij J, Marsh S, Sparreboom A (2007) Irinotecan-induced diarrhea: functional significance of the polymorphic ABCC2 transporter protein. *Clin Pharmacol Ther*, 81, 42–49.

Dean M, Rzhetsky A, Allikmets R (2001) The human ATP-binding cassette (ABC) transporter superfamily. *Genome Res*, 11(7), 1156–1166.

Erdman AR, Mangravite LM, Urban TJ, Lagpacan LL, Castro RA, de la Cruz M, Chan W, Huang CC, Johns SJ, Kawamoto M, et al. (2006). The human organic anion transporter 3 (OAT3; SLC22A8): genetic variation and functional genomics. *Am J Physiol Renal Physiol*, 290(4), F905–F912.

Filippo CA, Ardon O, Longo N (2011) Glycosylation of the OCTN2 carnitine transporter: study of natural mutations identified in patients with primary carnitine deficiency. *Biochimica et biophysica acta*, 1812(3), 312–320.

Fujita T, Brown C, Carlson EJ, Taylor T, de la Cruz M, Johns SJ, Stryke D, Kawamoto M, Fujita K, Castro R, et al. (2005) Functional analysis of polymorphisms in the organic anion transporter, SLC22A6 (OAT1). *Pharmacogenetics and Genomics*, 15(4), 201–209.

Fujita T, Urban TJ, Leabman MK, Fujita K, Giacomini KM (2006) Transport of drugs in the kidney by the human organic cation transporter, OCT2 and its genetic variants. *J Pharm Sci*, 95, 25–36.

Germann UA (1996) P-glycoprotein–a mediator of multidrug resistance in tumour cells. *Eur J Cancer*, 32A, 927–944.

Glaeser H, Bailey DG, Dresser GK, Gregor JC, Schwarz UI, McGrath JS, Jolicoeur E, Lee W, Leake BF, Tirona RG, Kim RB (2007) Intestinal drug transporter expression and the impact of grapefruit juice in humans. *Clin Pharmacol Ther*, 81, 362–370.

Gottesman MM, Pastan I, Ambudkar SV (1996) P-glycoprotein and multidrug resistance. *Curr Opin Genet Develop*, 6, 610–617.

Gow JM, Chinn LW, Kroetz DL (2008) The effects of ABCB1 3'-untranslated region variants on mRNA stability. *Drug Metabol Dispos*, 36, 10–15.

Gow JM, Hodges LM, Chinn LW, Kroetz DL (2008a) Substrate-dependent effects of human ABCB1 coding polymorphisms. *J Pharmacol Exp Ther*, 325, 435–442.

Gow JM, Hodges LM, Chinn LW, Kroetz DL (2008b) Substrate-dependent effects of human ABCB1 coding polymorphisms. *J Pharmacol Exp Ther*, 325(2), 435–442.

Gray JH, Mangravite LM, Owen RP, Urban TJ, Chan W, Carlson EJ, Huang CC, Kawamoto M, Johns SJ, Stryke D, Ferrin TE, Giacomini KM (2004) Functional and genetic diversity in the concentrative nucleoside transporter, CNT1, in human populations. *Mol Pharmacol*, 65, 512–519.

Grundemann D, Babin-Ebell J, Martel F, Ording N, Schmidt A, Schomig E (1997) Primary structure and functional expression of the apical organic cation transporter from kidney epithelial LLC-PK1 cells. *J Biol Chem*, 272, 10408–10413.

Grundemann D, Harlfinger S, Golz S, Geerts A, Lazar A, Berkels R, Jung N, Rubbert A, Schomig E (2005) Discovery of the ergothioneine transporter. *Proc Natl Acad Sci USA*, 102(14), 5256–5261.

Grundemann D, Schechinger B, Rappold GA, Schomig E (1998) Molecular identification of the corticosterone-sensitive extraneuronal catecholamine transporter. *Nature Neurosci*, 1, 349–351.

Grundemann D, Schomig E (2000) Gene structures of the human non-neuronal monoamine transporters EMT and OCT2. *Human Genet*, 106, 627–635.

Hahn MK, Blakely RD (2002) Monoamine transporter gene structure and polymorphisms in relation to psychiatric and other complex disorders. *Pharmacogenomics J*, 2, 217–235.

Haimeur A, Conseil G, Deeley RG, Cole SP (2004) The MRP-related and BCRP/ABCG2 multidrug resistance proteins: biology, substrate specificity and regulation. *Curr Drug Metab*, 5, 21–53.

Hartkoorn RC, Kwan WS, Shallcross V, Chaikan A, Liptrott N, Egan D, Sora ES, James CE, Gibbons S, Bray PG, Back DJ, Khoo SH, Owen A (2010) HIV protease inhibitors are substrates for OATP1A2, OATP1B1 and OATP1B3 and lopinavir plasma concentrations are influenced by SLCO1B1 polymorphisms. *Pharmacogenet Genomics*, 20, 112–120.

Hediger MA, Romero MF, Peng JB, Rolfs A, Takanaga H, Bruford EA (2004) The ABCs of solute carriers: physiological, pathological and therapeutic implications of human membrane transport proteinsIntroduction. *Pflugers Archiv: European J Physiol*, 447, 465–468.

Hesselson SE, Matsson P, Shima JE, Fukushima H, Yee SW, Kobayashi Y, Gow JM, Ha C, Ma B, Poon A, Johns SJ, Stryke D, Castro RA, Tahara H, Choi JH, Chen L, Picard N, Sjodin E, Roelofs MJ, Ferrin TE, Myers R, Kroetz DL, Kwok PY, Giacomini KM (2009) Genetic variation in the proximal promoter of ABC and SLC superfamilies: liver and kidney specific expression and promoter activity predict variation. *PloS ONE*, 4, e6942.

Hirouchi M, Suzuki H, Itoda M, Ozawa S, Sawada J, Ieiri I, Ohtsubo K, Sugiyama Y (2004) Characterization of the cellular localization, expression level, and function of SNP variants of MRP2/ABCC2. *Pharm Res*, 21, 742–748.

Ho RH, Leake BF, Kilkenny DM, Meyer Zu Schwabedissen HE, Glaeser H, Kroetz DL, Kim RB (2010) Polymorphic variants in the human bile salt export pump (BSEP; ABCB11): functional characterization and interindividual variability. *Pharmacogenet Genomics*, 20, 45–57.

Ianculescu AG, Friesema EC, Visser TJ, Giacomini KM, Scanlan TS (2010) Transport of thyroid hormones is selectively inhibited by 3-iodothyronamine. *Mol BioSystems*, 6, 1403–1410.

Innocenti F, Kroetz DL, Schuetz E, Dolan ME, Ramirez J, Relling M, Chen P, Das S, Rosner GL, Ratain MJ (2009) Comprehensive pharmacogenetic analysis of irinotecan neutropenia and pharmacokinetics. *J Clin Oncol*, 27, 2604–2614.

International Transporter Consortium, Giacomini KM, Huang SM, Tweedie DJ, Benet LZ, Brouwer KL, Chu X, Dahlin A, Evers R, Fischer V, Hillgren KM, et al. (2010) Membrane transporters in drug development. *Nature Rev Drug Discovery*, 9, 215–236.

Iyer L, Das S, Janisch L, Wen M, Ramirez J, Karrison T, Fleming GF, Vokes EE, Schilsky RL, Ratain MJ (2002) UGT1A1*28 polymorphism as a determinant of irinotecan disposition and toxicity. *Pharmacogenomics J*, 2, 43–47.

Iyer L, Hall D, Das S, Mortell MA, Ramirez J, Kim S, Di Rienzo A, Ratain MJ (1999) Phenotype-genotype correlation of in vitro SN-38 (active metabolite of irinotecan) and bilirubin glucuronidation in human liver tissue with UGT1A1 promoter polymorphism. *Clin Pharmacol Ther*, 65, 576–582.

Jeong H, Herskowitz I, Kroetz DL, Rine J (2007) Function-altering SNPs in the human multidrug transporter gene ABCB1 identified using a Saccharomyces-based assay. *PLoS Genet*, 3, e39.

Kajiwara M, Terada T, Ogasawara K, Iwano J, Katsura T, Fukatsu A, Doi T, Inui K (2009) Identification of multidrug and toxin extrusion (MATE1 and MATE2-K) variants with complete loss of transport activity. *J Hum Genet*, 54, 40–46.

Kalliokoski A, Niemi M (2009) Impact of OATP transporters on pharmacokinetics. *Brit J Pharmacol*, 158, 693–705.

Kelly L, Fukushima H, Karchin R, Gow JM, Chinn LW, Pieper U, Segal MR, Kroetz DL, Sali A (2010) Functional hot spots in human ATP-binding cassette transporter nucleotide binding domains. *Protein Sci*, 19, 2110–2121.

Kim MJ, Skewes-Cox P, Fukushima H, Hesselson S, Yee SW, Ramsey LB, Nguyen L, Eshragh JL, Castro RA, Wen CC, Stryke D, Johns SJ, Ferrin TE, Kwok PY, Relling MV, Giacomini KM, Kroetz DL, Ahituv N (2011) Functional characterization of liver enhancers that regulate drug-associated transporters. *Clin Pharmacol Ther*, 89, 571–578.

Kim RB, Leake BF, Choo EF, Dresser GK, Kubba SV, Schwarz UI, Taylor A, Xie HG, McKinsey J, Zhou S, Lan LB, Schuetz JD, Schuetz EG, Wilkinson GR (2001) Identification of functionally variant MDR1 alleles among European Americans and African Americans. *Clin Pharmacol Ther*, 70, 189–199.

Kimchi-Sarfaty C, Gribar JJ, Gottesman MM (2002) Functional characterization of coding polymorphisms in the human MDR1 gene using a vaccinia virus expression system. *Mol Pharmacol*, 62, 1–6.

Kimchi-Sarfaty C, Oh JM, Kim IW, Sauna ZE, Calcagno AM, Ambudkar SV, Gottesman MM (2007) A "silent" polymorphism in the MDR1 gene changes substrate specificity. *Science*, 315, 525–528.

Klucken J, Buchler C, Orso E, Kaminski WE, Porsch-Ozcurumez M, Liebisch G, Kapinsky M, Diederich W, Drobnik W, Dean M, Allikmets R, Schmitz G (2000) ABCG1 (ABC8), the human homolog of the Drosophila white gene, is a regulator of macrophage cholesterol and phospholipid transport. *Proc Natl Acad Sci USA*, 97, 817–822.

Koehler MR, Wissinger B, Gorboulev V, Koepsell H, Schmid M (1997) The two human organic cation transporter genes SLC22A1 and SLC22A2 are located on chromosome 6q26. *Cytogenet Cell Genet*, 79, 198–200.

Koepsell H, Lips K, Volk C (2007) Polyspecific organic cation transporters: structure, function, physiological roles, and biopharmaceutical implications. *Pharm Res*, 24, 1227–1251.

Konig J, Zolk O, Singer K, Hoffmann C, Fromm M (2010) Double-transfected MDCK cells expressing human OCT1/MATE1 or OCT2/MATE1: determinants of uptake and transcellular translocation of organic cations. *Br J Pharmacol*, 163, 546–555.

Kroetz DL, Pauli-Magnus C, Hodges LM, Huang CC, Kawamoto M, Johns SJ, Stryke D, Ferrin TE, DeYoung J, Taylor T, Carlson EJ, Herskowitz I, Giacomini KM, Clark AG (2003) Sequence diversity and haplotype structure in the human ABCB1 (MDR1, multidrug resistance transporter) gene. *Pharmacogenetics*, 13, 481–494.

Kruh GD, Belinsky MG (2003) The MRP family of drug efflux pumps. *Oncogene*, 22, 7537–7552.

Leabman MK, Huang CC, DeYoung J, Carlson EJ, Taylor TR, de la Cruz M, Johns SJ, Stryke D, Kawamoto M, Urban TJ, et al., Pharmacogenetics of Membrane Transporters Investigators (2003) Natural variation in human membrane transporter genes reveals evolutionary and functional constraints. *Proc Natl Acad Sci USA*, 100, 5896–5901.

Lee MH, Lu K, Hazard S, Yu H, Shulenin S, Hidaka H, Kojima H, Allikmets R, Sakuma N, Pegoraro R, Srivastava AK, Salen G, Dean M, Patel SB (2001) Identification of a gene, ABCG5, important in the regulation of dietary cholesterol absorption. *Nature Genet*, 27, 79–83.

Lee W, Glaeser H, Smith LH, Roberts RL, Moeckel GW, Gervasini G, Leake BF, Kim RB (2005) Polymorphisms in human organic anion-transporting polypeptide 1A2 (OATP1A2): implications for altered drug disposition and central nervous system drug entry. *J Biol Chem*, 280, 9610–9617.

Leslie EM, Deeley RG, Cole SP (2005) Multidrug resistance proteins: role of P-glycoprotein, MRP1, MRP2, and BCRP (ABCG2) in tissue defense. *Toxicol Appl Pharmacol*, 204, 216–237.

Li L, Fridley BL, Kalari K, Jenkins G, Batzler A, Weinshilboum RM, Wang L (2009) Gemcitabine and arabinosylcytosin pharmacogenomics: genome-wide association and drug response biomarkers. *PloS One*, 4, e7765.

Lin JH, Yamazaki M (2003) Clinical relevance of P-glycoprotein in drug therapy. *Drug Metabol Rev*, 35, 417–454.

Lombard-Platet G, Savary S, Sarde CO, Mandel JL, Chimini G (1996) A close relative of the adrenoleukodystrophy (ALD) gene codes for a peroxisomal protein with a specific expression pattern. *Proc Natl Acad Sci USA*, 93, 1265–1269.

Loubiere LS, Vasilopoulou E, Bulmer JN, Taylor PM, Stieger B, Verrey F, McCabe CJ, Franklyn JA, Kilby MD, Chan SY (2010) Expression of thyroid hormone transporters in the human placenta and changes associated with intrauterine growth restriction. *Placenta*, 31, 295–304.

Ma MK, McLeod HL (2003) Lessons learned from the irinotecan metabolic pathway. *Curr Med Chem*, 10, 41–49.

Masuda S, Terada T, Yonezawa A, Tanihara Y, Kishimoto K, Katsura T, Ogawa O, Inui K (2006) Identification and functional characterization of a new human kidney-specific H+/organic cation antiporter, kidney-specific multidrug and toxin extrusion 2. *J Am Soc Nephrology*, 17, 2127–2135.

Matsson P, Yee SW, Markova S, Morrissey K, Jenkins G, Xuan J, Jorgenson E, Kroetz DL, Giacomini KM (2011) Discovery of regulatory elements in human ATP-binding cassette transporters through expression quantitative trait mapping. *Pharmacogenomics J*, doi:10.1038/tpj.2011.8.

Meyer ZU, Schwabedissen HE, Verstuyft C, Kroemer HK, Becquemont L, Kim RB (2010) Human multidrug and toxin extrusion 1 (MATE1/SLC47A1) transporter: functional characterization, interaction with OCT2 (SLC22A2), and single nucleotide polymorphisms. *Am J Physiol Renal Physiol*, 298, F997–F1005.

Mills RE, Walter K, Stewart C, Handsaker RE, Chen K, Alkan C, Abyzov A, Yoon SC, Ye K, Cheetham RK, et al., 1000 Genomes Project (2011) Mapping copy number variation by population-scale genome sequencing. *Nature*, 470, 59–65.

Molina-Arcas M, Casado FJ, Pastor-Anglada M (2009) Nucleoside transporter proteins. *Current Vascular Pharmacol*, 7, 426–434.

Molina-Arcas M, Trigueros-Motos L, Casado FJ, Pastor-Anglada M (2008) Physiological and pharmacological roles of nucleoside transporter proteins. *Nucleosides, Nucleotides & Nucleic Acids*, 27, 769–778.

More SS, Li S, Yee SW, Chen L, Xu Z, Jablons DM, Giacomini KM (2010) Organic cation transporters modulate the uptake and cytotoxicity of picoplatin, a third-generation platinum analogue. *Mol Cancer Ther*, 9, 1058–1069.

Morello J, Cuenca L, Soriano V, Medrano J, Madejon A, Vispo E, Barreiro P, Labarga P, Jimenez-Nacher I, Rodriguez-Novoa S (2010) Influence of a single nucleotide polymorphism at the main ribavirin transporter gene on the rapid virological response to pegylated interferon-ribavirin therapy in patients with chronic hepatitis C virus infection. *J Infectious Diseases*, 202, 1185–1191.

Morita N, Yasumori T, Nakayama K (2003) Human MDR1 polymorphism: G2677T/A and C3435T have no effect on MDR1 transport activities. *Biochem Pharmacol*, 65, 1843–1852.

Moriyama Y, Hiasa M, Matsumoto T, Omote H (2008) Multidrug and toxic compound extrusion (MATE)-type proteins as anchor transporters for the excretion of metabolic waste products and xenobiotics. *Xenobiotica*, 38, 1107–1118.

Naito T, Tokashiki S, Mino Y, Otsuka A, Ozono S, Kagawa Y, Kawakami J (2010) Impact of concentrative nucleoside transporter 1 gene polymorphism on oral bioavailability of mizoribine in stable kidney transplant recipients. *Basic Clin Pharmacol Toxicol*, 106, 310–316.

Nakamura T, Yonezawa A, Hashimoto S, Katsura T, Inui K (2010) Disruption of multidrug and toxin extrusion MATE1 potentiates cisplatin-induced nephrotoxicity. *Biochem Pharmacol*, 80, 1762–1767.

Nishimura M, Naito S (2005) Tissue-specific mRNA expression profiles of human ATP-binding cassette and solute carrier transporter superfamilies. *Drug Metabol Pharmacokin*, 20, 452–477.

Ohashi R, Tamai I, Nezu Ji J, Nikaido H, Hashimoto N, Oku A, Sai Y, Shimane M, Tsuji A (2001) Molecular and physiological evidence for multifunctionality of carnitine/organic cation transporter OCTN2. *Molecular Pharmacology*, 59(2), 358–366.

Ohashi R, Tamai I, Yabuuchi H, Nezu JI, Oku A, Sai Y, Shimane M, Tsuji A (1999) Na(+)-dependent carnitine transport by organic cation transporter (OCTN2): its pharmacological and toxicological relevance. *J Pharmacol Exp Ther*, 291, 778–774.

Osato DH, Huang CC, Kawamoto M, Johns SJ, Stryke D, Wang J, Ferrin TE, Herskowitz I, Giacomini KM (2003) Functional characterization in yeast of genetic variants in the human equilibrative nucleoside transporter, ENT1. *Pharmacogenetics*, 13, 297–301.

Otsuka M, Matsumoto T, Morimoto R, Arioka S, Omote H, Moriyama Y (2005) A human transporter protein that mediates the final excretion step for toxic organic cations. *Proc Natl Acad Sci USA*, 102, 17923–17928.

Owen RP, Gray JH, Taylor TR, Carlson EJ, Huang CC, Kawamoto M, Johns SJ, Stryke D, Ferrin TE, Giacomini KM (2005) Genetic analysis and functional characterization of polymorphisms in the human concentrative nucleoside transporter, CNT2. *Pharmacogenet Genomics*, 15, 83–90.

Owen RP, Lagpacan LL, Taylor TR, De La Cruz M, Huang CC, Kawamoto M, Johns SJ, Stryke D, Ferrin TE, Giacomini KM (2006) Functional characterization and haplotype analysis of polymorphisms in the human equilibrative nucleoside transporter, ENT2. *Drug Netbol Dispos*, 34, 12–15.

Pastor-Anglada M, Casado FJ, Valdes R, Mata J, Garcia-Manteiga J, Molina M (2001) Complex regulation of nucleoside transporter expression in epithelial and immune system cells. *Mol Membrane Biol*, 18, 81–85.

Pastor-Anglada M, Errasti-Murugarren E, Aymerich I, Casado FJ (2007) Concentrative nucleoside transporters (CNTs) in epithelia: from absorption to cell signaling. *J Physiol Biochem*, 63, 97–110.

Peltekova VD, Wintle RF, Rubin LA, Amos CI, Huang Q, Gu X, Newman B, Van Oene M, Cescon D, Greenberg G, Griffiths AM, St George-Hyslop PH, Siminovitch KA (2004) Functional variants of OCTN cation transporter genes are associated with Crohn disease. *Nature Genet*, 36, 471–475.

Picard N, Yee SW, Woillard JB, Lebranchu Y, Le Meur Y, Giacomini KM, Marquet P (2010) The role of organic anion-transporting polypeptides and their common genetic variants in mycophenolic acid pharmacokinetics. *Clin Pharmacol Ther*, 87, 100–108.

Rodrigues AC (2010) Efflux and uptake transporters as determinants of statin response. *Expert Opin Drug Metabol & Toxicol*, 6, 621–632.

Rose JB, Naydenova Z, Bang A, Eguchi M, Sweeney G, Choi DS, Hammond JR, Coe IR (2010) Equilibrative nucleoside transporter 1 plays an essential role in cardioprotection. *Am J Physiol Heart and Circulatory Physiology*, 298, H771–H777.

Sachidanandam R, Weissman D, Schmidt SC, Kakol JM, Stein LD, Marth G, Sherry S, Mullikin JC, Mortimore BJ, Willey DL, et al., International SNP Map Working Group (2001) A map of human genome sequence variation containing 1.42 million single nucleotide polymorphisms. *Nature*, 409, 928–933.

Sato T, Masuda S, Yonezawa A, Tanihara Y, Katsura T, Inui K (2008) Transcellular transport of organic cations in double-transfected MDCK cells expressing human organic cation transporters hOCT1/hMATE1 and hOCT2/hMATE1. *Biochem Pharmacol*, 76, 894–903.

Schinkel AH, Jonker JW (2003) Mammalian drug efflux transporters of the ATP binding cassette (ABC) family: an overview. *Adv Drug Delivery Rev*, 55, 3–29.

Schloss P, Williams DC (1998) The serotonin transporter: a primary target for antidepressant drugs. *J Psychopharmacology*, 12, 115–121.

SEARCH Collaborative Group, Link E, Parish S, Armitage J, Bowman L, Heath S, Matsuda F, Gut I, Lathrop M, Collins R (2008) SLCO1B1 variants and statin-induced myopathy–a genomewide study. *New Engl J Med*, 359, 789–799.

Shima JE, Komori T, Taylor TR, Stryke D, Kawamoto M, Johns SJ, Carlson EJ, Ferrin TE, Giacomini KM (2010) Genetic variants of human organic anion transporter 4 demonstrate altered transport of endogenous substrates. *Am J Physiol Renal Physiology*, 299, F767–F775.

Shu Y, Brown C, Castro RA, Shi RJ, Lin ET, Owen RP, Sheardown SA, Yue L, Burchard EG, Brett CM, Giacomini KM (2008) Effect of genetic variation in the organic cation transporter 1, OCT1, on metformin pharmacokinetics. *Clin Pharmacol Ther*, 83, 273–280.

Shu Y, Leabman MK, Feng B, Mangravite LM, Huang CC, Stryke D, Kawamoto M, Johns SJ, DeYoung J, Carlson E, Ferrin TE, Herskowitz I, Giacomini KM, Pharmacogenetics of Membrane Transporters Investigators (2003) Evolutionary conservation predicts function of variants of the human organic cation transporter, OCT1. *Proc Natl Acad Sci USA*, 100, 5902–5907.

Shu Y, Sheardown SA, Brown C, Owen RP, Zhang S, Castro RA, Ianculescu AG, Yue L, Lo JC, Burchard EG, Brett CM, Giacomini KM (2007) Effect of genetic variation in the organic cation transporter 1 (OCT1) on metformin action. *J Clin Invest*, 117, 1422–1431.

Stieger B (2009) Recent insights into the function and regulation of the bile salt export pump (ABCB11). *Curr Opin Lipidol*, 20, 176–181.

Stieger B, Meier Y, Meier PJ (2007) The bile salt export pump. *Pflugers Arch*, 453, 611–620.

Tahara H, Yee SW, Urban TJ, Hesselson S, Castro RA, Kawamoto M, Stryke D, Johns SJ, Ferrin TE, Kwok PY, Giacomini KM (2009) Functional genetic variation in the basal promoter of the organic cation/carnitine transporters OCTN1 (SLC22A4) and OCTN2 (SLC22A5). *J Pharmacol Exp Ther*, 329, 262–271.

Tamai I, Ohashi R, Nezu J, Yabuuchi H, Oku A, Shimane M, Sai Y, Tsuji A (1998) Molecular and functional identification of sodium ion-dependent, high affinity human carnitine transporter OCTN2. *J Biol Chem*, 273, 20378–20382.

Tanihara Y, Masuda S, Sato T, Katsura T, Ogawa O, Inui K (2007) Substrate specificity of MATE1 and MATE2-K, human multidrug and toxin extrusions/H(+)-organic cation antiporters. *Biochem Pharmacol*, 74, 359–371.

Trevino LR, Shimasaki N, Yang W, Panetta JC, Cheng C, Pei D, Chan D, Sparreboom A, Giacomini KM, Pui CH, Evans WE, Relling MV (2009) Germline genetic variation in an organic anion transporter polypeptide associated with methotrexate pharmacokinetics and clinical effects. *J Clin Oncol*, 27, 5972–5978.

Tsuda M, Terada T, Mizuno T, Katsura T, Shimakura J, Inui K (2009) Targeted disruption of the multidrug and toxin extrusion 1 (mate1) gene in mice reduces renal secretion of metformin. *Mol Pharmacol*, 75, 1280–1286.

Urban TJ, Brown C, Castro RA, Shah N, Mercer R, Huang Y, Brett CM, Burchard EG, Giacomini KM (2008) Effects of genetic variation in the novel organic cation transporter, OCTN1, on the renal clearance of gabapentin. *Clin Pharmacol Ther*, 83, 416–421.

Urban TJ, Gallagher RC, Brown C, Castro RA, Lagpacan LL, Brett CM, Taylor TR, Carlson EJ, Ferrin TE, Burchard EG, Packman S, Giacomini KM (2006a) Functional genetic diversity in the high-affinity carnitine transporter OCTN2 (SLC22A5). *Mol Pharmacol*, 70, 1602–1611.

Urban TJ, Giacomini KM, Risch N (2005) Haplotype structure and ethnic-specific allele frequencies at the OCTN locus: implications for the genetics of Crohn's disease. *Inflamm Bowel Dis*, 11, 78–79.

Urban TJ, Sebro R, Hurowitz EH, Leabman MK, Badagnani I, Lagpacan LL, Risch N, Giacomini KM (2006b) Functional genomics of membrane transporters in human populations. *Genome Res*, 16, 223–230.

Urban TJ, Yang C, Lagpacan LL, Brown C, Castro RA, Taylor TR, Huang CC, Stryke D, Johns SJ, Kawamoto M, Carlson EJ, Ferrin TE, Burchard EG, Giacomini KM (2007) Functional effects of protein sequence polymorphisms in the organic cation/ergothioneine transporter OCTN1 (SLC22A4). *Pharmacogenet Genomics*, 17, 773–782.

Urquhart BL, Kim RB (2009) Blood-brain barrier transporters and response to CNS-active drugs. *Eur J Clin Pharmacol*, 65, 1063–1070.

van de Steeg E, Wagenaar E, van der Kruijssen CM, Burggraaff JE, de Waart DR, Elferink RP, Kenworthy KE, Schinkel AH (2010) Organic anion transporting polypeptide 1a/1b-knockout mice provide insights into hepatic handling of bilirubin, bile acids, and drugs. *J Clin Invest*, 120, 2942–2952.

VanWert AL, Gionfriddo MR, Sweet DH (2010) Organic anion transporters: discovery, pharmacology, regulation and roles in pathophysiology. *Biopharma Drug Dispos*, 31, 1–71.

Vlaming ML, Lagas JS, Schinkel AH (2009) Physiological and pharmacological roles of ABCG2 (BCRP): recent findings in Abcg2 knockout mice. *Adv Drug Delivery Reviews*, 61, 14–25.

Wang D, Johnson AD, Papp AC, Kroetz DL, Sadee W (2005) Multidrug resistance polypeptide 1 (MDR1, ABCB1) variant 3435C>T affects mRNA stability. *Pharmacogenet Genomics*, 15, 693–704.

Wang ZJ, Yin OQ, Tomlinson B, Chow MS (2008) OCT2 polymorphisms and in-vivo renal functional consequence: studies with metformin and cimetidine. *Pharmacogenet Genomics*, 18, 637–645.

Watanabe S, Tsuda M, Terada T, Katsura T, Inui K (2010) Reduced renal clearance of a zwitterionic substrate cephalexin in MATE1-deficient mice. *J Pharmacol Exp Ther*, 334, 651–656.

Watkins PA, Gould SJ, Smith MA, Braiterman LT, Wei HM, Kok F, Moser AB, Moser HW, Smith KD (1995) Altered expression of ALDP in X-linked adrenoleukodystrophy. *Am J Hum Genet*, 57, 292–301.

Yee SW, Shima JE, Hesselson S, Nguyen L, De Val S, Lafond RJ, Kawamoto M, Johns SJ, Stryke D, Kwok PY, Ferrin TE, Black BL, Gurwitz D, Ahituv N, Giacomini KM (2009) Identification and characterization of proximal promoter polymorphisms in the human concentrative nucleoside transporter 2 (SLC28A2). *J Pharmacol Exp Ther*, 328, 699–707.

Zhang L, Dresser MJ, Gray AT, Yost SC, Terashita S, Giacomini KM (1997) Cloning and functional expression of a human liver organic cation transporter. *Mol Pharamcol*, 51, 913–921.

Zhou M, Xia L, Engel K, Wang J (2007) Molecular determinants of substrate selectivity of a novel organic cation transporter (PMAT) in the SLC29 family. *J Biol Chem*, 282, 3188–3195.

CHAPTER 5

EMERGING NEW TECHNOLOGY OF SNP TYPING

Toshihisa Ishikawa
Yoshihide Hayashizaki

5.1 INTRODUCTION

Single-nucleotide polymorphisms (SNPs) widely distributed throughout the genome have been used as markers of genetic variation. SNPs in drug-metabolizing enzyme genes have long been recognized as useful biomarkers for guiding therapeutic decisions regarding dose selection. SNPs occur within coding regions and noncoding regions of genes, or in inter-gene regions. Nonsynonymous SNPs occurring within coding regions result in the substitution of an amino acid residue in the gene product. In some cases, a nonsynonymous SNP can greatly affect the activity and/or protein stability of the gene product (Wang and Weinshilboum, 2006; Nakagawa et al., 2011). In addition, SNPs within noncoding regions can affect the transcription of genes. Identification and characterization of "clinically significant SNPs" are important to aid our understanding of the impact of genetic polymorphisms on drug action. To date, several SNP databases are available to the public, including HapMap (http://hapmap.ncbi.nlm.nih.gov), the website for the International HapMap project containing information on genotype data; dbSNP (http://www.ncbi.nlm.nih.gov/SNP), the portal to NCBI's SNP database; JSNP (http://snp.ims.u-tokyo.ac.jp), a database of Japanese SNPs; and HGVbase (http://www.hgvbaseg2p.org), a database of human genome variants for genotype–phenotype association analyses.

During the past two decades, the role of carrier-mediated transport in determining the pharmacokinetics of drugs has becoming increasingly evident with the discovery of genetic variants that affect the expression and/or function of a given drug transporter. Drug transporters are expressed in different cells, such as intestinal epithelial cells, hepatocytes, renal tubular cells, and cells of the blood–brain barrier to play a pivotal role in the absorption, distribution, and elimination of drugs (Giacomini et al., 2010). Pharmacogenomic studies have been elucidating the inherited nature of differences in pharmacokinetic and pharmacodynamic effects, thereby enhancing drug discovery and providing a stronger scientific basis for optimizing

Pharmacogenomics of Human Drug Transporters: Clinical Impacts, First Edition.
Edited by Toshihisa Ishikawa, Richard B. Kim and Jörg König

drug therapy on the basis of each patient's genetic constitution. Hitherto, researchers have associated alterations in drug uptake and/or disposition with the presence of genetic polymorphisms. In fact, the Pharmacogenomics of Membrane Transporter (PMT) project has identified sequence variants in different membrane transporter genes in the solute carrier (SLC) and ATP-binding cassette (ABC) transporter families (Kroetz et al., 2010). The PMT has been contributing new information about genetic variants in transporters, the functional activities of variant transporters, and the clinical relationships between variant transporters and drug disposition and response (http://pharmacogenetics.ucsf.edu).

There may be many different causes for personal variations in the response to a drug, such as genetic variations and expression levels of drug-targeted molecules, including membrane receptors, nuclear receptors, signal transduction components, and enzymes, as well as those of drug-metabolizing enzymes and drug transporters (Evans and Johnson, 2001; Kalow et al., 2001; Kim, 2002; Ishikawa et al., 2004; Andersson et al., 2005; Wilke et al., 2007; Weiss et al., 2008; Sadee, 2008). Recently, tools such as next-generation sequencing technologies and genome-wide association studies (GWAS) have been used to uncover a number of variants that affect drug toxicity and efficacy as well as potential risk of diseases. The costs involved in carrying out GWAS and sequencing have been dropping dramatically, while data are being provided at unprecedented rates. The GWAS approach has been applied for identifying genetic contributions to variations in drug response (Cooper et al., 2008; Daly et al., 2009; Ge et al., 2009; Schuldiner et al., 2009; Kamatani et al., 2010). As a result, there have been dramatic increases in our understanding of the mechanisms of drug action and of the genetic determinants responsible for variable responses to both rarely and widely used drugs, such as warfarin, tamoxifen, and clopidogrel.

The next important challenge is to incorporate pharmacogenomics data into routine clinical practice. Technologies are evolving to transform diagnostic devices for rapid genetic testing. Portable devices are being engineered for use in a range of settings to perform robust assays for the diagnosis of disease that will improve patient management and result in greater convenience and more rapid results. Genetic diagnostics is a growing field that is gradually becoming more user-friendly with the introduction of portable devices and faster nucleic acid detection. Successful genetic diagnostics requires four major elements: rapid reaction, low cost, low energy consumption, and simple analysis (requiring minimal technical training and the inclusion of controls but no off-instrument processing or reagent preparation). In this context, we need to develop a point-of-care "POC" technology and apply it to medical advances.

5.2 CURRENT TECHNOLOGIES OF SNP DETECTION

Basic technologies of molecular diagnostics play a pivotal role in pharmacogenomics, particularly with respect to SNP genotyping. Diagnosis is integrated with therapy for selecting treatments as well for monitoring results. Cost-effective methods should be developed for genotyping, and it would be desirable to include this information in each patient's record as guidance for physicians to provide individualized treatment. The accurate measurement of allele frequency variations among population groups with different sensitivities to diseases and/or different responses to drugs is fundamental to genetic epidemiology. Genotyping errors can markedly influence the clinical conclusions of pharmacotherapy. Thus, it is critically important to choose the appropriate method for accurate SNP detection.

There are many different methods for SNP typing, such as allele-specific hybridization (microarrays, AmpliChip CYP450), enzymatic cleavage (PCR-restriction fragment length polymorphisms (RFLP), invader assay), allele-specific PCR, mass-based detection (Sequenom), chemoluminescence (pyrosequencing), and fluorescence methods (direct sequencing and TaqMan). The central core components of these SNP typing methods rely primarily upon the stringency of DNA hybridization and/or the fidelity of enzyme reactions. Recent articles (Heim and Meyer, 1990; Saperstein and Nickerson, 1991; Bassler et al., 1995; Ross et al., 1998; Livak, 1999; Lymichev et al., 1999; Fan et al., 2000; Griffin and Smith, 2000; See et al., 2000) describe various SNP typing methods in detail. Therefore, we herein briefly summarize some of those technologies.

5.2.1 Allele-specific Hybridization

Many chip-based hybridization methods make use of universal sequence tags attached to SNP-specific genotyping oligonucleotides (Fan et al., 2000). AmpliChip CYP450 (Roche–Affymetrix), for example, is a chip-based detection method that enables the detection of a total of 31 polymorphisms in the *CYP2D6* gene as well as two polymorphisms in the *CYP2C19* gene. The AmpliChip method requires multiple steps before detection; however, including DNA extraction, amplification of those genes by PCR, and hybridization. Therefore, its overall detection time is about 8 h.

5.2.2 Enzymatic Cleavage

Enzyme-based approaches to SNP genotyping generally possess a higher degree of assay fidelity than those primarily dependent upon hybridization for their specificity. PCR-RFLP is an example of the historical standard for genotyping of genetic variations (Saperstein and Nickerson, 1991). It relies on the exquisite selectivity of restriction endonuclease for short stretches of DNA sequence that act as recognition sites for DNA strand cleavage. The major drawback of this method is the requirement for electrophoretic separation of digested products, which severely limits the throughput and automatability.

5.2.3 Allele-specific PCR

Another method for genotyping in small-scale applications is allele-specific PCR amplification, which combines selective PCR primer hybridization with a subsequent PCR reaction (Heim and Meyer, 1990; See et al., 2000). Conditions are so optimized that hybridization and subsequent amplification occur only when the PCR priming oligonucleotide is perfectly matched with the target site (usually with the polymorphic site at the 3′ end of the oligonucleotide). Again, the major drawback of this method for high-throughput applications is the requirement for gel electrophoresis, a process that is not amenable to automation.

5.2.4 Mass-based Detection

Sequenom involves DNA extraction, multiplex PCR amplification, one-base extension using dideoxynucleotide triphosphates (ddNTPs), and matrix-assisted laser desorption/ionization time-of-flight mass spectrometry (MALDI-TOF MS) analysis. MALDI-TOF MS supports high resolution analysis capable of resolving two components differing by substitution of a single DNA base; therefore, a large number of species can be separated and detected in a single sample (Ross et al., 1998; Griffin and Smith, 2000). SNP genotypings are determined by MS analysis of single-nucleotide primer extension reactions. The mass of any given dideoxynucleotide monophosphate (ddNMP) differs by at least 9 Da from each of the other ddNMPs; hence, all possible genotypes and heterozygous combinations at a particular site can be resolved by using a single primer. One major drawback of this SNP detection method, however, is that MALDI-TOF MS systems are expensive.

5.2.5 Fluorescent Method

In the case of the TaqMan method, a fluorogenic probe, consisting of an oligonucleotide labeled with both a fluorescent reporter dye and a quencher dye, is included in a typical PCR. Amplification of the probe-specific product causes cleavage of the probe, generating an increase in reporter fluorescence. By the use of different reporter dyes, cleavage of allele-specific probes can be detected in a single PCR. The 5′ nuclease assay is successfully used to discriminate alleles that differ by a single base substitution (Bassler et al., 1995; Livak, 1999).

5.2.6 Invader Assay (Enzymatic Cleavage)

The invader assay lacks the requirement for a prior PCR amplification step. Cleavase (Flap endonuclease) recognizes and cleaves a structure formed when two overlapping oligonucleotides hybridize to a target DNA strand. Following the cleavage reactions, the 5′ fragment is released from the signal probe. The use of appropriate probes that either fully match or contain a single mismatch at the site of cleavage allows discrimination of SNPs directly from human genomic DNA (Lymichev et al., 1999). However, due to the requirement for large quantities of input genomic DNA and the replacement of one expensive enzyme approach (*Taq* DNA polymerase) for another (Cleavase), the cost-effectiveness of this method is a practical issue.

5.3 SmartAmp METHOD

To effectively advance personalized medicine, it is necessary to be able to rapidly and conveniently test for patients' genetic polymorphisms and/or mutations. To address this clinical need, we have recently developed a rapid and cost-effective method, called the smart amplification process (SmartAmp), which enables us to detect genetic polymorphisms or mutations in target genes within 30–45 min under isothermal conditions that do not require DNA isolation and PCR amplification (Mitani et al., 2007; Watanabe et al., 2007; Aomori et al., 2009; Mitani et al., 2009; Toyoda et al., 2009; Ishikawa et al., 2010; Lezhava et al., 2010; Okada et al., 2010; Ota et al., 2010; Aw et al., 2011; Azuma et al., 2011).

The SmartAmp method was developed based on the principal concept that DNA amplification itself is the signal for detection of a genetic mutation or SNP. Differing from the widely used PCR, the SmartAmp method is an isothermal DNA amplification reaction (Mitani et al., 2007, 2009). In the SmartAmp method, the entire DNA amplification process requires five primers: turnback primer (TP), folding primer (FP), boost primer (BP), and two outer primers (OP1 and OP2) (Fig. 5.1). Primers are selected based on those algorithms considering the free energy, probability of base

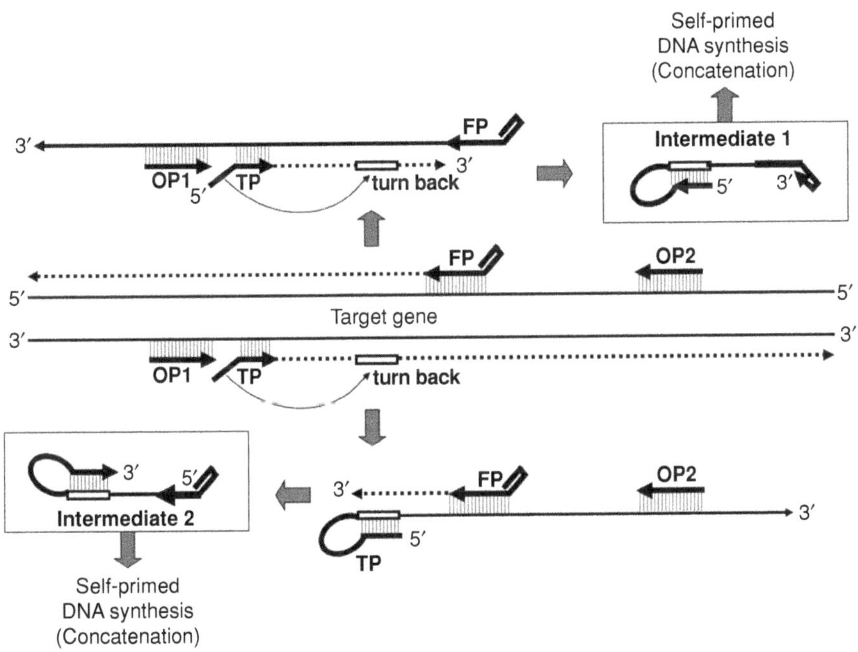

Figure 5.1 Schematic illustration of the SmartAmp reaction using five primers: turnback primer (TP), boost primer (BP), forward primer (FP), and two outer primers (OP1 and OP2). As an initial step, priming by the primers and the DNA polymerase reaction generates two intermediates 1 and 2. The inner primer set with FP and TP initiates the reaction by hybridizing to opposite strands of a target region. Linear primer extension products from the FP and TP primers are then released from their templates in a second primer extension reaction driven by the set of outer primers (OP1 and OP2) that hybridize downstream of the FP and TP primers. Because of the special features of the FP and TP primers, single-stranded primer extension products from those primers will refold at their 3′ and 5′ ends to form new priming sites that maintain self-amplification in a cautious process driven by the DNA strand-displacement activity of *Aac* DNA polymerase.

pairing, product size range, and optimal melting temperature. The design of these primers contributes to the specificity of SmartAmp. In particular, two primers (TP and FP) are critically important for the amplification process. The genomic sequence between the annealing sites of the TP and FP primers is the target region that will be amplified by the SmartAmp reaction. The other primers (BP, OP1, and OP2) are additionally employed to accelerate the process and enhance specificity.

5.4 MOLECULAR MECHANISM UNDERLYING ISOTHERMAL DNA AMPLIFICATION

In isothermal DNA amplification by the SmartAmp method, the initial step of copying a target sequence from the genomic DNA is a prerequisite. FP and TP hybridize the template genomic DNA. Next, both products primed for the FP and TP are detached from the template genomic DNA by strand-displacing DNA polymerase, whose extensions are primed by OP1 and OP2. The single-stranded DNA products, thus displaced, become templates in the second step for the opposing FP and TP. These single-stranded DNA products are generated by the strand-displacement activity of the DNA polymerase, being primed from the flanking region of OP primers adjacent to the target sequence. The resulting DNA products are referred to as "intermediate products" that play key roles in the subsequent amplification steps (Figs. 5.1 and 5.2).

The formation of those intermediate products is the rate-limiting step in SmartAmp-based isothermal DNA amplification. Intermediate 1 (IM1) has the TP sequence at the 5′ end and the FP complementary sequence at the 3′ end; and intermediate 2 (IM2) is complementary to IM1 (Fig. 5.2). The initial self-priming site on IM1 is the 3′ end of the FP sequence of IM1. Concatenated products of IM1 are synthesized by an elongation process termed pathway A. The characteristic feature of the products of pathway A is that the free 5′ and 3′ ends carry TP and its complementary sequence to form long double-stranded hairpin DNA. The initial self-priming elongation site on IM2 is located at the 3′ end of the TP sequence of IM2. Long concatenated DNA products are synthesized as in pathway A, but end products in pathway B are different. The long-hairpin DNA products of pathway B carry FP and its complementary sequence at the free 5′ and 3′ ends, respectively. There is another elongation pathway that starts from the 3′ end of a free TP primer, which hybridizes to the looping structure of the TP complementary sequence located at the intermediate region of the long products of pathway A. Thus, concatenated DNA products are formed in the SmartAmp reaction. The resulting DNA products could be detected by conventional agarose gel electrophoresis, where DNA ladder patterns represented the formation of concatenated DNA products (Mitani et al., 2007) (Fig. 5.2).

5.5 MOLECULAR MECHANISM UNDERLYING SNP DETECTION

To ensure high fidelity of SNP detection by the SmartAmp method, exponential amplification of mis-primed DNA must be suppressed. In the original SmartAmp

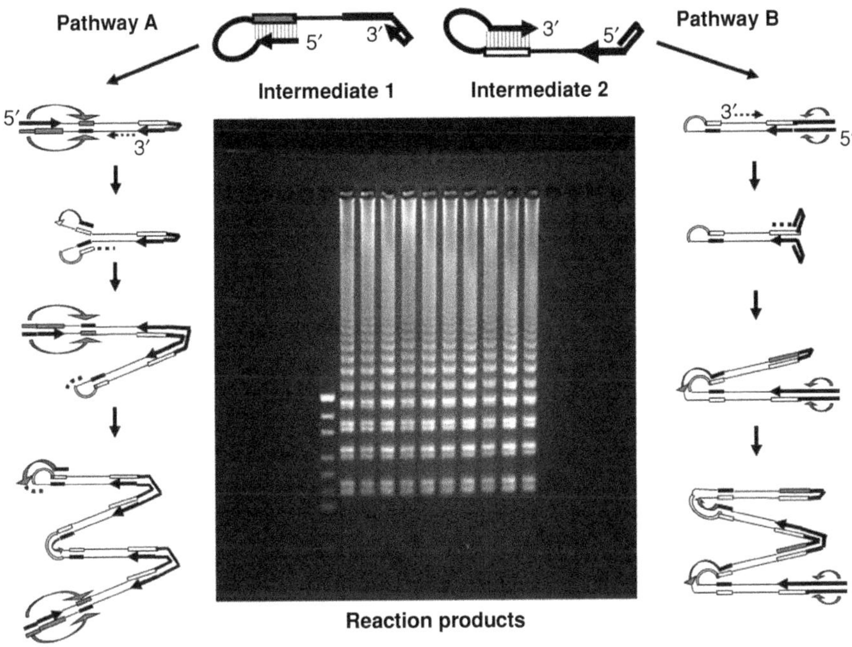

Figure 5.2 Formation of concatenated DNA products in the SmartAmp reaction. Self-priming DNA synthesis from each of intermediates 1 and 2 creates hairpin molecules via pathways A and B. These structures lead to further self-primed DNA synthesis to create dimeric amplicons and then subsequently concatenated DNA products.

method, this was achieved by adding either the mismatch-binding protein (MutS) *Thermus aquaticus* (Mitani et al., 2007, 2009) or a competitive probe (Toyoda et al., 2009) to the reaction mixture. MutS inhibits background DNA from entering the amplification cycle by specifically binding to mis-primed amplification products (Fig. 5.3). In addition, a combination of the asymmetrical primers, that is, TP and FP is used to minimize alternative mis-amplification pathways (Mitani et al., 2007).

The SmartAmp method utilizes *Aac* polymerase as a DNA polymerase with strand-displacement activity. This DNA polymerase is highly resistant to cellular contaminants and hence works directly on blood samples, just after a simple heat treatment (98°C, 3 min) to degrade RNA and denature proteins. This is a great advantage of the SmartAmp method over the commonly used PCR-based techniques that require careful DNA extraction. Actually, in the conventional method, the enzymatic activity of *Taq* DNA polymerase is easily inhibited by impurities.

5.6 EXAMPLE OF SNP DETECTION BY SmartAmp METHOD

Clinical application of SmartAmp to practical SNP detection should be evaluated with clinical samples (either blood or genomic DNA) according to the principle of

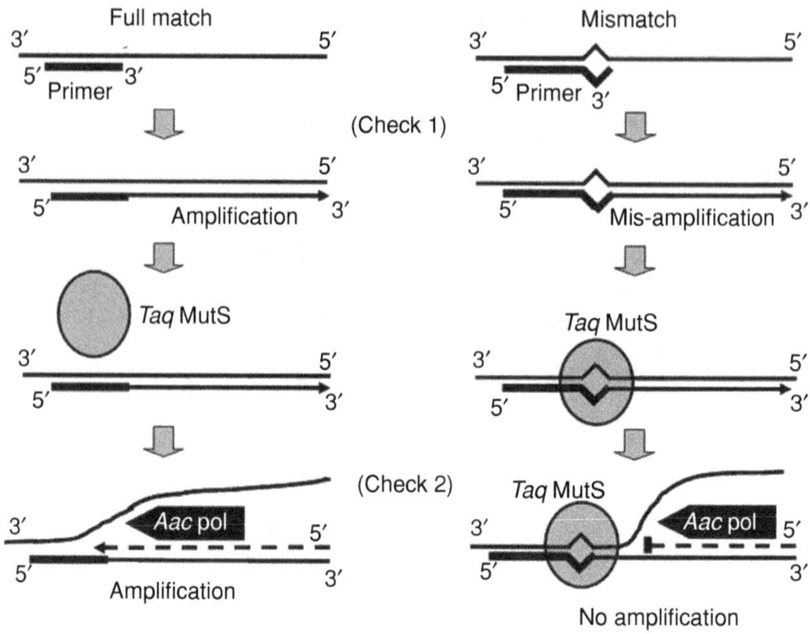

Figure 5.3 The mechanism of allele discrimination as exercised by *Taq* MutS. SNP typing with a wild-type allele detection primer, using the wild-type allele (left) and the mutant-type allele (right) as templates. The wild-type allele detection primer is designed to encompass the SNP nucleotide site at each 3′ position. Amplification is not allowed when the primer mismatches with the mutant-type allele (Check 1). If Check 1 fails, *Taq* MutS strongly binds to mismatched nucleotides and *Aac* DNA polymerase cannot strand-displace or extend the newly synthesized strand (Check 2).

amplification versus non-amplification as compared to threshold values. The amount of DNA-intercalating SYBR Green I dye generated during the reaction can be monitored in a real-time PCR system, for example, M×3000P (Stratagene, La Jolla, CA, USA), whereby SNP typing can be determined by referring to the intensity of fluorescence.

Each SmartAmp reaction is performed in a 25 μL volume tube at 60°C. The standard reaction mixture contains 3.2 μM each of TP and FP, 0.4 μM each of OP1 and OP2, 1.6 μM BP, 1.4 nM dNTPs, 5% DMSO, 20 mM tris-HCl (pH 8.8), 10 mM KCl, 10 mM $(NH_4)_2SO_4$, 8 mM $MgSO_4$, 0.1% Tween 20, SYBR Green I (1/100,000 diluted), 40 units of *Aac* DNA polymerase, 1.5–2.4 μg of *Taq* MutS (optional), and 1 μL of blood or genomic DNA sample. Each reaction mixture should be incubated at 60°C for 40–60 min under isothermal conditions in a real-time PCR model M×3000P system whereby changes in the fluorescence intensity of SYBR Green I dye are monitored to detect the DNA amplification. Figure 5.4 depicts the SmartAmp-based detection of a clinically important SNP, 460G>A, in the human thiopurine S-methyltransferase (*TPMT*) gene.

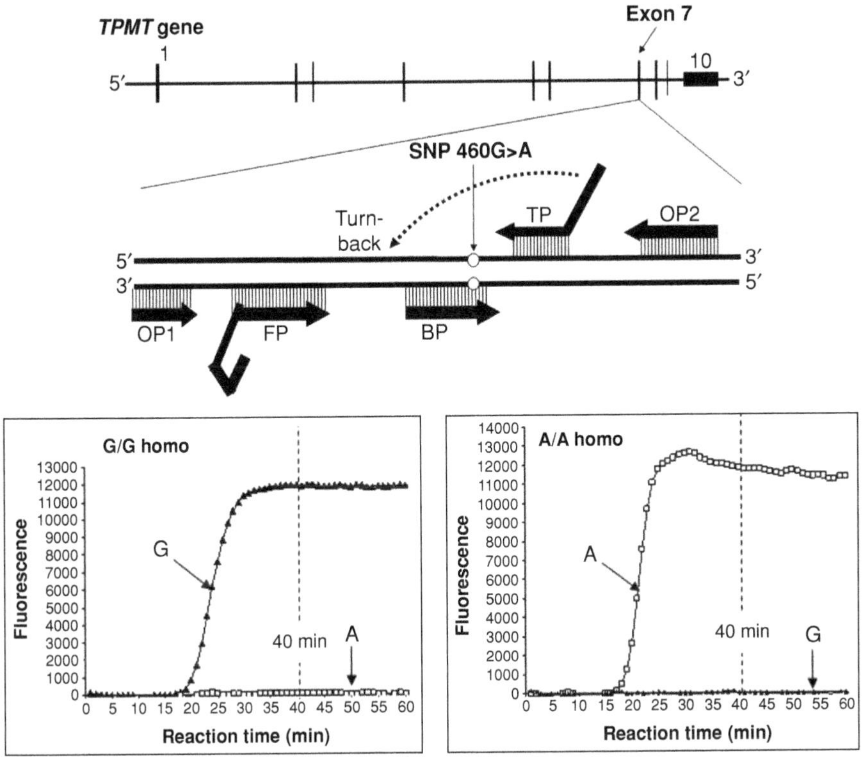

Figure 5.4 Schematic illustration of the human *TPMT* gene and detection of the SNP 460G>A by the SmartAmp method. The *TPMT* gene comprising 10 exons is located on chromosome 6p22.3. The two bottom panels depict the time courses of the SmartAmp assay reactions with *TPMT*-specific primers carrying WT (460G) or SNP (460A) alleles; namely, G/G homozygote and A/A homozygote. In this case, two different BP primers were used to detect the SNP 460G>A; one for detecting the WT (460G) and another for the SNP (460A). Detection of the SNP 460G>A was achieved by the SmartAmp method within 40 min.

5.7 THIOPURINE TOXICITY AND GENETIC POLYMORPHISMS IN THE *TPMT* GENE

Thiopurines are effective immunosuppressants and anticancer agents used for treating childhood acute lymphoblastic leukemia, acute myeloblastic leukemia, autoimmune disease, rheumatoid arthritis, and inflammatory bowel diseases (IBDs). The intracellular accumulation of such active metabolites as 6-thioguanine nucleotides (6-TGNs), however, causes dose-limiting hematopoietic toxicity (Weinshilboum and Sladek, 1980; Tai et al., 1997). The enzyme *TPMT* operates in the main inactivation pathway for thiopurine drugs, and *TPMT* deficiency has been reported to exacerbate thiopurine toxicity (Fig. 5.5). *TPMT* activity is inversely correlated to 6-TGN levels in erythrocytes and other hematopoietic tissues (Anstey et al., 1992; Krynetski et al., 1995; Yates et al., 1997; Black et al., 1998; Stolk et al., 1998; Clunie and Lennard,

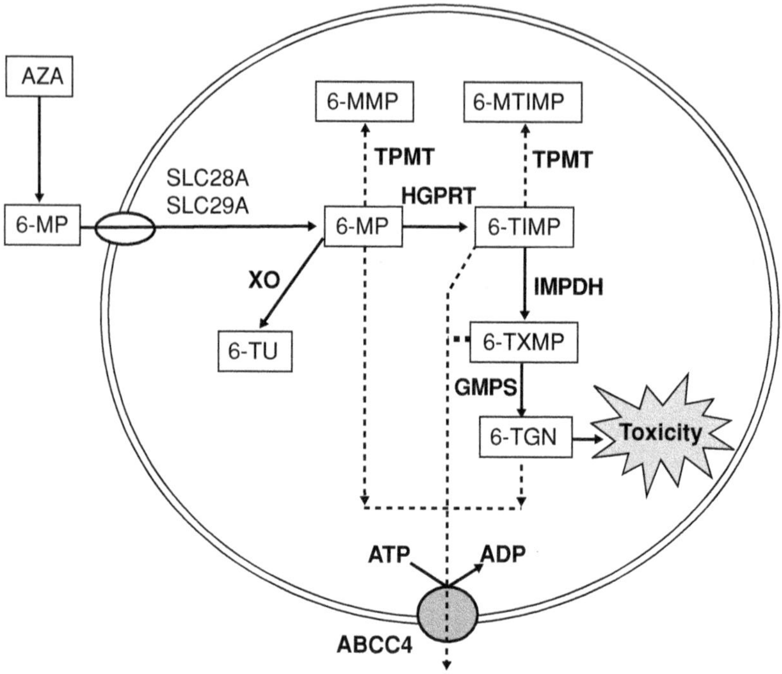

Figure 5.5 Cellular metabolism of azathioprine (AZA) and membrane transport of its metabolite. AZA is converted to 6-MP and transported into the cell via nucleoside transporters (SLC28A, SLC29A). HGPRT is responsible for the conversion of 6-MP to 6-TIMP. Thiopurine toxicity is caused by the cellular accumulation of 6-TGN. Human ABC transporter *ABCC4* plays a role of extruding the cytotoxic 6-TGN as well as 6-MP, 6-TIMP, and 6-TXMPfrom the cells. HGPRT, hypoxanthine-guanine phosphoribosyl transferase; IMPDH, inosine monophosphate dehydrogenase; XO, xanthine oxidase; 6-MP, 6-mercaptopurine; 6-TU, 6-thiouric acid; 6-MMP, 6-methylmercaptopurine; 6-TIMP, 6-thioinosine 5′-monophosphate; 6-MTIMP, 6-methylthioinosine monophosphate; 6-TXMP, 6-thioxanthosine monophosphate; GMPS, guanosine monophosphate synthetase; 6-TGN, 6-thioguanine nucleotides

2004; Evans, 2004). Certain genetic polymorphisms in the *TPMT* gene cause low or no *TPMT* activity in certain patients, leading to thiopurine-induced life-threatening hematologic toxicity. The *TPMT*3A* allele, which is frequently found in the Caucasian population, carries two nonsynonymous SNPs of both 460G>A (Ala154Thr) and 719A>G (Tyr240Lys). The TPMT*3A protein undergoes rapid degradation to exhibit no *TPMT* activity (Wang et al., 2003; Wang et al., 2005; Wang and Weinshilboum, 2006). A total of 24 functionally related alleles have been reported to date (Schütz et al., 2000; Lee et al., 2008; Schaeffeler et al., 2008). Therefore, based on the *TPMT* genotype of a patient, the thiopurine dose should be optimized to reduce the toxicity risk, for example, a 50% reduced dose for an intermediate metabolizer or an 80–90% reduced dose for a poor metabolizer.

5.8 SNP 2269G>A (Glu757Lys) IN THE *ABCC4* GENE AND THIOPURINE TOXICITY

For largely unknown reasons, there are subsets of Japanese patients who suffer from dose-limiting hematopoietic toxicity of thiopurine drugs, but are not *TPMT* deficient (Takatsu et al., 2009). Recent studies have revealed that *ABCC4* plays a protective role against thiopurine-induced hematopoietic toxicity by actively exporting thiopurine nucleotides (Krishnamurthy et al., 2008; Ban et al., 2010). *ABCC4* is reportedly involved in the transport of antiviral agents, such as azidothymidine, adefovir, tenofovir, lamivudine, and ganciclovir (Schuetz et al., 1999; Adachi et al., 2002; Anderson et al., 2006; Imaoka et al., 2007), as well as anticancer drugs including 6-MP, 6-TG, methotrexate, and camptothecins (Lee et al., 2000; Chen et al., 2002; Wielinga et al., 2002; Tian et al., 2005).

Krishnamurthy et al. have shown that patients carrying SNP 2269G>A (Glu757Lys) in the human *ABCC4* gene have severely reduced *ABCC4* function resulting from an impairment of its cell membrane localization (Krishnamurthy et al., 2008). *ABCC4* protects against thiopurine-induced hematologic toxicity by actively exporting 6-TGN, a toxic metabolite in the thiopurine drug metabolic pathway. Interestingly, the *ABCC4* 2269G>A SNP is common in the Japanese population (15–18% frequency), which suggests that this nonsynonymous SNP could provide an explanation for the unsolved thiopurine toxicity that is not associated with genetic polymorphisms of *TPMT* (Ando et al., 2001; Takatsu et al., 2009; Ban et al., 2010). We, therefore, prepared one set of SmartAmp primers to detect the *ABCC4* SNP 2269G>A (Fig. 5.6) and validated it with clinical samples from a total of 282 Japanese patients with IBD. Results obtained by SmartAmp-based detection of the SNP revealed a 100% consistency with the sequence analysis and TaqMan data (data not shown).

Ban et al. investigated an association between the 2269G>A polymorphism in the *ABCC4* gene and thiopurine sensitivity in Japanese IBD patients (Ban et al., 2010). A total of 235 IBD patients were treated with azathioprine (AZA). The 6-TGN levels in red blood cells were significantly higher in patients with the allele of *ABCC4* SNP 2269G>A than in patients with the wild-type allele ($p = 0.049$). Furthermore, the white blood cell count was significantly lower in patients with the SNP 2269G>A allele than in patients with the wild-type allele. Among 15 patients with leucopenia ($<3 \times 10^9$/L), 7 carried the SNP 2269G>A allele (Ban et al., 2010). The odds ratio of carrying the SNP allele and having leucopenia was 3.33 (95% confidence interval: 1.03–10.57, $p = 0.036$) (Ban et al., 2010). Compared with the AZA dose of 2–3 mg/kg recommended in Western countries (Lichtenstein et al., 2006), lower doses of AZA (0.6–1.2 mg/kg) are used in Japan because of the relatively higher sensitivity to AZA (Hibi et al., 2003). It is therefore suggested that the *ABCC4* SNP 2269G>A is a new diagnostic marker indicative of thiopurine toxicity/sensitivity in Japanese patients with IBD. In this context, the SmartAmp primer set for rapid detection of the *ABCC4* SNP 2269G>A (Fig. 5.6) would provide a clinically practical tool for prediction of thiopurine toxicity/sensitivity in Japanese patients with IBD and other diseases, such as childhood acute lymphoblastic leukemia, acute myeloblastic leukemia,

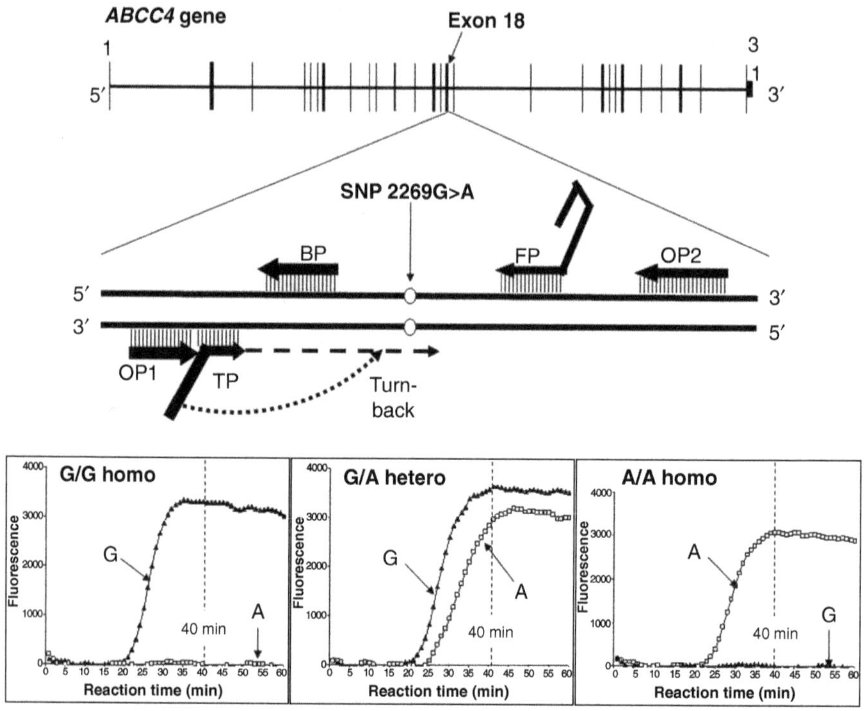

Figure 5.6 Detection of SNP 538G>A in the human *ABCC4* gene by the SmartAmp method. The human *ABCC4* gene is located on chromosome 13q32.1, where the SNP 2269G>A resides in exon 18. The three lower panels depict the time courses of the SmartAmp reaction with *ABCC4* allele-specific primers carrying WT (2269G) or SNP (2269A) alleles; namely, WT/WT homozygote, WT/SNP heterozygote, and SNP/SNP homozygote.

autoimmune disease, and rheumatoid arthritis. The advantage of SmartAmp-based genotyping is that there is no need for pretreatments, such as DNA extraction from a patient's blood or PCR-based DNA amplification steps.

5.9 CONCLUSION

One of the key issues affecting the advancement of personalized medicine is the speed by which we can develop cost-effective, rapid, and accurate genotyping tools for use in routine clinical practice. The isothermal amplification technologies have great potential to cover different applications for molecular genetic tests. As discussed above, the SmartAmp method comprises an isothermal DNA amplification reaction with background suppression technology based on mismatched binding proteins or competitive primers. The SmartAmp method is based on the concept that DNA amplification itself is the signal for detection of a specific target sequence, which can benefit other types of isothermal amplifications. This method makes use of *Aac*

DNA polymerase, which is highly resistant to cellular contaminants and hence works directly on blood samples

In the near future, the SmartAmp method is expected to translate into less complex and less expensive instrumentation. Although we have shown data generated on real-time PCR detection systems in this article, a method for end-point detection that uses a CCD camera-linked digital processor has recently been developed (Mitani et al., 2009). Furthermore, by using 96-well or 384-well plates and automated dispenser units, the throughput of SmartAmp reactions could be markedly increased. In addition, the potential also exists for utilizing very small-scale microfluidic designs for on-site detection of genotypes.

ACKNOWLEDGMENTS

The authors thank Prof. Akira Andoh (Shiga University of Medical Science, Otsu, Japan) and Prof. John D. Schuetz (St. Jude Children's Hospital, Memphis, TN, USA) for their fruitful discussions about genetic polymorphisms of the human *ABCC4* gene. In addition, thanks go to Drs. Alexander Lezhava and Wanping Aw (RIKEN Omics Science Center) for their generous support in the SmartAmp experiments. The authors' study was supported by a research project by Japan Science and Technology Agency (JST) named "Development of the world's fastest SNP detection system" (to TI) and a Research Grant for RIKEN Omics Science Center from the Ministry of Education, Culture, Sports, Science, and Technology (to YH).

REFERENCES

Adachi M, Sampath J, Lan LB, Sun D, Hargrove P, Flatley R, Tatum A, Edwards MZ, Wezeman M, Matherly L, Drake R, Schuetz J (2002) Expression of MRP4 confers resistance to ganciclovir and compromises bystander cell killing. *J Biol Chem*, 277, 38998–39004.

Anderson PL, Lamba J, Aquilante CL, Schuetz E, Fletcher CV (2006) Pharmacogenetic characteristics of indinavir, zidovudine, and lamivudine therapy in HIV-infected adults: a pilot study. *J Acquir Immune Defic Syndr*, 42, 441–449.

Andersson T, Flockhart DA, Goldstein DB, Huang SM, Kroetz DL, Milos PM, Ratain MJ, Thummel K (2005) Drug-metabolizing enzymes: evidence for clinical utility of pharmacogenomic tests. *Clin Pharmacol Ther*, 78, 559–581.

Ando M, Ando Y, Hasegawa Y, Sekido Y, Shimokata K, Horibe K (2001) Genetic polymorphisms of thiopurine S-methyltransferase and 6-mercaptopurine toxicity in Japanese children with acute lymphoblastic leukaemia. *Pharmacogenetics*, 11, 269–273.

Anstey A, Lennard L, Mayou SC, Kirby, JD (1992) Pancytopenia related to azathioprine—an enzyme deficiency caused by a common genetic polymorphism: a review. *J Royal Soc Med*, 85, 752–756.

Aomori T, Yamamoto K, Oguchi-Katayama A, Kawai Y, Ishidao T, Mitani Y, Kogo Y, Lezhava A, Fujita Y, Obayashi K, et al. (2009) Rapid SNP detection of the cytochrome P-450 (CYP) 2C9 and the vitamin K oxide reductase (VKORC1) gene for the warfarin dose adjustment by Smart-Amplification process version 2. *Clin Chem*, 55, 804–812.

Aw W, Lezhava A, Hayashizaki Y, Ishikawa, T (2011) A new trend in personalized medicine: rapid detection of SNPs in drug transporter genes by SmartAmp method. *Clin Pharmacol Ther*, 89, 617–620.

Azuma K, Lezhava A, Shimizu M, Kimura Y, Ishizu Y, Ishikawa T, Kamataki T, Hayashizaki Y, Yamazaki H (2011) Direct genotyping of cytochrome P450 2A6 whole gene deletion from human blood samples by SmartAmp method. *Clin Chim Acta*, 412, 1249–1251.

Ban H, Andoh A, Imaeda H, Kobori A, Bamba S, Tsujikawa T, Sasaki M, Saito Y, Fujiyama Y (2010) The multidrug-resistance protein 4 polymorphism is a new factor accounting for thiopurine sensitivity in Japanese patients with inflammatory bowel disease. *J Gastroenterol*, 45, 1014–1021.

Bassler HA, Flood SJ, Livak KJ, Marmaro J, Knorr R, Batt CA (1995) Use of a fluorogenic probe in a PCR-based assay for the detection of *Listeria monocytogenes*. *Appl Environ Microbiol*, 61, 3724–3728.

Black AJ, McLeod HL, Capell HA, Powrie RH, Matowe LK, Pritchard SC, Collie-Duguid ES, Reid DM (1998) Thiopurine methyltransferase genotype predicts therapy—limiting severe toxicity from azathioprine. *Ann Internal Med*, 129, 716–718.

Chen ZS, Lee K, Walther S, Raftogianis RB, Kuwano M, Zeng H, Kruh GD (2002) Analysis of methotrexate and folate transport by multidrug resistance protein 4 (ABCC4): MRP4 is a component of the methotrexate efflux system. *Cancer Res*, 62, 3144–3150.

Clunie GP, Lennard L (2004) Relevance of thiopurine methyltransferase status in rheumatology patients receiving azathioprine. *Rheumatology*, 43, 13–18.

Cooper GM, Johnson JA, Langaee TY, Feng H, Stanaway IB, Schwarz UI, Ritchie MD, Stein CM, Roden DM, Smith JD, Veenstra DL, Rettie AE, Rieder MJ (2008) A genome-wide scan for common genetic variants with a large influence on warfarin maintenance dose. *Blood*, 112, 1022–1027.

Daly AK, Donaldson PT, Bhatnagar P, Shen Y, Pe'er I, Floratos A, Daly MJ, Goldstein DB, John S, Nelson MR, et al., DILIGEN Study, International SAE Consortium (2009) HLA-B*5701 genotype is a major determinant of drug-induced liver injury due to flucloxacillin. *Nat Genet*, 41, 816–819.

Evans WE (2004) Pharmacogenetics of thiopurine S-methyltransferase and thiopurine therapy. *Ther Drug Monit*, 26, 185–191.

Evans WE, Johnson JA (2001) Pharmacogenomics: the inherited basis for interindividual differences in drug response. *Annu Rev Genomics Hum Genet*, 2, 9–39.

Fan JB, Chen X, Halushka MK, Berno A, Huang X, Ryder T, Lipshutz RJ, Lockhart DJ, Chakravarti A (2000) Parallel genotyping of human SNPs using generic high-density oligonucleotide tag arrays. *Genome Res*, 10, 853–860.

Ge D, Fellay J, Thompson AJ, Simon JS, Shianna KV, Urban TJ, Heinzen EL, Qiu P, Bertelsen AH, Muir AJ, Sulkowski M, McHutchison JG, Goldstein DB (2009) Genetic variation in IL28B predicts hepatitis C treatment-induced viral clearance. *Nature*, 461, 399–401.

Giacomini KM, Huang SM, Tweedie DJ, Benet LZ, Brouwer KL, Chu X, Dahlin A, Evers R, Fischer V, Hillgren KM, et al. (2010) Membrane transporters in drug development: report from the FDA critical path initiative-sponsored workshop. *Nat Rev Drug Discov*, 9, 215–236.

Griffin TJ, Smith LM, (2000) Single nucleotide polymorphism analysis by MALDI-TOF mass spectrometry. *Trends Biotechnol*, 18, 77–84.

Heim M, Meyer UA (1990) Genotyping of poor metabolisers of debrisoquine by allele-specific PCR amplification. *Lancet*, 336, 529–532.

Hibi T, Naganuma M, Kitahora T, Kinjyo F, Shimoyama T (2003) Low-dose azathioprine is effective and safe for maintenance of remission in patients with ulcerative colitis. *J Gastroenterol*, 38, 740–746.

Imaoka T, Kusuhara H, Adachi M, Schuetz JD, Takeuchi K, Sugiyama Y (2007) Functional involvement of multidrug resistance-associated protein 4 (MRP4/ABCC4) in the renal elimination of the antiviral drugs adefovir and tenofovir. *Mol Pharmacol*, 71, 619–627.

Ishikawa T, Sakurai A, Hirano H, Lezhava A, Sakurai M, Hayashizaki Y (2010) Emerging new technologies in pharmacogenomics: rapid SNP detection, molecular dynamic simulation, and QSAR analysis methods to validate clinically important genetic variants of human ABC Transporter ABCB1 (P-gp/MDR1). *Pharmacol Ther*, 126, 69–81.

Ishikawa T, Tsuji A, Inui K, Sai Y, Anzai N, Wada M, Endou H, Sumino, Y (2004) The genetic polymorphism of drug transporters: functional analysis approaches. *Pharmacogenomics*, 5, 67–99.

Kalow W, Meyer UA, Tyndale RF, editors. (2001) *Pharmacogenomics*. New York: Marcel Dekker.

Kamatani Y, Matsuda K, Okada Y, Kubo M, Hosono N, Daigo Y, Nakamura Y, Kamatani N (2010) Genome-wide association study of hematological and biochemical traits in a Japanese population. *Nat Genet*, 42, 210–215.

Kim RB (2002) Pharmacogenetics of CYP enzymes and drug transporters: remarkable recent advances. *Adv Drug Deliv Rev*, 54, 1241–1242.

Krishnamurthy P, Schwab M, Takenaka K, Nachagari D, Morgan J, Leslie M, Du W, Boyd K, Cheok M, Nakauchi H, et al. (2008) Transporter-mediated protection against thiopurine-induced hematopoietic toxicity. *Cancer Res*, 68, 4983–4989.

Kroetz DL, Yee SW, Giacomini GK (2010) The pharmacogenomics of membrane transporters project: research at the interface of genomics and transporter pharmacology. *Clin Pharmacol Ther*, 87, 109–116.

Krynetski EY, Schuetz JD, Galpin AJ, Pui C, Relling MV, Evans WE (1995) A single point mutation leading to loss of catalytic activity in human thiopurine S-methyltransferase. *Proc Natl Acad Sci USA*, 92, 949–953.

Lee K, Klein-Szanto AJ, Kruh GD (2000) Analysis of the MRP4 drug resistance profile in transfected NIH3T3 cells. *J Natl Cancer Inst*, 92, 1934–1940.

Lee SS, Kim WY, Jang YJ, Schin LG (2008) Duplex pyrosequencing of the TPMT*3C and TPMT*6 alleles in Korean and Vietnamese populations. *Clin Chem*, 398, 82–85.

Lezhava A, Ishidao T, Ishizu Y, Naito K, Hanami T, Katayama A, Kogo Y, Soma T, Ikeda S, Murakami K, et al. (2010) Exciton primer-mediated SNP detection in SmartAmp2 reactions. *Hum Mutat*, 31, 208–217.

Lichtenstein GR, Abreu MT, Cohen R, Tremaine W (2006) American Gastroenterological Association Institute technical review on corticosteroids, immunomodulators, and infiximab in inflammatory bowel disease. *Rev Gastroenterol Mex*, 71, 351–401.

Livak KJ (1999) Allelic discrimination using fluorogenic probes and the 5′ nuclease assay. *Genet Anal*, 14, 143–149.

Lymichev V, Mast AL, Hall JG, Prudent JR, Kaiser MW, Takova T, Kwiatkowski RW, Sander TJ, de Arruda M, Arco DA, Neri BP, Brow MA (1999) Polymorphism identification and quantitative detection of genomic DNA by invasive cleavage of oligonucleotide probes. *Nat Biotechnol*, 17, 292–296.

Mitani Y, Lezhava A, Kawai Y, Kikuchi T, Oguchi-Katayama A, Kogo Y, Itoh M, Miyagi T, Takakura H, Hoshi K, et al. (2007) Rapid SNP diagnostics using asymmetric isothermal amplification and a new mismatch-suppression technology. *Nat Methods*, 4, 257–262.

Mitani Y, Lezhava A, Sakurai A, Horikawa A, Nagakura M, Hayashizaki Y, Ishikawa T (2009) A rapid and cost-effective SNP detection method: application of SmartAmp2 to pharmacogenomics research. *Pharmacogenomics*, 10, 1187–1197.

Nakagawa H, Toyoda Y, Wakabayashi-Nakao K, Tamaki H, Osumi M, Ishikawa T (2011) Ubiquitin-mediated proteasomal degradation of ABC transporters: a new aspect of genetic polymorphisms and clinical impacts. *J Pharm Sci*, 100, 3602–3619.

Okada R, Ishizu Y, Endo R, Lezhava A, Ieiri I, Kusuhara H, Sugiyama Y, Hayashizaki Y (2010) Direct rapid genotyping of glutathione-S-transferase M1 and T1 from human blood specimens using the SmartAmp2 method. *Drug Metab Dispos*, 38, 1636–1639.

Ota I, Sakurai A, Toyoda Y, Morita S, Sasaki T, Chishima T, Yamakado M, Kawai Y, Ishidao T, Lezhava A, et al. (2010) Association between breast cancer risk and the wild-type allele of human ABC transporter ABCC11. *Anticancer Res*, 30, 5189–5194.

Ross P, Hall L, Smirnov I, Haff L (1998) High level multiplex genotyping by MALDI-TOF mass spectrometry. *Nat Biotechnol*, 16, 1347–1351.

Sadee W (2008) Drug therapy and personalized health care: pharmacogenomics in perspective. *Pharm Res*, 25, 2713–2719.

Saperstein DA, Nickerson JM (1991) Restriction fragment length polymorphism analysis using PCR coupled to restriction digests. *Biotechniques*, 10, 488–489.

Schaeffeler E, Zanger UM, Eichelbaum M, Asante-Poku S, Shin J-G, Schwab M (2008) Highly multiplexed genotyping of thiopurine S-methyltransferase variants using MALDI-TOF mass spectrometry: reliable genotyping in different ethnic groups. *Clin Chem*, 54, 1637–1647.

Schuetz JD, Connelly MC, Sun D, Paibir SG, Flynn PM, Srinivas RV, Kumar A, Fridland A (1999) MRP4: a previously unidentified factor in resistance to nucleoside-based antiviral drugs. *Nat Med*, 5, 1048–51.

Schuldiner AR, O'Connell JR, Bliden KP, Gandhi A, Ryan K, Horenstein RB, Damcott CM, Pakyz R, Tantry US, Gibson Q, et al. (2009) Association of cytochrome P450 2C19 genotype with the antiplatelet effect and clinical efficacy of clopidogrel therapy. *JAMA*, 302, 849–857.

Schütz E, von Ahsen N, Oellerich M (2000) Genotyping of eight thiopurine methyltransferase mutations: three-color multiplexing, "two-color/shared" anchor, and fluorescence-quenching hybridization probe assays based on thermodynamic nearest-neighbor probe design. *Clin Chem*, 46, 1728–1737.

See D, Kanazin V, Talbert H, Blake T (2000) Electrophoretic detection of single-nucleotide polymorphisms. *Biotechniques*, 28, 710–714.

Stolk JN, Boerbooms AM, de Abreu RA, de Koning DG, van Beusekom HJ, Muller WH, van de Putte LB (1998) Reduced thiopurine methyltransferase activity and development of side effects of azathioprine treatment in patients with rheumatoid arthritis. *Arthritis Rheum*, 41, 1858–1866.

Tai HL, Krynetski EY, Schuetz EG, Yanishevski Y, Evans WE (1997) Enhanced proteolysis of thiopurine S-methyltransferase (TPMT) encoded by mutant alleles in humans (TPMT*3A,TPMT*2): mechanism for genetic polymorphisms of TPMT activity. *Proc Natl Acad Sci USA*, 94, 6444–6449.

Takatsu N, Matsui T, Murakami Y, Ishihara H, Hisabe T, Nagahama T, Maki S, Beppu T, Takaki Y, Hirai F, Yao K (2009) Adverse reactions to azathioprine cannot be predicted by thiopurine S-methyltransferase genotype in Japanese patients with inflammatory bowel disease. *J Gastroenterol Hepatol*, 24, 1258–1264.

Tian Q, Zhang J, Tan TM, Chan E, Duan W, Chan SY, Boelsterli UA, Ho PC, Yang H, Bian JS, et al. (2005) Human multidrug resistance associated protein 4 confers resistance to camptothecins. *Pharm Res*, 22, 1837–1853.

Toyoda Y, Sakurai A, Mitani Y, Nakashima M, Yoshiura K, Nakagawa H, Sakai Y, Ota I, Lezhava A, Hayashizaki Y, Niikawa N, Ishikawa, T (2009) Earwax, osmidrosis, and breast cancer: why does one SNP (538G>A) in the human ABC transporter ABCC11 gene determine earwax type? *FASEB J*, 23, 2001–2013.

Wang L, Nguyen TV, McLaughlin RW, Sikkink LA, Ramirwez-Alvarado M, Weinshilboum R (2005) Human thiopurine S-methyltansferase pharmacogenetics: variant allozyme misfolding and aggresome formation. *Proc Natl Acad Sci U S A*, 102, 9394–9399.

Wang L, Sullivan W, Toft D, Weinshilboum R (2003) Thiopurine S-methyltransferase pharmacogenetics: chaperone protein association and allozyme degradation. *Pharmacogenetics*, 13, 555–564.

Wang L, Weinshilboum R (2006) Thiopurine S-methyltransferase pharmacogenetics: insights, challenges and future directions. *Oncogene*, 25, 1629–1638.

Watanabe J, Mitani Y, Kawai Y, Kikuchi T, Kogo Y, Oguchi-Katayama A, Kanamori H, Usui K, Itoh M, Cizdziel PE, et al. (2007) Use of a competitive probe in assay design for genotyping of the UGT1A1*28 microsatellite polymorphism by the smart amplification process. *Biotechniques*, 43, 479–484.

Weinshilboum RM, Sladek SL (1980) Mercaptopurine pharmacogenetics: monogenic inheritance of erythrocyte thiopurine methyltransferase activity. *Am J Hum Genet*, 32, 651–662.

Weiss ST, McLeod HL, Flockhart DA, Dolan ME, Benowitz NL, Johnson JA, Ratain MJ, Giacomini KM (2008) Creating and evaluating genetic tests predictive of drug response. *Nat Rev Drug Discov*, 7, 568–574.

Wielinga PR, Reid G, Challa EE, van der Heijden I, van Deemter L, de Haas M, Mol C, Kuil AJ, Groeneveld E, Schuetz JD, et al. (2002) Thiopurine metabolism and identification of the thiopurine metabolites transported by MRP4 and MRP5 overexpressed in human embryonic kidney cells. *Mol Pharmacol*, 62, 1321–1331.

Wilke RA, Lin DW, Roden DM, Watkins PB, Flockhart D, Zineh I, Giacomini KM, Krauss RM (2007) Identifying genetic risk factors for serious adverse drug reactions: current progress and challenges. *Nat Rev Drug Discov*, 6, 904–916.

Yates CR, Krynetski EY, Loennechen T, Fessing MY, Tai HL, Pui CH, Relling MV, Evans WE (1997) Molecular diagnosis of thiopurine S-methyltransferase deficiency: genetic basis for azathioprine and mercaptopurine intolerance. *Ann Internal Med*, 126, 608–614.

CHAPTER 6

OATP1A2, OAT1, AND OAT3

Rommel G. Tirona

6.1 INTRODUCTION

Organic anion transporting polypeptide (OATP) 1A2 and the organic anion transporters 1 (OAT1) and 3 (OAT3) are considered among the most important human uptake drug transporters in nonhepatic tissues. Since their discoveries about 15 years ago, there has been progress in delineating much of the molecular biology, biochemistry, and pharmacology of these membrane proteins (Vanwert et al., 2010; Burckhardt and Burckhardt, 2011; König, 2011). The clinical pharmacology of OATP1A2, OAT1, and OAT3 has similarly developed over the years, highlighted by the studies exposing their significant roles in the mechanisms of well-established and recently uncovered drug–drug interactions (Greenblatt, 2009; Hagenbuch, 2010). At the same time, much attention has been focused on the pharmacogenetics of these OATs as possible determinants of interindividual variability in drug responses (Franke et al., 2009). In this chapter, the clinical relevance of OATP1A2, OAT1, and OAT3 are reviewed, focusing on the roles in drug disposition and the emerging impact of genetic variation in these transporters in determining drug response.

6.2 OATP1A2 GENE EXPRESSION AND FUNCTION

The OATP1A2 cDNA was the first member of the OATP superfamily cloned in humans (Kullak-Ublick et al., 1995). Although it was initially cloned from human liver, OATP1A2 expression appears restricted to the apical membrane of cholangiocytes (Lee et al., 2005). OATP1A2 mRNA expression is highest in the brain where it has been localized to the capillary endothelial cells of the blood–brain barrier (Gao et al., 2000; Lee et al., 2005) as well as in the ciliary body epithelium of the eye (Gao et al., 2005). In the kidney, OATP1A2 is expressed in the abluminal membrane of the distal collecting ducts (Lee et al., 2005). Lastly, an apical localization of OATP1A2 has been described in enterocytes of the duodenum (Glaeser et al., 2007). The *SLCO1A2* gene, encoding OATP1A2, was localized to chromosome 12p12 and initial analysis of the regulatory region of the gene demonstrated cell-independent promoter activity, consistent with gene expression in multiple tissues (Kullak-Ublick

Pharmacogenomics of Human Drug Transporters: Clinical Impacts, First Edition.
Edited by Toshihisa Ishikawa, Richard B. Kim and Jörg König

TABLE 6.1 Transport Substrates of OATP1A2

Class	Substrate	Reference
Drugs	Fexofenadine	Cvetkovic et al. (1999)
	Imatinib	Hu et al. (2008)
	Rocuronium	van Montfoort et al. (2001)
	Methotrexate	Badagnani et al. (2006)
	Erythromycin	Franke et al. (2008)
	Statins	Ho et al. (2006); Knauer et al. (2010); Shirasaka et al. (2010)
	Levofloxacin	Maeda et al. (2007)
	HIV protease inhibitors	Su et al. (2004); Hartkoorn et al. (2010)
Xenobiotics	Bromosulfophthalein	Kullak-Ublick et al. (1995)
	N-methylquinine, N-methylquinidine	van Montfoort et al. (2001)
	Microcystin	Fischer et al. (2005)
	Deltorphin II	Gao et al. (2000); Kullak-Ublick et al. (2001))
	BQ-123	Kullak-Ublick et al. (2001)
	DPDPE	Gao et al. (2000); Fischer et al. (2005)
Bile acids	Taurocholate, cholate, glycocholate	Kullak-Ublick et al. (1995); Kullak-Ublick et al. (2001)
	Taurochenodeoxycholate, tauroursodeoxycholate	
Hormones	E_1S, E_2G	Kullak-Ublick et al. (1998); Kullak-Ublick et al. (2001)
	DHEAS	Kullak-Ublick et al. (1998)
	T_4, T_3, rT_3	Fujiwara et al. (2001)
Other endobiotics	PGE_2	Kullak-Ublick et al. (2001)

et al., 1997). OATP1A2 has been localized in breast carcinoma cells (Miki et al., 2006) where expression is thought to be driven by the actions of the nuclear hormone receptor, Pregnane X Receptor (Meyer zu Schwabedissen et al., 2008). Increased expression of OAT1A2 is found in bone cancer (Liedauer et al., 2009), whereas it is decreased in primary and metastatic liver cancer (Wlcek et al., 2011).

Studies in a number of cell systems and laboratories have defined the broad substrate specificity of OATP1A2 (Table 6.1). Of particular note, OATP1A2 is unique among OATPs in that it transports anions, cations, and neutral compounds. Whereas originally characterized as a transporter for endogenous compounds such as bile acids and hormones (steroid conjugates and thyroid hormones), it was soon found to accept xenobiotic substrates including drugs in common use such as fexofenadine, statins, HIV protease inhibitors, imatinib, and methotrexate (Table 6.1).

Given the diversity of transport substrates and tissue localization, OATP1A2 is considered to play important roles in drug disposition. For example, OATP1A2 is thought to be involved in the gastrointestinal absorption of drugs. Indeed, fexofenadine bioavailability is reduced when coadministered with OATP1A2 inhibitors, grapefruit juice, or its constituent, naringenin (Dresser et al., 2005; Bailey et al., 2007). A role for OATP1A2 in facilitating brain entry or renal tubular absorption of drugs has not been formally examined in clinical studies.

6.3 *SLCO1A2* (OATP1A2) GENETIC VARIATION

A number of genetic polymorphisms in the coding region of *SLCO1A2* have been described (Lee et al., 2005; Badagnani et al., 2006). Several nonsynonymous single nucleotide polymorphisms (SNPs) such as 38T>C, 516A>C, and 2003G>A are relatively common (5–11%) in ethnic populations (Table 6.2). When studied *in vitro*, the 516C (172D) variant of OATP1A2 has reduced transport activity for several substrates (E_1S, DPDPE, Deltorphin II, methotrexate) in comparison to the 516A (172E) wildtype form (Lee et al., 2005; Badagnani et al., 2006). The mechanism for altered transport of this variant appears to be reduction in cell surface transporter expression. Genetic variations in the *SLCO1A2* intron region have also been detailed (Kim et al., 2011). The common C>T variant (rs4148981) in the intronic evolutionarily conserved region (ECR) 35, termed *3, is associated with reduced hepatic mRNA expression of OATP1A2 (Kim et al., 2011).

TABLE 6.2 Selected Genetic Variants of *SLCO1A2* (OATP1A2)

				Minor allele frequency (%)			
Variation	dbSNP ID	Location	Effect	Caucasian	African	Asian	Global
−1105G>A	rs4148977	5′-Flanking					35
−1032G>A	rs4148978	5′-Flanking					34
−361G>A	rs3764043	5′-Flanking		0.05	0	24	6
−189 insA	rs3834939	5′-Flanking					36
ECR35*3	rs4148981	Intron		15	43	34	
38T>C	rs10841795	Exon	13I>T	11	2	0	6
382A>T	rs11568567	Exon	128N>Y	0	1.3	0	1
404A>T	rs45502302	Exon	135N>I	0	1.3	0	
502C>T	rs11568564	Exon	168R>C	0.6	0	0	
516A>C	rs11568563	Exon	172E>D	5	2	0	5
559G>A	NA	Exon	187A>T	0	0	0	
830C>A	NA	Exon	277T>N	0	0.6	0	
833delA	NA	Exon	278N>del	0	0.6	0	
841A>G	rs11568551	Exon	281I>V	0	0	0.8	
968T>C	rs11568579	Exon	323L>P	0.6	0	0	
1063>A>G	rs45628437	Exon	355I>V	1.9	0	0	
2003A>G	rs11568557	Exon	668T>S	0	4.4	0	3

Source: From Lee et al. (2005); Badagnani et al. (2006); Yamakawa et al. (2011), and dbSNP 09/2011.

TABLE 6.3 Clinical Studies Examining the Impact of *SLCO1A2* Genetic Variation on Drug Response

Drug/substrate	Variant/haplotype	Impact	Reference
Lopinavir	37T>C	No effect on C_{min} or $C_{2\text{-}6}$	Hartkoorn et al. (2010)
	516A>C	No effect on C_{min} or $C_{2\text{-}6}$	
Imatinib	38T>C	No difference in $C_{ss\ avg}$	Eechoute et al. (2011))
	516A>C	No difference in $C_{ss\ avg}$	
Imatinib	−1105G>A/ −1032G>A	Decreased clearance	Yamakawa et al. (2011)
	−361G>A	Increased clearance	
Methotrexate	ECR*1*3	Increased clearance	Kim et al. (2011)
Estrone sulfate	5 SNPs in *SLCO1A2*	No effect on breast cancer risk	Justenhoven et al. (2011)

ECR, evolutionarily conserved region.

6.4 CLINICAL RELEVANCE OF *SLCO1A2* GENETIC VARIATION

Few studies have now examined a potential role of *SLCO1A2* genetic variation in pharmacokinetics (Table 6.3). The first study examined the relationship between *SLCO1A2* nonsynonymous polymorphisms and plasma lopinavir levels in a cohort of 400 patients with routine therapeutic drug monitoring (Hartkoorn et al., 2010). Here, though there was a lack of association between *SLCO1A2* 38T>C and 516A>C variations with lopinavir concentrations, the *SLCO1B1* 521T>C polymorphism was linked to higher drug levels. Two other reports have focused on the OATP1A2 substrate and tyrosine kinase inhibitor, imatinib. In one study, common promoter polymorphisms in *SLCO1A2* were associated with increased (−1105GG/−1032GG) or decreased (−361GG) imatinib clearance in a cohort of Japanese patients ($n = 34$) with chronic myeloid leukemia (Yamakawa et al., 2011). In the second report, steady-state plasma imatinib concentrations were not different in Caucasian patients (cohort, $n = 94$), harboring either 38T>C or 516A>C variations in *SLCO1A2* (Eechoute et al., 2011). This was despite *in vitro* studies that demonstrated decreased imatinib transport by the 516A>C OATP1A2 variant.

6.5 OAT1 GENE EXPRESSION AND FUNCTION

OAT1 is expressed primarily in kidney (Reid et al., 1998; Cihlar et al., 1999; Hosoyamada et al., 1999; Lu et al., 1999; Race et al., 1999) but also detected at low level in brain. In the kidneys, OAT1 is localized to the basolateral membrane of proximal tubule cells (Hosoyamada et al., 1999), whereas in brain it is found in the cytoplasm of choroid plexus epithelia (Alebouyeh et al., 2003). The expression of OAT1 in kidney appears to be partly controlled by the transcription factors, hepatocyte nuclear factor 1α (Saji et al., 2008), and hepatocyte nuclear factor 4α (Ogasawara et al., 2007). In renal diseases, OAT1 mRNA expression in kidney was lower than in healthy individuals (Sakurai et al., 2004). Acting as an α-ketoglutarate exchanger, OAT1 in the kidney is involved in the renal tubular uptake of solutes from blood (Hosoyamada et al., 1999; Lu et al., 1999), with *p*-aminohippurate being the prototypical substrate. Since its discovery, a wide variety of drugs have been found to be transport substrates, most of which are efficiently cleared from the body by renal secretion (Table 6.4). Among drug classes, many members of the angiotensin converting enzyme (ACE) inhibitors, angiotensin receptor blockers (ARBs), diuretics, cephalosporin antibiotics, nucleoside antiviral drugs, and nonsteroidal anti-inflammatory drugs (NSAID) are substrates of OAT1. The broad substrate specificity of OAT1 is not only toward anionic drugs but also toward cationic compounds such as the histamine 2 (H2) receptor blockers, cimetidine and ranitidine.

6.6 *SLC22A6* (OAT1) GENETIC VARIATION

A number of genetic variations have been described in the coding regions of *SLC22A6* (Bleasby et al., 2005; Fujita et al., 2005) (Table 6.5). Interestingly, the allelic frequencies of observed nonsynonymous polymorphisms are low. Among those with highest prevalence are the 149G>A (50R>H) and 1361G>A (454R>Q) variations that occur at 3% and 2% allelic frequencies, respectively, in the African-American population (Fujita et al., 2005). When studied in the *Xenopus laevis* oocyte system, the 50R>H variant of OAT1 was not different to wildtype OAT1 with respect to transport kinetics for substrates *p*-aminohippurate and ochratoxin A (Fujita et al., 2005). However, for antiviral substrates adefovir, tenofovir, and cidofovir, the K_m for OAT1-mediated uptake was lower in the 50R>H variant than the wildtype form (Bleasby et al., 2005). No differences in transport activity were observed for the other most common 454R>Q variant of OAT1 (Fujita et al., 2005). Regulatory region polymorphisms in *SLC22A6* have also been reported. For example, the −126G>A polymorphism is rather common with allelic frequencies of 7%, 40%, and 27% in the Caucasian, African, and Asian populations (Kiser et al., 2008). Another polymorphism in the 5′ untranslated region of *SLC22A6*, −3655G>A, is found only in Asians with a frequency of 7% (Bhatnagar et al., 2006). Finally, a common, far upstream 5′ flanking region SNP of *SLC22A6* (rs10792367), in proximity to the *SLC22A8* gene, has recently been studied (see Sections 6.8 and 6.9) with global allelic frequencies of 45% (Han et al., 2011).

6.7 OAT3 GENE EXPRESSION AND FUNCTION

OAT3 was cloned from human kidney (Race et al., 1999), where mRNA expression of OAT3 is greater than OAT1 (Motohashi et al., 2002). Similar to OAT1, OAT3 is localized on the basolateral membrane of renal proximal tubular cells (Motohashi et al., 2002). Other tissues where OAT3 mRNA has been detected are adrenals (Asif et al., 2005), skeletal muscle (Takeda et al., 2004), and choroid plexus (Alebouyeh et al., 2003). A role for OAT3 in the blood–brain barrier efflux of solutes has been suggested (Kusch-Poddar et al., 2005) similar to what has been established

TABLE 6.4 Transport Substrates of OAT1 and OAT3

Transporter	Class	Substrates	Reference
OAT1	ACE inhibitors	Captoprilat, quinaprilat	Ueo et al. (2005)
	ARBs	Olmesartan	Yamada et al. (2007)
	Diuretics	Bumetanide, furosemide	Hasannejad et al. (2004)
	Cephalosporins	Cefdinir, ceftibuten, ceftizoxime, cephaloridine	Takeda et al. (2002a)
	Antivirals	Acyclovir, adefovir, cidofovir, ganciclovir, tenofovir, zalcitabine, zidovudine	Cihlar et al. (1999); Takeda et al. (2002b)
	Anticancer	Methotrexate	Ueo et al. (2005)
	H2 blockers	Cimetidine, ranitidine	Burckhardt et al. (2003)
	NSAIDs	Ibuprofen, indomethacin, ketoprofen, salicylate	Khamdang et al. (2002)
	Amino acids	Citrulline	Nakakariya et al. (2009)
	Hormones	E_1S	Ueo et al. (2005)
	Miscellaneous	Dimesna	Cutler et al. (2012)
		p-aminohippurate	Hosoyamada et al. (1999); Lu et al. (1999)
		Tetracycline	Babu et al. (2002)
OAT3	ACE inhibitors	Captoprilat, quinaprilat	Yuan et al. (2009)
	ARBs	Olmesartan	Yamada et al. (2007)
	Diuretics	Bumetanide, furosemide	Hasannejad et al. (2004)
	Statins	Pravastatin, rosuvastatin	Nishizato et al. (2003); Windass et al. (2007)
	Penicillins	Benzylpenicillin	Nozaki et al. (2007)
	Cephalosporins	Cefaclor, cefazolin, ceftibuten, ceftizoxime, cefaloridine	Takeda et al. (2002a); Ueo et al. (2005))
	Antivirals	Adefovir, cidofovir, tenofovir, valacyclovir, zidovudine	Takeda et al. (2002a); Uwai et al. (2007a)
	Anticancer	Methotrexate	Cha et al. (2001)
		Topotecan	Matsumoto et al. (2007)
		Erlotinib	Elmeliegy et al. (2011)
	H2 blockers	Cimetidine, Famotidine, Ranitidine	Cha et al. (2001)
	NSAIDs	Ibuprofen, Indomethacin, Ketoprofen, Salicylate	Cha et al. (2001); Asif et al. (2005)
	Bile acids	Cholate, Taurocholate	Cha et al. (2001)
	Second messengers	cAMP	Cha et al. (2001))
	Metabolites	Urate	Cha et al. (2001)
	Hormones	E_1S, E_2G	Cha et al. (2001)
		Cortisol	Kimura et al. (2002); Asif et al. (2005)
		DHEAS	Cha et al. (2001)
		PGE_2	Cha et al. (2001)
		$PGF_{2\alpha}$	Kimura et al. (2002)
	Miscellaneous	*p*-aminohippurate	Cha et al. (2001)
		Mycophenolate-glucuronide	Uwai et al. (2007b)
		Tetracycline	Babu et al. (2002)

TABLE 6.5 Selected Genetic Variations in *SLC22A6* (OAT1)

				Minor allele frequency (%)			
Variation	dbSNP ID	Location	Effect	Caucasian	African	Asian	Global
62758799C>G	rs10792367	Intergenic		50	43	58	45
−3655G>A	NA	5′-Flanking		0	0	7.2	
−126G>A	rs4149170	5′-Flanking		6.8	40	27	21
20T>C	NA	Exon	7L>P	0	0	5	
149G>A	rs11568626	Exon	50R>H	0	3.2	0	
180C>T	NA	Exon	Synonymous	0	0	0.8	
186G>A	rs11568625	Exon	Synonymous	0.6	0	0	
252G>T	rs11568628	Exon	Synonymous	7.5	2.5	1.7	3.3
311C>T	rs11568627	Exon	104P>L	0	0.6	1	
351A>G	rs11568629	Exon	Synonymous	7.5	2.5	1.7	3
677T>C	rs11568623	Exon	226I>T	0.6	0	0	
767C>T	rs11568624	Exon	256A>V	0	0.6	0	
877C>T	rs45607933	Exon	293R>W	0	1.9	0	
1233C>T	rs11568619	Exon	Synonymous	0	0	0.8	
1361G>A	rs11568634	Exon	454R>Q	0	0.6	0	
1470C>T	rs11568635	Exon	Synonymous	0	1.3	0	1.4
1574A>T	NA	Exon	525K>I	0	0	0	0.5

Source: From Bleasby et al. (2005); Xu et al. (2005); Fujita et al. (2005); Bhatnagar et al. (2006); Han et al. (2011), and dbSNP 09/2011.

in rodents (Kusuhara et al., 1999; Sugiyama et al., 2001). Mechanisms of OAT3 transcription have not been well studied and differential gene expression with OAT1 suggests that these homologous proteins are regulated by differing factors. In the kidney, OAT3 expression is controlled by hepatocyte nuclear factor 1α. OAT3, like OAT1, acts to facilitate drug uptake into cells with exchange with α-ketoglutarate (Bakhiya et al., 2003). Moreover, OAT3 shares many drug substrates with OAT1 (Table 6.4) including ACE inhibitors, ARBs, antibiotics, diuretics, and NSAIDs. However, some drugs appear to be substrates of OAT3 but not OAT1 such as benzylpenicillin. Drug–drug interactions at the level of OAT3 are of particular clinical relevance. For instance, the uricosuric agent, probenecid, has been used clinically to decrease the renal clearance of penicillins and cephalosporins. Although probenecid is an inhibitor of both OAT1 and OAT3, the relatively higher affinity of drug substrates with OAT3, together with its presumed greater expression in kidney, suggests that such drug–drug interaction largely involves OAT3. Moreover, deleterious drug–drug interactions such as methotrexate–NSAIDs are thought to occur through inhibition of OAT3-mediated renal secretion (Maeda et al., 2008).

6.8 *SLC22A8* (OAT3) GENETIC VARIATION

Table 6.6 summarizes genetic variations of *SLC22A8* (OAT3). Coding region variation in *SLC22A8* appears to be relatively uncommon with the exception of one synonymous polymorphism, 723A>T (rs2276299), which occurs at a global frequency

TABLE 6.6 Selected Genetic Variations in *SLC22A8* (OAT3)

				Minor allele frequency (%)			
Variation	dbSNP ID	Location	Effect	Caucasian	African	Asian	Global
153G>A	rs4149180	Exon	Synonymous	0.8	59	4	17
387C>A	rs11568479	Exon	129F>L	0	0	0	
445C>A	rs45566039	Exon	149R>S	0.6	0	0.8	
523A>G	NA	Exon	175I>V				
715C>T	rs11568496	Exon	239Q>X	0	0	0.8	
723A>T	rs2276299	Exon	Synonymous	18	24	0.4	17
779T>G	rs11568493	Exon	260I>R	0	0	0.8	
829C>T	rs11568492	Exon	277R>W	0	0.7	0	
842T>C	rs45438191	Exon	281V>A	0	6	0	3.5
913A>T	rs11568482	Exon	305I>F	0	0	3.5	3.1
929C>T	rs11568481	Exon	310A>V	0.6	0	0	
1166C>T	NA	Exon	389A>V			0.8	
1195G>T	rs11568497	Exon	399A>S	0	0	0	
1342G>A	rs11568486	Exon	448V>I	1.3	0.6	0	0.5

Source: From Nishizato et al. (2003); Xu et al. (2005); Erdman et al. (2006), and dbSNP 09/2011.

of 17% (Erdman et al., 2006). The 842T>C, 913A>T, and the 1342G>A nonsynonymous polymorphisms can be found at greater than 1% in various ethnic populations. Of these genetic polymorphisms, only the 913A>T (305T>F) transporter variant appeared to behave differently from the wildtype form, in a substrate-dependent fashion, when studied in a mammalian cell expression system (Erdman et al., 2006). Several rare nonsynonymous variants (445C>A, 715C>T, and 779T>G) had significant transport deficiency when compared to the wildtype transporter (Erdman et al., 2006). Genetic variation has also been examined in the *SLC22A8* regulatory regions (Bhatnagar et al., 2006; Ogasawara et al., 2008). Among five haplotypes comprising five promoter SNPs (−659G>del, −578C>G, −515A>C, −461T>C, −19C>A), none were found to be associated with OAT3 expression in kidney (Ogasawara et al., 2008).

6.9 CLINICAL RELEVANCE OF *SLC22A6* AND *SLC22A8* GENETIC VARIATION ON DRUG RESPONSE

Few studies have examined the role of *SLC22A6* and *SLC22A8* genetic variants on drug response (Table 6.7). The first study involved a study of pravastatin pharmacokinetics in 120 healthy Japanese subjects (Nishizato et al., 2003). The authors found that the 723T>A and the 1166C>T genetic variations in *SLC22A8* did not relate to the renal clearance of pravastatin. However, Nishizato and colleagues were the first to find an association between pravastatin pharmacokinetics and *SLCO1B1* genetic variants.

TABLE 6.7 Clinical Studies Examining the Impact of *SLC22A6* (OAT1) and *SLC22A8* (OAT3) Genetic Variation on Drug Response

Drug/substrate	Variant	Impact	Reference
Pravastatin	723T>A and 1166C>T in *SLC22A8*	No differences in renal clearance	Nishizato et al. (2003)
Torsemide	22 SNPs in *SLC22A6*, *SLC22A8*, and *SLC22A11*	Increased clearance with Haplotype H10 of *SLC22A11* (OAT4). Decreased clearance with homogzygous carriers of the rs11231809 TT genotype. No effect of *SLC22A6* and *SLC22A8* variants	Vormfelde et al. (2006)
Adefovir	1361G>A (454R>Q), rs11568634 in *SLC22A6*	No effect on overall renal clearance or secretory clearance	Fujita et al. (2005)
Tenofovir	rs11568626 (−401G>A) and rs4149170 (−126G>A) in *SLC22A6*	No association with intracellular tenofovir diphosphate levels	Kiser et al. (2008)
Tenofovir	rs4149170 (−126G>A) in *SLC22A6*	No association with tenofovir-induced renal damage	Rodriguez-Novoa et al. (2009)
Hydrochlorothiazide	rs10792367 (G>C; intergenic of *SLC22A6* and *SLC22A8*)	Decreased blood pressure response in C carriers	Han et al. (2011)

Two groups have studied the influence of OAT SNPs with diuretic response. In one study, the pharmacokinetics of the loop diuretic, torsemide, was characterized in 95 Caucasian Germans (Vormfelde et al., 2006). Whereas there was a 6.6-fold variation in renal torsemide clearance, no genetic variation in *SLC22A6* or *SLC22A8* was found to determine drug elimination. However, genetic polymorphisms in *SLC22A11* (OAT4), the OAT implicated in the renal reabsorption of solutes, were associated with torsemide pharmacokinetics. Here, subjects with the rs1123809 TT genotype were found to have approximately 30% decreased torsemide renal clearance compared to those with the AA genotype. Carriers with two copies of the *SLC22A11* H10 haplotype were found to have increased torsemide renal clearance in comparison to non-H10 carriers. In a recent study, Han and colleagues examined the role of OAT polymorphisms with hypertension and drug response to the diuretic, hydrochlorothiazide (Han et al., 2011). From two case–control studies totaling over 1000 hypertensives and control subjects, it was found that a common polymorphism in the *SLC22A6*–*SLC22A8* intergenic region (rs10792367; G>C) was not associated with hypertension risk. This was despite the studies with OAT3 knockout mice suggesting a role for OAT3 in blood pressure regulation (Vallon et al., 2008). In two clinical trials of hydrochlorothiazide in 574 and 274 hypertensive patients, it was found that carriers of the C allele of the intergenic polymorphism (rs10792367) had decreased systolic blood pressure response and this effect was gene-dose dependent (Han et al., 2011). The mechanisms involved in the genetic control of hydrochlorothiazide response remain to be determined.

Genetic polymorphisms in OATs have also been examined as a contributor to antiviral drug response. In a small family-based cohort of two individuals heterozygous for the *SLC22A6* 1361G>A (454R>Q) polymorphism and two noncarriers, the renal clearance of adefovir was not associated with this genetic variant (Fujita et al., 2005). In the second study, the levels of active tenofovir diphosphate in peripheral blood mononuclear cells were measured in a cohort ($n = 30$) of HIV-infected patients treated with tenofovir and lopinaivr/ritonavir (Kiser et al., 2008). It was determined that two promoter polymorphisms (−401G>A and −126G>A) in *SLC22A6* were not associated with tenofovir diphosphate concentrations. However, the 13-fold variation observed in tenofovir diphosphate levels was related to renal function and *ABCC4* (MRP4) 3463A>G genotype (Kiser et al., 2008). In the third study, investigators explored the role of drug-transporter genetics on renal dysfunction caused by tenofovir treatment (Rodriguez-Novoa et al., 2009). In a cohort of HIV-infected, tenofovir-treated patients with kidney tubular dysfunction ($n = 19$) or normal tubular function ($n = 96$), the investigators found a lack of association between the *SLC22A6* promoter polymorphism (−126G>A) and drug-related renal damage. However, there was a link between *ABCC2* (MRP2) −24C genotype with tenofovir renal tubular dysfunction.

6.10 CONCLUSION

The roles of OATP1A2, OAT1, and OAT3 in drug disposition and response have been developed since their discoveries over 15 years ago. For OATP1A2, there are some findings that support a role in intestinal drug absorption. However, less is known regarding the relevance of OATP1A2 in other pharmacokinetic processes such as tissue distribution or elimination. Difficulty in addressing these issues lies in a lack of understanding of the relative contribution of transporters and drug-metabolizing enzymes in an OATP1A2 substrate drug's disposition. Clearly, better OATP1A2-specific drug probes are required to better clarify the influence of this transporter to drug response. What is certain at the moment is that *SLCO1A2* genetic polymorphisms have so far not consistently been shown to influence substrate pharmacokinetics. As for OAT1 and OAT3, there is much compelling evidence for their key roles in the renal tubular secretion of drugs largely based on significant drug–drug interactions at the transporter level affecting renal clearance. Interestingly, there is a current lack of evidence to show that *SLC22A6* or *SLC22A8* genetic polymorphisms affect the renal elimination of substrate drugs. The relatively low frequency of genetic variation within the *SLC22A6* and *SLC22A8* coding regions affecting transport activity suggests evolutionary pressures maintaining function. Furthermore, substrate overlap and tissue co-localization among different OAT transporters may contribute to the blunting of overall effects of gene-specific polymorphisms on drug disposition. Taken together, there remains much to be explored regarding OATP1A2, OAT1, and OAT3, and their associated genetic polymorphisms as contributors to drug response.

REFERENCES

Alebouyeh M, Takeda M, Onozato ML, Tojo A, Noshiro R, Hasannejad H, Inatomi J, Narikawa S, Huang XL, Khamdang S, Anzai N, Endou H (2003) Expression of human organic anion transporters in the choroid plexus and their interactions with neurotransmitter metabolites. *J Pharmacol Sci*, 93, 430–436.

Asif AR, Steffgen J, Metten M, Grunewald RW, Muller GA, Bahn A, Burckhardt G, Hagos Y (2005) Presence of organic anion transporters 3 (OAT3) and 4 (OAT4) in human adrenocortical cells. *Pflugers Arch*, 450, 88–95.

Babu E, Takeda M, Narikawa S, Kobayashi Y, Yamamoto T, Cha SH, Sekine T, Sakthisekaran D, Endou H (2002) Human organic anion transporters mediate the transport of tetracycline. *Jpn J Pharmacol*, 88, 69–76.

Badagnani I, Castro RA, Taylor TR, Brett CM, Huang CC, Stryke D, Kawamoto M, Johns SJ, Ferrin TE, Carlson EJ, et al. (2006) Interaction of methotrexate with organic-anion transporting polypeptide 1A2 and its genetic variants. *J Pharmacol Exp Ther*, 318, 521–529.

Bailey DG, Dresser GK, Leake BF, Kim RB (2007) Naringin is a major and selective clinical inhibitor of organic anion-transporting polypeptide 1A2 (OATP1A2) in grapefruit juice. *Clin Pharmacol Ther*, 81, 495–502.

Bakhiya A, Bahn A, Burckhardt G, Wolff N (2003) Human organic anion transporter 3 (hOAT3) can operate as an exchanger and mediate secretory urate flux. *Cell Physiol Biochem*, 13, 249–256.

Bhatnagar V, Xu G, Hamilton BA, Truong DM, Eraly SA, Wu W, Nigam SK (2006) Analyses of 5′ regulatory region polymorphisms in human SLC22A6 (OAT1) and SLC22A8 (OAT3). *J Hum Genet*, 51, 575–580.

Bleasby K, Hall LA, Perry JL, Mohrenweiser HW, Pritchard JB (2005) Functional consequences of single nucleotide polymorphisms in the human organic anion transporter hOAT1 (SLC22A6). *J Pharmacol Exp Ther*, 314, 923–931.

Burckhardt BC, Brai S, Wallis S, Krick W, Wolff NA, Burckhardt G (2003) Transport of cimetidine by flounder and human renal organic anion transporter 1. *Am J Physiol Renal Physiol*, 284, F503–509.

Burckhardt G, Burckhardt BC (2011) In vitro and in vivo evidence of the importance of organic anion transporters (OATs) in drug therapy. In: *Handbook of Experimental Pharmacology*, Vol. 201. Heidelberg, Germany: Springer. pp 29–104.

Cha SH, Sekine T, Fukushima JI, Kanai Y, Kobayashi Y, Goya T, Endou H (2001) Identification and characterization of human organic anion transporter 3 expressing predominantly in the kidney. *Mol Pharmacol*, 59, 1277–1286.

Cihlar T, Lin DC, Pritchard JB, Fuller MD, Mendel DB, Sweet DH (1999) The antiviral nucleotide analogs cidofovir and adefovir are novel substrates for human and rat renal organic anion transporter 1. *Mol Pharmacol*, 56, 570–580.

Cutler MJ, Urquhart BL, Velenosi TJ, Meyer Zu Schwabedissen HE, Dresser GK, Leake BF, Tirona RG, Kim RB, Freeman DJ (2012) In vitro and in vivo assessment of renal drug transporters in the disposition of mesna and dimesna. *J Clin Pharmacol*, 52, 530–542.

Cvetkovic M, Leake B, Fromm MF, Wilkinson GR, Kim RB (1999) OATP and P-glycoprotein transporters mediate the cellular uptake and excretion of fexofenadine. *Drug Metab Dispos*, 27, 866–871.

Dresser GK, Kim RB, Bailey DG (2005) Effect of grapefruit juice volume on the reduction of fexofenadine bioavailability: possible role of organic anion transporting polypeptides. *Clin Pharmacol Ther*, 77, 170–177.

Eechoute K, Franke RM, Loos WJ, Scherkenbach LA, Boere I, Verweij J, Gurney H, Kim RB, Tirona RG, Mathijssen RH, Sparreboom A (2011) Environmental and genetic factors affecting transport of imatinib by OATP1A2. *Clin Pharmacol Ther*, 89, 816–820.

Elmeliegy MA, Carcaboso AM, Tagen M, Bai F, Stewart CF (2011) Role of ATP-binding cassette and solute carrier transporters in erlotinib CNS penetration and intracellular accumulation. *Clin Cancer Res*, 17, 89–99.

Erdman AR, Mangravite LM, Urban TJ, Lagpacan LL, Castro RA, De La Cruz M, Chan W, Huang CC, Johns SJ, Kawamoto M, et al. (2006) The human organic anion transporter 3 (OAT3; SLC22A8): genetic variation and functional genomics. *Am J Physiol Renal Physiol*, 290, F905–F912.

Fischer WJ, Altheimer S, Cattori V, Meier PJ, Dietrich DR, Hagenbuch B (2005) Organic anion transporting polypeptides expressed in liver and brain mediate uptake of microcystin. *Toxicol Appl Pharmacol*, 203, 257–263.

Franke RM, Baker SD, Mathijssen RH, Schuetz EG, Sparreboom A (2008) Influence of solute carriers on the pharmacokinetics of CYP3A4 probes. *Clin Pharmacol Ther*, 84, 704–709.

Franke RM, Scherkenbach LA, Sparreboom A (2009) Pharmacogenetics of the organic anion transporting polypeptide 1A2. *Pharmacogenomics*, 10, 339–344.

Fujita T, Brown C, Carlson EJ, Taylor T, De La Cruz M, Johns SJ, Stryke D, Kawamoto M, Fujita K, Castro R, et al. (2005) Functional analysis of polymorphisms in the organic anion transporter, SLC22A6 (OAT1). *Pharmacogenet Genomics*, 15, 201–209.

Fujiwara K, Adachi H, Nishio T, Unno M, Tokui T, Okabe M, Onogawa T, Suzuki T, Asano N, Tanemoto M, et al. (2001) Identification of thyroid hormone transporters in humans: different molecules are involved in a tissue-specific manner. *Endocrinology*, 142, 2005–2012.

Gao B, Hagenbuch B, Kullak-Ublick GA, Benke D, Aguzzi A, Meier PJ (2000) Organic anion-transporting polypeptides mediate transport of opioid peptides across blood-brain barrier. *J Pharmacol Exp Ther*,294, 73–79.

Gao B, Huber RD, Wenzel A, Vavricka SR, Ismair MG, Reme C, Meier PJ (2005) Localization of organic anion transporting polypeptides in the rat and human ciliary body epithelium. *Exp Eye Res*, 80, 61–72.

Glaeser H, Bailey DG, Dresser GK, Gregor JC, Schwarz UI, Mcgrath JS, Jolicoeur E, Lee W, Leake BF, Tirona RG, Kim RB (2007) Intestinal drug transporter expression and the impact of grapefruit juice in humans. *Clin Pharmacol Ther*, 81, 362–370.

Greenblatt DJ (2009) Analysis of drug interactions involving fruit beverages and organic anion-transporting polypeptides. *J Clin Pharmacol*, 49, 1403–1407.

Hagenbuch B (2010) Drug uptake systems in liver and kidney: a historic perspective. *Clin Pharmacol Ther*, 87, 39–47.

Han YF, Fan XH, Wang XJ, Sun K, Xue H, Li WJ, Wang YB, Chen JZ, Zhen YS, Zhang WL, et al. (2011) Association of intergenic polymorphism of organic anion transporter 1 and 3 genes with hypertension and blood pressure response to hydrochlorothiazide. *Am J Hypertens*, 24, 340–346.

Hartkoorn RC, Kwan WS, Shallcross V, Chaikan A, Liptrott N, Egan D, Sora ES, James CE, Gibbons S, Bray PG, et al. (2010) HIV protease inhibitors are substrates for OATP1A2, OATP1B1 and OATP1B3 and lopinavir plasma concentrations are influenced by SLCO1B1 polymorphisms. *Pharmacogenet Genomics*, 20, 112–120.

Hasannejad H, Takeda M, Taki K, Shin HJ, Babu E, Jutabha P, Khamdang S, Aleboyeh M, Onozato ML, Tojo A, et al. (2004) Interactions of human organic anion transporters with diuretics. *J Pharmacol Exp Ther*, 308, 1021–1029.

Ho RH, Tirona RG, Leake BF, Glaeser H, Lee W, Lemke CJ, Wang Y, Kim RB (2006) Drug and bile acid transporters in rosuvastatin hepatic uptake: function, expression, and pharmacogenetics. *Gastroenterology*, 130, 1793–1806.

Hosoyamada M, Sekine T, Kanai Y, Endou H (1999) Molecular cloning and functional expression of a multispecific organic anion transporter from human kidney. *Am J Physiol*, 276, F122–F128.

Hu S, Franke RM, Filipski KK, Hu C, Orwick SJ, De Bruijn EA, Burger H, Baker SD, Sparreboom A (2008) Interaction of imatinib with human organic ion carriers. *Clin Cancer Res*, 14, 3141–3148.

Justenhoven C, Schaeffeler E, Winter S, Baisch C, Hamann U, Harth V, Rabstein S, Spickenheuer A, Pesch B, Bruning T, Ko YD, Schwab M, Brauch H (2011) Polymorphisms of the nuclear receptor pregnane X receptor and organic anion transporter polypeptides 1A2, 1B1, 1B3, and 2B1 are not associated with breast cancer risk. *Breast Cancer Res Treat*, 125, 563–569.

Khamdang S, Takeda M, Noshiro R, Narikawa S, Enomoto A, Anzai N, Piyachaturawat P, Endou H (2002) Interactions of human organic anion transporters and human organic cation transporters with nonsteroidal anti-inflammatory drugs. *J Pharmacol Exp Ther*, 303, 534–539.

Kim MJ, Skewes-Cox P, Fukushima H, Hesselson S, Yee SW, Ramsey LB, Nguyen L, Eshragh JL, Castro RA, Wen CC, et al. (2011) Functional characterization of liver enhancers that regulate drug-associated transporters. *Clin Pharmacol Ther*, 89, 571–578.

Kimura H, Takeda M, Narikawa S, Enomoto A, Ichida K, Endou H (2002) Human organic anion transporters and human organic cation transporters mediate renal transport of prostaglandins. *J Pharmacol Exp Ther*, 301, 293–298.

Kiser JJ, Aquilante CL, Anderson PL, King TM, Carten ML, Fletcher CV (2008) Clinical and genetic determinants of intracellular tenofovir diphosphate concentrations in HIV-infected patients. *J Acquir Immune Defic Syndr*, 47, 298–303.

Knauer MJ, Urquhart BL, Meyer Zu Schwabedissen HE, Schwarz UI, Lemke CJ, Leake BF, Kim RB, Tirona RG (2010) Human skeletal muscle drug transporters determine local exposure and toxicity of statins. *Circ Res*, 106, 297–306.

König J (2011) Uptake transporters of the human OATP family: molecular characteristics, substrates, their role in drug-drug interactions, and functional consequences of polymorphisms. In: *Handbook of Experimental Pharmacology*, Vol. 201. Heidelberg, Germany: Springer. pp 1–28.

Kullak-Ublick GA, Beuers U, Fahney C, Hagenbuch B, Meier PJ, Paumgartner G (1997) Identification and functional characterization of the promoter region of the human organic anion transporting polypeptide gene. *Hepatology*, 26, 991–997.

Kullak-Ublick GA, Fisch T, Oswald M, Hagenbuch B, Meier PJ, Beuers U, Paumgartner G (1998) Dehydroepiandrosterone sulfate (DHEAS): identification of a carrier protein in human liver and brain. *FEBS Lett*, 424, 173–176.

Kullak-Ublick GA, Hagenbuch B, Stieger B, Schteingart CD, Hofmann AF, Wolkoff AW, Meier PJ (1995) Molecular and functional characterization of an organic anion transporting polypeptide cloned from human liver. *Gastroenterology*, 109, 1274–1282.

Kullak-Ublick GA, Ismair MG, Stieger B, Landmann L, Huber R, Pizzagalli F, Fattinger K, Meier PJ, Hagenbuch B (2001) Organic anion-transporting polypeptide B (OATP-B) and its functional comparison with three other OATPs of human liver. *Gastroenterology*, 120, 525–533.

Kusch-Poddar M, Drewe J, Fux I, Gutmann H (2005) Evaluation of the immortalized human brain capillary endothelial cell line BB19 as a human cell culture model for the blood-brain barrier. *Brain Res*, 1064, 21–31.

Kusuhara H, Sekine T, Utsunomiya-Tate N, Tsuda M, Kojima R, Cha SH, Sugiyama Y, Kanai Y, Endou H (1999) Molecular cloning and characterization of a new multispecific organic anion transporter from rat brain. *J Biol Chem*, 274, 13675–13680.

Lee W, Glaeser H, Smith LH, Roberts RL, Moeckel GW, Gervasini G, Leake BF, Kim RB (2005) Polymorphisms in human organic anion-transporting polypeptide 1A2 (OATP1A2): implications for altered drug disposition and central nervous system drug entry. *J Biol Chem*, 280, 9610–9617.

Liedauer R, Svoboda M, Wlcek K, Arrich F, Ja W, Toma C, Thalhammer T (2009) Different expression patterns of organic anion transporting polypeptides in osteosarcomas, bone metastases and aneurysmal bone cysts. *Oncol Rep*, 22, 1485–1492.

Lu R, Chan BS, Schuster VL (1999) Cloning of the human kidney PAH transporter: narrow substrate specificity and regulation by protein kinase C. *Am J Physiol*, 276, F295–303.

Maeda A, Tsuruoka S, Kanai Y, Endou H, Saito K, Miyamoto E, Fujimura A (2008) Evaluation of the interaction between nonsteroidal anti-inflammatory drugs and methotrexate using human organic anion transporter 3-transfected cells. *Eur J Pharmacol*, 596, 166–172.

Maeda T, Takahashi K, Ohtsu N, Oguma T, Ohnishi T, Atsumi R, Tamai I (2007) Identification of influx transporter for the quinolone antibacterial agent levofloxacin. *Mol Pharm*, 4, 85–94.

Matsumoto S, Yoshida K, Ishiguro N, Maeda T, Tamai I (2007) Involvement of rat and human organic anion transporter 3 in the renal tubular secretion of topotecan [(S)-9-dimethylaminomethyl-10-hydroxy-camptothecin hydrochloride]. *J Pharmacol Exp Ther*, 322, 1246–1252.

Meyer Zu Schwabedissen HE, Tirona RG, Yip CS, Ho RH, Kim RB (2008) Interplay between the nuclear receptor pregnane X receptor and the uptake transporter organic anion transporter polypeptide 1A2 selectively enhances estrogen effects in breast cancer. *Cancer Res*, 68, 9338–9347.

Miki Y, Suzuki T, Kitada K, Yabuki N, Shibuya R, Moriya T, Ishida T, Ohuchi N, Blumberg B, Sasano H (2006) Expression of the steroid and xenobiotic receptor and its possible target gene, organic anion transporting polypeptide-A, in human breast carcinoma. *Cancer Res*, 66, 535–542.

Motohashi H, Sakurai Y, Saito H, Masuda S, Urakami Y, Goto M, Fukatsu A, Ogawa O, Inui K (2002) Gene expression levels and immunolocalization of organic ion transporters in the human kidney. *J Am Soc Nephrol*, 13, 866–874.

Nakakariya M, Shima Y, Shirasaka Y, Mitsuoka K, Nakanishi T, Tamai I (2009) Organic anion transporter OAT1 is involved in renal handling of citrulline. *Am J Physiol Renal Physiol*, 297, F71–F79.

Nishizato Y, Ieiri I, Suzuki H, Kimura M, Kawabata K, Hirota T, Takane H, Irie S, Kusuhara H, Urasaki Y, et al. (2003) Polymorphisms of OATP-C (SLC21A6) and OAT3 (SLC22A8) genes: consequences for pravastatin pharmacokinetics. *Clin Pharmacol Ther*, 73, 554–565.

Nozaki Y, Kusuhara H, Kondo T, Hasegawa M, Shiroyanagi Y, Nakazawa H, Okano T, Sugiyama Y (2007) Characterization of the uptake of organic anion transporter (OAT) 1 and OAT3 substrates by human kidney slices. *J Pharmacol Exp Ther*, 321, 362–369.

Ogasawara K, Terada T, Asaka J, Katsura T, Inui K (2007) Hepatocyte nuclear factor-4{alpha} regulates the human organic anion transporter 1 gene in the kidney. *Am J Physiol Renal Physiol*, 292, F1819–F1826.

Ogasawara K, Terada T, Motohashi H, Asaka J, Aoki M, Katsura T, Kamba T, Ogawa O, Inui K (2008) Analysis of regulatory polymorphisms in organic ion transporter genes (SLC22A) in the kidney. *J Hum Genet*, 53, 607–614.

Race JE, Grassl SM, Williams WJ, Holtzman EJ (1999) Molecular cloning and characterization of two novel human renal organic anion transporters (hOAT1 and hOAT3). *Biochem Biophys Res Commun*, 255, 508–514.

Reid G, Wolff NA, Dautzenberg FM, Burckhardt G (1998) Cloning of a human renal p-aminohippurate transporter, hROAT1. *Kidney Blood Press Res*, 21, 233–237.

Rodriguez-Novoa S, Labarga P, Soriano V, Egan D, Albalater M, Morello J, Cuenca L, Gonzalez-Pardo G, Khoo S, Back D, Owen A (2009) Predictors of kidney tubular dysfunction in HIV-infected patients treated with tenofovir: a pharmacogenetic study. *Clin Infect Dis*, 48, e108–e116.

Saji T, Kikuchi R, Kusuhara H, Kim I, Gonzalez FJ, Sugiyama Y (2008) Transcriptional regulation of human and mouse organic anion transporter 1 by hepatocyte nuclear factor 1 alpha/beta. *J Pharmacol Exp Ther*, 324, 784–790.

Sakurai Y, Motohashi H, Ueo H, Masuda S, Saito H, Okuda M, Mori N, Matsuura M, Doi T, Fukatsu A, Ogawa O, Inui K (2004) Expression levels of renal organic anion transporters (OATs) and their correlation with anionic drug excretion in patients with renal diseases. *Pharm Res*, 21, 61–67.

Shirasaka Y, Suzuki K, Nakanishi T, Tamai I (2010) Intestinal absorption of HMG-CoA reductase inhibitor pravastatin mediated by organic anion transporting polypeptide. *Pharm Res*, 27, 2141–2149.

Su Y, Zhang X, Sinko PJ (2004) Human organic anion-transporting polypeptide OATP-A (SLC21A3) acts in concert with P-glycoprotein and multidrug resistance protein 2 in the vectorial transport of Saquinavir in Hep G2 cells. *Mol Pharm*, 1, 49–56.

Sugiyama D, Kusuhara H, Shitara Y, Abe T, Meier PJ, Sekine T, Endou H, Suzuki H, Sugiyama Y (2001) Characterization of the efflux transport of 17beta-estradiol-D-17beta-glucuronide from the brain across the blood-brain barrier. *J Pharmacol Exp Ther*, 298, 316–322.

Takeda M, Babu E, Narikawa S, Endou H (2002a) Interaction of human organic anion transporters with various cephalosporin antibiotics. *Eur J Pharmacol*, 438, 137–142.

Takeda M, Khamdang S, Narikawa S, Kimura H, Kobayashi Y, Yamamoto T, Cha SH, Sekine T, Endou H (2002b) Human organic anion transporters and human organic cation transporters mediate renal antiviral transport. *J Pharmacol Exp Ther*, 300, 918–924.

Takeda M, Noshiro R, Onozato ML, Tojo A, Hasannejad H, Huang XL, Narikawa S, Endou H (2004) Evidence for a role of human organic anion transporters in the muscular side effects of HMG-CoA reductase inhibitors. *Eur J Pharmacol*, 483, 133–138.

Ueo H, Motohashi H, Katsura T, Inui K (2005) Human organic anion transporter hOAT3 is a potent transporter of cephalosporin antibiotics, in comparison with hOAT1. *Biochem Pharmacol*, 70, 1104–1113.

Uwai Y, Ida H, Tsuji Y, Katsura T, Inui K (2007a) Renal transport of adefovir, cidofovir, and tenofovir by SLC22A family members (hOAT1, hOAT3, and hOCT2). *Pharm Res*, 24, 811–815.

Uwai Y, Motohashi H, Tsuji Y, Ueo H, Katsura T, Inui K (2007b) Interaction and transport characteristics of mycophenolic acid and its glucuronide via human organic anion transporters hOAT1 and hOAT3. *Biochem Pharmacol*, 74, 161–168.

Vallon V, Eraly SA, Wikoff WR, Rieg T, Kaler G, Truong DM, Ahn SY, Mahapatra NR, Mahata SK, Gangoiti JA, et al. (2008) Organic anion transporter 3 contributes to the regulation of blood pressure. *J Am Soc Nephrol*, 19, 1732–1740.

Van Montfoort JE, Muller M, Groothuis GM, Meijer DK, Koepsell H, Meier PJ (2001) Comparison of "type I" and "type II" organic cation transport by organic cation transporters and organic anion-transporting polypeptides. *J Pharmacol Exp Ther*, 298, 110–115.

Vanwert AL, Gionfriddo MR, Sweet DH (2010) Organic anion transporters: discovery, pharmacology, regulation and roles in pathophysiology. *Biopharm Drug Dispos*, 31, 1–71.

Vormfelde SV, Schirmer M, Hagos Y, Toliat MR, Engelhardt S, Meineke I, Burckhardt G, Nurnberg P, Brockmoller J (2006) Torsemide renal clearance and genetic variation in luminal and basolateral organic anion transporters. *Br J Clin Pharmacol*, 62, 323–335.

Windass AS, Lowes S, Wang Y, Brown CD (2007) The contribution of organic anion transporters OAT1 and OAT3 to the renal uptake of rosuvastatin. *J Pharmacol Exp Ther*, 322, 1221–1227.

Wlcek K, Svoboda M, Riha J, Zakaria S, Olszewski U, Dvorak Z, Sellner F, Ellinger I, Jager W, Thalhammer T (2011) The analysis of organic anion transporting polypeptide (OATP) mRNA and protein patterns in primary and metastatic liver cancer. *Cancer Biol Ther*, 11, 801–811.

Xu G, Bhatnagar V, Wen G, Hamilton BA, Eraly SA, Nigam SK (2005) Analyses of coding region polymorphisms in apical and basolateral human organic anion transporter (OAT) genes [OAT1 (NKT), OAT2, OAT3, OAT4, URAT (RST)]. *Kidney Int*, 68, 1491–1499.

Yamada A, Maeda K, Kamiyama E, Sugiyama D, Kondo T, Shiroyanagi Y, Nakazawa H, Okano T, Adachi M, Schuetz JD, et al. (2007) Multiple human isoforms of drug transporters contribute to the hepatic and renal transport of olmesartan, a selective antagonist of the angiotensin II AT1-receptor. *Drug Metab Dispos*, 35, 2166–2176.

Yamakawa Y, Hamada A, Shuto T, Yuki M, Uchida T, Kai H, Kawaguchi T, Saito H (2011) Pharmacokinetic impact of SLCO1A2 polymorphisms on imatinib disposition in patients with chronic myeloid leukemia. *Clin Pharmacol Ther*, 90, 157–163.

Yuan H, Feng B, Yu Y, Chupka J, Zheng JY, Heath TG, Bond BR (2009) Renal organic anion transporter-mediated drug-drug interaction between gemcabene and quinapril. *J Pharmacol Exp Ther*, 330, 191–197.

CHAPTER 7

OATP1B1, OATP1B3, AND OATP2B1

Jörg König
Martin F. Fromm

7.1 INTRODUCTION

7.1.1 Characteristics of the Human OATP Family

The liver plays an important role in the metabolism and elimination of endogenous substances and drugs (Petzinger, 1994; Okudaira and Sugiyama, 1996). Whereas in the past decades most studies have focused on the role of metabolizing enzymes as key determinants of drug disposition, the importance of membrane transporters for the pharmacokinetics of drugs has been increasingly recognized in the last years. Drug-metabolizing reactions can be grouped into phase I reactions, mediated by cytochrome P450 enzymes and phase II reactions, mediated by conjugating enzymes (Batt et al., 1994). Since metabolic enzymes are located intracellularly, substances have to be taken up from blood across the basolateral membrane into hepatocytes before subsequent metabolism. Prior to the identification of basolaterally localized uptake transporters, it was generally accepted that drugs enter the hepatocytes mainly by diffusion (Schwenk, 1987). However, in recent years, studies investigating the molecular mechanisms of the hepatobiliary excretion of xenobiotics in more detail revealed that transport proteins located in plasma membranes are responsible for the uptake and the export of substance in and out of cells (Suzuki and Sugiyama, 2000; Kusuhara and Sugiyama, 2002).

Transport proteins can be divided into two major groups: uptake transporters mediating the uptake of substances from the environment into cells and export pumps actively secreting drugs or drug metabolites out of cells. These export pumps belong to the superfamily of ATP-binding cassette (ABC) transporters mediating the ATP-driven export of substances out of cells mostly against a concentration gradient (Allikmets et al., 1996; Keppler et al., 2000; Kruh et al., 2001). Uptake transporters mostly belong to the solute carriers (SLC) superfamily of transport proteins (Hediger et al., 2004). Today, this superfamily is comprised of 51 families (SLC1–SLC51) with more than 375 individual members.

Pharmacogenomics of Human Drug Transporters: Clinical Impacts, First Edition.
Edited by Toshihisa Ishikawa, Richard B. Kim and Jörg König

In the focus of this chapter are uptake transporters of the OATP family, expressed in human hepatocytes. The human OATP family consists of 10 different members which are grouped into six different subfamilies (OATP1–OATP6). Most of them are expressed in multiple tissues including the brain (Kusuhara and Sugiyama, 2005), intestine (Kunta and Sinko, 2004; Zair et al., 2008), placenta (St-Pierre et al., 2002), heart (Grube et al., 2006b), and kidney (van Montfoort et al., 2003). A short summary of the characteristics of the human OATP family members is shown in Table 7.1. Interestingly, the OATP family members OATP1B1 (formerly termed OATP-C and OATP2; gene symbol *SLCO1B1*) and OATP1B3 (formerly termed OATP8; gene symbol *SLCO1B3*) are predominantly, if not exclusively, expressed in human liver mediating the hepatic uptake of substance from blood into hepatocytes (Abe et al., 1999; Hsiang et al., 1999; König et al., 2000a, 2000b). In addition to OATP1B1 and OATP1B3, OATP2B1 (formerly termed OATP-B; gene symbol *SLCO2B1*) is the third human OATP family member highly expressed in hepatocytes. Distinct from the first two, expression of OATP2B1 was also reported from several other tissues (Kobayashi et al., 2003; Gao et al., 2005; Grube et al., 2006b) suggesting that this uptake transporter is ubiquitously expressed.

Rat Oatp1a1 (formerly termed Oatp or Oatp1; gene symbol *Slco1a1*) was the first characterized member of the OATP/Oatp family (Jacquemin et al., 1994) and it was demonstrated that this transport protein mediates the sodium-independent uptake of several organic anions including conjugated and unconjugated bile salts (Kullak-Ublick et al., 1994). Based on this sequence information several related transport proteins could be identified and characterized in the following years and today, more than 40 different family members from human, rat, mouse, and other species are known (Hagenbuch and Meier, 2003). Interestingly, all OATP family members share a very similar topology with 12 predicted transmembrane helices (TMH) and a large fifth extracellular loop between helices 9 and 10. As shown in the alignment of the three hepatic OATPs OATP1B1, OATP1B3, and OATP2B1 (Fig. 7.1), the arrangement of all helices is comparable, even if the amino acid identity is low (between OATP1B1 and OATP2B1 only 34.7%). The large fifth extracellular loop contains several conserved cysteine residues, N-glycosylation sites are found in the extracellular loops 2 and 5 and the OATP family signature D-X-RW-(I,V)-GAWW-X-(G,N)-(F,L), a highly conserved peptide sequence found in all OATP/Oatp family members, is located at the border between extracellular loop 3 and TMH 6 (Fig. 7.1) (Hagenbuch and Meier, 2003).

The phylogenetic classification of human, mouse, and rat OATP/Oatp family members demonstrates (Fig. 7.2) that most of the human OATPs have direct orthologous proteins in mouse and rat as exemplified by closely related protein sequences found in both species. Interestingly, human OATP1A2 (formerly termed OATP-A; gene symbol *SLCO1A2*) and both hepatic OATPs, OATP1B1 and OATP1B3, have no direct rodent orthologous. For OATP1B1, the closest related protein is OATP1B3 (79.8% amino acid identity) and the closest rodent protein shares only 64.7% identity to this transporter. In contrast, OATP2B1 has closely related rodent orthologous with an amino acid identity (to mouse Oatp2b1) of 77.1%. Orthologous proteins in rodents showing a comparable substrate spectrum and the same tissue expression pattern are a prerequisite for using animals as model systems for the functional characterization

TABLE 7.1 Characteristics of Human OATP Family Members

Gene symbol	Sequence accession ID	Protein name	Amino acids	Chromosomal localization	Tissue distribution	Predominant substrates
SLCO1A2	NM_021094	OATP1A2	670	12p12	Brain, kidney	Bile salts, org. anions, drugs
SLCO1B1	NM_006446	OATP1B1	691	12p	Liver	Bile salts, org. anions, drugs
SLCO1B3	NM_019844	OATP1B3	702	12p12	Liver	Bile salts, org. anions, drugs
SLCO1C1	NM_017435	OATP1C1	712	12p12.2	Brain, testis	Thyroid hormones
SLCO2A1	NM_005630	OATP2A1	643	3q21	Ubiquitous	Prostaglandin
SLCO2B1	NM_007256	OATP2B1	709	11q13	Ubiquitous	E3S, DHEAS, BSP, drugs
SLCO3A1	NM_013272	OATP3A1	710	15q26	Ubiquitous	E3S, Prostaglandins
SLCO4A1	NM_016354	OATP4A1	722	20q13.33	Ubiquitous	Taurocholate, T3
SLCO4C1	NM_180991	OATP4C1	724	5q21.2	Kidney	Digoxin, thyroid hormones
SLCO5A1	NM_030958	OATP5A1	848	8q13.3	–[a]	–[a]
SLCO6A1	NM_173488	OATP6A1	719	5q21.1	Testis[a]	–[a]

Data compiled from NCBI (www.ncbi.nml.nih.gov) and SLC tables (www.bioparadigms.org/slc) databases.

T3, triiodothyronine; BSP, bromosulfophthalein; DHEAS, dehydroepiandrosterone sulfate.

[a]Unknown tissue distribution for OATP5A1, unknown substrates for OATP5A1 and OATP6A1 and doubtful expression of OATP6A1 in testis.

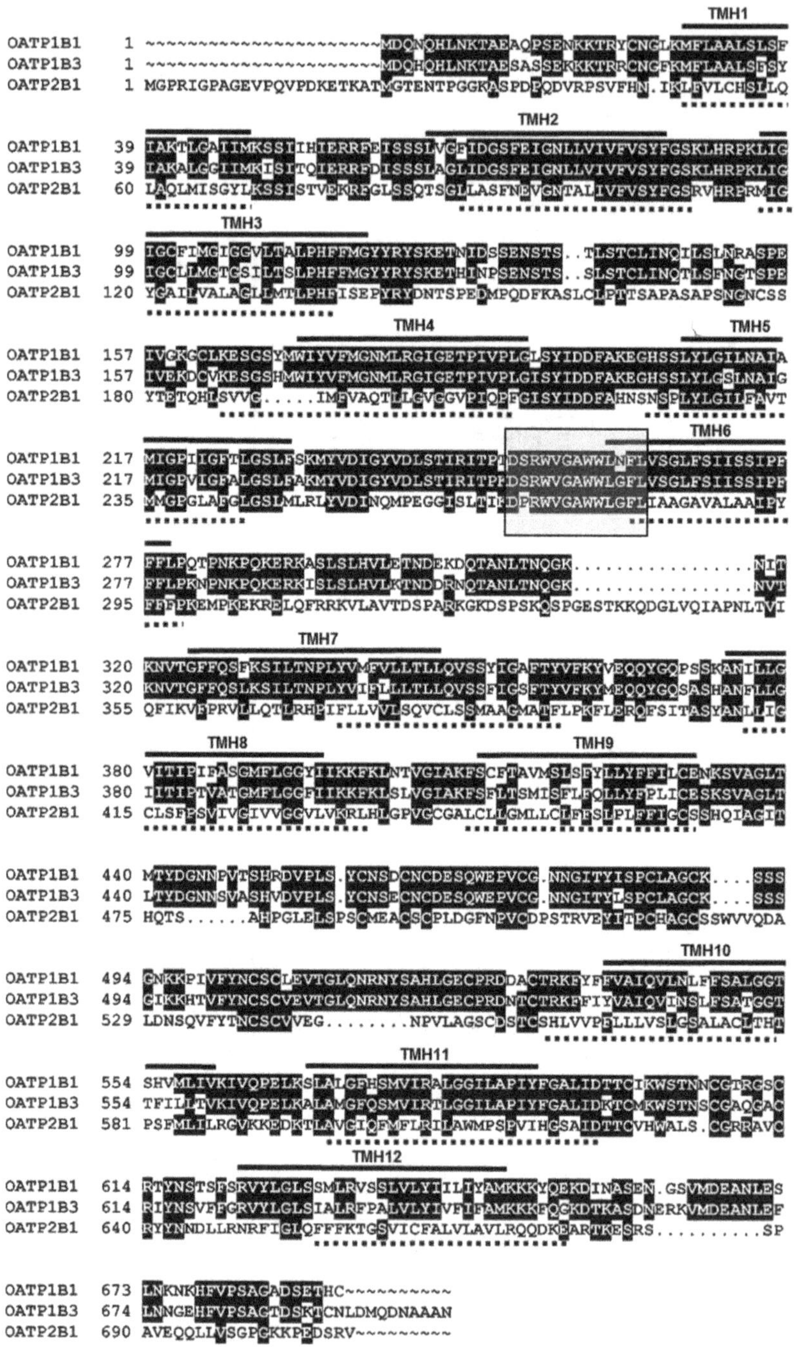

Figure 7.1 Alignment of the three hepatic OATP family members. The bold lines above the alignments represent the predicted TMHs (TMH1–TMH12) for the closely related proteins OATP1B1 and OATP1B3, the dotted lines below the alignments represent the TMHs predicted for the OATP2B1 protein. The box shows the localization of the OATP family signature.

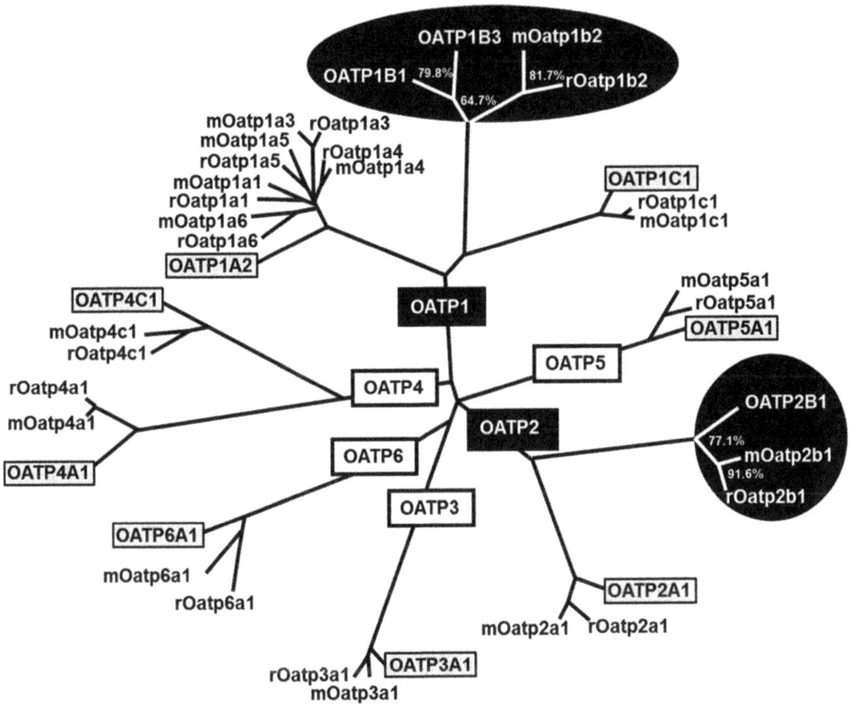

Figure 7.2 Phylogenetic tree of human, mouse, and rat OATP family members. The human OATP proteins are boxed in grey, hepatic OATP family members OATP1B1, OATP1B3, and OATP2B1, which are in the focus of this chapter, are highlighted in black.

of human proteins. Since OATP1B1 and OATP1B3 have no orthologous proteins in other species, it become evident that the transferability of animal data (e.g., analyzing the hepatobiliary elimination of drugs in mice or rats) with regard to the role of the OATPs to the human situation is limited. Furthermore, there are no defined knockout animals for these uptake transporters available. Therefore, the availability of genetically engineered cellular model systems (e.g., stably transfected cells recombinantly overexpressing transport proteins) is important for the functional analysis of these uptake transporters or for the investigation of the functional consequences of genetic variations.

As determined so far, all human OATP family members have a broad substrate spectrum transporting a variety of endogenously synthesized organic anions as well as several widely prescribed drugs (Hsiang et al., 1999; Kullak-Ublick et al., 2001; König et al., 2006; Fahrmayr et al., 2010). Best characterized OATPs with respect to their substrate spectra are the founding member OATP1A2 and both hepatic OATPs, OATP1B1 and OATP1B3. They share a very similar substrate spectrum and are capable of transporting a variety of different endogenous compounds including hormone conjugates (Kullak-Ublick et al., 2001; König et al., 2006), bilirubin and bilirubin glucuronides (Cui et al., 2001), and bile salts such as taurocholate (Maeda et al., 2006b). For drug treatment, the ability of OATP family members in transporting drugs is important. Drugs identified as substrates for human OATP1B1 and OATP1B3 include HMG-CoA reductase inhibitors (statins) such as pravastatin (Hsiang et al., 1999; Seithel et al., 2007) and fluvastatin (Kopplow et al., 2005), the antihistaminic drug fexofenadine (Shimizu et al., 2005), anticancer drugs such as methotrexate, and antibiotics such as rifampicin (Vavricka et al., 2002; Tirona et al., 2003). Table 7.2 summarizes selected endogenous and xenobiotic substrates and affinity constants (K_m values) for OATP1B1-mediated substrate transport. In contrast to OATP1B1 and OATP1B3, only a few drugs have been identified as substrates for OATP2B1. These include some statins (e.g., fluvastatin) (Kopplow et al., 2005), fexofenadine (Shimizu et al., 2005), and benzylpenicillin (Grube et al., 2006b).

Because the hepatic OATP family members mediate the uptake of drugs from blood into hepatocytes, it became evident that alterations in transporter function may result in a change of the plasma levels of the given drug substrates. Two major molecular mechanisms may account for such alterations. One possibility is that one given drug influences the transport kinetics of a second drug, which is also a substrate of the uptake transporter. These so-called transporter-mediated drug–drug interactions finally may result in altered transport kinetics of one or both drugs (for review see Fahrmayr et al., 2010). The second possibility of altered transport kinetics is due to the functional consequences of genetic variations in transporter genes (so-called polymorphisms or haplotypes; for review see (König et al., 2006; Niemi et al., 2011)). So far, several genetic variations have been identified in different transporter genes and characterized with respect to their functional consequences. In the focus of this chapter are genetic variations in the transporters genes *SLCO1B1* (encoding OATP1B1), *SLCO1B3* (encoding OATP1B3), and *SLCO2B1* (encoding OATP2B1).

7.2 GENETIC VARIATIONS IN THE *SLCO1B1* GENE ENCODING OATP1B1

7.2.1 Pharmacogenetics of OATP1B1—*In Vitro* Data

Considerable effort has been made over the recent years to identify and functionally characterize genetic variations in the *SLCO1B1* gene and to determine their ethnic frequencies. Base pair exchanges in the *SLCO1B1* gene leading to amino acid exchanges in the OATP1B1 protein together with the respective allele nomenclature are summarized in Table 7.3, the localizations of the amino acid exchanges in the predicted protein topology are shown Figure 7.3. The first pioneering work regarding the functional consequences of base pair exchanges in the *SLCO1B1* gene has been published by Tirona and coworkers (Tirona et al., 2001). They identified 14 nonsynonymous polymorphisms in a population of Africans and European-Americans and found that some of them affect transport function and/or the subcellular protein localization. In particular, they demonstrated that the variants *SLCO1B1*3* (OATP1B1p.V82A), *5 (OATP1B1p.V174A), *6 (OATP1B1p.I353T), and *13 (haplotype: OATP1B1p.V82A+p.E156G+p.E667G) showed reduced transport activity for the substrates 17ß-estradiol glucuronide, estrone-3-sulfate, and rifampicin. The variants *SLCO1B1*2* (OATP1B1p.F73L) and *12 (OATP1B1p.F73L+p.D655G)

TABLE 7.2 Selected Substrates of OATP1B1

Endogenous Substrates			
Substrate	Class	K_m value	Reference
Bilirubin	Porphyrin	0.16 μM	Cui et al. (2001)
Bilirubin monoglucuronide	Heme metabolite	0.1 μM	Cui et al. (2001)
Bilirubin diglucuronide	Heme metabolite	0.3 μM	Cui et al. (2001))
CCK-8	Peptide hormone		Hirano et al. (2004)
Cholic acid	Bile acid	11 μM	Cui et al. (2001)
DHEAS	Hormone conjugate	22 μM	Abe et al. (1999)
Estradiol-17ß-glucuronide	Hormone conjugate	8.2 μM	König et al. (2000b)
Estrone-3-sulfate	Hormone conjugate	13 μM	Cui et al. (2001)
Glycocholic acid	Bile acid		Kullak-Ublick et al. (2001)
Glycoursodeoxycholic acid	Bile acid		Maeda et al. (2006b)
Iodothyronine sulfate	Thyroid hormone		van der Deure et al. (2008)
Leukotriene C_4	Eicosanoide		Abe et al. (1999)
Leukotriene E_4	Eicosanoide		Abe et al. (1999)
Prostaglandin E_2	Eicosanoide		Abe et al. (1999)
Taurocholic acid	Bile acid	10 μM	Hsiang et al. (1999)
Taurolithocholic acid 3-sulfate	Bile acid		Sasaki et al. (2002)
Tauroursodeoxycholic acid	Bile acid		Maeda et al. (2006b)
Thromboxane B_2	Eicosanoide		Abe et al. (1999)
Thyroxine	Thyroid hormone	3 μM	Hsiang et al. (1999)
Triiodothyronine	Thyroid hormone	2.7 μM	Abe et al. (1999)
Xenobiotic Substrates			
Substrate	Class	K_m value	Reference
Atorvastatin	Antihyperlipidemic drug	12.4 μM	(Kameyama et al. (2005)
Arsenate	Toxin		Lu et al. (2006)
Arsenite	Toxin		Lu et al. (2006)
Bamet-UD2	Anticancer drug	10 μM	Briz et al. (2002)
Bamet-R2	Anticancer drug	10 μM	Briz et al. (2002)
BDE99	Toxin	0.91 μM	Pacyniak et al. (2010)
BDE153	Toxin	1.91 μM	Pacyniak et al. (2010)
Benzylpenicillin	Antibiotic		Tamai et al. (2001)
Bosentan	Antihypertensive drug	44 μM	Treiber et al. (2007)
BQ-123	Endothelin receptor antagonist		Kullak-Ublick et al. (2001)
BSP	Model substrate	0.1 μM	Cui et al. (2001)
Caspofungin	Fungicide		Sandhu et al. (2005)

(continued)

TABLE 7.2 (*Continued*)

Xenobiotic Substrates			
Substrate	Class	K_m value	Reference
Cefditoren	Antibiotic	3.45 μM	Nakakariya et al. (2008)
Cefoperazone	Antibiotic	4.84 μM	Nakakariya et al. (2008))
Cerivastatin	Antihyperlipidemic drug		Shitara et al. (2003)
DADLE	Synthetic opioide peptide	12.4 μM	Nozawa et al. (2003)
Darunavir	Protease inhibitor		Hartkoorn et al. (2010)
Dimethylphalloidin	Toxin		Meier-Abt et al. (2004)
DPDPE	Synthetic opioide peptide	10 μM	Kullak-Ublick et al. (2001)
Enalapril	Cardiovascular drug	10 μM	Liu et al. (2006)
Ezetimibe glucuronide	Antihyperlipidemic drug metabolite	0.91 μM	Oswald et al. (2008)
Fexofenadine	Antihistaminic drug	1.91 μM	Matsushima et al. (2008)
Flavopiridol	Kinase inhibitor		Ni et al. (2010)
Flavopiridol glucuronide	Kinase inhibitor metabolite	44 μM	Ni et al. (2010)
Fluvastatin	Antihyperlipidemic drug	2.4 μM	Kopplow et al. (2005)
Gd-EOB-DTPA	Contrast agent	0.1 μM	Leonhardt et al. (2010)
Gimatecan	Anticancer drug		Oostendorp et al. (2009)
Lopinavir	Protease inhibitor	3.45 μM	Hartkoorn et al. (2010)
Methotrexate	Anticancer drug	4.84 μM	Abe et al. (1999)
Microcystin-LR	Toxin	39 μM	Fischer et al. (2005)
Mycophenolic acid glucuronide	Immunosuppressant		Picard et al. (2010)
Nafcillin	Antibiotic	11.1 μM	Nakakariya et al. (2008)
Olmesartan	Cardiovascular drug	12.8 μM	Yamada et al. (2007)
Pazopanib	Anticancer drug		Keisner and Shah (2011)
Phalloidin	Toxin	39 μM	Fehrenbach et al. (2003)
Pitavastatin	Antihyperlipidemic drug	3 μM	Hirano et al. (2004)
Pravastatin	Antihyperlipidemic drug	35 μM	Hsiang et al. (1999)
Rifampin	Antibiotic	13 μM	Vavricka et al. (2002)
Rosuvastatin	Antihyperlipidemic drug	7.3 μM	Brown (2001)

(continued)

TABLE 7.2 (*Continued*)

Xenobiotic Substrates			
Substrate	Class	K_m value	Reference
Saquinavir	Protease inhibitor		Hartkoorn et al. (2010)
Simvastatin	Antihyperlipidemic drug		Neuvonen et al. (2008)
Sirolimus	Immunosuppressant		Oswald et al. (2010)
SN-38	Anticancer drug		Nozawa et al. (2005)
Temocapril	Cardiovascular drug		Maeda et al. (2006a)
Torasemide	Diuretic	6.2 μM	Werner et al. (2010)
Troglitazone sulfate	Oral antidiabetic drug		Nozawa et al. (2004b)
Valsartan	Cardiovascular drug	17.8 μM	Poirier et al. (2009)

Bamet-UD2, bile acid cisplatin derivative [cis-diammine-bisursodeoxycholate-platinum(II)]; Bamet-R2, Cis-diammineplatinum-(II)-chlorocholylglycinate; BDE99, 2,2′,4,4′,5-pentabromodiphenylether; BDE153, 2,2′,4,4′,5,5′-hexabromodiphenylether; BQ123, cyclic pentapeptide (cyclo[D-Trp-D-Asp-L-Pro-D-Val-L-Leu]); BSP, Bromosulfophthalein; DADLE, [D-Ala^2,D-Leu^2]-enkephalin; DPDPE, [D-penicillamine^{2,5}]enkephalin; GDF-EOB-DTPA, gadolinium-ethoxy-benzyl-diethylenetriamine pentaacetic acid; SN-38, 7-ethyl-10-hydroxycamptothecin.

increased the K_m value for estrone-3-sulfate. Interestingly, many of the amino acid exchanges having functional consequences on protein function or protein localization are located within TMHs or in the second or fifth extracellular loop (Tirona et al., 2001). Exchanges located in the second loop were also in the focus of the analysis by Michalski and colleagues (Michalski et al., 2002). Analyzing the OATP1B1 protein expression in 81 human liver samples, they identified one liver sample with a significantly reduced amount of OATP1B1 protein. Sequencing the *SLCO1B1* cDNA amplified from this liver sample demonstrated that three base pair substitutions leading to three amino acid exchanges were present within this gene. Two (*SLCO1B1**1b*; OATP1B1p.N130D and *SLCO1B1**4*; OATP1B1p.P155T) of the exchanges were previously identified by Tirona et al. (2001), whereas the third exchange (*SLCO1B1**18*; OATP1B1p.L193R) was the first naturally occurring mutation (frequency below 1%) identified in the *SLCO1B1* gene. Using stably transfected MDCK cells recombinantly overexpressing the single mutations or the haplotype with all three mutations in one cDNA, they demonstrated that the effect of the variation depends on the substrate used for the analysis. Whereas the prototypic substrate bromosulfophthalein (BSP) was transported by the *1b and the *4 variant with similar transport rates obtained with the protein encoded by the reference sequence (*SLCO1B1**1a*), transport of 17ß-estradiol glucuronide was significantly reduced for the *4 variant. Using the bile salt taurocholate as substrate, the transport of the *1b variant was significantly reduced and totally abolished by the *4 variant. The base pair exchange leading to the first naturally occurring mutation (OATP1B1p.L193R) totally abolished transport for all investigated substrates (Michalski et al., 2002). Furthermore, this mutation led to an altered protein localization as observed by a high amount of intracellularly

TABLE 7.3 Nonsynonymous Sequence Variations in the *SLCO1B1* Gene Encoding Human OATP1B1

rsnumber	Exon	Nucleotide exchange	Amino acid exchange	Alleles
rs11557087	1	c.28A>G	p.T10A	n.a.
rs61760183	2	c.170G>A	p.R57Q	n.a.
rs56101265	2	c.217T>C	p.F73L	*2;,*12
rs56061368	3	c.245T>C	p.V82A	*3;,*13
rs2306283	4	c.388A>G	p.N130D	*1b; *14, *15, *16(II), *17, *18* *20, *21
rs2306282	4	c.452A>G	p.N151S	*16(I)
n.a.	4	c.455G>A	p.R152K	*1c
rs11045819	4	c.463C>A	p.P155T	*4, *14, *18
rs72559745	4	c.467A>G	p.E156G	*3, *13
rs4149056	5	c.521T>C	p.V174A	*5, *15, *16(II), *17
rs72559746	5	c.578T>G	p.L193R	*18
rs4603354	5	c.608G>A	p.G203E	n.a.
n.a.	6	c.721G>A	p.D241N	*1c
rs11045852	7	c.733A>G	p.I245V	n.a.
rs11045853	7	c.758G>A	p.R253Q	n.a.
rs72559742	7	c.841C>A	p.Q281K	n.a.
rs77871475	8	c.1000A>T	p.T334S	n.a.
rs72559747	8	c.1007C>G	p.P336R	n.a.
rs79109623	8	c.1034C>T	p.T345M	n.a.
rs55901008	8	c.1058T>C	p.I353T	*6
rs59113707	9	c.1200C>G	p.F400L	n.a.
rs77468276	9	c.1246G>C	p.V416L	n.a.
rs56387224	9	c.1294A>G	p.N432D	*7
rs72559748	10	c.1385A>G	p.D462G	*8
n.a.	10	c.1388C>G	p.L463F	*20
n.a.	10	c.1454G>A	p.C485F	n.a.
rs59502379	10	c.1463G>C	p.G488A	*9
rs74064213	10	c.1495A>G	p.I499V	n.a.
rs71581987	11	c.1622A>T	p.Q541L	n.a.
rs71581988	11	c.1628T>G	p.L543W	n.a.
rs74700754	12	c.1724A>T	p.H575L	n.a.
rs34671512	14	c.1929A>C	p.L643F	*19, *20, *21
rs56199088	14	c.1964A>G	p.D655G	*10, *12
rs55737008	14	c.2000A>G	p.E667G	*11, *13

retained OATP1B1 protein compared to the *1a wild-type protein. These results for the first time demonstrated that the variations in the second extracellular loop of the OATP1B1 protein may influence the substrate spectrum or transport kinetics of selected substrates and that for a detailed analysis of the functional consequences of genetic variations the transport of more than one substrate should be investigated.

Since most of the widely prescribed HMG-CoA reductase inhibitors (statins) are substrates of hepatic OATP family members, the functional consequences of

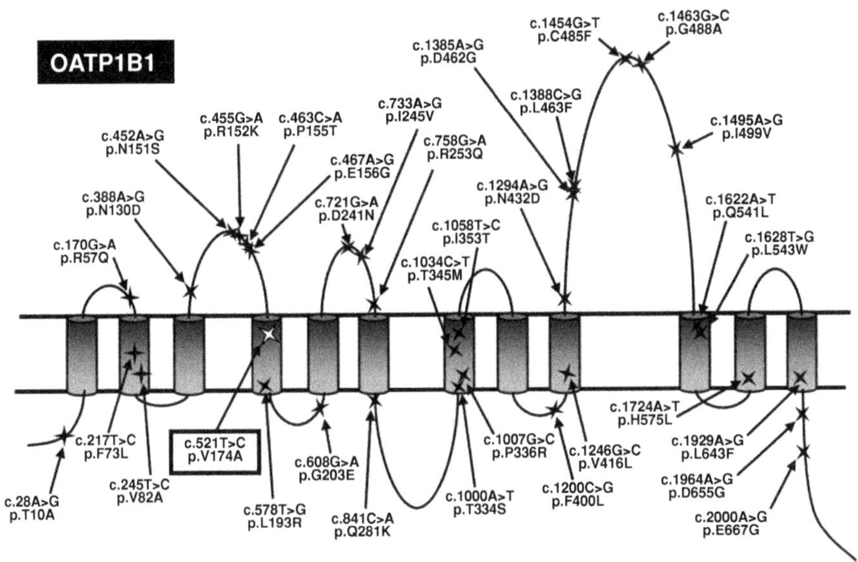

Figure 7.3 Localization of amino acid exchanges originating from of genetic variations (polymorphisms) in the *SLCO1B1* gene in the predicted protein topology of the human OATP1B1 protein. Highlighted in the box is the variation OATP1B1p.V174A (c.521T>C; (designated in the text as OATP1B1*5 variant)).

genetic variations in these transporter genes on statin transport have been investigated in detail using *in vitro* as well as *in vivo* studies (see the next paragraph). Especially, the *SLCO1B1*5* variant (*SLCO1B1c.521T>C*; OATP1B1p.V174A) has been investigated over the last few years and it has been demonstrated for several drugs that this exchange is associated with a reduced transport activity of the OATP1B1 protein (Table 7.4). Using cell systems expressing OATP1B1 reduced uptake activity has been shown for atorvastatin, cerivastatin, and rosuvastatin (Kameyama et al., 2005) as well as for ezetimibe glucuronide, the metabolite of the antihyperlipidemic drug ezetimibe (Oswald et al., 2008). Furthermore, reduced transport kinetics by the *SLCO1B1*5* variant could be detected for the endothelin-receptor antagonist, atrasentan, (Katz et al., 2006) and the antibiotic rifampicin (Tirona et al., 2003). Using pravastatin as substrate and human embryonic kidney (HEK) 293 cells recombinantly overexpressing the *SLCO1B1*1a* and the *5 variant, no significant alteration of the affinity constant (K_m value) could be detected whereas the maximal transport velocity (V_{max} value) was reduced in the *5 variant suggesting that this amino acid exchange influences transport kinetics rather than substrate affinity (Kameyama et al., 2005).

Interesting observations were made analyzing the haplotype *SLCO1B1*15* carrying both the single-nucleotide polymorphisms, *SLCO1B1*1b* (OATP1B1p.N130D) and *SLCO1B1*5*. Using estrone-3-sulfate as substrate and stably transfected cells recombinantly expressing this haplotype protein it could be demonstrated that no alterations in K_m or V_{max} values were detected compared to the transport mediated by

TABLE 7.4 Influence of Polymorphisms in the *SLCO1B1* Gene Encoding Human OATP1B1 on *In Vitro* Substrate Uptake and Pharmacokinetics or Pharmacodynamics of Selected Drug Substrates

Allele	Amino acid exchange	Drug	*In vivo/ in vitro*	Effect	Reference
*1b	p.N130D	Ezetimibe	*In vivo*	Decreased serum concentrations	Oswald et al. (2008)
*1b	p.N130D	Irinotecan	*In vivo*	Association with grade 3 diarrhea in NSCLC patients	Han et al. (2008)
*1b	p.N130D	Rifampin	*in vitro*	↓ Uptake	Tirona et al. (2003)
*1b	p.N130D	Pravastatin	*In vivo*	Values of AUC more than 60 % lower *; significantly decreased amount of pravastatin excreted into urine	Mwinyi et al. (2004)
*1b	p.N130D	Pravastatin	*In vivo*	↓ AUC	Maeda et al. (2006a)
*1b	p.N130D	Pravastatin	*In vivo*	Stronger decrease in lathosterol levels	Gerloff et al. (2006)
*1b	p.N130D	Rosuvastatin	*In vivo*	↑ C_{max}	Lee et al. (2005)
*1b	p.N130D	Tacrolimus	*In vivo*	Lower blood concentration in heterozygous carriers	Elens et al. (2007)
*1b	p.N130D	Talinolol	*In vivo*	Shorter half-lives	Bernsdorf et al. (2006)
*1b	p.N130D	Torasemide	*In vivo*	Increased oral clearance	Vormfelde et al. (2008)
*1c	p.P152K+ D241N	Rosuvastatin	*in vitro*	↓ Uptake	Ho et al. (2006)
*2	p.F73L	Rifampicin	*in vitro*	↓ Uptake	Tirona et al. (2003)
*2	p.F73L	Rosuvastatin	*in vitro*	↓ Uptake	Ho et al. (2006)
*3	p.V82A+ E156G	Rifampicin	*in vitro*	↓ Uptake	Tirona et al. (2003)
*3	p.V82A+ E156G	Rosuvastatin	*in vitro*	↓ Uptake	Ho et al. (2006)
*5	p.V174A	Atorvastatin	*in vitro*	↓ Uptake	Kameyama et al. (2005)
*5	p.V174A	Atrasentan	*in vitro*	↓ Uptake	Katz et al. (2006)
*5	p.V174A	Cerivastatin	*in vitro*	↓ Uptake	Kameyama et al. (2005)
*5	p.V174A	Ezetimibe-glucuronide	*in vitro*	↓ Uptake	Oswald et al. (2008)
*5	p.V174A	Pravastatin	*in vitro*	↓ Uptake	Kameyama et al. (2005)
*5	p.V174A	Rifampicin	*in vitro*	↓ Uptake	Tirona et al. (2003)
*5	p.V174A	Rosuvastatin	*in vitro*	↓ Uptake	Ho et al. (2006)
*5	p.V174A	Atorvastatin	*In vivo*	↑ AUC of atorvastatin and 2-hydroxyatorva-statin	Pasanen et al. (2007)

(*continued*)

TABLE 7.4 (*Continued*)

Allele	Amino acid exchange	Drug	*In vivo/ in vitro*	Effect	Reference
*5	p.V174A	Ezetimibe	*In vivo*	Decreased fecal excretion	Oswald et al. (2008)
*5	p.V174A	Fexofenadine	*In vivo*	↑ AUC	Niemi et al. (2005b)
*5	p.V174A	Irinotecan	*In vivo*	↑ AUC of SN-38; association with grade 4 neutropenia	Han et al. (2008)
*5	p.V174A	Nateglinide	*In vivo*	↑ AUC and C_{max}	Zhang et al. (2006)
*5	p.V174A	Pravastatin	*In vivo*	↑ AUC	Niemi et al. (2005c)
*5	p.V174A	Pravastatin	*In vivo*	↑ AUC and C_{max}	Niemi et al. (2006b)
*5	p.V174A	Pravastatin	*In vivo*	↑ AUC	Mwinyi et al. (2004)
*5	p.V174A	Pravastatin	*In vivo*	↑ AUC and C_{max}	Ho et al. (2006)
*5	p.V174A	Pravastatin	*In vivo*	↓ AUC and shorter half-life for heterozygous transplant recipients; smaller decrease in total and LDL cholesterol	Hedman et al. (2006)
*5	p.V174A	Pravastatin	*In vivo*	Lower reduction in total cholesterol	Zhang et al. (2007)
*5	p.V174A	Pravastatin	*In vivo*	Lower decrease in lathosterol levels	Gerloff et al. (2006)
*5	p.V174A	Pravastatin	*In vivo*	Lower decrease in total cholesterol	Tachibana-Iimori et al. (2004)
*5	p.V174A	Repaglinide	*In vivo*	↑ AUC of repaglinide and the repaglinide metabolites M2 + M4; correlation of maximum decrease in blood concentrations and repaglinide AUC	Kalliokoski et al. (2008a)
*5	p.V174A	Repaglinide	*In vivo*	↑ AUC	Niemi et al. (2005a)
*5	p.V174A	Repaglinide	*In vivo*	↑ AUC	Kalliokoski et al. (2008b)
*5	p.V174A	Rosuvastatin	*In vivo*	↑ AUC and C_{max}	Pasanen et al. (2007)
*5	p.V174A	Rosuvastatin	*In vivo*	↑ AUC + C_{max}	Lee et al. (2005)
*5	p.V174A	Simvastatin acid	*In vivo*	↑ AUC and C_{max}	Pasanen et al. (2006)
*5	p.V174A	Tacrolimus	*In vivo*	↑ Blood concentration in heterozygous carriers	Elens et al. (2007)
*5	p.V174A	Torasemide	*In vivo*	↓ Oral clearance	Vormfelde et al. (2008)
*5	p.V174A	Torasemide	*In vivo*	↑ AUC	Werner et al. (2010)

(*continued*)

TABLE 7.4 (*Continued*)

Allele	Amino acid exchange	Drug	*In vivo/ in vitro*	Effect	Reference
*6	p.I353T	Rifampicin	*in vitro*	↓ Uptake	Tirona et al. (2003)
*6	p.I353T	Rosuvastatin	*in vitro*	↓ Uptake	Ho et al. (2006)
*7	p.N432D	Rifampicin	*in vitro*	↓ Uptake	Tirona et al. (2003)
*7	p.N432D	Rosuvastatin	*in vitro*	↓ Uptake	Ho et al. (2006)
*9	p.G488A	Atrasentan	*in vitro*	↓ Uptake	Katz et al. (2006)
*9	p.G488A	Rifampicin	*in vitro*	↓ Uptake	Tirona et al. (2003)
*9	p.G488A	Rosuvastatin	*in vitro*	↓ Uptake	Ho et al. (2006)
*11	p.E667G	Rifampicin	*in vitro*	↓ Uptake	Tirona et al. (2003)
*12	p.F73l+ D655G	Rifampicin	*in vitro*	↓ Uptake	Tirona et al. (2003)
*12	p.F73l+ D655G	Rosuvastatin	*in vitro*	↓ Uptake	Ho et al. (2006)
*13	p.V82A+ E156G+ E667G	Rifampicin	*in vitro*	↓ Uptake	Tirona et al. (2003)
*13	p.V82A+ E156G+ E667G	Rosuvastatin	*in vitro*	↓ Uptake	Ho et al. (2006)
*14	p.N130D+ P155T	Fluvastatin	*In vivo*	Stronger reduction of LDL and total cholesterol	Couvert et al. (2008)
*14	p.N130D+ P155T	Rifampicin	*in vitro*	↓ Uptake	Tirona et al. (2003)
*15	p.N130D+ V174A	Atorvastatin	*in vitro*	↓ Uptake	Kameyama et al. (2005)
*15	p.N130D+ V174A	Atrasentan	*in vitro*	↓ Uptake	Katz et al. (2006)
*15	p.N130D+ V174A	Cerivastatin	*in vitro*	↓ Uptake	Kameyama et al. (2005)
*15	p.N130D+ V174A	SN-38	*in vitro*	↓ Uptake	Nozawa et al. (2005)
*15	p.N130D+ V174A	Pitavastatin	*in vitro*	↓ Uptake	Deng et al. (2008)
*15	p.N130D+ V174A	Pravastatin	*in vitro*	↓ Uptake	Kameyama et al. (2005)
*15	p.N130D+ V174A	Pravastatin	*in vitro*	↓ Uptake	Nozawa et al. (2005)
*15	p.N130D+ V174A	Rosuvastatin	*in vitro*	↓ Uptake	Ho et al. (2006)
*15	p.N130D+ V174A	TR-14035	*in vitro*	↓ Uptake	Tsuda-Tsukimoto et al. (2006)
*15	p.N130D+ V174A	Atrasentan	*In vivo*	↑ AUC	Katz et al. (2006)
*15	p.N130D+ V174A	Ezetimibe	*In vivo*	Reduced fecal excretion	Oswald et al. (2008)

(*continued*)

TABLE 7.4 (*Continued*)

Allele	Amino acid exchange	Drug	*In vivo/ in vitro*	Effect	Reference
*15	p.N130D+ V174A	Irinotecan	*In vivo*	↑ AUC of irinotecan and SN-38 and lower clearance	Han et al. (2008)
*15	p.N130D+ V174A	Irinotecan	*In vivo*	↑ AUC of irinotecan and SN-38	Ieiri et al. (2009)
*15	p.N130D+ V174A	Irinotecan	*In vivo*	↑ AUC of SN-38	Xiang et al. (2006)
*15	p.N130D+ V174A	Olmesartan	*In vivo*	↑ AUC and C_{max}, significantly lower clearance	Suwannakul et al. (2008)
*15	p.N130D+ V174A	Pitavastatin	*In vivo*	↑ AUC and C_{max}	Deng et al. (2008)
*15	p.N130D+ V174A	Pitavastatin	*In vivo*	↑ AUC and C_{max}	Chung et al. (2005)
*15	p.N130D+ V174A	Pitavastatin	*In vivo*	↑ AUC and C_{max}	Ieiri et al. (2007)
*15	p.N130D+ V174A	Pravastatin	*In vivo*	↑ AUC and C_{max}	Deng et al. (2008)
*15	p.N130D+ V174A	Pravastatin	*In vivo*	↑ AUC	Niemi et al. (2004)
*15	p.N130D+ V174A	Pravastatin	*In vivo*	↑ AUC	Ho et al. (2006)
*15	p.N130D+ V174A	Pravastatin	*In vivo*	↑ AUC and C_{max}	Igel et al. (2006)
*15	p.N130D+ V174A	Pravastatin	*In vivo*	Lower LDL-cholesterol reduction	Takane et al. (2006)
*15	p.N130D+ V174A	Pravastatin	*In vivo*	Reduction of total and nonrenal clearance compared with *1b	Nishizato et al. (2003)
*15	p.N130D+ V174A	Rosuvastatin	*In vivo*	↑ AUC and C_{max}	Lee et al. (2005)
*15	p.N130D+ V174A	Rosuvastatin	*In vivo*	↑ AUC and C_{max}	Choi et al. (2008)
*17	-11187G>A^{+} + p.N130D+ pV174A	Pravastatin	*In vivo*	↑ AUC	Niemi et al. (2004)
*17	-11187G>A^{+} + p.N130D+ pV174A	Pravastatin	*In vivo*	Smaller effects on plasma lathosterol concentration and lathosterol-to-cholesterol ratio	Niemi et al. (2005c)

(*continued*)

TABLE 7.4 (*Continued*)

Allele	Amino acid exchange	Drug	*In vivo/ in vitro*	Effect	Reference
*17	-11187G>A⁺+ p.N130D+ pV174A	Pravastatin	*In vivo*	↑ AUC and C_{max}	Igel et al. (2006)
*18	pN130D+ S137S+ P155T+ L235L+ L193R	Atrasentan	*in vitro*	↓ Uptake	Katz et al. (2006)
*18	pN130D+ S137S+ P155T+ L235L+ L193R	Rosuvastatin	*in vitro*	↓ Uptake	Ho et al. (2006)
	p.P155T	Fluvastatin	*In vivo*	Stronger reduction of LDL and total cholesterol	Couvert et al. (2008)
	p.L587W	Pravastatin	*In vivo*	Association with pravastatin-induced myopathy	Morimoto et al. (2004)
	-11187G>A⁺	Pravastatin	*In vivo*	↑ AUC	Niemi et al. (2004)
		Pravastatin	*In vivo*	↓ C_{max} and AUC; greater increase in HDL cholesterol in patients with heterozygous familial hyperc-holesterolemia	Hedman et al. (2006)
		Repaglinide	*In vivo*	Enhanced effect on blood glucose	Niemi et al. (2005a)
		Irinotecan	*In vivo*	↑ AUC of SN-38	Han et al. (2008)

↓ Uptake Reduced uptake compared to uptake mediated by the *SLCO1B1** 1a* variant (OATP1B1-WT) measured in *in vitro* cell systems or in oocytes expressing the respective protein.
↓ Decrease in AUC (area under the concentration time curve) and decrease in C_{max} (maximum peak concentration in plasma).
↑ Increase in AUC (area under the concentration time curve) and decrease in C_{max} (maximum peak concentration in plasma).
NSCLC, non-small-cell lung cancer; SN-38, active metabolite of the topoisomerase inhibitor irinotecan.
+ = variation within the *SLCO1B1* gene promoter region = base pair exchange.

the protein encoded by the *SLCO1B1* 1a* gene (Nozawa et al., 2002). These results were confirmed using 17ß-estradiol glucuronide as substrate (Iwai et al., 2004). Interestingly, for SN38, the active metabolite of the topoisomerase inhibitor, irinotecan, as well as for atrasentan, atorvastatin, cerivastatin, pitavastatin, and rosuvastatin, reduced transport activity of the *15 variant compared to the *1a variant could be detected demonstrating again the necessity of using different substrates for analyzing the functional consequences of genetic variations.

7.2.2 Pharmacogenetics of OATP1B1—*In Vivo* Data

Polymorphisms in the *SLCO1B1* gene show distinct differences in their frequencies between geographical regions and ethnicities. Pasanen et al. (Pasanen et al., 2008) investigated the allelic frequencies of 12 different polymorphisms including 5 nonsynonymous and 2 promoter variants in 941 persons from 52 populations. The most investigated variant today, *SLCO1B1*5,* has an allelic frequency of 8–20% in the European population, 1–8% in African-Americans, and 8–16% in East Asians. The *SLCO1B1* 1b* variant has a frequency of about 26% in Europeans, 63% in East Asians, and 77% in sub-Saharan Africans (Pasanen et al., 2008).

The effect of *SLCO1B1* polymorphisms and haplotypes on drug pharmacokinetics has been investigated for more than 15 clinically used drugs (Table 7.4). The first drug investigated in detail was the HMG-CoA reductase inhibitor pravastatin. In spite of the fact that pravastatin is not significantly metabolized in hepatocytes, pharmacokinetic studies reported a large interindividual variability in pravastatin plasma levels (Neuvonen et al., 1998). Based on the findings that pravastatin is a substrate for OATP1B1 (Hsiang et al., 1999) one would expect that reduced OATP1B1 uptake function, as shown *in vitro* for the *SLCO1B1*5* variant, would limit the access of pravastatin into hepatocytes resulting in increased plasma concentrations. Interestingly, this was demonstrated in several studies (Nishizato et al., 2003; Mwinyi et al., 2004; Niemi et al., 2006b) despite the fact that pravastatin is also a substrate for the second hepatic OATP family member, OATP1B3 (Seithel et al., 2007). In detail, Niemi and coworkers found that carriers with the 521 TC genotype (heterozygous) have a 106% higher pravastatin plasma concentration than homozygous TT carriers (Niemi et al., 2004). Since the target enzyme of statins is located intracellular in hepatocytes it was speculated that reduced uptake leads to a decreased effect on HMG-CoA reductase. Indeed, this was demonstrated in another study by Niemi and colleagues and they reported significantly reduced effects of pravastatin on cholesterol synthesis in carriers of the *SLCO1B1*17* haplotype compared with non carriers (Niemi et al., 2005c). This *SLCO1B1*17* haplotype contains, in addition to the *SLCO1B1c.521T>C* (*SLCO1B1*5*) exchange, a variation in the promoter region (*SLCO1B1g.-11187G>A*) and the *SLCO1B1c.388A>G* (*SLCO1B1* 1b*) exchange. Despite the fact that some results analyzing the effect of the *SLCO1B1c.388A>G* variation (*SLCO1B1* 1b*) are controversial and could be due to different substrates used for the analysis, one *in vitro* study demonstrated that the transport of pravastatin was reduced by the *SLCO1B1c.521T>C* (*5) variation irrespective of this exchange was in the same haplotype with the *SLCO1B1c.388A>G* (*1b) variation or not (Kameyama et al., 2005). This suggests that the observed pharmacodynamic effect of the *SLCO1B1*17* allele was a consequence only of the amino acid exchange OATP1B1p.V174A (*SLCO1B1*5*). If the polymorphism in the *SLCO1B1* promoter region (*SLCO1B1g.-11187G>A*), which is strongly linked to the *SLCO1B1*5* variant, has also an effect on pravastatin transport has to be elucidated in the future. The pharmacodynamic effect of the *5 variant was also observed in a study using pravastatin multiple dosing in healthy volunteers (Igel et al., 2006). In this study, 4 carriers heterozygous for the *17 allele, 3 homozygous for the *15 allele, 1 carrier homozygous for the *17 haplotype compared to 8 carriers of the *SLCO1B1* 1a* wild-type allele received 40 mg pravastatin once daily for 3 weeks. The maximum peak concentration time curve (AUC) was 100% higher in the variant group compared with the control group (Igel et al., 2006).

Similar to pravastatin, several *in vivo* studies have demonstrated that the *SLCO1B1*5* allele is also associated with elevated plasma levels of atorvastatin (Pasanen et al., 2007), rosuvastatin (Lee et al., 2005), and simvastatin acid (Pasanen et al., 2006). Interestingly, this variant seems to have no effect on the plasma concentrations of fluvastatin (Neuvonen et al., 2008). The major importance of this variant on statin pharmacokinetics has been demonstrated in a genome-wide association study. In this study using 300,000 markers in 85 patients with definite or incipient myopathy compared to 90 control volunteers, the *SLCO1B1*5* variant was identified to be strongly associated with the risk for simvastatin-induced myopathy (Link et al., 2008). A second genome-wide association meta-analysis combining the results from three association studies (Rotterdam study with $n = 3847$; Age, Gene, Environment and Susceptibility-Reykjavik with $n = 2193$, and the Framingham heart study with $n = 3424$) demonstrated that the *5 allele is also associated with variations in serum bilirubin levels (Johnson et al., 2009), in agreement with the fact that bilirubin is a substrate for OATP1B1 (Cui et al., 2001).

In contrast to the effect of the *SLCO1B1*5* allele, the *SLCO1B1*1b* variant has been associated with a decreased AUC of pravastatin (Mwinyi et al., 2004; Maeda et al., 2006a) suggesting an increase in uptake transport function of this protein. The AUC in volunteers receiving 10 mg pravastatin was 35% lower in persons with the **1b/*1b* allele than in those with the **1a/*1a* genotype (Maeda et al., 2006a). In contrast, for rosuvastatin this variant seems not to have an influence (Lee et al., 2005) suggesting that the effects may be substrate specific.

Apart from statins, the influence of the *SLCO1B1*5* variant on the pharmacokinetics of several other endogenous compounds or drugs has been studied. In most cases, reduced transport function resulting in elevated plasma levels (higher AUC) and decreased hepatic uptake of the investigated drug could be observed (Table 7.4). For the oral antidiabetic drug repaglinide, it was shown in 56 healthy volunteers that the *SLCO1B1c.521CC* genotype is associated with a nearly 3-fold increase in the AUC (Niemi et al., 2005a). Interestingly, repaglinide has not been identified as a substrate for OATP1B1 but it inhibits OATP1B1-mediated pravastatin uptake *in vitro* (Bachmakov et al., 2008). The effect of the *SLCO1B1*5* variant on repaglinide pharmacokinetics has been confirmed in further studies and found to be consistent throughout a wide dose range (Kalliokoski et al., 2008a). Moreover, the *SLCO1B1*1b* variant also has been demonstrated to affect the kinetics of repaglinide as demonstrated by a 32% decrease in persons with the *1b genotype compared to persons with the *SLCO1B1*1a* genotype (Kalliokoski et al., 2008a). These results show that variations in the *SLCO1B1* gene affect the pharmacokinetics of repaglinide and that for the *5 variant this reduced uptake could be correlated with a lower decrease in blood-glucose levels (Niemi et al., 2005c) and also a pharmacodynamic effect of these polymorphisms.

In healthy Chinese volunteers the AUC of nateglinide, another oral antidiabetic drug, was found to be increased 2-fold in persons homozygous for the

SLCO1B1c.521CC allele and 1.8-fold in the heterozygous *SLCO1B1c.521TC* persons compared to the *SLCO1B1**1a* homozygous carriers (Zhang et al., 2006). Interestingly, these results could not be confirmed in larger panel studies (Kalliokoski et al., 2008b). Furthermore, the *SLCO1B1**5* allele has no effect on the pharmacokinetics of either rosiglitazone or pioglitazone or their metabolites (Aquilante et al., 2008; Kalliokoski et al., 2008b) despite the fact that it has been demonstrated *in vitro* that rosiglitazone interacts with the OATP1B1 protein (Bachmakov et al., 2008). These results demonstrate that *in vitro* observed results may not be easily transferable to the *in vivo* situation. Theoretically, hepatocytes expressing mutated OATP1B1 proteins might develop adaptive mechanisms such as compensatory upregulation of another uptake transporter. For OATP1B1, a potential candidate could be the closely related OATP family member OATP1B3. Both transporters share an overlapping substrate spectrum and for some drugs one can speculate that the observed *in vitro* effect of an OATP1B1 variant could be compensated *in vivo* by transport mediated by OATP1B3.

In the case of the antihyperlipidemic drug, ezetimibe, it was found that in carriers of the *SLCO1B1**5* variant, the fecal excretion was increased after single-dose administration (Oswald et al., 2008). These results were also obtained from analyzing carriers of the *SLCO1B1**1b* allele. Further drug pharmacokinetics affected by the *SLCO1B1**5* allele are the immunosuppressive drug tacrolimus (Elens et al., 2007), and the loop diuretic torasemide (Werner et al., 2010). In both cases reduced hepatic uptake resulting in increased plasma levels of the given drug were observed in carriers of the *5 allele compared to carriers of the *SLCO1B1**1a* allele. Despite the effect on drug pharmacokinetics, genetic variations in the *SLCO1B1* gene also affect the disposition of several bile salts and bile acid synthesis markers. This was shown by analyzing the effect of the *1b, the *5, and the *15 alleles on the plasma concentrations of chenodeoxycholic acid (CDCA), glycoursodeoxycholic acid (GUDCA), and ursodeoxycholic acid (UDCA). This analysis (Xiang et al., 2009) demonstrated that persons carrying the *5 or the *15 allele have 50–240% higher plasma concentrations of these bile salts than persons with the *SLCO1B1**1a* allele.

In a study published recently, it was shown that the variation *SLCO1B1c.463C>A* (present in the variants *SLCO1B1**4*, *14 and *18; see Table 7.3) is associated with a reduced AUC of the antibiotic rifampicin (Weiner et al., 2010). Interestingly, the *SLCO1B1**5* allele has no effect on the induction of hepatic CYP3A4 by rifampicin (Niemi et al., 2006a) suggesting that the *5 allele has only a minor effect on the liver exposure of rifampicin.

Taken together these *in vivo* analyses demonstrated that the kinetics of endogenous compounds as well as the pharmacokinetics of drugs might be altered due to genetic variations in the *SLCO1B1* gene encoding the hepatic uptake transporter, OATP1B1. Since statins are a major class of drug substrates for this transporter, genotyping specific variants associated with reduced or increased uptake may help to achieve the benefits of statin therapy more efficiently and, as demonstrated for the effect of the variant OATP1B1*5, more safely.

7.3 GENETIC VARIATIONS IN THE *SLCO1B3* GENE ENCODING OATP1B3

7.3.1 Pharmacogenetics of OATP1B3—*In Vitro* and *In Vivo* Data

Compared to the comprehensive knowledge on the functional consequences of polymorphisms in the *SLCO1B1* gene, data describing functional consequences in the *SLCO1B3* gene encoding the closely related family member OATP1B3 are limited. The uptake transporter OATP1B3 is the second OATP family member which seems to be liver specific (König et al., 2000a) and the protein shares 80% amino acid identity to OATP1B1. Furthermore, they have an overlapping and similar substrate spectrum. As for OATP1B1, endogenous substrates of OATP1B3 include hormone metabolites and bile salts (König et al., 2000a). Drugs that are substrates for OATP1B3 (for review see Fahrmayr et al., 2010) include antibiotics (e.g., rifampicin), anticancer drugs (e.g., paclitaxel and methotrexate), cardioactive drugs (e.g., digoxin), and HMG-CoA reductase inhibitors (e.g., fluvastatin, pitavastatin, and pravastatin).

The first identification of polymorphisms in the *SLCO1B3* gene was published by Iida et al. (2001). They screened the DNA from 48 Japanese individuals and identified a total of 250 polymorphisms among different transporter genes including three OATP family members without further analyzing the functional consequences of these variations. Regarding OATP1B3, they identified the nonsynonymous variation *SLCO1B3c.334T>G* (OATP1B3p.S112A) as well as several base pair exchanges in noncoding regions of the *SLCO1B3* gene. The first detailed analysis about the functional consequences of variations in the *SLCO1B3* gene was published by Letschert and colleagues (Letschert et al., 2004). They found that the polymorphism OATP1B3p.S112A as well as the polymorphisms OATP1B3p.M233I (*SLCO1B3c.699G>A*) are frequent within the Caucasian population with allelic frequencies of 74% and 71%, respectively, whereas the third variation analyzed in this study (OATP1B3p.G522C; *SLCO1B3c.1564G>T*) has a relatively rare frequency of about 1.9% in the same population. Using stably transfected MDCKII and HEK293 cells they investigated the functional consequences and demonstrated that effects of this base pair exchanges on transport and protein localization were cell line and substrate dependent. Using the substrates BSP, 17ß-estradiol glucuronide, cholecystokinin-8, taurocholate, estrone-3-sulfate, and dehydroepiandrosterone-3-sulfate it was shown that polymorphisms OATP1B3p.S112A and OATP1B3p.M233I did not result in significant changes in protein expression or transport function whereas the variation OATP1B1p.G522C revealed cell-line and substrate-dependent changes in transport activity (Letschert et al., 2004). Using 17ß-estradiol glucuronide as substrate and MDCKII cells as expression system this variant shows transport rates higher than the transport rates obtained with the OATP1B1-WT protein whereas in HEK293 cells using the same substrate the transport of the OATP1B3p.G522C variant was totally abolished. For OATP1B3p.S112A and OATP1B3p.M233I these results were confirmed later in a study using Xenopus laevis oocytes as expression system and paclitaxel as substrate (Smith et al., 2007). In an *in vivo* study the variant OATP1B3p.S112A was identified as affecting the pharmacokinetics of the immunosuppressant mycophenolic acid (MPA) (Miura et al., 2007). In this study the authors showed that variations in the *SLCO1B3* gene, together with polymorphisms in the *ABCC2* gene encoding the apically localized export pump MRP2, may affect the interindividual variability in MPA plasma concentrations. Interestingly, it has been demonstrated that the *SLCO1B1*5* allele also might influence MPA treatment as shown in a study with 218 renal transplant recipients receiving MPA from the first day of transplantation (Michelon et al., 2010).

As for OATP1B1, bilirubin and bilirubin glucuronide have been identified as substrates for OATP1B3 (Briz et al., 2003). A genome-wide association study scanning 4300 Sardinian individuals for identifying genetic variations responsible for alterations in bilirubin plasma levels (Sanna et al., 2009) identified a strong association for the regulation of serum bilirubin levels located on chromosome 12 within the *SLCO1B3* gene ($P = 3.9 \times 10^9$). This strong association was successfully confirmed in an independent Sardinian sample cohort and in a cohort of Amish individuals demonstrating that OATP1B3 is a plausible candidate gene responsible for alterations in bilirubin plasma levels. Interestingly, in this study, weaker associations for OATP1B1 were detected and the authors found that variations in the *SLCO1B3* gene are associated with altered unconjugated bilirubin levels, whereas polymorphism in the *SLCO1B1* gene might influence conjugated bilirubin transport (Sanna et al., 2009). It should be noted in this connection that a second genome-wide association study published in 2009 demonstrated that the *SLCO1B1*5* allele is associated with alterations in serum bilirubin levels (Johnson et al., 2009).

Interestingly, and despite the fact that expression of OATP1B3 has not been analyzed in leukocytes, a study has been published recently showing an association of the OATP1B3p.S112A polymorphism with the intracellular accumulation of imatinib in patients with chronic myeloid leukemia (Nambu et al., 2011). In this study, the authors demonstrated that the intracellular imatinib concentration was significantly higher in patients with the *SLCO1B3c.TT* genotype compared to those in the heterozygous *SLCO1B3c.TG/GG* genotype. These results demonstrated that possibly the so-called liver-specific transport might be expressed or upregulated in tissues or cell types not investigated so far.

7.4 GENETIC VARIATIONS IN THE *SLCO2B1* GENE ENCODING OATP2B1

7.4.1 Pharmacogenetics of OATP2B1—*In Vitro* and *In Vivo* Data

OATP2B1 is the third OATP family member highly expressed in human hepatocytes but unlike OATP1B1 and OATP1B3, OATP2B1 is also expressed in several other tissues and it is assumed that it is ubiquitously expressed (Grube et al., 2006a, 2006b). Drugs indentified as substrates for OATP2B1 include antibiotics (e.g., benzylpenicillin) (Tamai et al., 2001), the antihistaminic drug fexofenadine (Nozawa et al., 2004a), the leukotriene-receptor antagonist montelukast (Mougey et al., 2009), ezetimibe glucuronide, the metabolite of the antihyperlipidemic drug ezetimibe (Oswald et al., 2008) as well as several statins, for example atorvastatin (Grube et al., 2006b) and fluvastatin (Kopplow et al., 2005).

To this day, only a few variants in the *SLCO2B1* gene encoding human OATP2B1 have been described. The two variants *SLCO2B1*2* (*SLCO2B1c.1175T>C*; OATP2B1p.T392I) and *SLCO2B1*3* (*SLCO2B1c.1457C>T*; OATP2B1p.S486P) were identified in a study by Tamai et al. (Tamai et al., 2001) followed by the analysis of their allelic frequencies (Nozawa et al., 2002). This study also included the first characterization of the functional consequences of these variations using HEK293 cells expressing the mutated proteins. They found that the allelic frequencies in the Japanese population were 69.1% for the wild-type *SLCO2B1*1* protein, 0% for the *2 variant, and 30.9% for the *SLCO2B1*3*. Using estrone-3-sulfate as substrate it was shown that the V_{max} value of the *SLCO2B1*3* variant was decreased to 42.5% of *SLCO2B1*1*, whereas the K_m values of both proteins were comparable. A third variation in the *SLCO2B1* gene was identified by database analysis (rs12422149, resulting in *SLCO2B1c.G935A*; OATP2B1p.R312Q) and further functionally characterized. It was demonstrated (Mougey et al., 2009) that this variation is associated with altered pharmacokinetics of montelukast as analyzed *in vitro* by using stably transfected MDCKII cells as well as *in vivo* shown in patients expressing the *SLCO2B1*1* wild-type or the mutated OATP2B1pR312Q protein. This analysis has been extended by *in vivo* studies analyzing the effect of citrus juice on the pharmacokinetics of montelukast (Mougey et al., 2010). In this study the authors conducted a single-dose, pharmacokinetic analysis of montelukast co-ingested with citrus juice and they found that under these conditions the variant shows a significant reduction in AUC compared to the wild-type protein. In a second study analyzing the variant *SLCO2B1*3* and using apple juice as interacting compound for OATP2B1-mediated fexofenadine transport, Imanga et al. (2011) demonstrated that when fexofenadine was administered with water, subjects with *SLCO2B1c.1457C>T* allele showed a significant decrease in fexofenadine AUC compared with *SLCO2B1c.1457CC* subjects. When administered with apple juice, a significant decrease in fexofenadine AUC compared to the administration with water was observed and this decrease was lower in subjects with the *SLCO2B1c.1457C>T* allele. These results demonstrated that the pharmacokinetics of fexofenadine may be altered as a result of the OATP2B1 genotype.

The impact of regulatory polymorphisms in the genes encoding the three hepatic OATP family members has been analyzed by Aoki and colleagues (Aoki et al., 2009). Interestingly, this study demonstrates that the polymorphism *SLCO2B1g.-282G>A* is a major factor affecting the expression of the *SLCO2B1* mRNA suggesting that this variation contributes to interindividual differences in the protein levels of OATP2B1.

Taken together, the presented *in vitro* as well as *in vivo* studies demonstrated that several genetic variations in the transporter genes encoding the three hepatic OATP family members OATP1B1, OATP1B3, and OATP2B1 may influence the pharmacokinetics of drugs that are substrates for these uptake transporters. Continued investigations in this important research field will have an important impact on further optimizing drug therapy and for explaining the interindividual response to drug treatment.

REFERENCES

Abe T, Kakyo M, Tokui T, Nakagomi R, Nishio T, Nakai D, Nomura H, Unno M, Suzuki M, Naitoh T, et al. (1999) Identification of a novel gene family encoding human liver-specific organic anion transporter LST-1. *J Biol Chem*, 274, 17159–17163.

Allikmets R, Gerrard B, Hutchinson A, Dean M (1996) Characterization of the human ABC superfamily: isolation and mapping of 21 new genes using the expressed sequence tags database. *Hum Mol Genet*, 5, 1649–1655.

Aoki M, Terada T, Ogasawara K, Katsura T, Hatano E, Ikai I, Inui KI (2009) Impact of regulatory polymorphisms in organic anion transporter genes in the human liver. *Pharmacogenet Genomics*, 19, 647–656.

Aquilante CL, Bushman LR, Knutsen SD, Burt LE, Rome LC, Kosmiski LA (2008) Influence of SLCO1B1 and CYP2C8 gene polymorphisms on rosiglitazone pharmacokinetics in healthy volunteers. *Hum Genomics*, 3, 7–16.

Bachmakov I, Glaeser H, Fromm MF, König J (2008) Interaction of oral antidiabetic drugs with hepatic uptake transporters: focus on organic anion transporting polypeptides and organic cation transporter 1. *Diabetes*, 57, 1463–1469.

Batt AM, Magdalou J, Vincent-Viry M, Ouzzine M, Fournel-Gigleux S, Galteau MM, Siest G (1994) Drug metabolizing enzymes related to laboratory medicine: cytochromes P-450 and UDP-glucuronosyltransferases. *Clinica Chimica Acta*, 226, 171–190.

Bernsdorf A, Giessmann T, Modess C, Wegner D, Igelbrink S, Hecker U, Haenisch S, Cascorbi I, Terhaag B, Siegmund W (2006) Simvastatin does not influence the intestinal P-glycoprotein and MPR2, and the disposition of talinolol after chronic medication in healthy subjects genotyped for the ABCB1, ABCC2 and SLCO1B1 polymorphisms. *Br J Clin Pharmacol*, 61, 440–450.

Briz O, Serrano MA, MacIas RI, Gonzalez-Gallego J, Marin JJ (2003) Role of organic anion-transporting polypeptides, OATP-A, OATP-C and OATP-8, in the human placenta-maternal liver tandem excretory pathway for foetal bilirubin. *Biochem J*, 371, 897–905.

Briz O, Serrano MA, Rebollo N, Hagenbuch B, Meier PJ, Koepsell H, Marin JJ (2002) Carriers involved in targeting the cytostatic bile acid-cisplatin derivatives cis-diammine-chloro-cholylglycinate-platinum(II) and cis-diammine-bisursodeoxycholate-platinum(II) toward liver cells. *Mol Pharmacol*, 61, 853–860.

Brown WV (2001) Therapies on the horizon for cholesterol reduction. *Clinical Cardiology*, 24, III24–III27.

Choi JH, Lee MG, Cho JY, Lee JE, Kim KH, Park K (2008) Influence of OATP1B1 genotype on the pharmacokinetics of rosuvastatin in Koreans. *Clin Pharmacol Ther*, 83, 251–257.

Chung JY, Cho JY, Yu KS, Kim JR, Oh DS, Jung HR, Lim KS, Moon KH, Shin SG, Jang IJ (2005) Effect of OATP1B1 (SLCO1B1) variant alleles on the pharmacokinetics of pitavastatin in healthy volunteers. *Clin Pharmacol Ther*, 78, 342–350.

Couvert P, Giral P, Dejager S, Gu J, Huby T, Chapman MJ, Bruckert E, Carrie A (2008) Association between a frequent allele of the gene encoding OATP1B1 and enhanced LDL-lowering response to fluvastatin therapy. *Pharmacogenomics*, 9, 1217–1227.

Cui Y, König J, Leier I, Buchholz U, Keppler D (2001) Hepatic uptake of bilirubin and its conjugates by the human organic anion transporter SLC21A6. *J Biol Chem*, 276, 9626–9630.

Deng JW, Song IS, Shin HJ, Yeo CW, Cho DY, Shon JH, Shin JG (2008) The effect of SLCO1B1*15 on the disposition of pravastatin and pitavastatin is substrate dependent: the contribution of transporting activity changes by SLCO1B1*15. *Pharmacogenet Genomics*, 18, 424–433.

Elens L, Capron A, Kerckhove VV, Lerut J, Mourad M, Lison D, Wallemacq P, Haufroid V (2007) 1199G>A and 2677G>T/A polymorphisms of ABCB1 independently affect tacrolimus concentration in hepatic tissue after liver transplantation. *Pharmacogenet Genomics*, 17, 873–883.

Fahrmayr C, Fromm MF, König J (2010) Hepatic OATP and OCT uptake transporters: their role for drug-drug interactions and pharmacogenetic aspects. *Drug Metab Rev*, 42, 380–401.

Fehrenbach T, Cui Y, Faulstich H, Keppler D (2003) Characterization of the transport of the bicyclic peptide phalloidin by human hepatic transport proteins. *Naunyn Schmiedebergs Arch Pharmacol*, 368, 415–420.

Fischer WJ, Altheimer S, Cattori V, Meier PJ, Dietrich DR, Hagenbuch B (2005) Organic anion transporting polypeptides expressed in liver and brain mediate uptake of microcystin. *Toxicol Appl Pharmacol*, 203, 257–263.

Gao B, Huber RD, Wenzel A, Vavricka SR, Ismair MG, Reme C, Meier PJ (2005) Localization of organic anion transporting polypeptides in the rat and human ciliary body epithelium. *Exp Eye Res*, 80, 61–72.

Gerloff T, Schaefer M, Mwinyi J, Johne A, Sudhop T, Lutjohann D, Roots I, von Bergmann K (2006) Influence of the SLCO1B1*1b and *5 haplotypes on pravastatin's cholesterol lowering capabilities and basal sterol serum levels. *Naunyn Schmiedebergs Arch Pharmacol*, 373, 45–50.

Grube M, Kock K, Karner S, Reuther S, Ritter CA, Jedlitschky G, Kroemer HK (2006a) Modification of OATP2B1-mediated transport by steroid hormones. *Mol Pharmacol*, 70, 1735–1741.

Grube M, Kock K, Oswald S, Draber K, Meissner K, Eckel L, Bohm M, Felix SB, Vogelgesang S, Jedlitschky G, et al. (2006b) Organic anion transporting polypeptide 2B1 is a high-affinity transporter for atorvastatin and is expressed in the human heart. *Clin Pharmacol Ther*, 80, 607–620.

Hagenbuch B, Meier PJ (2003) The superfamily of organic anion transporting polypeptides. *Biochim Biophys Acta*, 1609, 1–18.

Han JY, Lim HS, Shin ES, Yoo YK, Park YH, Lee JE, Kim HT, Lee JS (2008) Influence of the organic anion-transporting polypeptide 1B1 (OATP1B1) polymorphisms on irinotecan-pharmacokinetics and clinical outcome of patients with advanced non-small cell lung cancer. *Lung Cancer*, 59, 69–75.

Hartkoorn RC, Kwan WS, Shallcross V, Chaikan A, Liptrott N, Egan D, Sora ES, James CE, Gibbons S, Bray PG, et al. (2010) HIV protease inhibitors are substrates for OATP1A2, OATP1B1 and OATP1B3 and lopinavir plasma concentrations are influenced by SLCO1B1 polymorphisms. *Pharmacogenet Genomics*, 20, 112–120.

Hediger MA, Romero MF, Peng JB, Rolfs A, Takanaga H, Bruford EA (2004) The ABCs of solute carriers: physiological, pathological and therapeutic implications of human membrane transport proteins. Introduction. *Pflugers Arch*, 447, 465–468.

Hedman M, Antikainen M, Holmberg C, Neuvonen M, Eichelbaum M, Kivisto KT, Neuvonen PJ, Niemi M (2006) Pharmacokinetics and response to pravastatin in paediatric patients with familial hypercholesterolaemia and in paediatric cardiac transplant recipients in relation to polymorphisms of the SLCO1B1 and ABCB1 genes. *Br J Clin Pharmacol*, 61, 706–715.

Hirano M, Maeda K, Shitara Y, Sugiyama Y (2004) Contribution of OATP2 (OATP1B1) and OATP8 (OATP1B3) to the hepatic uptake of pitavastatin in humans. *J Pharmacol Exp Ther*, 311, 139–146.

Ho RH, Tirona RG, Leake BF, Glaeser H, Lee W, Lemke CJ, Wang Y, Kim RB (2006) Drug and bile acid transporters in rosuvastatin hepatic uptake: function, expression, and pharmacogenetics. *Gastroenterology*, 130, 1793–1806.

Hsiang B, Zhu Y, Wang Z, Wu Y, Sasseville V, Yang WP, Kirchgessner TG (1999) A novel human hepatic organic anion transporting polypeptide (OATP2). Identification of a liver-specific human organic anion transporting polypeptide and identification of rat and human hydroxymethylglutaryl-CoA reductase inhibitor transporters. *J Biol Chem*, 274, 37161–37168.

Ieiri I, Higuchi S, Sugiyama Y (2009) Genetic polymorphisms of uptake (OATP1B1, 1B3) and efflux (MRP2, BCRP) transporters: implications for inter-individual differences in the pharmacokinetics and pharmacodynamics of statins and other clinically relevant drugs. *Expert Opin Drug Metab Toxicol*, 5, 703–729.

Ieiri I, Suwannakul S, Maeda K, Uchimaru H, Hashimoto K, Kimura M, Fujino H, Hirano M, Kusuhara H, Irie S, Higuchi S, Sugiyama Y (2007) SLCO1B1 (OATP1B1, an uptake transporter) and ABCG2 (BCRP, an efflux transporter) variant alleles and pharmacokinetics of pitavastatin in healthy volunteers. *Clin Pharmacol Ther*, 82, 541–547.

Igel M, Arnold KA, Niemi M, Hofmann U, Schwab M, Lutjohann D, von Bergmann K, Eichelbaum M, Kivisto KT (2006) Impact of the SLCO1B1 polymorphism on the pharmacokinetics and lipid-lowering efficacy of multiple-dose pravastatin. *Clin Pharmacol Ther*, 79, 419–426.

Iida A, Saito S, Sekine A, Mishima C, Kondo K, Kitamura Y, Harigae S, Osawa S, Nakamura Y (2001) Catalog of 258 single-nucleotide polymorphisms (SNPs) in genes encoding three organic anion transporters, three organic anion-transporting polypeptides, and three NADH:ubiquinone oxidoreductase flavoproteins. *J Hum Genet*, 46, 668–683.

Imanaga J, Kotegawa T, Imai H, Tsutsumi K, Yoshizato T, Ohyama T, Shirasaka Y, Tamai I, Tateishi T, Ohashi K (2011) The effects of the SLCO2B1 c.1457C > T polymorphism and apple juice on the pharmacokinetics of fexofenadine and midazolam in humans. *Pharmacogenet Genomics*, 21, 84–93.

Iwai M, Suzuki H, Ieiri I, Otsubo K, Sugiyama Y (2004) Functional analysis of single nucleotide polymorphisms of hepatic organic anion transporter OATP1B1 (OATP-C). *Pharmacogenetics*, 14, 749–757.

Jacquemin E, Hagenbuch B, Stieger B, Wolkoff AW, Meier PJ (1994) Expression cloning of a rat liver Na(+)-independent organic anion transporter. *Proc Natl Acad Sci U S A*, 91, 133–137.

Johnson AD, Kavousi M, Smith AV, Chen MH, Dehghan A, Aspelund T, Lin JP, van Duijn CM, Harris TB, Cupples LA, et al. (2009) Genome-wide association meta-analysis for total serum bilirubin levels. *Hum Mol Genet*, 18, 2700–2710.

Kalliokoski A, Backman JT, Neuvonen PJ, Niemi M (2008a) Effects of the SLCO1B1*1B haplotype on the pharmacokinetics and pharmacodynamics of repaglinide and nateglinide. *Pharmacogenet Genomics*, 18, 937–942.

Kalliokoski A, Neuvonen M, Neuvonen PJ, Niemi M (2008b) Different effects of SLCO1B1 polymorphism on the pharmacokinetics and pharmacodynamics of repaglinide and nateglinide. *J Clin Pharmacol*, 48, 311–321.

Kameyama Y, Yamashita K, Kobayashi K, Hosokawa M, Chiba K (2005) Functional characterization of SLCO1B1 (OATP-C) variants, SLCO1B1*5, SLCO1B1*15 and SLCO1B1*15+C1007G, by using transient expression systems of HeLa and HEK293 cells. *Pharmacogenet Genomics*, 15, 513–522.

Katz DA, Carr R, Grimm DR, Xiong H, Holley-Shanks R, Mueller T, Leake B, Wang Q, Han L, Wang PG, et al. (2006) Organic anion transporting polypeptide 1B1 activity classified by SLCO1B1 genotype influences atrasentan pharmacokinetics. *Clin Pharmacol Ther*, 79, 186–196.

Keisner SV, Shah, SR (2011) Pazopanib: the newest tyrosine kinase inhibitor for the treatment of advanced or metastatic renal cell carcinoma. *Drugs*, 71, 443–454.

Keppler D, Kamisako T, Leier I, Cui Y, Nies AT, Tsujii H, König J (2000) Localization, substrate specificity, and drug resistance conferred by conjugate export pumps of the MRP family. *Adv Enzyme Regul*, 40, 339–349.

Kobayashi D, Nozawa T, Imai K, Nezu J, Tsuji A, Tamai I (2003) Involvement of human organic anion transporting polypeptide OATP-B (SLC21A9) in pH-dependent transport across intestinal apical membrane. *J Pharmacol Exp Ther*, 306, 703–708.

König J, Cui Y, Nies AT, Keppler D (2000a) Localization and genomic organization of a new hepatocellular organic anion transporting polypeptide. *J Biol Chem*, 275, 23161–23168.

König J, Cui Y, Nies AT, Keppler D (2000b) A novel human organic anion transporting polypeptide localized to the basolateral hepatocyte membrane. *Am J Physiol Gastrointest Liver Physiol*, 278, G156 G164.

König J, Seithel A, Gradhand U, Fromm MF (2006) Pharmacogenomics of human OATP transporters. *Naunyn Schmiedebergs Arch Pharmacol*, 372, 432–443.

Kopplow K, Letschert K, König J, Walter B, Keppler D (2005) Human hepatobiliary transport of organic anions analyzed by quadruple-transfected cells. *Mol Pharmacol*, 68, 1031–1038.

Kruh GD, Zeng H, Rea PA, Liu G, Chen ZS, Lee K, Belinsky MG (2001) MRP subfamily transporters and resistance to anticancer agents. *J Bioenerg Biomembr*, 33, 493–501.

Kullak-Ublick GA, Hagenbuch B, Stieger B, Wolkoff AW, Meier PJ (1994) Functional characterization of the basolateral rat liver organic anion transporting polypeptide. *Hepatology*, 20, 411–416.

Kullak-Ublick GA, Ismair MG, Stieger B, Landmann L, Huber R, Pizzagalli F, Fattinger K, Meier PJ, Hagenbuch B (2001) Organic anion-transporting polypeptide B (OATP-B) and its functional comparison with three other OATPs of human liver. *Gastroenterology*, 120, 525–533.

Kunta JR, Sinko PJ (2004) Intestinal drug transporters: in vivo function and clinical importance. *Curr Drug Metab*, 5, 109–124.

Kusuhara H, Sugiyama Y (2002) Role of transporters in the tissue-selective distribution and elimination of drugs: transporters in the liver, small intestine, brain and kidney. *J Control Release*, 78, 43–54.

Kusuhara H, Sugiyama Y (2005) Active efflux across the blood-brain barrier: role of the solute carrier family. *NeuroRx*, 2, 73–85.

Lee E, Ryan S, Birmingham B, Zalikowski J, March R, Ambrose H, Moore R, Lee C, Chen Y, Schneck D (2005) Rosuvastatin pharmacokinetics and pharmacogenetics in white and Asian subjects residing in the same environment. *Clin Pharmacol Ther*, 78, 330–341.

Leonhardt M, Keiser M, Oswald S, Kuhn J, Jia J, Grube M, Kroemer HK, Siegmund W, Weitschies W (2010) Hepatic uptake of the magnetic resonance imaging contrast agent Gd-EOB-DTPA: role of human organic anion transporters. *Drug Metab Dispos*, 38, 1024–1028.

Letschert K, Keppler D, König J (2004) Mutations in the SLCO1B3 gene affecting the substrate specificity of the hepatocellular uptake transporter OATP1B3 (OATP8). *Pharmacogenetics*, 14, 441–452.

Link E, Parish S, Armitage J, Bowman L, Heath S, Matsuda F, Gut I, Lathrop M, Collins R (2008) SLCO1B1 variants and statin-induced myopathy–a genomewide study. *N Engl J Med*, 359, 789–799.

Liu L, Cui Y, Chung AY, Shitara Y, Sugiyama Y, Keppler D, Pang KS (2006) Vectorial transport of enalapril by Oatp1a1/Mrp2 and OATP1B1 and OATP1B3/MRP2 in rat and human livers. *J Pharmacol Exp Ther*, 318, 395–402.

Lu WJ, Tamai I, Nezu J, Lai ML, Huang JD (2006) Organic anion transporting polypeptide-C mediates arsenic uptake in HEK-293 cells. *J Biomed Sci*, 13, 525–533.

Maeda K, Ieiri I, Yasuda K, Fujino A, Fujiwara H, Otsubo K, Hirano M, Watanabe T, Kitamura Y, Kusuhara H, et al. (2006a) Effects of organic anion transporting polypeptide 1B1 haplotype on pharmacokinetics of pravastatin, valsartan, and temocapril. *Clin Pharmacol Ther*, 79, 427–439.

Maeda K, Kambara M, Tian Y, Hofmann AF, Sugiyama Y (2006b) Uptake of ursodeoxycholate and its conjugates by human hepatocytes: role of Na(+)-taurocholate cotransporting polypeptide (NTCP), organic anion transporting polypeptide (OATP) 1B1 (OATP-C), and oatp1B3 (OATP8). *Mol Pharm*, 3, 70–77.

Matsushima S, Maeda K, Hayashi H, Debori Y, Schinkel AH, Schuetz JD, Kusuhara H, Sugiyama Y (2008) Involvement of multiple efflux transporters in hepatic disposition of fexofenadine. *Mol Pharmacol*, 73, 1474–1483.

Meier-Abt F, Faulstich H, Hagenbuch B (2004) Identification of phalloidin uptake systems of rat and human liver. *Biochim Biophys Acta*, 1664, 64–69.

Michalski C, Cui Y, Nies AT, Nuessler AK, Neuhaus P, Zanger UM, Klein K, Eichelbaum M, Keppler D, König J (2002) A naturally occurring mutation in the SLC21A6 gene causing impaired membrane localization of the hepatocyte uptake transporter. *J Biol Chem*, 277, 43058–43063.

Michelon H, König J, Durrbach A, Quteineh L, Verstuyft C, Furlan V, Ferlicot S, Letierce A, Charpentier B, Fromm MF, et al. (2010) SLCO1B1 genetic polymorphism influences mycophenolic acid tolerance in renal transplant recipients. *Pharmacogenomics*, 11, 1703–1713.

Miura M, Satoh S, Inoue K, Kagaya H, Saito M, Inoue T, Suzuki T, Habuchi T (2007) Influence of SLCO1B1, 1B3, 2B1 and ABCC2 genetic polymorphisms on mycophenolic acid pharmacokinetics in Japanese renal transplant recipients. *Eur J Clin Pharmacol*, 63, 1161–1169.

Morimoto K, Oishi T, Ueda S, Ueda M, Hosokawa M, Chiba K (2004) A novel variant allele of OATP-C (SLCO1B1) found in a Japanese patient with pravastatin-induced myopathy. *Drug Metab Pharmacokinet*, 19, 453–455.

Mougey EB, Feng H, Castro M, Irvin CG, Lima JJ (2009) Absorption of montelukast is transporter mediated: a common variant of OATP2B1 is associated with reduced plasma concentrations and poor response. *Pharmacogenet Genomics*, 19, 129–138.

Mougey EB, Lang JE, Wen X, Lima JJ (2010) Effect of citrus juice and SLCO2B1 genotype on the pharmacokinetics of montelukast. *J Clin Pharmacol*, 51, 751–760.

Mwinyi J, Johne A, Bauer S, Roots I, Gerloff T (2004) Evidence for inverse effects of OATP-C (SLC21A6) 5 and 1b haplotypes on pravastatin kinetics. *Clin Pharmacol Ther*, 75, 415–421.

Nakakariya M, Shimada T, Irokawa M, Maeda T, Tamai I (2008) Identification and species similarity of OATP transporters responsible for hepatic uptake of beta-lactam antibiotics. *Drug Metab Pharmacokinet*, 23, 347–355.

Nambu T, Hamada A, Nakashima R, Yuki M, Kawaguchi T, Mitsuya H, Saito H (2011) Association of SLCO1B3 polymorphism with intracellular accumulation of imatinib in leukocytes in patients with chronic myeloid leukemia. *Biol Pharm Bull*, 34, 114–119.

Neuvonen PJ, Backman JT, Niemi M (2008) Pharmacokinetic comparison of the potential over-the-counter statins simvastatin, lovastatin, fluvastatin and pravastatin. *Clin Pharmacokinet*, 47, 463–474.

Neuvonen PJ, Kantola T, Kivisto KT (1998) Simvastatin but not pravastatin is very susceptible to interaction with the CYP3A4 inhibitor itraconazole. *Clin Pharmacol Ther*, 63, 332–341.

Ni W, Ji J, Dai Z, Papp A, Johnson AJ, Ahn S, Farley KL, Lin TS, Dalton JT, Li X, et al. (2010) Flavopiridol pharmacogenetics: clinical and functional evidence for the role of SLCO1B1/OATP1B1 in flavopiridol disposition. *PLoS One*, 5, e13792.

Niemi M, Backman JT, Kajosaari LI, Leathart JB, Neuvonen M, Daly AK, Eichelbaum M, Kivisto KT, Neuvonen PJ (2005a) Polymorphic organic anion transporting polypeptide 1B1 is a major determinant of repaglinide pharmacokinetics. *Clin Pharmacol Ther*, 77, 468–478.

Niemi M, Kivisto KT, Diczfalusy U, Bodin K, Bertilsson L, Fromm MF, Eichelbaum M (2006a) Effect of SLCO1B1 polymorphism on induction of CYP3A4 by rifampicin. *Pharmacogenet Genomics*, 16, 565–568.

Niemi M, Kivisto KT, Hofmann U, Schwab M, Eichelbaum M, Fromm MF (2005b) Fexofenadine pharmacokinetics are associated with a polymorphism of the SLCO1B1 gene (encoding OATP1B1). *Br J Clin Pharmacol*, 59, 602–604.

Niemi M, Neuvonen PJ, Hofmann U, Backman JT, Schwab M, Lutjohann D, von Bergmann K, Eichelbaum M, Kivisto KT (2005c) Acute effects of pravastatin on cholesterol synthesis are associated with SLCO1B1 (encoding OATP1B1) haplotype *17. *Pharmacogenet Genomics*, 15, 303–309.

Niemi M, Pasanen MK, Neuvonen PJ (2006b) SLCO1B1 polymorphism and sex affect the pharmacokinetics of pravastatin but not fluvastatin. *Clin Pharmacol Ther*, 80, 356–366.

Niemi M, Pasanen MK, Neuvonen PJ (2011) Organic anion transporting polypeptide 1B1: a genetically polymorphic transporter of major importance for hepatic drug uptake. *Pharmacol Rev*, 63, 157–181.

Niemi M, Schaeffeler E, Lang T, Fromm MF, Neuvonen M, Kyrklund C, Backman JT, Kerb R, Schwab M, Neuvonen PJ, et al. (2004) High plasma pravastatin concentrations are associated with single nucleotide polymorphisms and haplotypes of organic anion transporting polypeptide-C (OATP-C, SLCO1B1). *Pharmacogenetics*, 14, 429–440.

Nishizato Y, Ieiri I, Suzuki H, Kimura M, Kawabata K, Hirota T, Takane H, Irie S, Kusuhara H, Urasaki Y, et al. (2003) Polymorphisms of OATP-C (SLC21A6) and OAT3 (SLC22A8) genes: consequences for pravastatin pharmacokinetics. *Clin Pharmacol Ther*, 73, 554–565.

Nozawa T, Imai K, Nezu J, Tsuji A, Tamai I (2004a) Functional Characterization of pH-sensitive organic anion transporting polypeptide OATP-B in human. *J Pharmacol Exp Ther*, 308, 438–445.

Nozawa T, Minami H, Sugiura S, Tsuji A, Tamai I (2005) Role of organic anion transporter OATP1B1 (OATP-C) in hepatic uptake of irinotecan and its active metabolite, 7-ethyl-10-hydroxycamptothecin: in vitro evidence and effect of single nucleotide polymorphisms. *Drug Metab Dispos*, 33, 434–439.

Nozawa T, Nakajima M, Tamai I, Noda K, Nezu J, Sai Y, Tsuji A, Yokoi T (2002) Genetic polymorphisms of human organic anion transporters OATP-C (SLC21A6) and OATP-B (SLC21A9): allele frequencies in the Japanese population and functional analysis. *J Pharmacol Exp Ther*, 302, 804–813.

Nozawa T, Sugiura S, Nakajima M, Goto A, Yokoi T, Nezu J, Tsuji A, Tamai I (2004b) Involvement of organic anion transporting polypeptides in the transport of troglitazone sulfate: implications for understanding troglitazone hepatotoxicity. *Drug Metab Dispos*, 32, 291–294.

Nozawa T, Tamai I, Sai Y, Nezu J, Tsuji A (2003) Contribution of organic anion transporting polypeptide OATP-C to hepatic elimination of the opioid pentapeptide analogue [D-Ala2, D-Leu5]-enkephalin. *J Pharm Pharmacol*, 55, 1013–1020.

Okudaira N, Sugiyama Y (1996) Use of an isolated perfused kidney to assess renal clearance of drugs: information obtained in steady-state and non-steady-state experimental systems. *Pharm Biotechnol*, 8, 211–238.

Oostendorp RL, van de Steeg E, van der Kruijssen CM, Beijnen JH, Kenworthy KE, Schinkel AH, Schellens JH (2009) Organic anion-transporting polypeptide 1B1 mediates transport of gimatecan and BNP1350 and can be inhibited by several classic ATP-binding cassette (ABC) B1 and/or ABCG2 inhibitors. *Drug Metab Dispos*, 37, 917–923.

Oswald S, König J, Lutjohann D, Giessmann T, Kroemer HK, Rimmbach C, Rosskopf D, Fromm MF, Siegmund W (2008) Disposition of ezetimibe is influenced by polymorphisms of the hepatic uptake carrier OATP1B1. *Pharmacogenet Genomics*, 18, 559–568.

Oswald S, Nassif A, Modess C, Keiser M, Hanke U, Engel A, Lutjohann D, Weitschies W, Siegmund W (2010) Pharmacokinetic and pharmacodynamic interactions between the immunosuppressant sirolimus and the lipid-lowering drug ezetimibe in healthy volunteers. *Clin Pharmacol Ther*, 87, 663–667.

Pacyniak E, Roth M, Hagenbuch B, Guo GL (2010) Mechanism of polybrominated diphenyl ether uptake into the liver: PBDE congeners are substrates of human hepatic OATP transporters. *Toxicol Sci*, 115, 344–353.

Pasanen MK, Fredrikson H, Neuvonen PJ, Niemi M (2007) Different effects of SLCO1B1 polymorphism on the pharmacokinetics of atorvastatin and rosuvastatin. *Clin Pharmacol Ther*, 82, 726–733.

Pasanen MK, Neuvonen M, Neuvonen PJ, Niemi M (2006) SLCO1B1 polymorphism markedly affects the pharmacokinetics of simvastatin acid. *Pharmacogenet Genomics*, 16, 873–879.

Pasanen MK, Neuvonen PJ, Niemi M (2008) Global analysis of genetic variation in SLCO1B1. *Pharmacogenomics*, 9, 19–33.

Petzinger E (1994) Transport of organic anions in the liver. An update on bile acid, fatty acid, monocarboxylate, anionic amino acid, cholephilic organic anion, and anionic drug transport. *Rev Physiol Biochem Pharmacol*, 123, 47–211.

Picard N, Yee SW, Woillard JB, Lebranchu Y, Le Meur Y, Giacomini KM, Marquet P (2010) The role of organic anion-transporting polypeptides and their common genetic variants in mycophenolic acid pharmacokinetics. *Clin Pharmacol Ther*, 87, 100–108.

Poirier A, Cascais AC, Funk C, Lave T (2009) Prediction of pharmacokinetic profile of valsartan in human based on in vitro uptake transport data. *J Pharmacokinet Pharmacodyn*, 36, 585–611.

Sandhu P, Lee W, Xu X, Leake BF, Yamazaki M, Stone JA, Lin JH, Pearson PG, Kim RB (2005) Hepatic uptake of the novel antifungal agent caspofungin. *Drug Metab Dispos*, 33, 676–682.

Sanna S, Busonero F, Maschio A, McArdle PF, Usala G, Dei M, Lai S, Mulas A, Piras MG, Perseu L, et al. (2009) Common variants in the SLCO1B3 locus are associated with bilirubin levels and unconjugated hyperbilirubinemia. *Hum Mol Genet*, 18, 2711–2718.

Sasaki M, Suzuki H, Ito K, Abe T, Sugiyama Y (2002) Transcellular transport of organic anions across a double-transfected Madin-Darby canine kidney II cell monolayer expressing both human organic anion-transporting polypeptide (OATP2/SLC21A6) and multidrug resistance-associated protein 2 (MRP2/ABCC2). *J Biol Chem*, 277, 6497–6503.

Schwenk M (1987) Drug transport in intestine, liver and kidney. *Archives of Toxicology*, 60, 37–42.

Seithel A, Eberl S, Singer K, Auge D, Heinkele G, Wolf NB, Dorje F, Fromm MF, König J (2007) The influence of macrolide antibiotics on the uptake of organic anions and drugs mediated by OATP1B1 and OATP1B3. *Drug Metab Dispos*, 35, 779–786.

Shimizu M, Fuse K, Okudaira K, Nishigaki R, Maeda K, Kusuhara H, Sugiyama Y (2005) Contribution of OATP (organic anion-transporting polypeptide) family transporters to the hepatic uptake of fexofenadine in humans. *Drug Metab Dispos*, 33, 1477–1481.

Shitara Y, Itoh T, Sato H, Li AP, Sugiyama Y (2003) Inhibition of transporter-mediated hepatic uptake as a mechanism for drug-drug interaction between cerivastatin and cyclosporin A. *J Pharmacol Exp Ther*, 304, 610–616.

Smith NF, Marsh S, Scott-Horton TJ, Hamada A, Mielke S, Mross K, Figg WD, Verweij J, McLeod HL, Sparreboom A (2007) Variants in the SLCO1B3 gene: interethnic distribution and association with paclitaxel pharmacokinetics. *Clin Pharmacol Ther*, 81, 76–82.

St-Pierre MV, Hagenbuch B, Ugele B, Meier PJ, Stallmach T (2002) Characterization of an organic anion-transporting polypeptide (OATP-B) in human placenta. *J Clin Endocrinol Metab*, 87, 1856–1863.

Suwannakul S, Ieiri I, Kimura M, Kawabata K, Kusuhara H, Hirota T, Irie S, Sugiyama Y, Higuchi S (2008) Pharmacokinetic interaction between pravastatin and olmesartan in relation to SLCO1B1 polymorphism. *J Hum Genet*, 53, 899–904.

Suzuki H, Sugiyama Y (2000) Transport of drugs across the hepatic sinusoidal membrane: sinusoidal drug influx and efflux in the liver. *Semin Liver Dis*, 20, 251–263.

Tachibana-Iimori R, Tabara Y, Kusuhara H, Kohara K, Kawamoto R, Nakura J, Tokunaga K, Kondo I, Sugiyama Y, Miki T (2004) Effect of genetic polymorphism of OATP-C (SLCO1B1) on lipid-lowering response to HMG-CoA reductase inhibitors. *Drug Metab Pharmacokinet*, 19, 375–380.

Takane H, Miyata M, Burioka N, Shigemasa C, Shimizu E, Otsubo K, Ieiri I (2006) Pharmacogenetic determinants of variability in lipid-lowering response to pravastatin therapy. *J Hum Genet*, 51, 822–826.

Tamai I, Nozawa T, Koshida M, Nezu J, Sai Y, Tsuji A (2001) Functional characterization of human organic anion transporting polypeptide B (OATP-B) in comparison with liver-specific OATP-C. *Pharm Res*, 18, 1262–1269.

Tirona RG, Leake BF, Merino G, Kim RB (2001) Polymorphisms in OATP-C: identification of multiple allelic variants associated with altered transport activity among European- and African-Americans. *J Biol Chem*, 276, 35669–35675.

Tirona RG, Leake BF, Wolkoff AW, Kim RB (2003) Human organic anion transporting polypeptide-C (SLC21A6) is a major determinant of rifampin-mediated pregnane X receptor activation. *J Pharmacol Exp Ther*, 304, 223–228.

Treiber A, Schneiter R, Hausler S, Stieger B (2007) Bosentan is a substrate of human OATP1B1 and OATP1B3: inhibition of hepatic uptake as the common mechanism of its interactions with cyclosporin A, rifampicin, and sildenafil. *Drug Metab Dispos*, 35, 1400–1407.

Tsuda-Tsukimoto M, Maeda T, Iwanaga T, Kume T, Tamai I (2006) Characterization of hepatobiliary transport systems of a novel alpha4beta1/alpha4beta7 dual antagonist, TR-14035. *Pharm Res*, 23, 2646–2656.

van der Deure WM, Hansen PS, Peeters RP, Kyvik KO, Friesema EC, Hegedus L, Visser TJ (2008) Thyroid hormone transport and metabolism by organic anion transporter 1C1 and consequences of genetic variation. *Endocrinology*, 149, 5307–5314.

van Montfoort JE, Hagenbuch B, Groothuis GM, Koepsell H, Meier PJ, Meijer DK (2003) Drug uptake systems in liver and kidney. *Curr Drug Metab*, 4, 185–211.

Vavricka SR, Van Montfoort J, Ha HR, Meier PJ, Fattinger K (2002) Interactions of rifamycin SV and rifampicin with organic anion uptake systems of human liver. *Hepatology*, 36, 164–172.

Vormfelde SV, Toliat MR, Schirmer M, Meineke I, Nurnberg P, Brockmoller J (2008) The polymorphisms Asn130Asp and Val174Ala in OATP1B1 and the CYP2C9 allele *3 independently affect torsemide pharmacokinetics and pharmacodynamics. *Clin Pharmacol Ther*, 83, 815–817.

Weiner M, Peloquin C, Burman W, Luo CC, Engle M, Prihoda TJ, Mac Kenzie WR, Bliven-Sizemore E, Johnson JL, Vernon A (2010) Effects of tuberculosis, race, and human gene SLCO1B1 polymorphisms on rifampin concentrations. *Antimicrobial Agents and Chemotherapy*, 54, 4192–4200.

Werner U, Werner D, Heinbuchner S, Graf B, Ince H, Kische S, Thurmann P, König J, Fromm MF, Zolk O (2010) Gender is an important determinant of the disposition of the loop diuretic torasemide. *J Clin Pharmacol*, 50, 160–168.

Xiang X, Han Y, Neuvonen M, Pasanen MK, Kalliokoski A, Backman JT, Laitila J, Neuvonen PJ, Niemi M (2009) Effect of SLCO1B1 polymorphism on the plasma concentrations of bile acids and bile acid synthesis marker in humans. *Pharmacogenet Genomics*, 19, 447–457.

Xiang X, Jada SR, Li HH, Fan L, Tham LS, Wong CI, Lee SC, Lim R, Zhou QY, Goh BC, et al. (2006) Pharmacogenetics of SLCO1B1 gene and the impact of *1b and *15 haplotypes on irinotecan disposition in Asian cancer patients. *Pharmacogenet Genomics*, 16, 683–691.

Yamada A, Maeda K, Kamiyama E, Sugiyama D, Kondo T, Shiroyanagi Y, Nakazawa H, Okano T, Adachi M, Schuetz JD, et al. (2007) Multiple human isoforms of drug transporters contribute to the hepatic and renal transport of olmesartan, a selective antagonist of the angiotensin II AT1-receptor. *Drug Metab Dispos*, 35, 2166–2176.

Zair ZM, Eloranta JJ, Stieger B, Kullak-Ublick GA (2008) Pharmacogenetics of OATP (SLC21/SLCO), OAT and OCT (SLC22) and PEPT (SLC15) transporters in the intestine, liver and kidney. *Pharmacogenomics*, 9, 597–624.

Zhang W, Chen BL, Ozdemir V, He YJ, Zhou G, Peng DD, Deng S, Xie QY, Xie W, Xu LY, et al. (2007) SLCO1B1 521T->C functional genetic polymorphism and lipid-lowering efficacy of multiple-dose pravastatin in Chinese coronary heart disease patients. *Br J Clin Pharmacol*, 64, 346–352.

Zhang W, He YJ, Han CT, Liu ZQ, Li Q, Fan L, Tan ZR, Zhang WX, Yu BN, Wang D, et al. (2006) Effect of SLCO1B1 genetic polymorphism on the pharmacokinetics of nateglinide. *Br J Clin Pharmacol*, 62, 567–572.

CHAPTER 8

OCT (SLC22A) AND OCTN FAMILY

Sophie L. Stocker
Arian Emami Riedmaier
Matthias Schwab
Kathleen M. Giacomini

8.1 INTRODUCTION

The organic ion transporter family, *SLC22*, includes 22 transporters in the human genome. Within the family, there are several subfamilies, which consist of members that cluster together based on sequence homology, including the organic anion transporters (OATs), the organic cation transporters (OCTs), and the novel organic cation transporters (OCTNs). The focus of this chapter is on the OCTs and the OCTNs. These two subfamilies are distinct not only based on sequence similarities but also on characteristic substrate and inhibitor specificities. In particular, OCTs favor positively charged substrates, specifically organic bases and quaternary amines, whereas OCTNs accept zwitterions as well as cationic substrates. Inhibitors of OCTs and OCTNs tend to have similar chemical structures as substrates but may be more hydrophobic.

Both common and rare genetic variants in the OCT and OCTN families have been identified and characterized. Importantly, a number of association studies have revealed that variants in the two major families are associated with risk for disease as well as variation in drug response. In particular, polymorphisms in OCTN1 and OCTN2 have been associated with risk for inflammatory bowel syndrome and systemic carnitine deficiency. In the OCT family, polymorphisms have generally been associated with variation in drug levels and response, though some polymorphisms have been associated with disease. This chapter is divided into two major sections, focused on the OCTs and the OCTNs, respectively. Within each section, following a description of the transporters and their polymorphisms, studies focusing on functional genomic and clinical implications of the transporter polymorphisms are presented. These studies clearly show that genetic variants in the OCTs and OCTNs are major contributors to variation in drug disposition, therapeutic and adverse drug response as well as risk for human disease.

Pharmacogenomics of Human Drug Transporters: Clinical Impacts, First Edition.
Edited by Toshihisa Ishikawa, Richard B. Kim and Jörg König

8.2 ORGANIC CATION TRANSPORTERS

8.2.1 Structure

In humans, OCT1, OCT2, and OCT3 are encoded by genes, *SLC22A1–3*, which are localized within a cluster on chromosome 6 (q26–27) (Koehler et al., 1997; Grundemann et al., 1998b; Grundemann and Schomig, 2000). Each of the three genes consists of 11 exons and 10 introns (Hayer et al., 1999; Grundemann and Schomig, 2000) and encodes proteins of 553 (OCT1), 555 (OCT2), and 556 (OCT3) amino acids. Hydropathy analyses suggest that the secondary structure of all three proteins consists of 12 transmembrane spanning α-helices, C- and N-termini inside the cell, an intracellular loop with phosphorylation sites between the sixth and seventh transmembrane domains (TMDs), and a large extracellular loop between the first and second TMDs containing glycosylation sites (Fig. 8.1). Based on pairwise comparisons of the human amino acid sequences, OCT1 and OCT2 are markedly homologous (70% identity, 84% similarity), whereas, OCT3 shows less sequence homology to both OCT1 (50%, 70%) and OCT2 (50%, 73%) (Grundemann et al., 1999).

8.2.2 Tissue Distribution and Membrane Localization

In humans, OCT1, OCT2, and OCT3 exhibit broad tissue distribution and are expressed in epithelial cells and neurons (Koepsell et al., 2003; Muller et al., 2005). OCT1 and OCT3 are also expressed in skeletal muscle and OCT3 is expressed in glial cells. In humans, OCT1 is most highly expressed in the liver but it is also expressed in the kidney, brain, and to a lesser extent in heart, smooth muscle, and skeletal muscle (Gorboulev et al., 1997; Zhang et al., 1997; Chen et al., 2010a). The tissue distribution of OCT2 is more restricted compared to that of OCT1 and OCT3. OCT2 is most strongly expressed in (all three segments) the kidney and also in the small intestine, lung, skin, brain, and heart (Gorboulev et al., 1997; Busch et al., 1998; Motohashi et al., 2002; Chen et al., 2010a). OCT3 has the broadest tissue distribution and is strongly expressed in placenta and blood vessels, as well as, skeletal and smooth muscle, liver, kidney, placenta, brain, and heart (Wu et al., 2000; Muller et al., 2005; Chen et al., 2010a). DNA methylation has been shown to modulate the expression of OCT2 in human kidney (Aoki et al., 2008) and OCT1 in human liver (Schaeffeler et al., 2011). In terms of subcellular expression, generally, OCT1, OCT2, and OCT3 are located on the basolateral membrane of epithelial cells, except in trachea and bronchial epithelial cells where they appear to be located on the luminal (apical) membrane (Lips et al., 2005). OCT3 is also expressed in the luminal membrane of enterocytes (Muller et al., 2005).

8.2.3 Substrate and Inhibitor Specificity

In general, OCTs transport positively charged compounds with a relative molecular mass below 500, in an electrogenic and sodium-independent manner (Koepsell et al.,

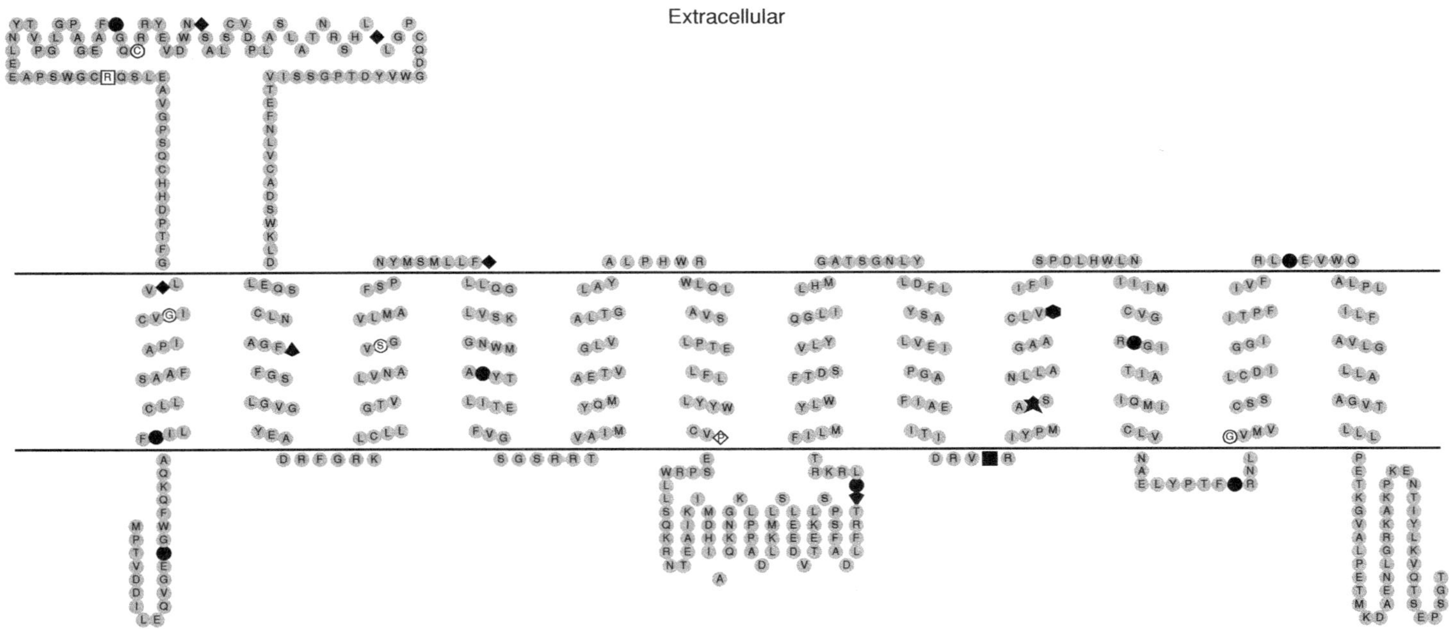

Figure 8.1 The secondary structure of OCT1 and coding polymorphisms. The different shapes represent genetic variants specific to an ethnic group. No SNP, light grey with white residue names; African-American only, black circle with white residue name ($n = 8$); Caucasian only, black circle outline with black residue name ($n = 4$); Japanese only, black diamond with white residue name ($n = 4$); Korean only, black diamond outline with black residue name ($n = 1$); African-Americans and Caucasian, black square with white residue name ($n = 1$); Caucasian and Mexican-American, black square outline with black residue name ($n = 1$); African-American and Caucasian and Mexican-American, black hexagon with white residue name ($n = 1$); African-American and Asian and Japanese and Korean, black down arrow with white residue name ($n = 1$); African-American and Caucasian and Asian and Japanese and Korean and Mexican-American, black up arrow with white residue name ($n = 1$); African-American and Caucasian and Asian and Japanese and Korean and Mexican-American and Pacific Islander, black star with white residue name ($n = 1$).

2007). OCTs are bidirectional, that is, they are able to translocate organic cations across the plasma membrane in either direction, that is, influx as well as efflux. The driving force is supplied by the electrochemical gradient of the transported organic cation. Representative substrates and inhibitors along with kinetic constants are listed in Table 8.1.

TABLE 8.1 Substrates and Inhibitors of Human OCT Transporters

Protein (Gene)	Model substrate (K_m μM)	Endogenous substrate (K_m μM)	Xenobiotics (K_m μM)	Inhibitors (K_i^a or IC_{50} μM)
OCT1	ASP+ (9.21)	Agmatine (1900)	Acyclovir (151)	Amantadine (**236**)
(SLC22A1)	MPP+ (15–25)	Prostaglandin E2 (0.66)	Gancyclovir (516)	Cimetidine (149)
	N-methylquinine (20)	Prostaglandin F2α (0.48)	Metformin (1060–5450)	Cocaine (**85**)
	TBuMA (53)		Ranitidine (70)	Decynium-22 (**2.73**)
	TEA (69–566)			Metformin (**3.7–493**)
				Midazolam (**3.7**)
				Omeprazole (15.7)
				Phenformin (**13–23**)
				Quinidine (**17.5**)
				Ranitidine (28–33)
OCT2	ASP+ (24)	Acetylcholine (117)	Amantadine (27)	Amantadine (**23–28**)
(SLC22A2)	MPP+ (8–25)	Agmatine (1800)	Amiloride (95)	Cimetidine (110–373)
	TEA (48–500)	Choline (102, 210)	Aminoguandine (4100)	Cocaine (**113**)
		Creatinine (4000, 5640)	Cimetidine (8.6, 73)	Decynium-22 (**0.1**)
		Dopamine (330–1400)	Famotidine (56)	MDMA (**1.63**)
		Epinephrine (420)	Memantine (34)	Metformin (**289–1380**)
				Omeprazole (6.7)
		Norepinephrine (1500–5452)	Metformin (680–3356)	Phenformin (**54–111**)
		Histamine (940, 1300)	Quinine (2.6)	Procainamide (**50**)
		5-Hydroxytryptamine (80, 290)	Paraquat (95–114)	Quinine (**2.9–3.4**)
		Prostaglandin E2 (0.03)	Propranolol (11.8)	Ranitidine (**31–79**)
		Prostaglandin F2α (0.33)	Ranitidine (265)	Trimethoprim (**49.7**)
		Ranitidine (65)	Varenicline (370)	Verapamil (13–85)
		Serotonin (80, 290)		
OCT3	MPP+ (47–114)	Dopamine (1200)	Etilefrine (180–2800)	Amantadine (>**1000**)
(SLC22A3)		Epinephrine (240)	Metformin (2260)	Cocaine (>**1000**)
		Histamine (140–220)		Desipramine (**14**)
		Norepinephrine (510, 2630)		MDMA (**73.6**)
		Serotonin (2500)		Metformin (**904**)
				Phenformin (**134**)
				Omeprazole (22)
				Procainamide (**738**)

Source: Data from http://bts.ucsf.edu/fdatransportal and Gorboulev et al., 1997; Zhang et al., 1997, 2006; Busch et al., 1998; Grundemann et al., 1999, 2003; Wu et al., 2000; Pietig et al., 2001; Van Montfoort et al., 2001; Barendt and Wright, 2002; Dresser et al., 2002; Leabman et al., 2002; Urakami et al., 2004; Bourdet et al., 2005; Muller et al., 2005; Sata et al., 2005; Tahara et al., 2005; Amphoux et al., 2006; Biermann et al., 2006; Chen et al., 2007, 2010b; Choi et al., 2007; Shu et al., 2007; Umehara et al., 2007; Feng et al., 2008; Kimura et al., 2009; Nies et al., 2009, 2011a; Ahlin et al., 2010; Burger et al., 2010; Imamura et al., 2010; Meyer zu Schwabedissen et al., 2010; More et al., 2010; Tzvetkov et al., 2012.

ASP+, 4-(4-dimethylamino)styryl-methylpyridinium; MPP+, 1-methyl-4-phenylpyridinium; TBuMA, tributylammonium; MDMA 3,4=methylenedioxy-N-methylamphetamin; TEA, tetraethylammonium.

[a]The K_i values are given in bold.

8.3 POLYMORPHISMS IN ORGANIC CATION TRANSPORTERS

The response of different people to the same medication varies substantially. It has been estimated that genetics can account for 20–95% of variability in drug disposition and effects (Kalow et al., 1999). Since OCTs play an important role in the metabolism of many medications, polymorphisms of OCTs can be of relevance in determining patient response to pharmacotherapy.

Several studies have reported polymorphisms in the genes encoding human OCT1, OCT2, and OCT3 from different ethnic groups (Table 8.2). Some of the single nucleotide polymorphisms (SNPs) have been characterized *in vitro* using various model organic cations (Table 8.3) and a select few have subsequently been characterized in human clinical studies (Table 8.4; Shu et al., 2007, 2008; Shikata et al., 2007; Song et al., 2008a; Chen et al., 2009). SNPs and deletion polymorphisms have been identified in promoter, intronic, and 3′ and 5′ untranslated and flanking regions. In comparison to coding regions, polymorphisms in noncoding regions are more abundant and minor allele frequencies are higher (Kang et al., 2007; Hesselson et al., 2009).

8.3.1 OCT1

Genotyping studies have identified over 200 SNPs in OCT1, some of which are ethnic-specific. Currently, 24 nonsynonymous variants have been identified in OCT1 with four (F160L, P341L, M420del, M408V) being present in multiple ethnic groups (Fig. 8.1). Overall, M408V is the most common nonsynonymous genetic variant. In Caucasian individuals, 14 nonsynonymous and six synonymous SNPs of OCT1 have been identified (Table 8.2). Of these, four nonsynonymous (G38A, C88R, S189L, and G465R), as well as two synonymous SNPs are specific to this ethnic group. Several studies have identified SNPs in OCT1 in Asian populations by direct sequencing of genomic DNA from 78 unrelated Japanese individuals (Saito et al., 2002; Chen et al., 2010b), 116 arrhythmic Japanese patients (Fukushima-Uesaka et al., 2004; Itoda et al., 2004), 99 Japanese diabetic patients (Shikata et al., 2007; Chen et al., 2010b), 150 unrelated healthy Koreans and 30 unrelated healthy Chinese (Chen et al., 2010b). To date, eight nonsynonymous and three synonymous SNPs have been identified in East Asian populations. (Table 8.2) of which the *in vitro* function of 21 SNPs has been determined (Table 8.3). Some of the nonsynonymous SNPs found in Japanese, Korean, and Chinese populations have also been detected in other populations, including Caucasian (P160L and M408V) (Kerb et al., 2002) and Asian-American (F160L, P341L, and M480V) (Shu et al., 2003), although the allele frequencies of most polymorphisms differ among the various ethnic groups. By contrast, some nonsynonymous SNPs appear to be found specifically in Japanese F41L, Q97K, P114L and R206C and Korean populations, (P283L) (Table 8.2).

Many of the nonsynonymous polymorphisms in OCT1 occur in conserved regions of the gene. For example, F160 is conserved for both OCT1 and OCT2, and P341 is conserved among OCT1, OCT2, and OCT3. Polymorphisms resulting

TABLE 8.2 Nonsynonymous Genetic Variants and Allele Frequencies in the Genes Encoding OCT1 (*SLC22A1*), OCT2 (*SLC22A2*) and OCT3 (*SLC22A3*) in Various Populations

					Allele frequency							
Gene	SNP ID	Location	Nucleotide change	Amino acid change	AA	CA	AS	CH	JP	KO	ME	PA
OCT1	rs3447885	Exon 1	41C>T	Ser14Phe	0.031	0	0		0		0	
	rs34570655	Exon 1	67C>G	Leu23Val	0.005	0	0				0	
	rs35888596	Exon 1	113G>A	Gly38Asp	0	0.005	0				0	
	rs2297373	Exon 1	123C>G	Phe41Leu					0.004			
	rs12208357	Exon 1	181C>T	Arg61Cys	0	0.072	0		0		0.056	0
	rs35546288	Exon 1	253C>T	Leu85Phe	0.010	0	0				0	0
	rs55918055	Exon 1	262T>C	Cys88Arg		0.003						
	–	Exon 1	289C>A	Gln97Lys					0.017			
	–	Exon 1	350C>T	Pro117Leu					0.004–0.023			
	rs683369	Exon 2	8237C>G	Phe160Leu	0.005	0.065	0.017		0.086–0.88	0.130	0.050	0
	rs34104736	Exon 3	566C>T	Ser189Leu	0	0.005	0				0	0
	–	Exon 3	616C>T	Arg206Cys					0.008			
	rs36103319	Exon 3	659G>T	Gly220Val	0.005	0	0				0	0
	rs4646277	Exon 5	848C>T	Pro283Leu						0.020		
	rs2282143	Exon 6	1022A>G	Pro341Leu	0.082	0	0.117		0.168–0.19	0.160	0	0
	rs34205214	Exon 6	1025G>A	Arg342His	0.031	0	0				0	0
	rs34130495	Exon 7	1201G>A	Gly401Ser	0.007	0.011	0				0	0
	rs628031	Exon 7	1222A>G	Met408Val	0.735	0.598	0.762		0.15–0.810	0.740	0.786	0.929
	rs35167514	Exon 7	1258delA	Met420del	0.029	0.185	0				0.214	0
	rs34305973	Exon 7	1258delT	Met420del	0.029	0.185	0				0.214	0

	rs35191146	Exon 7	1258delG	Met420del	0.029	0.185	0				0.214	0
	rs35956182	Exon 8	1320G>A	Met440Ile	0.005	0	0				0	0
	rs34295611	Exon 8	1381G>A	Val461Ile	0.010	0	0				0	0
	rs34059508	Exon 9	1393G>A	Gly465Arg	0	0.04	0				0	0
	rs35270274	Exon 9	1463G>T	Arg488Met	0.05	0	0				0	0
OCT2	–	Exon 1	133T>insTA	Phe45INS	0	0.005	0				0	0
	rs8177504	Exon 1	160C>T	Pro54Ser	0.005	0	0				0	0
	rs8177509	Exon 2	481T>C	Phe161Leu	0	0.005	0				0	0
	rs8177508	Exon 2	493A>G	Met165Val	0.005	0	0		0		0	0
	rs8177507	Exon 2	495G>A	Met165Ile	0.015	0	0				0	0
	–	Exon 3	596C>T	Thr199Ile					0.009	0.010		
	–	Exon 3	602C>T	Thr201Met				0.004	0.013	0.020		
	rs316019	Exon 4	808G>T	Ala270Ser	0.110	0.158	0.086	0.133	0.168	0.140	0.150	0.071
	rs8177513	Exon 5	890C>G	Ala297Gly	0.010	0.010	0				0	0.071
	rs8177516	Exon 7	1198C>T	Arg400Cys	0.015	0	0		0		0	0
	rs8177517	Exon 8	1294A>C	Lys432Gln	0.010	0	0		0		0.050	0
OCT3	rs8187715	Exon 1	131C>T	Thr44Met	0.006	0.006	0				0	0.143
	rs8187717	Exon 1	346G>T	Ala116Ser	0.017	0	0				0	0
	rs8187725	Exon 7	1199C>T	Thr400Ile	0	0.005	0				0	0

Columns refer to the frequencies of each variable site in each ethnic group.

Source: Data from http://pharmacogenetics.ucsf.edu/ and Shu et al., 2003; Fukushima-Uesaka et al., 2004; Itoda et al., 2004; Sakata et al., 2004; Kang et al., 2007; Chen et al., 2010a.

AA, African-American; CA, Caucasian; AS, Asian-American; CH, Chinese; JP, Japanese; KO, Korean; ME, Mexican-American; PA, Pacific Islander.

TABLE 8.3 The *In Vitro* Function of Polymorphisms in the OCT1 (*SLC22A1*), OCT2 (*SLC22A2*) and OCT3 (*SLC22A3*) Genes

		Transport activity compared to reference[a]				
NCBI dbSNP	Cell system	MPP$^+$	Metformin	TEA	Transporter kinetics, surface expression, localization	Reference
OCT1						
S14F	Oocytes	190%	–	–	Lower K_m; higher V_{max}	(Shu et al., 2003)
	HEK293	–	60%	–	Similar K_m; lower V_{max}	(Shu et al., 2007)
R61C	HEK293	–	9%	–	Low expression, intracellular localization	(Shu et al., 2007; Nies et al., 2009; Ahlin et al., 2010)
	Oocytes	30%	–	–	Higher K_m; lower V_{max}	(Kerb et al., 2002; Shu et al., 2003)
L85F	Oocytes	Similar	–	–	ND	(Shu et al., 2003)
C88L	Oocytes	1.4%	–	6%	ND	(Kerb et al., 2002)
Q97K	HEK293	–	65%	–	Normal localization	(Chen et al., 2010b)
P117L	HEK293	–	55%	–	Normal localization	(Chen et al., 2010b)
F160L	Oocytes	Similar	–	130%	ND	(Kerb et al., 2002; Shu et al., 2003; Sakata et al., 2004)
	HEK293	–	Similar	–	ND	(Shu et al., 2007)
S189L	HEK293	–	20%	–	Similar K_m; lower V_{max}	(Shu et al., 2007)
	Oocytes	Similar	–	–	ND	(Shu et al., 2003)
R206C	HEK293	–	25%	–	Low expression	(Chen et al., 2010b)
G220V	HEK293	–	0%	–	ND	(Shu et al., 2007)
	Oocytes	0%	–	–	ND	(Shu et al., 2003)
P283L	HEK293	–	–	12%	Normal localization	(Takeuchi et al., 2003)
	Oocytes	30%	–	30%	ND	(Sakata et al., 2004)
R287G	HEK293	–	–	8%	Normal localization	(Takeuchi et al., 2003)
	Oocytes	30%	–	30%	ND	(Sakata et al., 2004)

P341L	HEK293	–	Similar	67%	Normal localization; lower K_m and V_{max} (TEA)	(Takeuchi et al., 2003; Shu et al., 2007)
	Oocytes	20–100%	Similar	57%	Higher K_m; lower V_{max} (MPP^+)	(Shu et al., 2003; Sakata et al., 2004; Song et al., 2008b)
R342H	HEK293	–	Similar	–	ND	(Shu et al., 2007)
	Oocytes	Similar	–	–	ND	(Shu et al., 2003)
G401S	Oocytes	0%	–	8%	ND	(Kerb et al., 2002; Shu et al., 2003, 2007)
	HEK293	–	11%	–	Lower V_{max}; similar K_m	(Shu et al., 2007)
M408V	HEK293	Similar	100–135%	–	Normal localization; higher hepatic mRNA expression; higher K_m and V_{max} (metformin)	(Shikata et al., 2007; Shu et al., 2007; Ahlin et al., 2010)
M420del	Oocytes	Similar	–	–	ND	(Kerb et al., 2002; Shu et al., 2003)
	HEK293	–	25–30%	–	Normal localization and expression; similar K_m; lower V_{max}	(Shu et al., 2007; Ahlin et al., 2010)
M440I	Oocytes	Similar	–	–	ND	(Shu et al., 2003)
V461I	Oocytes	Similar	–	–	ND	(Shu et al., 2003)
G465R	HEK293	–	0%	–	Low expression	(Shu et al., 2007; Ahlin et al., 2010)
	Oocytes	0%	–	–	Low expression	(Shu et al., 2003)
R488M	HEK293	Similar	Similar	–	ND	(Shu et al., 2007)
	Oocytes	Similar	–	–	ND	(Shu et al., 2003)

(continued)

TABLE 8.3 (*Continued*)

NCBI dbSNP	Cell system	Transport activity compared to reference[a]			Transporter kinetics, surface expression, localization	Reference
		MPP^+	Metformin	TEA		
OCT2						
F45ins	Oocytes	0%	–	–	ND	(Leabman et al., 2002; Fujita et al., 2006)
P54S	Oocytes	130%	–	–	ND	(Leabman et al., 2002; Fujita et al., 2006)
F161L	Oocytes	Similar	–	–	ND	(Leabman et al., 2002; Fujita et al., 2006)
M165I	Oocytes	Similar	–	–	Lower V_{max} (MPP^+)	(Leabman et al., 2002; Fujita et al., 2006)
M165V	Oocytes	Similar	–	–	ND	(Leabman et al., 2002; Fujita et al., 2006)
T199I	Oocytes	0%	0%	–	ND	(Song et al., 2008b)
T201M	Oocytes	0%	0%	–	ND	(Song et al., 2008b)
A270S	Oocytes	10%	10%	–	Higher K_i with TBA (MPP^+); higher K_m; lower V_{max}	(Leabman et al., 2002; Fujita et al., 2006; Kang et al., 2007; Song et al., 2008b)
	HEK293	–	150%	–	Normal localization; similar K_m but higher V_{max} (metformin) or lower V_{max} (metformin); 28% higher protein expression in variant	(Chen et al., 2009; Zolk et al., 2009a)
A297G	Oocytes	90%	–	–	ND	(Leabman et al., 2002; Fujita et al., 2006)

R400C	Oocytes	80%	–	–	Lower V_{max}; lower K_i with TBA (MPP^+)	(Leabman et al., 2002; Fujita et al., 2006)
K432Q	Oocytes	Similar	–	–	Lower K_m; lower K_i with TBA (MPP^+)	(Leabman et al., 2002; Fujita et al., 2006)
OCT3						
T44M	HEK293	150%	260%	–	Similar K_m; higher V_{max} (MPP+ and metformin); normal localization; 30% higher expression	(Chen et al., 2010a)
A116S	HEK293	Similar	90%	–	Similar K_m and V_{max}; normal localization; similar expression	(Chen et al., 2010a)
L186F	HEK293	Similar	Similar	–	Similar K_m and V_{max}; normal localization; similar expression	(Chen et al., 2010a)
V388M	HEK293	Similar	Similar	–	Similar K_m and V_{max}; normal localization; similar expression	(Chen et al., 2010a)
T400I	HEK293	Similar	25%	–	Higher K_m; similar V_{max}; normal localization; similar expression	(Chen et al., 2010a)
V423F	HEK293	Similar	50%	–	Higher K_m; similar V_{max}; normal localization; similar expression	(Chen et al., 2010a)

ins, insertion; MPP^+, 1-methyl-4-phenylpyridinium; TBA, tetrabutylammonium; ND, not determined; HEK293, human embryonic kidney 293 cells.
[a]Transport compared to reference cell line.

TABLE 8.4 Pharmacogenomic Studies Investigating the Genes and Genetic Polymorphisms Influencing Response to Metformin

Study population	Metformin dose	Gene(s) and variants studied	Pharmacokinetic/ pharmacodyanmic differences[a]
Diabetics treated with metformin for at least 1 month ($N = 33$; $M = 9$, $F = 24$) (Shikata et al., 2007)	ND[b]	OCT1, OCT2	PK: ND PD: M480V positive predictor of metformin efficacy (not significant), that is, decrease in HbA1c by more than 0.5% from baseline after 3 mo of metformin therapy
Caucasian-European diabetics started on metformin ($N = 132$; $M = 40$, $F = 62$) (Becker et al., 2009)	677 mg (average)[b]	OCT1-12 tagging SNPs	PK: ND PD: rs622342 higher HbA1c levels, that is, reduced glucose-lowering effect
Caucasian-European diabetics started on metformin ($N = 152$; $M = 38$, $F = 60$) (Becker et al., 2010a)	521 mg (average)[b]	OCT1-rs622342 MATE1-rs2289669	PK: ND PD: Higher HbA1c levels in homozygous variant rs622342 (OCT1) and reference rs228669 (MATE1)
Caucasian-European diabetics started on metformin ($N = 1{,}014$; $M = 534$, $F = 480$) (Zhou et al., 2009)	ND	OCT1-R61C, M420del	PK: ND PD: No effect on reduction in HbA1c after 18 mo of metformin treatment
Healthy Caucasian-Americans ($N = 12$) (Shu et al., 2008)	1000 mg single dose	OCT1-R61C, G401S, M420del, G465R	PK: Higher AUC; Higher C_{max}; Lower V_d/F PD: Higher plasma glucose levels
Healthy Koreans ($N = 26$) (Song et al., 2008a)	500 mg single dose	OCT2-T199I, T201M, A270S	PK: A270S caused higher AUC, C_{max} and lower V_d/F, renal clearance PD: ND
Healthy Caucasian and African-Americans ($N = 23$) (Chen et al., 2009)	850 mg single dose	OCT2 and A270S variant	PK: Higher renal clearance PD: ND

TABLE 8.4 *(Continued)*

Study population	Metformin dose	Gene(s) and variants studied	Pharmacokinetic/ pharmacodyanmic differences[a]
Healthy European-Caucasians ($N = 103$; $M = 103$, $F = 0$) (Tzvetkov et al., 2009)	500 mg single dose	OCT1, OCT2, OCT3	PK: Combination of R61C, G401S, M420del and G465R increases renal clearance of metformin PD: ND
Healthy Chinese ($N = 15$; $M = 15$, $F = 0$) (Wang et al., 2008b)	500 mg single dose	OCT2	PK: A270S lower renal clearance PD: ND

ND, not determined; *M*, male; *F*, female; PK, pharmacokinetic; PD, pharmacodynamic.
[a]Comparison with reference.
[b]Subjects being coadministered other oral hypoglycemic agents included in the study.

in substitutions of leucine at both these positions have been observed. In contrast, both P283 and R287 are conserved for OCT1-3 as well as for OCTN1-2. Amino acid substitutions of leucine and glycine, respectively, occur at these positions in OCT1. Interestingly, R287 is conserved not only in the OCT and OCTN families but also in the OAT family, suggesting a fundamental role of this residue for transport activities. Overall, as with other genes, the greatest level of genetic variation in the coding region of OCT1 is present in African-Americans, with eight nonsynonymous (S14F, L23V, L85F, G220V, R342H, M440I, V461I and R488M) and 14 synonymous SNPs only found in this population.

8.3.2 OCT2

In comparison to the liver-specific OCT1, OCT2 is less diverse in terms of nonsynonymous genetic variation, and resultant functional changes (Koepsell et al., 2003). The ratio of nonsynonymous to synonymous nucleotide changes is significantly lower than the reported genetic variations in a population of more than 75 other genes (Cargill et al., 1999; Halushka et al., 1999). The higher degree of amino acid conservation in OCT2 implicates a higher selective pressure, underscoring the biological importance of OCT2, which is probably related to the elimination and detoxification of important endogenous compounds or compounds derived from the diet. Indeed, OCT2 has been shown to play a major role in brain choline homeostasis by mediating choline transport across the ventricular membrane of the choroid plexus (Sweet et al., 2001). Despite the limited amino acid variation, OCT2 appears to be more diverse in terms of both functional activity and heterozygosity than the organic anion transporter OAT1 (*SLC22A6*), which is also localized to the basolateral membrane of the human proximal tubule. Several genetic variants of OCT2 have been identified with very ethnic-specific frequencies (Leabman et al., 2002; Fukushima-Uesaka et al., 2004; Kang et al., 2007). For example, P54S, M165V, M165I, R400C polymorphisms in OCT2 have only been found in African-Americans (Leabman et al., 2002). Similarly, T199I and T201M are only present in Japanese, Korean, and Chinese populations (Kang et al., 2007; Song et al., 2008a; Chen et al., 2010b) and K432Q has been found in both African-American and Mexican-American populations. By contrast, A270S is present in all populations studied and has allele frequencies in the various populations ranging from 10% to 30% (Table 8.2). Interestingly, this polymorphism is the only one identified, thus far, at a frequency >1% in Caucasians.

In an ethnically diverse population, Leabman et al. (2002) identified 28 SNPs in OCT2, including eight nonsynonymous amino acid changes (P54S, P161L, M165V, and A297G), of which four (M165I, A270S, R400C, and K432Q) were present at a frequency of >1%. These four most common SNPs also exhibited altered transporter function (Table 8.3). In Japanese, 3 nonsynonymous and 20 synonymous SNPs have been identified (Fukushima-Uesaka et al., 2004). The remaining SNPs, insertion and deletion polymorphisms appeared in intronic, 3′ untranslated regions and 3′ flanking regions of the OCT2 gene. As with other genes, variation in African-American populations in OCT2 is greater than any other ethnic group. Currently, there are four nonsynonymous (P54S, M165V, M165I, and R400C) and seven synonymous SNPs, which are only found in the African-American population. One deletion polymorphism (-578-576delAAG) with significantly reduced promoter activity was identified in Japanese and carriers of the variant tended (not significant) to have lower OCT2 mRNA levels (Ogasawara et al., 2008).

8.3.3 OCT3

Recently, genetic polymorphisms in OCT3 have been identified in Caucasian (Lazar et al., 2003), Korean (Kang et al., 2007), and an ethnically diverse population (Chen et al., 2010a). Two synonymous SNPs of OCT3, R120R and A411A, occur at higher frequencies (10–60%) in various ethnic groups. By contrast, the six nonsynonymous variants identified occur at very low frequencies in all ethnic groups (Table 8.2). Interestingly, three (T44M, T400I, and V423F) of the six nonsynonymous polymorphisms exhibit altered transport function with respect to metformin and other selected endogenous amine uptake (Chen et al., 2010a; Table 8.3). The reason for the limited protein sequence variation in OCT3 is unclear but may reflect the important endogenous role of OCT3 in humans (Tajima, 1989) but could also be caused by demographic factors such as population subdivision (Simonsen et al., 1995). The investigation of patients with candidate disorders might reveal deleterious polymorphisms that occur at low prevalence and that have not been found in healthy individuals.

8.3.4 Summary

In summary, the number and allele frequencies of nonsynonymous SNPs differ among the three paralogs: OCT1, OCT2, and OCT3. In general, OCT1 is more variable, followed by OCT2 and then OCT3. The reduced nonsynonymous variation in OCT3, compared to OCT1 and OCT2 and normalized to the synonymous variation, suggests that OCT3 is under a high degree of selective pressure and may play a critical role in survival and/or reproduction. Different ethnic groups have their own specific variants and also share more cosmopolitan variants (Table 8.2). As with many genes, cosmopolitan variants or variants shared among several ethnic groups are generally more frequent; however, some population specific variants reached high allele frequency, for example, OCT1 Met420del, which appears to be a European specific variant, and is present at about 20% in European American populations. Coding region insertions or deletions are generally rare, but OCT1 Met420del results from a 3-bp deletion in the coding sequence of OCT1. Information gained about the function of these polymorphisms will help in understanding interindividual differences in drug response and the pharmacokinetics of cationic drugs important for particular ethnic groups.

8.4 FUNCTIONAL GENOMICS OF TRANSPORTER VARIANTS

Functionally relevant genetic variants in OCT1–3 can alter the handling of certain drugs in the body and account for some of the interindividual variability in drug response and toxicity. Previous studies have suggested that rare variants may be more likely to affect function than more common variants (Leabman et al., 2003). Several nonsynonymous variants have shown reduced or loss of uptake capacity of metformin without altered expression of mRNA.

As shown in Table 8.3, some polymorphisms in OCTs are associated with altered substrate selectivity. For example, the OCT1 M420del polymorphism, which demonstrated normal transport of model substrates such as 1-methyl-4-phenylpyridinium (MPP^+) MPP^+, exhibits reduced transport of metformin (Shu et al., 2007). Accordingly, the potential pharmacokinetic consequences of OCT genetic variants may be different depending on the substrate administered. Such substrate specific effects of genetic polymorphisms of OCTs highlight the need to interpret functional studies with caution and not extend results of *in vitro* studies beyond the substrates and inhibitors that have been characterized.

8.4.1 OCT1

Recently genetic variants in OCT1 have been shown to reduce the transport of model substrates such as MPP^+ and tetraethylammonium (TEA) and also drugs such as metformin in cellular assays. Currently, seven nonsynonymous polymorphisms in OCT1 (R61C, C88L, G220V, P283L, R287G, G401S, and G465R), from multiple ethnic populations, have been shown to exhibit reduced or negligible uptake of MPP^+ and/or TEA (Kerb et al., 2002; Table 8.3). Of these, four SNPs (R61C, G220V, G401S, and G465R) also have demonstrated reduced or negligible uptake of metformin. An additional six polymorphisms show reduced uptake of metformin whereas the transport of MPP^+ or TEA remained unaltered or were not investigated (Ahlin et al., 2010). Studies using another model substrate 4-(4-(dimethylamino)styryl)-*N*-methylpyridinium (ASP^+) found two variants (V408M and M420del) which transported ASP^+ at the same rate as the OCT1-reference whereas the transport by the R61C variant was reduced and the G465R variant was nonfunctional (Ahlin et al., 2010). The common M420del, G465R, and R61C polymorphisms also show reduced uptake and DNA adduct formation of picoplatin (More et al., 2010), a third generation chemotherapeutic agent. The reduced membrane localization and mRNA expression of R61C and in particular G465R, most likely contribute to their lower transport efficiency (Nies et al., 2009; Ahlin et al., 2010). Four rare polymorphisms in OCT1 (C88L, G220V, G401S, and G465R), primarily found in Caucasians and African-Americans, show little to no uptake of MPP^+ and/or metformin. Interestingly, the G220V found only in African-Americans is in TMD 4 and very close to the proposed substrate binding sites of OCT1. Therefore, perhaps the reason that uptake by G220V is negligible is because this polymorphism interferes with the binding site of OCT1 and results in altered binding and/or translocation of substrates of OCT1.

8.4.2 OCT2

As noted above, the gene encoding OCT2 appears to be under more selective pressure than the OCT1 gene. The selection against deleterious amino acid changes in OCT2 is supported by the fact that fewer nonsynonymous variants in the transporter have been identified and only five of the 11 nonsynonymous SNPs that were functionally characterized showed reduced uptake of MPP^+ and/or metformin. Interestingly, two variants (T199I and T201M), identified only in Japanese, Korean, and Chinese populations, also showed a significant decrease in the MPP^+ and metformin uptake (Kang et al., 2007; Song et al., 2008b), despite the protein expression and membrane localization remaining unaltered (Kang et al., 2007). Given that both the K_m and V_{max} were significantly reduced, it is suggested that these polymorphisms affect the substrate binding specificity of OCT2. The more common nonsynonymous variants (M165I, K432Q, and R400C) also exhibit functional differences from the reference OCT2, such as altered interactions with MPP^+ and metformin and reduced transport activity (Leabman et al., 2002; Fujita et al., 2006). The protein variant A270S, common in Caucasian, African-American, and Asian populations (Leabman et al., 2002; Table 8.2) has exhibited modest reduction in the uptake of MPP^+ with variable alterations of the transport function whereas uptake for picoplatin (More et al., 2010) and paraquat (Chen et al., 2007) appears to have increased. The variable transport function may be due to inherent differences in the cellular systems used. For example, in transiently transfected MDCK cells, the K_m of MPP^+ was increased (Kang et al., 2007) whereas in HEK cells and oocytes, the K_m was similar between the A270S variant and reference (Leabman et al., 2002; Zolk et al., 2009a). A lower V_{max} (Kang et al., 2007; Zolk et al., 2009a) and less sensitivity to inhibition of MPP^+ uptake by tetrabutylammonium, TBA, (increased K_i value for TBA) have, however, been consistently demonstrated (Leabman et al., 2002; Fujita et al., 2006).

OCT2 also transports metformin with a lower K_m than OCT1 (Kimura et al., 2005a) and three nonsynonymous variants have been associated with both reduced and increased uptake of metformin. Interestingly, the A270S variant demonstrates reduced metformin uptake in oocytes (Fujita et al., 2006; Kang et al., 2007; Song et al., 2008b) yet uptake in HEK293 cells is increased (Chen et al., 2009; Table 8.3). The reason for this discrepancy remains unclear; however, cellular studies using metformin show variable effects on metformin transport and protein expression compared to reference. In transiently transfected cells, the V_{max} and protein expression was increased (Chen et al., 2009) whereas in stably transfected cells V_{max} and protein expression was unaltered in comparison with cells expressing the reference OCT2 (Leabman et al., 2002; Kang et al., 2007; Zolk et al., 2009a). The K_m of metformin for OCT2 remains unaltered by the polymorphism (Chen et al., 2009; Zolk et al., 2009a). The A270S polymorphism is predicted to face the outer pore of the transporter, although the absolute position of the residue remains disputed (Zolk et al., 2009b). This raises the possibility that this polymorphism may alter substrate recognition and translocation by OCT2 (Chen et al., 2009; Zolk et al., 2009b). In general, the most frequent nonsynonymous variants exhibited more subtle functional differences from the reference forms of OCT2 when compared to the less frequent SNPs.

8.4.3 OCT3

Recently the functional significance of polymorphisms in OCT3 has been investigated in cellular studies. Only one variant, T44M has demonstrated altered (increased) MPP^+ transport. By contrast, three nonsynonymous variants displayed altered metformin transport, two polymorphisms (T400I and V423F) show decreased uptake whereas one polymorphism (T44M) had more than twofold increased uptake of metformin (Chen et al., 2010a). The amino acid substitutions of both T400I and V423F did not significantly affect protein expression or subcellular localization, suggesting that a disruption of the structure of OCT3 may contribute to altered metformin transport. Although the subcellular localization of T44M remained unaltered, the protein expression was significantly higher and thus may explain the increased transport of metformin and MPP^+. Interestingly, two SNPs, one synonymous (A411A) and one noncoding 3′UTR (rs3088442) polymorphism in OCT3 have been associated with decreased OCT3 mRNA and protein levels in Caucasian liver samples (Nies et al., 2009). The functional effects of other genetic variants previously identified in OCT3 remain to be explored.

8.4.4 Summary

In general, the effect of nonsynonymous genetic variants of OCTs on transport activity is substrate specific. Furthermore, rarer genetic variants in OCT1, OCT2, and OCT3 are more likely to result in deleterious consequences with regard to substrate-specific transporter interactions. Functional characterization may depend on cell systems used in *in vitro* studies. Because several variants in these transporters are common and functionally altered in cellular assays, functional genomic studies of OCTs suggest that the variants may play a role in the variation in drug disposition and response clinically.

8.5 CLINICAL STUDIES OF POLYMORPHISMS IN ORGANIC CATION TRANSPORTERS

Polymorphisms in the genes coding for drug metabolizing enzymes, transporters, receptors, and other structural and signaling proteins, which modulate drug action, may affect drug response. OCTs play an important role in the physiological and pathophysiological processes and in determining the interstitial concentrations of endogenous compounds, drugs, and xenobiotics in a variety of tissues. Polymorphisms in OCT1 and OCT3, in the liver, may reduce the hepatic uptake of substrates (including drugs) whereas polymorphisms in OCT2, in the kidney, may reduce renal excretion. The resulting higher plasma levels may cause adverse effects or allow a similar therapeutic effect at a lower dose. Therefore, polymorphisms in OCTs can be of relevance in determining the causes of variability in the pharmacokinetics and patient response to pharmacotherapy.

8.6 INFLUENCE ON THE PHARMACOKINETICS AND PHARMACODYNAMICS OF METFORMIN

Metformin, a biguanide, is used as first-line therapy for the treatment of type 2 diabetes mellitus and is among the most widely prescribed drugs in the United States (Kirpichnikov et al., 2002; Nathan et al., 2009). The major action of metformin is to decrease hepatic glucose production and gastrointestinal glucose absorption and facilitate glucose utilization in skeletal muscle and adipose tissue (Abbud et al., 2000; Zhou et al., 2001; Kirpichnikov et al., 2002). The molecular mechanisms underlying the action of metformin seem to be related to its activation (phosphorylation) of AMP-activated protein kinase (Zhou et al., 2001). The pharmacological response to metformin is governed in part by its pharmacokinetics, including plasma concentrations and tissue distribution (Shu et al., 2008). Metformin has been well characterized *in vitro* as a substrate of OCT1, OCT2, OCT3, the human multidrug and toxin extrusion proteins 1 (MATE1), MATE2K, and the plasma membrane monoamine transporter (PMAT) (Engel et al., 2004; Otsuka et al., 2005; Masuda et al., 2006; Tanihara et al., 2007; Nies et al., 2009). Differential expression or altered function of these transporters could alter the amount of drug taken up by the liver (the target tissue) and elimination through the kidney. The role of metformin transport in skeletal muscle or adipose tissue, and the subsequent implications for metformin pharmacodynamics are unknown (Shu et al., 2007). Recently, OCT1, OCT2 (and OCT3 to a limited extent) have been shown to play a role in the hepatic and renal transport of metformin, respectively (Wang et al., 2002; Kimura et al., 2005b; Fujita et al., 2006; Shu et al., 2007; Nies et al., 2009). Furthermore, genetic variants of these transporters are associated with variation in the pharmacokinetics and antidiabetic action of metformin (Stumvoll et al., 1995; Kimura et al., 2009; Becker et al., 2009, 2010a; Chen et al., 2010a; Table 8.4).

8.6.1 OCT1

There is a large variation in the response to metformin and it is estimated that more than one-third of patients receiving metformin monotherapy do not achieve acceptable control of fasting glucose levels (Hermann et al., 1994a, 1994b). Preliminary investigations into the genotype–phenotype relationship of OCTs and the efficacy of metformin found no significant difference in the prevalence of genetic polymorphisms in OCT1 and OCT2 between responders and nonresponders (Shikata et al., 2007). The authors concluded that polymorphisms in OCT1 and OCT2 contribute little to the efficacy of metformin (Shikata et al., 2007). Subsequently, however, genetic polymorphisms in OCT1 have been associated with variations in the pharmacokinetic and pharmacodynamic profile of metformin (Shu et al., 2007, 2008). A summary of these studies is provided in Table 8.4. A study in healthy volunteers demonstrated an increase in the area under the concentration-time curve (AUC) and maximal plasma concentration ($C_{\max}$) values and lower apparent volume of distribution (V_d/F) whereas the renal clearance of metformin was similar in subjects carrying reduced-function OCT1 alleles (R61C, G401S, M420del, and G465R) compared to those carrying the reference allele (Shu et al., 2008). In contrast, the combined effect of all four amino acid variants was associated with an incremental increase in the renal clearance of metformin in another study in healthy volunteers (Tzvetkov et al., 2009). The dose of metformin used, sex distribution, and number of subjects differ between the two studies and may account for the divergent findings. The effects of the same reduced-function alleles on the response to metformin have also been investigated in healthy volunteers. Although baseline glucose levels were similar, reduced-function OCT1 variants were associated with higher plasma glucose levels (after glucose tolerance test) following metformin administration than those subjects carrying the reference allele (Shu et al., 2007). However, the association of R61C and M420del with reduced metformin efficacy (defined as the reduction in HbA1c) was not confirmed in two large long-term follow-up observational cohorts of diabetic patients treated with metformin monotherapy (Zhou et al., 2009; Davis et al., 2010). The G465R variant was associated (not significantly) with a smaller absolute change in HbA1c (Davis et al., 2010) and the presence of one or more OCT1 polymorphisms reduced lipid and insulin responses to metformin in women with polycystic ovary syndrome (Gambineri et al., 2010). The reasons for these differences may be related to the populations studied, that is, healthy versus diabetic populations or to the difference in the pharmacodynamic measurements; plasma glucose levels in the healthy volunteer study versus HbA1c levels in diabetic patients. In another study, examining Caucasian-European diabetic patients commenced on metformin, the intronic polymorphism, rs622342, was associated with higher HbA1c levels, that is, reduced glucose-lowering effect when compared to those carrying the reference allele (Becker et al., 2009). Interestingly, in a study of metformin use in prevention of diabetes, a nonsynonymous SNP (F160L) in linkage disequilibrium (LD) with the rs622342 OCT1 polymorphism was associated with a reduced response to metformin (Jablonski et al., 2010). This finding was not reported in a recent GWAS study of diabetic patients of European ancestry which found no significant effects of polymorphisms in OCTs on the response to metformin; the only significant locus was that containing the ataxia telangiectasia mutated (ATM) gene (Zhou et al., 2010). Interestingly, in *Oct1* knockout mice, metformin levels were significantly lower in liver and intestine and plasma concentrations were higher (Wang et al., 2002; Shu et al., 2003). Moreover, the glucose-lowering effect was decreased compared to mice with normal functioning *Oct1* (Wang et al., 2002, 2003; Shu et al., 2007). Together, these data indicate that OCT1 plays an important role in the therapeutic action of metformin and genetic variation in OCT1 may explain some of the interindividual variation in response to metformin monotherapy.

8.6.2 OCT2

Metformin is eliminated from the body primarily by the kidney without any significant metabolism *in vivo* (Scheen, 1996). Variation in the renal clearance of metformin has a strong genetic component (Leabman et al., 2003). OCT2 plays a major role in the total clearance of metformin due to its high expression in the kidney and lower K_m for metformin compared to other OCTs (Kimura et al., 2005a, 2005b). Therefore, functional mutations in OCT2 are expected to decrease the renal clearance of metformin and subsequently increase plasma concentrations. In healthy Korean subjects, three nonsynonymous SNPs (T199I, T201M, and A270S) increased the plasma concentrations and decreased the renal clearance of metformin compared to the reference genotype (Song et al., 2008a). This is consistent with *in vitro* data showing reduced uptake of MPP^+ and metformin (Kang et al., 2007; Song et al., 2008b). In healthy Caucasian subjects, the A270S variant was associated with increased renal clearance of metformin (Chen et al., 2009). By contrast, in healthy Chinese, the renal metformin clearance was reduced with each copy of the variant allele and the A270S variant explained 28% of the interindividual variability in this parameter in the study population (Wang et al., 2008b). The reason for the discrepancy in results is not clear; however, ethnic-specific altered expression of OCT2 by SNPs in linkage with A270S may explain the differences in the studies (Chen et al., 2009). Interestingly, the inhibition of metformin renal clearance by cimetidine is dependent on the A270S polymorphism (Wang et al., 2008b). To date, the relationship of genetic polymorphisms in OCT2 on the glucose-lowering effect of metformin remains unknown.

8.6.3 OCT3

The effects of genetic variants in OCT3 on the pharmacokinetics and response to metformin have not been investigated.

8.6.4 Summary

Collectively, the clinical studies of polymorphisms in OCTs suggest that genetic variants in these transporters may play a role in variation in drug levels and response. Importantly, both rare and common variants have been studied and clinical results have been supported by a wealth of functional genomic studies. New areas that need study are the role of genetic variants in noncoding regions of transporters in variation in response to metformin. Further, OCTs collectively are responsible for tissue specific levels of metformin; therefore, gene × gene interactions may occur. For example, individuals with reduced-function variants of both OCT1 and OCT3 may have considerably reduced hepatic levels of metformin. Genetic variants in OCTs may also interact with variants in their functional partners, MATEs. In summary, genetic variants in OCTs and other genes should be studied in the context of nongenetic factors such as adherence, diet, and concomitant medications, to fully understand the factors that contribute to variation in response to metformin.

8.7 EFFECT OF OCT POLYMORPHISMS ON DISEASE RISK AND RESPONSE TO OTHER DRUGS

8.7.1 OCT1

Levodopa appears to be a substrate of OCTs. Recently, the minor allele of the intronic SNP rs622342 in OCT1 has been associated with higher prescribed doses of anti-Parkinsonian drugs and a shorter survival after the start of levodopa therapy (Becker et al., 2011).

The antiemetic drug, tropisetron has also been shown to be a substrate of OCT1 and the common genetic variants show reduced uptake of tropisetron (Tzvetkov et al., 2012). Tropisetron is primarily metabolized by CYP2D6 and genetic polymorphisms in CYP2D6 explain a small fraction of the interindividual variation in the efficacy of antiemetic treatment (Kaiser et al., 2002). However, before tropisetron can be metabolized it must enter the liver, therefore, genetic polymorphisms in OCT1 are postulated to contribute to variation in drug response. Indeed, plasma concentrations and correspondingly, the response to tropisetron were increased (i.e., decreased episodes of vomiting) in patients carrying reduced-functioning OCT1 alleles (R61C, C88R, G401S, M420del, and G465R) and receiving chemotherapy independent of CYP2D6 genotype (Tzvetkov et al., 2012).

Imatinib is a 2-phenylaminopyramidine derivative that inhibits the ABL tyrosine kinase activity and is used in the treatment of chronic myeloid leukemia (CML). Previously, OCT inhibitors, prazosin and verapamil, have been shown to decrease the uptake of imatinib in a concentration-dependent manner (Thomas et al., 2004; White et al., 2006). More recent *in vitro* studies indicate that imatinib is a weak substrate of OCT1 (Hu et al., 2008; Wang et al., 2008a; Minematsu and Giacomini, 2011). The weak association of imatinib with OCT1 has been confirmed in clinical studies, in which the reduced-functioning polymorphisms of OCT1, R61C, and G465R, were not found to alter steady-state plasma concentrations of imatinib (Hu et al., 2008) or response to imatinib (Zach et al., 2008) compared to the reference genotype. However, other studies have found that the pretreatment expression of the OCT1 gene and OCT1 activity may be the most important determinant for treatment outcome (time to complete cytogenetic remission or achieving a major molecular response) in CML patients treated with imatinib (Crossman et al., 2005; White et al., 2006; Wang et al., 2008a). Given the conflicting data regarding the importance of OCT1 in the pharmacokinetics and response to imatinib, further cellular and clinical studies are required to understand this transporter–drug interaction.

8.7.2 OCT2

Recently, OCT2 has been implicated in blood pressure control and associated with the development of essential hypertension in 607 Caucasian patients who underwent left heart catheterization (Lazar et al., 2006). In particular, the A270S polymorphism was associated with essential hypertension and the relationship was more pronounced in patients without diabetes mellitus, an independent factor for the development of hypertension. Although the mechanism remains unclear, OCT2 has been shown to play a role in the renal handling of dopamine (Grundemann et al., 1998a) and renal dopamine is known to be an important natriuretic factor involved in sodium homeostasis and blood pressure regulation (Aperia, 2000). Therefore, it is postulated that polymorphisms in OCT2 alter renal dopamine transport, disrupting the integrity of renal sodium homeostasis and hence blood pressure control (Lazar et al., 2006). Further, cellular and clinical studies are required to examine this hypothesis.

Cisplatin, one of the most widely used cytotoxic anticancer agents, is primarily eliminated by the kidneys. *In vitro* studies show that OCT2 transports cisplatin (Ciarimboli et al., 2005; Zhang et al., 2006; Filipski et al., 2008) and cisplatin-induced nephrotoxicity is reduced with coadministration of cimetidine (Ciarimboli et al., 2005). Cellular studies also suggest that OCT2 (and OCT1) are important in mediating the cytotoxicity of cisplatin and its analogs, oxaliplatin and carboplatin (Zhang et al., 2006). Furthermore, the urinary excretion of cisplatin is reduced in *Oct1/Oct2*(-/-) knockout mice (Filipski et al., 2008). However, in Caucasian cancer patients receiving cisplatin, the reduced-function A270S polymorphism had no effect on the pharmacokinetic parameters of cisplatin including urinary excretion (Filipski et al., 2008). Interestingly, patients carrying the A270S variant had a reduced risk of developing cisplatin-induced nephrotoxicity (Filipski et al., 2009), suggesting that OCT2 plays a major role in the cisplatin-induced nephrotoxicity and response whereas its effects on the pharmacokinetics of cisplatin are less prominent.

8.7.3 OCT3

Interestingly, functionally significant genetic polymorphisms of OCT3 in coding and proximal promoter regions have been identified and associated with an increased risk of diseases such as coronary artery disease (Tregouet et al., 2009), obsessive–compulsive disorder (Lazar et al., 2003, 2008), and prostate cancer (Eeles et al., 2008). In particular, in a case-control study of 84 Caucasian children and adolescents with obsessive–compulsive disorder, a nonsynonymous variant of OCT3, M370I, which had a 40% decline of transport capacity for norepinephrine and a genetic variant in the promoter region g.-81G>delGA of OCT3, which showed increased luciferase activity, were associated with obsessive–compulsive disorder (Lazar et al., 2008). Further studies are required to elucidate the mechanism behind the association.

8.7.4 Summary

In addition to the large number of studies focused on genetic polymorphisms in OCTs in response to metformin, current studies suggest that polymorphisms in OCTs play a role in therapeutic and adverse response to other drugs including levodopa and cisplatin. Moreover, polymorphisms in OCTs appear to play a role in the risk to diseases such as obsessive–compulsive disorders, coronary artery disease, and high blood pressure. Most studies have involved small patient populations or healthy volunteers. These proof-of-concept studies suggest the need for large prospective studies in patients to evaluate the clinical relevance of genetic polymorphisms and their impact of clinical pharmacokinetics, therapeutic effects, and adverse reactions in drug therapy as well as in risk for disease.

8.8 OCTNs

8.8.1 Structure

Extensive studies on OCT transporters had long suggested the need for an organic cation/proton antiporter in the renal apical membrane for the efficient elimination of organic cations into urine (Koepsell et al., 1999; Sweet and Pritchard, 1999; Nies et al., 2011b). Yet, only beginning in 1997, in an attempt to identify such transporters, Tamai et al. (1997) cloned and characterized several more remotely related members of the polyspecific OCT (*SLC22*) family involved in the transport of organic cations and carnitine, which were designated as the novel OCT subfamily or OCTNs. Hence, like the OCTs, the OCTNs play a significant role in the elimination of cationic drugs, but they are unique in their ability to transport carnitine, a function that is not shared by the OCTs (Tamai et al., 1997, 1998; Vermeire, 2004). This section will focus on members 4 and 5 of the *SLC22* family, which encode OCTN1 (*SLC22A4*) and OCTN2 (*SLC22A5*). However, it must also be mentioned that an additional member of this family, OCTN3, has also been characterized, yet to date, it has only been cloned from the mouse and the corresponding gene has not been identified in humans (Lamhonwah et al., 2003).

Similar to other members of the *SLC22* family, the membrane topology of OCTN1 and OCTN2 consists of 12 α-helical TMDs, an intracellular *N*-terminus domain, a large glycosylated extracellular loop between TMDs 1 and 2, a large intracellular loop with phosphorylated sites between TMDs 6 and 7, and an intracellular C-terminus domain (Grundemann et al., 1994; Koepsell et al., 2007). In humans, the genes coding for OCTN1 and OCTN2 are localized in a cluster on chromosome 5q31 and code for functional OCTs, which are involved in the transport of several endogenous, as well as pharmacologically relevant compounds (Rioux et al., 2001). OCTN1 contains 551 amino acids whereas OCTN2 consists of 557 amino acids. Homology studies suggest a close link between OCTN1 and OCTN2, with a 75.8% sequence similarity, compared to other members of the *SLC22* family, namely OCT1, OCT2, or OCT3, with which they display low structural similarities (Lahjouji et al., 2001; Ohashi et al., 2001).

8.8.2 Tissue Distribution and Membrane Localization

OCTN1 exhibits broad tissue distribution; in humans, the highest expression levels were identified in the proximal tubular cells of the kidney, skeletal muscle, bone marrow, and trachea. Lower levels have been reported in many other organs, but not in adult human liver (Tamai et al., 1997; Koepsell et al., 2007; Tahara et al., 2009). Recent studies have also located human OCTN1 in the mitochondria (Lamhonwah and Tein, 2006). OCTN2 displays relatively ubiquitous expression pattern; in humans, the strongest expression was observed in the kidney (brush-border membrane) with considerable expression in skeletal muscle, placenta, testis, prostate, liver, small intestine, and heart (Tamai et al., 1998; Wu et al., 1998).

8.8.3 Substrate and Inhibitor Specificity

The high expression of OCTN1 in renal proximal tubules suggests that it is primarily important in the tubular transport of substrate drugs. Initial studies by Tamai et al. (1997) suggested that OCTN1 may operate at the apical membrane of renal epithelial cells to carry out ATP-dependent reabsorption of endogenous cations and some xenobiotics in a metabolic inhibitor-susceptible and pH sensitive manner, suggesting that it may act as an organic cation/proton antiporter. Since these studies, OCTN1 has been shown to transport the zwitterions L-carnitine and stachydrine, the prototypical organic cation substrate TEA, as well as, several clinically important drugs, including quinidine, pyrilamine, and verapamil (Table 8.5; Tamai et al., 1997; Wu et al., 1998). More recently, studies by Grundemann et al. (2005) have shown that the endogenous substrate of OCTN1 is L-ergothioneine (EGT), which has long been recognized as an antioxidant amino acid that humans consume with plant food. It has been speculated that the uptake and concentration of EGT in the bone marrow and erythrocytes by OCTN1 may protect erythrocytes and monocytes against oxidative damage (Markova et al., 2009). Furthermore, OCTN1 activity may be inhibited by many additional compounds, including choline, L-carnitine, clonidine, and cimetidine (Table 8.6; Koepsell et al., 2007). The specific translocation of the above-mentioned substrates by OCTN1 displays different mechanisms and reaction kinetics for different substrates, such that this transporter is able to operate in both directions. For example, when measuring OCTN1-mediated TEA uptake in membrane vesicles, an energetic coupling between efflux of protons and influx of TEA was found, indicating an antiport transport mechanism whereas OCTN1-mediated uptake of EGT was found to be stimulated by extracellular sodium. Transcription of the *SLC22A4* gene is inhibited by the transcription factor RUNX1, which is mainly expressed in hematopoietic cells and binds to a consensus sequence in the first intron of this gene (Tamai et al., 1997; Koepsell et al., 2007; Urban et al., 2007).

The physiological role of OCTN2 expression in the kidney is primarily related to L-carnitine reabsorption in order to maintain the tightly regulated levels of carnitine stores. In addition, expression of OCTN2 in muscle cells acts to facilitate the uptake of this compound into cardiac and skeletal muscle, as these tissues are not capable of synthesizing carnitine but require it for the transfer of long-chain fatty acids into the mitochondrion for β-oxidation and energy generation (Grube et al., 2005; Urban et al., 2006). Subsequently, the affinity of OCTN2 for carnitine is markedly higher than that of OCTN1 (i.e., K_m of 24 μM for OCTN1 compared to 4.3 μM for OCTN2) (Koepsell et al., 2007). Other substrates of OCTN2 include the zwitterion β-lactam antibiotic cephaloridine, as well as, the cationic substrates TEA, choline, verapamil, and pyrilamine (Table 8.5). Inhibition of OCTN2-mediated uptake of these substrates has been found upon exposure to cimetidine, clonidine, procainamide, emetine,

TABLE 8.5 Substrate Specificity of OCTN Transporters

Protein (Gene)	Model substrate (K_m in μM)	Endogenous substrate (K_m or IC_{50} in μM)	Drugs (K_m or IC_{50} in μM)	Inhibitors (K_m or IC_{50} in μM)
OCTN1 (*SLC22A4*)	L-ergothioneine (**21**)	Choline (231) L-Carnitine (**24**) D-Carnitine (<2000) Stachydrine (**270**)	Pyrilamine (<100) Cimetidine (435) Clonidine (<5000) Verapamil (8.4) Lidocaine (0.8) Cephaloridine (<1000) Quinine (<500) Quinidine (69) Procainamide (<500)	TEA (**195**) N-methylnicotinamide (77) Tetra-butyl ammonium (535) Tetrapentyl ammonium (70)
OCTN2 (*SLC22A5*)	L-carnitine (**4.3**)	Acetyl-L-carnitine (**8.5**) Betaine (<500) Choline (<**2000**) D-carnitine (**11**) Aldosterone (<500) Corticosterone (<100)	Pyrilamine (<500) Cimetidine (<2000) Clonidine (<3000) Verapamil (<50) Procainamide (<3000) Quinidine (<100) Cefsulodine (<500) Emetine (<300) Quinine (<100) Actinomycin D (<500)	TEA (**300**) MPP^+ (>1000)

Source: Data according to Koepsell H. et al., 2007 and Ohashi R et al., 2001.
Note: The K_m values are given in bold. TEA, tetraethylammonium; MPP^+, 1-methyl-4-phenylpyridinium.

TABLE 8.6 Summary of Rare *SLC22A5* Variants and Their Frequency in Different Ethnic Population

				Allele frequency					
Location	SNP ID	Position (from translational starting site)	Amino acid change	Chinese	Indians	Asian (Japanese)	White (Caucasian)	Hispanic	African-Americans
5′ UTR	rs2631367	−207	–	0	0.81	0	0.475	0.02	0.373
Exon 1	rs2631365	285	Leu95Leu	0.58	0.64	0.8	0.35	–	–
Exon 2	rs10040427	430	Leu144Phe	0	–	0	0	0.01	0.075
Exon 4	rs274558	807	Leu269Leu	0.62	0.62	0.73	0.35	–	–
Exon 8	rs60376624	1400	Ser467Cys	0.01	0	–	–	–	–
Exon 9	rs28383481	1463	Arg488His	0	0.01	–	–	–	–
Exon 10	rs11568525	1645	Pro549Ser	0	–	0	0	0	0.1
3′ UTR	rs1045020	1721	–	0	–	0	0.094	0.03	0.013

Source: Data according to Urban T et al., 2006 and Toh DSL et al., 2010.

cefsulodine, and nicotine (Table 8.5; Tamai et al., 1998; Wu et al., 1998; Koepsell et al., 2007). Initial studies of OCTN2 indicated that this transporter is physiologically important in the high affinity uptake of carnitine and displays significant sodium ion dependence, such that sodium increased the affinity for L-carnitine uptake. Thus, it was concluded that OCTN2 is a sodium-L-carnitine cotransporter rather than a transporter that is activated by extracellular sodium. However, these initial studies also suggested that prototypical substrates for the previously known OCTs, including OCTN1, such as TEA were not good substitutes for L-carnitine, thus implicating OCTN2 in the sodium-independent transport of these organic cations (Tamai et al., 1998; Wu et al., 1998). Furthermore, recent observations have indicated that the influx of carnitine and the efflux of TEA are linked such that the sodium dependence of carnitine determines the direction of influx or efflux of carnitine and TEA, respectively, under physiological conditions. Taken together, these observations suggest that the reabsorption of carnitine is accompanied by accelerated elimination of organic cations in such a way that circulating organic cations are concentrated in the epithelial cells by basolaterally localized OCT transporters and are secreted into the urine across the apical membrane by OCTN2, among other apical-type transporters in the renal tubular epithelial cells (Ohashi et al., 2001).

8.9 POLYMORPHISMS IN OCTNs AND FUNCTIONAL GENOMICS OF TRANSPORTER VARIATION

Considering the central role of OCTNs in the maintenance of intracellular homeostasis and energy production in the cell, there has been intensive research in the past decades on the functional influence of genetic variations in the coding and noncoding regions of these transporters in different ethnic populations (Silverberg, 2006; Urban et al., 2006).

The first study on the functional effects of OCTN1 and OCTN2 polymorphisms were carried out by Peltekova et al. (2004). In this study, the authors identified two functional genetic variants in the OCTN cluster located on the inflammatory bowel disease (IBD) 5 locus (Peltekova et al., 2004). IBD5 is a cytokine-rich gene cluster on chromosome 5q31 that has consistently been linked to an increased risk for autoimmune diseases, such as Crohn's disease (CD), rheumatoid arthritis (RA), and/or ulcerative colitis (UC) (Rioux et al., 2001; Giallourakis et al., 2003). The first variant identified is a missense mutation in OCTN1 comprising a C→T change at position 1672 in exon 9, which leads to an amino acid substitution involving the exchange of leucine at position 503 by phenylalanine (L503F) (Peltekova et al., 2004). In fibroblasts and HeLa cells, this variation results in 2.7 times lower carnitine uptake compared to cells expressing the reference sequence (i.e., 503L); while both variants transport carnitine in a sodium and concentration dependent manner, the 503F variant displays a lower V_{max} and a higher K_m for carnitine uptake. Conversely, although both variants also transport TEA in a sodium-independent and pH-dependent manner, the 503F variant displays 3.7 times greater TEA uptake and a higher V_{max} and lower K_m for TEA uptake, thereby considerably altering the transport function of OCTN1. In fact, the characterization of OCTN1 as a low-affinity carnitine transporter has been based on analyses of the 503F variant rather than the more prevalent wild-type 503L variant, which has been shown by Peltekova et al. (2004) to be a better carnitine transporter. The second variant was identified in the promoter region of OCTN2 and consists of a G→C transversion at position -207, which disrupts a heat shock element in the promoter region of the SLC22A5 gene (Table 8.6; Rioux et al., 2001; Peltekova et al., 2004). In OCTN2-deficient fibroblasts expressing -207G, as well as in HeLa cells, the promoter activity induced by heat shock transcription factor was significantly greater than in those expressing -207C, indicating that the -207C variant disrupts a functional promoter element (Peltekova et al., 2004).

Several additional nucleotide substitutions have been reported in the SLC22A4 gene; including another common protein sequence variant T306I (Kawasaki et al., 2004). This substitution introduces a potential *N*-glycosylation site in the third extracellular loop of OCTN1, but has not been found to result in any significant alteration in its transport function. However, since this amino acid is conserved in nearly all mammalian OCTN1 orthologs, earlier studies suggest that it may have functional effects *in vivo* that are not detectable using cell-based assays alone. Hence, further *in vivo* studies still need to be carried out and will be of interest for determining whether this variation may also have functional implications, for example, by disrupting a potential regulatory site in the protein (Kawasaki et al., 2004; Urban et al., 2007).

Furthermore, additional rare variations in the SLC22A5 gene have been reported in various studies involving different population subsets and are summarized in Table 8.6. Although most of these variants have been reported at very low frequencies in the populations tested, the study of their effects on transport function may be of interest in the understanding of OCTN2's substrate specificity and population variations in its transport function (Tang et al., 1999; Toh et al., 2010).

Studies of OCTN1 and OCTN2 polymorphisms using ethnically diverse samples have suggested further variations among different ethnic groups. Results from studies by Peltekova et al. (2004) have been widely replicated and suggest that the two SNPs occur at particularly high allele frequencies (approximately 50%) in individuals of European ancestry (Urban et al., 2006). Furthermore, studies carried out among various Caucasian populations, including German, Greek, Canadian, and Spanish have shown an association of this haplotype with CD (Gazouli et al., 2005; Newman et al., 2005; Torok et al., 2005; Martinez et al., 2006). Yet, studies performed in African-Americans, Mexicans, and Asians have reported a reduced prevalence of these allelic variants (Li et al., 2008; Toh et al., 2010). Studies by Yamazaki et al. (2004) reported a complete absence of these SNPs in a Japanese population of 484 patients with CD and 345 healthy volunteers. Similarly, Li et al. (2008) reported a very low frequency of these variations in a Chinese population of 61 patients with CD, 151 patients with UC, and 200 healthy volunteers. Conversely, studies in a Asian populations have functionally characterized an exonic SNP in the SLC22A4 gene, involving an amino acid substitution (G462E) that results in complete loss of TEA transport activity that has not been reported at high frequencies in any other population (Kawasaki et al., 2004). These and results from various other studies suggest an apparent genetic heterogeneity among different ethnic populations for OCTN1 and OCTN2 polymorphisms, especially Caucasians and Asians.

8.10 CLINICAL STUDIES OF POLYMORPHISMS IN NOVEL ORGANIC CATION TRANSPORTERS

As previously described, initial studies by Peltekova et al. have identified two functional variants in the OCTN gene cluster located on the IBD5 locus. Furthermore, both of these variants display strong LD and together create a two-allele risk haplotype (TC haplotype) with a frequency of 54% in a Canadian population of CD patients compared to 42% in control subjects (Peltekova et al., 2004). Given the essential role of these transport proteins in the regulation of various homeostatic functions of the cell, several hypotheses have proposed a role for these functional SNPs in OCTNs in the context of CD (Roediger and Nance, 1986; Waller et al., 2006; Urban et al., 2007). Specifically, as described previously, OCTN2 has a high affinity for L-carnitine, which is essential in the passage of long-chain fatty acids from the cytosol into the mitochondria for β-oxidation. Thus, defects within this system have been linked to impaired pathogen killing by oxidation-burst mediated mechanisms, resulting in persistent infection, which has been proposed as a possible mechanism for the development of IBD (Roediger and Nance, 1986; Waller et al., 2006). Furthermore, although the precise physiological role of OCTN1 is less well defined, it is known to be expressed in immune active cells and variants in this gene have previously been associated with RA (Waller et al., 2006; Urban et al., 2007). Thus, the OCTN genes constitute plausible candidates for IBD and the findings by Peltekova and colleagues represent a novel and important association between alterations in transport function and IBD risk. As a result, numerous studies have been carried out to confirm the association of CD with the IBD5 risk haplotype. One such study, assessing 374 patients with CD, measured not only the OCTN variants but also a number of surrounding SNPs from the list of SNPs that comprise the IBD5 risk haplotype. From these studies, the authors concluded that while the OCTN1 and OCTN2 variants are associated with CD, this association is not independent of the background risk haplotype in the region (Noble et al., 2005). Additional studies have demonstrated similar findings, including a study carried out in a pediatric population, which confirmed the association of the OCTN variants with CD, but also demonstrated that the variants were not independently associated from the surrounding SNPs (Russell et al., 2006). In addition, studies from various ethnic populations, including Asians and African-Americans, show no or very little evidence for IBD5 as a contributing factor for IBD susceptibility (Urban et al., 2005; Tosa et al., 2006; Li et al., 2008). Additional information is needed to ascertain that these reduced-function variants are indeed responsible for the reported clinical phenotype, since without this information it may be possible that the clinical symptoms can be explained by undetected polymorphisms in the nearby genes that are in LD with the disease associated SNPs (Silverberg, 2006). Therefore, while the TC haplotype may thus far be the best candidate for the IBD5 association findings, much larger sample sizes are required in order to conclusively determine the significance of these alleles and the possibility of their independent association.

In addition, defects in OCTN2 function, which result from variations in the *SLC22A5* gene, have been associated with the development of primary systemic carnitine deficiency (PSCD) (Lahjouji et al., 2001; Urban et al., 2006; Koepsell et al., 2007). PSCD is an autosomal-recessive disorder that involves reduced renal L-carnitine reabsorption and entry of long-chain fatty acids into the mitochondrion. As a result, body carnitine stores are depleted, resulting in a clinical presentation of cardiomyopathy, muscle weakness, hyperammonemia, and hypoglycemia arising from insufficient β-oxidation and impaired energy production for muscle and cardiac functions. OCTN2 polymorphisms in patients with PSCD include both missense and nonsense mutations, as well as, insertion/deletion, and frameshift mutations (Lahjouji et al., 2001; Vermeire, 2004; Koepsell et al., 2007). However, studies in a Japanese population have also suggested that heterozygous expression of OCTN2, resulting in partial loss of function, may also have detrimental effects (Lahjouji et al., 2001; Tein, 2003). Hence, information on the incidences of OCTN2 polymorphisms may also be very helpful in predicting early susceptibility to PSCD.

ACKNOWLEDGMENT

This work was in part supported by the Robert Bosch Stiftung, Stuttgart, Germany and the 7FP EU Initial Training Network Program "Fighting Drug Failure" (GA-2009-238132). We gratefully acknowledge support from NIGMS GM61390.

REFERENCES

Abbud W, Habinowski S, Zhang JZ, Kendrew J, Elkairi FS, Kemp BE, Witters LA, Ismail-Beigi F (2000) Stimulation of AMP-activated protein kinase (AMPK) is associated with enhancement of Glut1-mediated glucose transport. *Arch Biochem Biophys*, 380, 347–352.

Ahlin G, Chen L, Lazorova L, Chen Y, Ianculescu AG, Davis RL, Giacomini KM, Artursson P (2010) Genotype-dependent effects of inhibitors of the organic cation transporter, OCT1: predictions of metformin interactions. *Pharmacogenomics J*, 11, 400–411.

Amphoux A, Vialou V, Drescher E, Bruss M, Mannoury La Cour C, Rochat C, Millan MJ, Giros B, Bonisch H, Gautron S (2006) Differential pharmacological in vitro properties of organic cation transporters and regional distribution in rat brain. *Neuropharmacology*, 50, 941–952.

Aoki M, Terada T, Kajiwara M, Ogasawara K, Ikai I, Ogawa O, Katsura T, Inui K (2008) Kidney-specific expression of human organic cation transporter 2 (OCT2/SLC22A2) is regulated by DNA methylation. *Am J Physiol Renal Physiol*, 295, F165–F170.

Aperia AC (2000) Intrarenal dopamine: a key signal in the interactive regulation of sodium metabolism. *Annu Rev Physiol*, 62, 621–647.

Barendt WM, Wright SH (2002) The human organic cation transporter (hOCT2) recognizes the degree of substrate ionization. *J Biol Chem*, 277, 22491-22496.

Becker ML, Visser LE, Van Schaik RH, Hofman A, Uitterlinden AG, Stricker BH (2009) Genetic variation in the organic cation transporter 1 is associated with metformin response in patients with diabetes mellitus. *Pharmacogenomics J*, 9, 242–247.

Becker ML, Visser LE, Van Schaik RH, Hofman A, Uitterlinden AG, Stricker BH (2010a) Interaction between polymorphisms in the OCT1 and MATE1 transporter and metformin response. *Pharmacogenet Genomics*, 20, 38–44.

Becker ML, Visser LE, Van Schaik RH, Hofman A, Uitterlinden AG, Stricker BH (2011) OCT1 polymorphism is associated with response and survival time in anti-Parkinsonian drug users. *Neurogenetics*, 12, 79–82.

Biermann J, Lang D, Gorboulev V, Koepsell H, Sindic A, Schroter R, Zvirbliene A, Pavenstadt H, Schlatter E, Ciarimboli G (2006) Characterization of regulatory mechanisms and states of human organic cation transporter 2. *Am J Physiol Cell Physiol*, 290, C1521–C1531.

Bourdet DL, Pritchard JB, Thakker DR (2005) Differential substrate and inhibitory activities of ranitidine and famotidine toward human organic cation transporter 1 (hOCT1; SLC22A1), hOCT2 (SLC22A2), and hOCT3 (SLC22A3). *J Pharmacol Exp Ther*, 315, 1288–1297.

Burger H, Zoumaro-Djayoon A, Boersma AW, Helleman J, Berns EM, Mathijssen RH, Loos WJ, Wiemer EA (2010) Differential transport of platinum compounds by the human organic cation transporter hOCT2 (hSLC22A2). *Br J Pharmacol*, 159, 898–908.

Busch AE, Karbach U, Miska D, Gorboulev V, Akhoundova A, Volk C, Arndt P, Ulzheimer JC, Sonders MS, Baumann C, et al. (1998) Human neurons express the polyspecific cation transporter hOCT2, which translocates monoamine neurotransmitters, amantadine, and memantine. *Mol Pharmacol*, 54, 342–352.

Cargill M, Altshuler D, Ireland J, Sklar P, Ardlie K, Patil N, Shaw N, Lane CR, Lim EP, Kalyanaraman N, et al. (1999) Characterization of single-nucleotide polymorphisms in coding regions of human genes. *Nat Genet*, 22, 231–238.

Chen L, Pawlikowski B, Schlessinger A, More SS, Stryke D, Johns SJ, Portman MA, Chen E, Ferrin TE, Sali A, Giacomini KM (2010a) Role of organic cation transporter 3 (SLC22A3) and its missense variants in the pharmacologic action of metformin. *Pharmacogenet Genomics*, 20, 687–699.

Chen L, Takizawa M, Chen E, Schlessinger A, Segenthelar J, Choi JH, Sali A, Kubo M, Nakamura S, Iwamoto Y, et al. (2010b) Genetic polymorphisms in organic cation transporter 1 (OCT1) in Chinese and Japanese populations exhibit altered function. *J Pharmacol Exp Ther*, 335, 42–50.

Chen Y, Li S, Brown C, Cheatham S, Castro RA, Leabman MK, Urban TJ, Chen L, Yee SW, Choi JH, et al. (2009) Effect of genetic variation in the organic cation transporter 2 on the renal elimination of metformin. *Pharmacogenet Genomics*, 19, 497–504.

Chen Y, Zhang S, Sorani M, Giacomini KM (2007) Transport of paraquat by human organic cation transporters and multidrug and toxic compound extrusion family. *J Pharmacol Exp Ther*, 322, 695–700.

Choi MK, Jin QR, Jin HE, Shim CK, Cho DY, Shin JG, Song IS (2007) Effects of tetraalkylammonium compounds with different affinities for organic cation transporters on the pharmacokinetics of metformin. *Biopharm Drug Dispos*, 28, 501–510.

Ciarimboli G, Ludwig T, Lang D, Pavenstadt H, Koepsell H, Piechota HJ, Haier J, Jaehde U, Zisowsky J, Schlatter E (2005) Cisplatin nephrotoxicity is critically mediated via the human organic cation transporter 2. *Am J Pathol*, 167, 1477–1484.

Crossman LC, Druker BJ, Deininger MW, Pirmohamed M, Wang L, Clark RE (2005) hOCT 1 and resistance to imatinib. *Blood*, 106, 1133–1134; author reply 1134.

Davis R, Giacomini K, Yee SW, Jenkins G, Mccarty C, Wilke R (2010) PS1-10: response to metformin and genetic variants of organic cation and multidrug and toxin extrusion transporters. *Clin Med Res*, 8, 191.

Dresser MJ, Xiao G, Leabman MK, Gray AT, Giacomini KM (2002) Interactions of n-tetraalkylammonium compounds and biguanides with a human renal organic cation transporter (hOCT2). *Pharm Res*, 19, 1244–1247.

Eeles RA, Kote-Jarai Z, Giles GG, Olama AA, Guy M, Jugurnauth SK, Mulholland S, Leongamornlert DA, Edwards SM, Morrison J, et al. (2008) Multiple newly identified loci associated with prostate cancer susceptibility. *Nat Genet*, 40, 316–321.

Engel K, Zhou M, Wang J (2004) Identification and characterization of a novel monoamine transporter in the human brain. *J Biol Chem*, 279, 50042–50049.

Feng B, Obach RS, Burstein AH, Clark DJ, De Morais SM, Faessel HM (2008) Effect of human renal cationic transporter inhibition on the pharmacokinetics of varenicline, a new therapy for smoking cessation: an in vitro-in vivo study. *Clin Pharmacol Ther*, 83, 567–576.

Filipski KK, Loos WJ, Verweij J, Sparreboom A (2008) Interaction of cisplatin with the human organic cation transporter 2. *Clin Cancer Res*, 14, 3875–3880.

Filipski KK, Mathijssen RH, Mikkelsen TS, Schinkel AH, Sparreboom A (2009) Contribution of organic cation transporter 2 (OCT2) to cisplatin-induced nephrotoxicity. *Clin Pharmacol Ther*, 86, 396–402.

Fujita T, Urban TJ, Leabman MK, Fujita K, Giacomini KM (2006) Transport of drugs in the kidney by the human organic cation transporter, OCT2 and its genetic variants. *J Pharm Sci*, 95, 25–36.

Fukushima-Uesaka H, Maekawa K, Ozawa S, Komamura K, Ueno K, Shibakawa M, Kamakura S, Kitakaze M, Tomoike H, Saito Y, Sawada J (2004) Fourteen novel single nucleotide polymorphisms in the SLC22A2 gene encoding human organic cation transporter (OCT2). *Drug Metab Pharmacokinet*, 19, 239–244.

Gambineri A, Tomassoni F, Gasparini DI, Di Rocco A, Mantovani V, Pagotto U, Altieri P, Sanna S, Fulghesu AM, Pasquali R (2010) Organic cation transporter 1 polymorphisms predict the metabolic response to metformin in women with the polycystic ovary syndrome. *J Clin Endocrinol Metab*, 95, E204–E208.

Gazouli M, Mantzaris G, Archimandritis AJ, Nasioulas G, Anagnou NP (2005) Single nucleotide polymorphisms of OCTN1, OCTN2, and DLG5 genes in Greek patients with Crohn's disease. *World J Gastroenterol*, 11, 7525–7530.

Giallourakis C, Stoll M, Miller K, Hampe J, Lander ES, Daly MJ, Schreiber S, Rioux JD (2003) IBD5 is a general risk factor for inflammatory bowel disease: replication of association with Crohn disease and identification of a novel association with ulcerative colitis. *Am J Hum Genet*, 73, 205–211.

Gorboulev V, Ulzheimer JC, Akhoundova A, Ulzheimer-Teuber I, Karbach U, Quester S, Baumann C, Lang F, Busch AE, Koepsell H (1997) Cloning and characterization of two human polyspecific organic cation transporters. *DNA Cell Biol*, 16, 871–881.

Grube M, Meyer Zu, Schwabedissen H, Draber K, Prager D, Moritz KU, Linnemann K, Fusch C, Jedlitschky G, Kroemer HK (2005) Expression, localization, and function of the carnitine transporter octn2 (slc22a5) in human placenta. *Drug Metab Dispos*, 33, 31–37.

Grundemann D, Gorboulev V, Gambaryan S, Veyhl M, Koepsell H (1994) Drug excretion mediated by a new prototype of polyspecific transporter. *Nature*, 372, 549–552.

Grundemann D, Hahne C, Berkels R, Schomig E (2003) Agmatine is efficiently transported by non-neuronal monoamine transporters extraneuronal monoamine transporter (EMT) and organic cation transporter 2 (OCT2). *J Pharmacol Exp Ther*, 304, 810–817.

Grundemann D, Harlfinger S, Golz S, Geerts A, Lazar A, Berkels R, Jung N, Rubbert A, Schomig E (2005) Discovery of the ergothioneine transporter. *Proc Natl Acad Sci U S A*, 102, 5256–5261.

Grundemann D, Koster S, Kiefer N, Breidert T, Engelhardt M, Spitzenberger F, Obermuller N, Schomig E (1998a) Transport of monoamine transmitters by the organic cation transporter type 2, OCT2. *J Biol Chem*, 273, 30915–30920.

Grundemann D, Liebich G, Kiefer N, Koster S, Schomig E (1999) Selective substrates for non-neuronal monoamine transporters. *Mol Pharmacol*, 56, 1–10.

Grundemann D, Schechinger B, Rappold GA, Schomig E (1998b) Molecular identification of the corticosterone-sensitive extraneuronal catecholamine transporter. *Nat Neurosci*, 1, 349–351.

Grundemann D, Schomig E (2000) Gene structures of the human non-neuronal monoamine transporters EMT and OCT2. *Hum Genet*, 106, 627–635.

Halushka MK, Fan JB, Bentley K, Hsie L, Shen N, Weder A, Cooper R, Lipshutz R, Chakravarti A (1999) Patterns of single-nucleotide polymorphisms in candidate genes for blood-pressure homeostasis. *Nat Genet*, 22, 239–247.

Hayer M, Bonisch H, Bruss M (1999) Molecular cloning, functional characterization and genomic organization of four alternatively spliced isoforms of the human organic cation transporter 1 (hOCT1/SLC22A1). *Ann Hum Genet*, 63, 473–482.

Hermann LS, Schersten B, Bitzen PO, Kjellstrom T, Lindgarde F, Melander A (1994a) Therapeutic comparison of metformin and sulfonylurea, alone and in various combinations. A double-blind controlled study. *Diabetes Care*, 17, 1100–1109.

Hermann LS, Schersten B, Melander A (1994b) Antihyperglycaemic efficacy, response prediction and dose-response relations of treatment with metformin and sulphonylurea, alone and in primary combination. *Diabet Med*, 11, 953–960.

Hesselson SE, Matsson P, Shima JE, Fukushima H, Yee SW, Kobayashi Y, Gow JM, Ha C, Ma B, Poon A, et al. (2009) Genetic variation in the proximal promoter of ABC and SLC superfamilies: liver and kidney specific expression and promoter activity predict variation. *PLoS One*, 4, e6942.

Hu S, Franke RM, Filipski KK, Hu C, Orwick SJ, De Bruijn EA, Burger H, Baker SD, Sparreboom A (2008) Interaction of imatinib with human organic ion carriers. *Clin Cancer Res*, 14, 3141–3148.

Imamura Y, Murayama N, Okudaira N, Kurihara A, Okazaki O, Izumi T, Inoue K, Yuasa H, Kusuhara H, Sugiyama Y (2010) Prediction of fluoroquinolone-induced elevation in serum creatinine levels: a case of drug-endogenous substance interaction involving the inhibition of renal secretion. *Clin Pharmacol Ther*, 89, 81–88.

Itoda M, Saito Y, Maekawa K, Hichiya H, Komamura K, Kamakura S, Kitakaze M, Tomoike H, Ueno K, Ozawa S, Sawada J (2004) Seven novel single nucleotide polymorphisms in the human SLC22A1 gene encoding organic cation transporter 1 (OCT1). *Drug Metab Pharmacokinet*, 19, 308–312.

Jablonski KA, Mcateer JB, De Bakker PI, Franks PW, Pollin TI, Hanson RL, Saxena R, Fowler S, Shuldiner AR, Knowler WC, Altshuler D, Florez JC (2010) Common variants in 40 genes assessed for diabetes incidence and response to metformin and lifestyle intervention in the diabetes prevention program. *Diabetes*, 59, 2672–2681.

Kaiser R, Sezer O, Papies A, Bauer S, Schelenz C, Tremblay PB, Possinger K, Roots I, Brockmoller J (2002) Patient-tailored antiemetic treatment with 5-hydroxytryptamine type 3 receptor antagonists according to cytochrome P-450 2D6 genotypes. *J Clin Oncol*, 20, 2805–2811.

Kalow W, Endrenyi L, Tang B (1999) Repeat administration of drugs as a means to assess the genetic component in pharmacological variability. *Pharmacology*, 58, 281–284.

Kang HJ, Song IS, Shin HJ, Kim WY, Lee CH, Shim JC, Zhou HH, Lee SS, Shin JG (2007) Identification and functional characterization of genetic variants of human organic cation transporters in a Korean population. *Drug Metab Dispos*, 35, 667–675.

Kawasaki Y, Kato Y, Sai Y, Tsuji A (2004) Functional characterization of human organic cation transporter OCTN1 single nucleotide polymorphisms in the Japanese population. *J Pharm Sci*, 93, 2920–2926.

Kerb R, Brinkmann U, Chatskaia N, Gorbunov D, Gorboulev V, Mornhinweg E, Keil A, Eichelbaum M, Koepsell H (2002) Identification of genetic variations of the human organic cation transporter hOCT1 and their functional consequences. *Pharmacogenetics*, 12, 591–595.

Kimura N, Masuda S, Katsura T, Inui K (2009) Transport of guanidine compounds by human organic cation transporters, hOCT1 and hOCT2. *Biochem Pharmacol*, 77, 1429–1436.

Kimura N, Masuda S, Tanihara Y, Ueo H, Okuda M, Katsura T, Inui K (2005a) Metformin is a superior substrate for renal organic cation transporter OCT2 rather than hepatic OCT1. *Drug Metab Pharmacokinet*, 20, 379–386.

Kimura N, Okuda M, Inui K (2005b) Metformin transport by renal basolateral organic cation transporter hOCT2. *Pharm Res*, 22, 255–259.

Kirpichnikov D, Mcfarlane SI, Sowers JR (2002) Metformin: an update. *Ann Intern Med*, 137, 25–33.

Koehler MR, Wissinger B, Gorboulev V, Koepsell H, Schmid M (1997) The two human organic cation transporter genes SLC22A1 and SLC22A2 are located on chromosome 6q26. *Cytogenet Cell Genet*, 79, 198–200.

Koepsell H, Gorboulev V, Arndt P (1999) Molecular pharmacology of organic cation transporters in kidney. *J Membr Biol*, 167, 103–117.

Koepsell H, Lips K, Volk C (2007) Polyspecific organic cation transporters: structure, function, physiological roles, and biopharmaceutical implications. *Pharm Res*, 24, 1227–1251.

Koepsell H, Schmitt BM, Gorboulev V (2003) Organic cation transporters. *Rev Physiol Biochem Pharmacol*, 150, 36–90.

Lahjouji K, Mitchell GA, Qureshi IA (2001) Carnitine transport by organic cation transporters and systemic carnitine deficiency. *Mol Genet Metab*, 73, 287–297.

Lamhonwah AM, Skaug J, Scherer SW, Tein I (2003) A third human carnitine/organic cation transporter (OCTN3) as a candidate for the 5q31 Crohn's disease locus (IBD5). *Biochem Biophys Res Commun*, 301, 98–101.

Lamhonwah AM, Tein I (2006) Novel localization of OCTN1, an organic cation/carnitine transporter, to mammalian mitochondria. *Biochem Biophys Res Commun*, 345, 1315–1325.

Lazar A, Grundemann D, Berkels R, Taubert D, Zimmermann T, Schomig E (2003) Genetic variability of the extraneuronal monoamine transporter EMT (SLC22A3). *J Hum Genet*, 48, 226–230.

Lazar A, Walitza S, Jetter A, Gerlach M, Warnke A, Herpertz-Dahlmann B, Grundemann D, Grimberg G, Schulz E, Remschmidt H, Wewetzer C, Schomig E (2008) Novel mutations of the extraneuronal monoamine transporter gene in children and adolescents with obsessive-compulsive disorder. *Int J Neuropsychopharmacol*, 11, 35–48.

Lazar A, Zimmermann T, Koch W, Grundemann D, Schomig A, Kastrati A, Schomig E (2006) Lower prevalence of the OCT2 Ser270 allele in patients with essential hypertension. *Clin Exp Hypertens*, 28, 645–653.

Leabman MK, Huang CC, Deyoung J, Carlson EJ, Taylor TR, De La Cruz M, Johns SJ, Stryke D, Kawamoto M, Urban TJ, et al. (2003) Natural variation in human membrane transporter genes reveals evolutionary and functional constraints. *Proc Natl Acad Sci U S A*, 100, 5896–5901.

Leabman MK, Huang CC, Kawamoto M, Johns SJ, Stryke D, Ferrin TE, Deyoung J, Taylor T, Clark AG, Herskowitz I, Giacomini KM (2002) Polymorphisms in a human kidney xenobiotic transporter, OCT2, exhibit altered function. *Pharmacogenetics*, 12, 395–405.

Li M, Gao X, Guo CC, Wu KC, Zhang X, Hu PJ (2008) OCTN and CARD15 gene polymorphism in Chinese patients with inflammatory bowel disease. *World J Gastroenterol*, 14, 4923–4927.

Lips KS, Volk C, Schmitt BM, Pfeil U, Arndt P, Miska D, Ermert L, Kummer W, Koepsell H (2005) Polyspecific cation transporters mediate luminal release of acetylcholine from bronchial epithelium. *Am J Respir Cell Mol Biol*, 33, 79–88.

Markova NG, Karaman-Jurukovska N, Dong KK, Damaghi N, Smiles KA, Yarosh DB (2009) Skin cells and tissue are capable of using L-ergothioneine as an integral component of their antioxidant defense system. *Free Radic Biol Med*, 46, 1168–1176.

Martinez A, Martin MC, Mendoza JL, Taxonera C, Diaz-Rubio M, De La Concha EG, Urcelay E (2006) Association of the organic cation transporter OCTN genes with Crohn's disease in the Spanish population. *Eur J Hum Genet*, 14, 222–226.

Masuda S, Terada T, Yonezawa A, Tanihara Y, Kishimoto K, Katsura T, Ogawa O, Inui K (2006) Identification and functional characterization of a new human kidney-specific H+/organic cation antiporter, kidney-specific multidrug and toxin extrusion 2. *J Am Soc Nephrol*, 17, 2127–2135.

Meyer Zu Schwabedissen HE, Verstuyft C, Kroemer HK, Becquemont L, Kim RB (2010) Human multidrug and toxin extrusion 1 (MATE1/SLC47A1) transporter: functional characterization, interaction with OCT2 (SLC22A2), and single nucleotide polymorphisms. *Am J Physiol Renal Physiol*, 298, F997–F1005.

Minematsu T, Giacomini KM (2011) Interactions of tyrosine kinase inhibitors with organic cation transporters and multidrug and toxic compound extrusion proteins. *Mol Cancer Ther*, 10, 531–539.

More SS, Li S, Yee SW, Chen L, Xu Z, Jablons DM, Giacomini KM (2010) Organic cation transporters modulate the uptake and cytotoxicity of picoplatin, a third-generation platinum analogue. *Mol Cancer Ther*, 9, 1058–1069.

Motohashi H, Sakurai Y, Saito H, Masuda S, Urakami Y, Goto M, Fukatsu A, Ogawa O, Inui K (2002) Gene expression levels and immunolocalization of organic ion transporters in the human kidney. *J Am Soc Nephrol*, 13, 866–874.

Muller J, Lips KS, Metzner L, Neubert RH, Koepsell H, Brandsch M (2005) Drug specificity and intestinal membrane localization of human organic cation transporters (OCT). *Biochem Pharmacol*, 70, 1851–1860.

Nathan DM, Buse JB, Davidson MB, Ferrannini E, Holman RR, Sherwin R, Zinman B (2009) Medical management of hyperglycaemia in type 2 diabetes mellitus: a consensus algorithm for the initiation and adjustment of therapy: a consensus statement from the American Diabetes Association and the European Association for the Study of Diabetes. *Diabetologia*, 52, 17–30.

Newman B, Gu X, Wintle R, Cescon D, Yazdanpanah M, Liu X, Peltekova V, Van Oene M, Amos CI, Siminovitch KA (2005) A risk haplotype in the Solute Carrier Family 22A4/22A5 gene cluster influences phenotypic expression of Crohn's disease. *Gastroenterology*, 128, 260–269.

Nies AT, Hofmann U, Resch C, Schaeffeler E, Rius M, Schwab M (2011a) Proton pump inhibitors inhibit metformin uptake by organic cation transporters (OCTs). *PLoS One*, 6, e22163.

Nies AT, Koepsell H, Damme K, Schwab M (2011b) Organic cation transporters (OCTs, MATEs), in vitro and in vivo evidence for the importance in drug therapy. In: *Handbook of Experimental Pharmacology*, Vol. 201. Heidelberg, Germany: Springer. pp 105–167.

Nies AT, Koepsell H, Winter S, Burk O, Klein K, Kerb R, Zanger UM, Keppler D, Schwab M, Schaeffeler E (2009) Expression of organic cation transporters OCT1 (SLC22A1) and OCT3 (SLC22A3) is affected by genetic factors and cholestasis in human liver. *Hepatology*, 50, 1227–1240.

Noble CL, Nimmo ER, Drummond H, Ho GT, Tenesa A, Smith L, Anderson N, Arnott ID, Satsangi J (2005) The contribution of OCTN1/2 variants within the IBD5 locus to disease susceptibility and severity in Crohn's disease. *Gastroenterology*, 129, 1854–1864.

Ogasawara K, Terada T, Motohashi H, Asaka J, Aoki M, Katsura T, Kamba T, Ogawa O, Inui K (2008) Analysis of regulatory polymorphisms in organic ion transporter genes (SLC22A) in the kidney. *J Hum Genet*, 53, 607–614.

Ohashi R, Tamai I, Nezu Ji J, Nikaido H, Hashimoto N, Oku A, Sai Y, Shimane M, Tsuji A (2001) Molecular and physiological evidence for multifunctionality of carnitine/organic cation transporter OCTN2. *Mol Pharmacol*, 59, 358–366.

Otsuka M, Matsumoto T, Morimoto R, Arioka S, Omote H, Moriyama Y (2005) A human transporter protein that mediates the final excretion step for toxic organic cations. *Proc Natl Acad Sci U S A*, 102, 17923–17928.

Peltekova VD, Wintle RF, Rubin LA, Amos CI, Huang Q, Gu X, Newman B, Van Oene M, Cescon D, Greenberg G, et al. (2004) Functional variants of OCTN cation transporter genes are associated with Crohn disease. *Nat Genet*, 36, 471–475.

Pietig G, Mehrens T, Hirsch JR, Cetinkaya I, Piechota H, Schlatter E (2001) Properties and regulation of organic cation transport in freshly isolated human proximal tubules. *J Biol Chem*, 276, 33741–33746.

Rioux JD, Daly MJ, Silverberg MS, Lindblad K, Steinhart H, Cohen Z, Delmonte T, Kocher K, Miller K, Guschwan S, et al. (2001) Genetic variation in the 5q31 cytokine gene cluster confers susceptibility to Crohn disease. *Nat Genet*, 29, 223–228.

Roediger WE, Nance S (1986) Metabolic induction of experimental ulcerative colitis by inhibition of fatty acid oxidation. *Br J Exp Pathol*, 67, 773–782.

Russell RK, Drummond HE, Nimmo ER, Anderson NH, Noble CL, Wilson DC, Gillett PM, Mcgrogan P, Hassan K, Weaver LT, Bisset WM, Mahdi G, Satsangi J (2006) Analysis of the influence of OCTN1/2 variants within the IBD5 locus on disease susceptibility and growth indices in early onset inflammatory bowel disease. *Gut*, 55, 1114–1123.

Saito S, Iida A, Sekine A, Ogawa C, Kawauchi S, Higuchi S, Nakamura Y (2002) Catalog of 238 variations among six human genes encoding solute carriers (hSLCs) in the Japanese population. *J Hum Genet*, 47, 576–584.

Sakata T, Anzai N, Shin HJ, Noshiro R, Hirata T, Yokoyama H, Kanai Y, Endou H (2004) Novel single nucleotide polymorphisms of organic cation transporter 1 (SLC22A1) affecting transport functions. *Biochem Biophys Res Commun*, 313, 789–793.

Sata R, Ohtani H, Tsujimoto M, Murakami H, Koyabu N, Nakamura T, Uchiumi T, Kuwano M, Nagata H, Tsukimori K, Nakano H, Sawada Y (2005) Functional analysis of organic cation transporter 3 expressed in human placenta. *J Pharmacol Exp Ther*, 315, 888–895.

Schaeffeler E, Hellerbrand C, Nies AT, Winter S, Kruck S, Hofmann U, Van Der Kuip H, Zanger UM, Koepsell H, Schwab M (2011) DNA methylation is associated with downregulation of the organic cation transporter OCT1 (SLC22A1) in human hepatocellular carcinoma. *Genome Med*, 3, 82.

Scheen AJ (1996) Clinical pharmacokinetics of metformin. *Clin Pharmacokinet*, 30, 359–371.

Shikata E, Yamamoto R, Takane H, Shigemasa C, Ikeda T, Otsubo K, Ieiri I (2007) Human organic cation transporter (OCT1 and OCT2) gene polymorphisms and therapeutic effects of metformin. *J Hum Genet*, 52, 117–122.

Shu Y, Brown C, Castro RA, Shi RJ, Lin ET, Owen RP, Sheardown SA, Yue L, Burchard EG, Brett CM, Giacomini KM (2008) Effect of genetic variation in the organic cation transporter 1, OCT1, on metformin pharmacokinetics. *Clin Pharmacol Ther*, 83, 273–280.

Shu Y, Leabman MK, Feng B, Mangravite LM, Huang CC, Stryke D, Kawamoto M, Johns SJ, Deyoung J, Carlson E, et al. (2003) Evolutionary conservation predicts function of variants of the human organic cation transporter, OCT1. *Proc Natl Acad Sci U S A*, 100, 5902–5907.

Shu Y, Sheardown SA, Brown C, Owen RP, Zhang S, Castro RA, Ianculescu AG, Yue L, Lo JC, Burchard EG, Brett CM, Giacomini KM (2007) Effect of genetic variation in the organic cation transporter 1 (OCT1) on metformin action. *J Clin Invest*, 117, 1422–1431.

Silverberg MS (2006) OCTNs: will the real IBD5 gene please stand up? *World J Gastroenterol*, 12, 3678–3681.

Simonsen KL, Churchill GA, Aquadro CF (1995) Properties of statistical tests of neutrality for DNA polymorphism data. *Genetics*, 141, 413–429.

Song IS, Shin HJ, Shim EJ, Jung IS, Kim WY, Shon JH, Shin JG (2008a) Genetic variants of the organic cation transporter 2 influence the disposition of metformin. *Clin Pharmacol Ther*, 84, 559–562.

Song IS, Shin HJ, Shin JG (2008b) Genetic variants of organic cation transporter 2 (OCT2) significantly reduce metformin uptake in oocytes. *Xenobiotica*, 38, 1252–1262.

Stumvoll M, Nurjhan N, Perriello G, Dailey G, Gerich JE (1995) Metabolic effects of metformin in non-insulin-dependent diabetes mellitus. *N Engl J Med*, 333, 550–554.

Sweet DH, Miller DS, Pritchard JB (2001) Ventricular choline transport: a role for organic cation transporter 2 expressed in choroid plexus. *J Biol Chem*, 276, 41611–41619.

Sweet DH, Pritchard JB (1999) The molecular biology of renal organic anion and organic cation transporters. *Cell Biochem Biophys*, 31, 89–118.

Tahara H, Kusuhara H, Endou H, Koepsell H, Imaoka T, Fuse E, Sugiyama Y (2005) A species difference in the transport activities of H2 receptor antagonists by rat and human renal organic anion and cation transporters. *J Pharmacol Exp Ther*, 315, 337–345.

Tahara H, Yee SW, Urban TJ, Hesselson S, Castro RA, Kawamoto M, Stryke D, Johns SJ, Ferrin TE, Kwok PY, Giacomini KM (2009) Functional genetic variation in the basal promoter of the organic cation/carnitine transporters OCTN1 (SLC22A4) and OCTN2 (SLC22A5). *J Pharmacol Exp Ther*, 329, 262–271.

Tajima F (1989) Statistical method for testing the neutral mutation hypothesis by DNA polymorphism. *Genetics*, 123, 585–595.

Takeuchi A, Motohashi H, Okuda M, Inui K (2003) Decreased function of genetic variants, Pro283Leu and Arg287Gly, in human organic cation transporter hOCT1. *Drug Metab Pharmacokinet*, 18, 409–412.

Tamai I, Ohashi R, Nezu J, Yabuuchi H, Oku A, Shimane M, Sai Y, Tsuji A (1998) Molecular and functional identification of sodium ion-dependent, high affinity human carnitine transporter OCTN2. *J Biol Chem*, 273, 20378–20382.

Tamai I, Yabuuchi H, Nezu J, Sai Y, Oku A, Shimane M, Tsuji A (1997) Cloning and characterization of a novel human pH-dependent organic cation transporter, OCTN1. *FEBS Lett*, 419, 107–111.

Tang NL, Ganapathy V, Wu X, Hui J, Seth P, Yuen PM, Wanders RJ, Fok TF, Hjelm NM (1999) Mutations of OCTN2, an organic cation/carnitine transporter, lead to deficient cellular carnitine uptake in primary carnitine deficiency. *Hum Mol Genet*, 8, 655–660.

Tanihara Y, Masuda S, Sato T, Katsura T, Ogawa O, Inui K (2007) Substrate specificity of MATE1 and MATE2-K, human multidrug and toxin extrusions/H(+)-organic cation antiporters. *Biochem Pharmacol*, 74, 359–371.

Tein I (2003) Carnitine transport: pathophysiology and metabolism of known molecular defects. *J Inherit Metab Dis*, 26, 147–169.

Thomas J, Wang L, Clark RE, Pirmohamed M (2004) Active transport of imatinib into and out of cells: implications for drug resistance. *Blood*, 104, 3739–3745.

Toh DS, Yee JY, Koo SH, Murray M, Lee EJ (2010) Genetic variations of the SLC22A5 gene in the Chinese and Indian populations of Singapore. *Drug Metab Pharmacokinet*, 25, 112–119.

Torok HP, Glas J, Tonenchi L, Lohse P, Muller-Myhsok B, Limbersky O, Neugebauer C, Schnitzler F, Seiderer J, Tillack C, et al. (2005) Polymorphisms in the DLG5 and OCTN cation transporter genes in Crohn's disease. *Gut*, 54, 1421–1427.

Tosa M, Negoro K, Kinouchi Y, Abe H, Nomura E, Takagi S, Aihara H, Oomori S, Sugimura M, Takahashi K, et al. (2006) Lack of association between IBD5 and Crohn's disease in Japanese patients demonstrates population-specific differences in inflammatory bowel disease. *Scand J Gastroenterol*, 41, 48–53.

Tregouet DA, Konig IR, Erdmann J, Munteanu A, Braund PS, Hall AS, Grosshennig A, Linsel-Nitschke P, Perret C, Desuremain M, et al. (2009) Genome-wide haplotype association study identifies the SLC22A3-LPAL2-LPA gene cluster as a risk locus for coronary artery disease. *Nat Genet*, 41, 283–285.

Tzvetkov MV, Saadatmand AR, Bokelmann K, Meineke I, Kaiser R, Brockmoller J (2012) Effects of OCT1 polymorphisms on the cellular uptake, plasma concentrations and efficacy of the 5-HT(3) antagonists tropisetron and ondansetron. *Pharmacogenomics J*, 12, 22–29.

Tzvetkov MV, Vormfelde SV, Balen D, Meineke I, Schmidt T, Sehrt D, Sabolic I, Koepsell H, Brockmoller J (2009) The effects of genetic polymorphisms in the organic cation transporters OCT1, OCT2, and OCT3 on the renal clearance of metformin. *Clin Pharmacol Ther*, 86, 299–306.

Umehara KI, Iwatsubo T, Noguchi K, Kamimura H (2007) Comparison of the kinetic characteristics of inhibitory effects exerted by biguanides and H2-blockers on human and rat organic cation transporter-mediated transport: Insight into the development of drug candidates. *Xenobiotica*, 37, 618–634.

Urakami Y, Kimura N, Okuda M, Inui K (2004) Creatinine transport by basolateral organic cation transporter hOCT2 in the human kidney. *Pharm Res*, 21, 976–981.

Urban TJ, Gallagher RC, Brown C, Castro RA, Lagpacan LL, Brett CM, Taylor TR, Carlson EJ, Ferrin TE, Burchard EG, Packman S, Giacomini KM (2006) Functional genetic diversity in the high-affinity carnitine transporter OCTN2 (SLC22A5). *Mol Pharmacol*, 70, 1602–1611.

Urban TJ, Giacomini KM, Risch N (2005) Haplotype structure and ethnic-specific allele frequencies at the OCTN locus: implications for the genetics of Crohn's disease. *Inflamm Bowel Dis*, 11, 78–79.

Urban TJ, Yang C, Lagpacan LL, Brown C, Castro RA, Taylor TR, Huang CC, Stryke D, Johns SJ, Kawamoto M, et al. (2007) Functional effects of protein sequence polymorphisms in the organic cation/ergothioneine transporter OCTN1 (SLC22A4). *Pharmacogenet Genomics*, 17, 773–782.

Van Montfoort JE, Muller M, Groothuis GM, Meijer DK, Koepsell H, Meier PJ (2001) Comparison of "type I" and "type II" organic cation transport by organic cation transporters and organic anion-transporting polypeptides. *J Pharmacol Exp Ther*, 298, 110–115.

Vermeire S (2004) DLG5 and OCTN. *Inflamm Bowel Dis*, 10, 888–890.

Waller S, Tremelling M, Bredin F, Godfrey L, Howson J, Parkes M (2006) Evidence for association of OCTN genes and IBD5 with ulcerative colitis. *Gut*, 55, 809–814.

Wang DS, Jonker JW, Kato Y, Kusuhara H, Schinkel AH, Sugiyama Y (2002) Involvement of organic cation transporter 1 in hepatic and intestinal distribution of metformin. *J Pharmacol Exp Ther*, 302, 510–515.

Wang DS, Kusuhara H, Kato Y, Jonker JW, Schinkel AH, Sugiyama Y (2003) Involvement of organic cation transporter 1 in the lactic acidosis caused by metformin. *Mol Pharmacol*, 63, 844–848.

Wang L, Giannoudis A, Lane S, Williamson P, Pirmohamed M, Clark RE (2008a) Expression of the uptake drug transporter hOCT1 is an important clinical determinant of the response to imatinib in chronic myeloid leukemia. *Clin Pharmacol Ther*, 83, 258–264.

Wang ZJ, Yin OQ, Tomlinson B, Chow MS (2008b) OCT2 polymorphisms and in-vivo renal functional consequence: studies with metformin and cimetidine. *Pharmacogenet Genomics*, 18, 637–645.

White DL, Saunders VA, Dang P, Engler J, Zannettino AC, Cambareri AC, Quinn SR, Manley PW, Hughes TP (2006) OCT-1-mediated influx is a key determinant of the intracellular uptake of imatinib but not nilotinib (AMN107): reduced OCT-1 activity is the cause of low in vitro sensitivity to imatinib. *Blood*, 108, 697–704.

Wu X, Huang W, Ganapathy ME, Wang H, Kekuda R, Conway SJ, Leibach FH, Ganapathy V (2000) Structure, function, and regional distribution of the organic cation transporter OCT3 in the kidney. *Am J Physiol Renal Physiol*, 279, F449–F458.

Wu X, Prasad PD, Leibach FH, Ganapathy V (1998) cDNA sequence, transport function, and genomic organization of human OCTN2, a new member of the organic cation transporter family. *Biochem Biophys Res Commun*, 246, 589–595.

Yamazaki K, Takazoe M, Tanaka T, Ichimori T, Saito S, Iida A, Onouchi Y, Hata A, Nakamura Y (2004) Association analysis of SLC22A4, SLC22A5 and DLG5 in Japanese patients with Crohn disease. *J Hum Genet*, 49, 664–668.

Zach O, Krieger O, Foedermayr M, Zellhofer B, Lutz D (2008) OCT1 (SLC22A1) R61C polymorphism and response to imatinib treatment in chronic myeloid leukemia patients. *Leuk Lymphoma*, 49, 2222–2223.

Zhang L, Dresser MJ, Gray AT, Yost SC, Terashita S, Giacomini KM (1997) Cloning and functional expression of a human liver organic cation transporter. *Mol Pharmacol*, 51, 913–921.

Zhang S, Lovejoy KS, Shima JE, Lagpacan LL, Shu Y, Lapuk A, Chen Y, Komori T, Gray JW, Chen X, Lippard SJ, Giacomini KM (2006) Organic cation transporters are determinants of oxaliplatin cytotoxicity. *Cancer Res*, 66, 8847–8857.

Zhou G, Myers R, Li Y, Chen Y, Shen X, Fenyk-Melody J, Wu M, Ventre J, Doebber T, Fujii N, Musi N, Hirshman MF, Goodyear LJ, Moller DE (2001) Role of AMP-activated protein kinase in mechanism of metformin action. *J Clin Invest*, 108, 1167–1174.

Zhou K, Bellenguez C, Spencer CC, Bennett AJ, Coleman RL, Tavendale R, Hawley SA, Donnelly LA, Schofield C, Groves CJ, et al. (2010) Common variants near ATM are associated with glycemic response to metformin in type 2 diabetes. *Nat Genet*, 43, 117–120.

Zhou K, Donnelly LA, Kimber CH, Donnan PT, Doney AS, Leese G, Hattersley AT, Mccarthy MI, Morris AD, Palmer CN, Pearson ER (2009) Reduced-function SLC22A1 polymorphisms encoding organic cation transporter 1 and glycemic response to metformin: a GoDARTS study. *Diabetes*, 58, 1434–1439.

Zolk O, Solbach TF, Konig J, Fromm MF (2009a) Functional characterization of the human organic cation transporter 2 variant p.270Ala>Ser. *Drug Metab Dispos*, 37, 1312–1318.

Zolk O, Solbach TF, Konig J, Fromm MF (2009b) Structural determinants of inhibitor interaction with the human organic cation transporter OCT2 (SLC22A2). *Naunyn Schmiedebergs Arch Pharmacol*, 379, 337–348.

CHAPTER 9

MATE (SLC47) FAMILY

Atsushi Yonezawa
Ken-ichi Inui

9.1 INTRODUCTION

The kidney plays an important role in the elimination of numerous compounds, including drugs, toxins, and endogenous metabolites. Renal elimination of drugs involves glomerular filtration and tubular secretion. The secretory process is performed by two distinct classes of transporters: One located at the basolateral membranes to mediate the cellular uptake of substrates from blood, and the other at the brush-border membranes to mediate the efflux of cellular substrates into the tubular lumen. More than 25 years ago, the transport mechanisms for a typical organic cation, tetraethylammonium (TEA), were characterized in rat renal brush-border and basolateral membranes by membrane vesicle technique (Takano et al., 1984; Inui and Okuda, 1998). TEA uptake by basolateral membrane vesicles was stimulated by an inside negative potential, whereas its uptake by brush-border membrane vesicles was driven by an outwardly directed H^+ gradient. These results indicated that the renal organic cation transport system consists of a membrane potential-dependent organic cation transporter (OCT) in the basolateral membrane and an H^+/organic cation antiporter in the brush-border membrane.

In 1994, a potential-dependent OCT, OCT1, was identified in the rat kidney (Grundemann et al., 1994). In 1996, kidney-specific OCT2 was also found (Okuda et al., 1996). Human OCT1/SLC22A1 and OCT2/SLC22A2 are expressed mainly in the liver and kidney, respectively, and mediate the uptake of cationic drugs (Inui et al., 2000; Koepsell et al., 2007). Ten years later, multidrug and toxin extrusion (MATE) was identified as H^+-coupled organic cation exporter (Otsuka et al., 2005). In this chapter, we summarize the importance of MATE1/SLC47A1 and MATE2-K/SLC47A2 to the pharmacokinetics, pharmacodynamics/toxicodynamics, and pharmacogenomics of cationic drugs.

9.2 DISCOVERY OF THE HUMAN *SLC47A* GENES

MATE transporters were originally identified in *Vibrio parahaemolyticus* and *Escherichia coli*, and named NorM and YdhE. In 2005, Otsuka et al., (Otsuka et al.,

Pharmacogenomics of Human Drug Transporters: Clinical Impacts, First Edition.
Edited by Toshihisa Ishikawa, Richard B. Kim and Jörg König

2005) searched for a mammalian ortholog of bacterial MATE-type multidrug transporters in genomic databases. Upon the screening of the entire working draft of the human genome database, human MATE1 was identified as a human ortholog of the bacterial NorM. The gene, which is located on chromosome 17 of p11.2, was designated as SLC47A1 (GenBank accession no. NM_018242). The gene product was also designated as MATE1 (GenBank accession no. NP_060712). Human MATE1 consists of 570 amino acids and shows 19.8% identity to NorM in *Vibrio parahaemolyticus*. It was suggested that MATE1 mediates the H^+-coupled electroneutral exchange of organic cations (Otsuka et al., 2005; Omote et al., 2006). Based on functional characterization, MATE was revealed to be an H^+/organic cation antiporter (Terada et al., 2006; Tsuda et al., 2007; Terada and Inui, 2008). The MATE family was assigned as the SLC47 family.

Human MATE2 (NP_690872) was also reported as a homolog of human MATE1 (Otsuka et al., 2005). Thereafter, two alternatively spliced variants of MATE2, MATE2-K (NP_001093116) and MATE2-B (BAF37007), were found (Masuda et al., 2006). They exhibit a different splicing pattern between exon 6 and exon 7. MATE2-K mRNA has a deletion of 108 bases in exon 7, compared with MATE2 mRNA. A 154-base intron between exon 6 and exon 7 is not spliced in MATE2-B, where there is a stop codon. MATE2, MATE2-K, and MATE2-B consist of 602, 566, and 219 amino acids, respectively. Among them, MATE2-K is predominantly expressed in human kidney, and is the active form of the *SLC47A2* gene (NM_001099646) (Masuda et al., 2006). Physiological roles of MATE2 and MATE2-B are unclear. Characteristics of human MATE1 and MATE2-K are summarized in Table 9.1 and previous reviews (Terada and Inui, 2008; Yonezawa and Inui, 2011a).

Animal orthologs of human MATE have also been found, although the nomenclature and classification are confusing. Human MATE2-K and rodent MATE2 exhibit

TABLE 9.1 Characteristics of Human MATE1/SLC47A1 and MATE2-K/SLC47A2

	MATE1	MATE2-K
Gene name	*SLC47A1*	*SLC47A2*
Nucleotide accession no.	NM_018242	NM_001099646
Chromosome location	17p11.2	17p11.2
Protein name	MATE1	MATE2-K
Protein accession no.	NP_060712	NP_001093116
Amino acids	570	566
Splicing variants	–	MATE2, MATE2-B (Inactive)
Tissue distribution	Kidney $\gg$ adrenal grand, testis > skeletal muscle, liver, uterus, etc.	Kidney
Membrane localization	Brush-border membrane	Brush-border membrane
Driving force	Oppositely directed H^+	Oppositely directed H^+
Transcription factor	Sp1, AP-1	–
Orthologs	Rat MATE1, Mouse MATE1, Rabbit MATE1	Rabbit MATE2-K

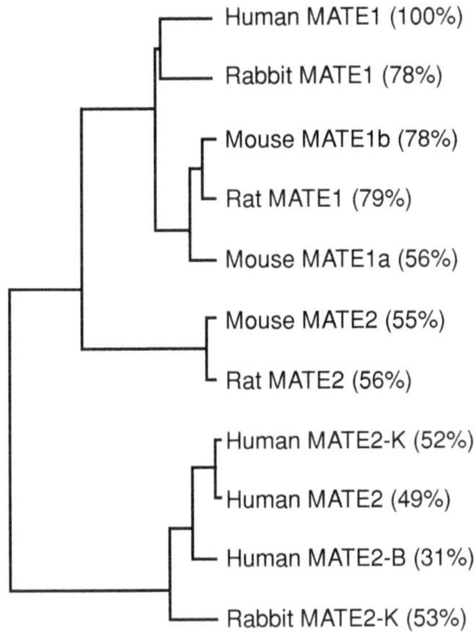

Figure 9.1 Phylogenetic tree of human, mouse, rat, and rabbit MATE. Amino acid identity compared with human MATE1 is shown.

only low mutual sequence identity (38.1%) and different expression patterns, although the characteristics of human MATE1 and rodent MATE1 are similar (Otsuka et al., 2005; Ohta et al., 2006; Omote et al., 2006; Terada et al., 2006; Terada and Inui, 2008; Yonezawa and Inui, 2011a). Rabbit MATE2-K shows similar features to human MATE2-K (Zhang et al., 2007). In fact, the counterparts of human MATE2-K have not been identified in rats and mice, and the counterpart of rodent MATE2 has not been found in humans (Omote et al., 2006; Terada and Inui, 2008; Yonezawa and Inui, 2011a). The phylogenetic tree of mammalian MATE-type transporters suggested that rodent MATE2 is classified into the MATE3 family, not the MATE2 family (Fig. 9.1) (Hiasa et al., 2007).

9.3 PHYSIOLOGICAL FUNCTION OF MATE

Human MATE1 is highly expressed in the kidney and moderately expressed in adrenal gland, liver, skeletal muscle, and several other tissues (Masuda et al., 2006) (Table 9.1). MATE2-K exhibits a kidney-specific expression. The MATE1 and MATE2-K mRNA are detected at similar levels in the kidney, and the proteins are similarly localized in the brush-border membrane of proximal tubules. Therefore, both MATE1 and MATE2-K could play a role in the renal tubular secretion of cationic drugs in human (Fig. 9.2). MATE1 could also act as an efflux transporter in other tissues. Species differences in the tissue distribution of MATEs have been demonstrated. In rats, MATE1 mRNA is expressed abundantly in the kidney and placenta, slightly in the spleen, but not in the liver (Terada et al., 2006). In mice, MATE1 is highly expressed in the kidney, liver, heart, and several tissues (Hiasa et al., 2007). The tissue distribution of MATE1 in mice is generally consistent

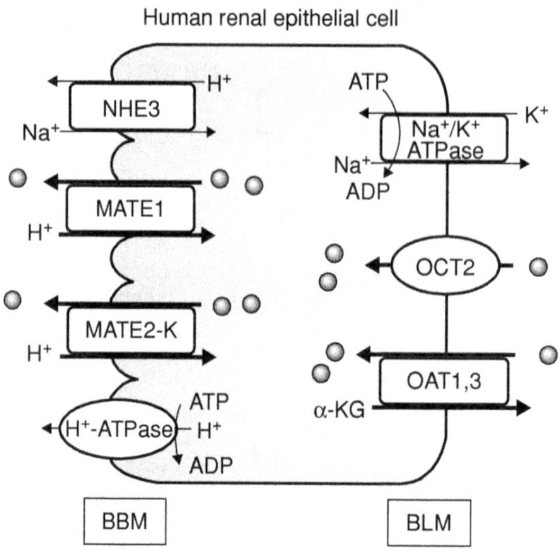

Figure 9.2 Organic cation transport systems in the proximal tubular epithelial cells of human kidney. BBM, brush-border membrane; BLM, basolateral membrane; OCT2, organic cation transporter 2; MATE, multidrug and toxin extrusion; NHE3, Na/H$^+$ exchanger 3; OAT, organic anion transporter: α-KG, α-Ketoglutarate.

with that in humans. However, MATE2-K is not expressed in rats and mice. Mate1 knockout (KO) mice could represent a model of MATE1 and MATE2-K deficiency in humans (Tsuda et al., 2009a).

Transport studies in brush-border and basolateral membrane vesicles prepared from the kidney have been successfully utilized to characterize a number of transport systems under well-defined *in vitro* conditions. This is because the ionic composition inside or outside the membrane vesicles is easily manipulated, and ion gradients and membrane potential can be provided artificially. In fact, it was clearly demonstrated that organic cation transport systems at the renal brush-border membranes are driven by an oppositely directed H$^+$ gradient (Takano et al., 1984). By using membrane vesicles from HEK293 cells expressing MATE, it is directly indicated that the driving force of TEA transport by MATE1 and MATE2-K is an oppositely directed H$^+$ gradient (Tsuda et al., 2007, 2009b). Furthermore, this stimulation disappeared in the presence of a protonophore, carbonyl cyanide *p*-trifluoromethoxyphenylhydrazone (FCCP). These results corresponded to previous findings using rat kidney brush-border membrane vesicles, and indicated that MATE1 and MATE2-K function as H$^+$/organic cation antiporters.

Substrates for MATE1 and MATE2-K are typical organic cations, TEA, 1-methyl-4-phenylpyridinium (MPP), metformin, cimetidine, procainamide, and so on (Table 9.2) (Tanihara et al., 2007). These compounds are also transported by OCT2 (Inui et al., 2000; Koepsell et al., 2007). K_m values of cationic drugs for MATE1 and MATE2-K are similar and higher than the plasma concentrations in clinical use. In addition, the anionic compounds—estrone sulfate, acyclovir, and ganciclovir—are also substrates for MATE1 and MATE2-K (Tanihara et al., 2007). However, most anions were not transported. Further study showed that 4′,6-diamidino-2-phenylindole (DAPI), a fluorescent probe, was a substrate for MATE1 and MATE2-K and useful for the assay of MATE activity (Yasujima et al., 2010). Zwitterionic drugs, cephalexin and cephradine, are mainly transported by MATE1 (Tanihara et al., 2007;

TABLE 9.2 Substrate Specificity of MATE1 and MATE2-K

	Classification	Transporter	Compounds
	Cationic compound	MATE1, MATE2-K	Tetraethylammonium, MPP, cimetidine, metformin, creatinine, guanidine, procainamide, thiamine, topotecan, paraquat, agmatine, 4′,6-diamidino-2-phenylindole (DAPI)
Substrate	Zwitterionic compound	MATE1	Cephalexin, cephradine, fexofenadine, PNU-288034
	Anionic compound	MATE1, MATE2-K	Estrone sulfate, acyclovir, ganciclovir
	Platinum compound	MATE1, MATE2-K	Cisplatin (MATE1, MATE2-K), oxaliplatin (MATE2-K)
Inhibitor	–	MATE1, MATE2-K	Cimetidine, pyrimethamine

Watanabe et al., 2010). As they were transported by organic anion transporters OAT1 and OAT3, but not OCT (Zhang et al., 2010), cephalexin and cephradine are likely eliminated by tubular secretion via OAT and MATE1. In addition, fluoroquinolones such as ciprofloxacin, enoxacin, gatifloxacin, levofloxacin, norfloxacin, pazufloxacin, and tosufloxacin were also transported by rat MATE1 (Ohta et al., 2009a). Thus, MATE1 can transport some substrates for OAT. On the other hand, oxaliplatin was a specific substrate for MATE2-K, although other platinum agents were hardly recognized by MATE2-K (Yonezawa et al., 2006; Yokoo et al., 2007). MATE1 and MATE2-K are very similar, but not completely the same, in substrate specificity (Table 9.1).

Some compounds were reported to be specific inhibitors of MATE, although OCT and MATE are similar in substrate specificity. The affinity of cimetidine for MATE1 and MATE2-K is much stronger than that for OCT2 (Matsushima et al., 2009; Ohta et al., 2009b; Tsuda et al., 2009b). IC_{50} values for MATE1 and MATE2-K are 1–10 μM, comparable to plasma concentrations (Tsuda et al., 2009b). Clinical drug–drug interactions have been previously reported between cimetidine and cationic drugs, metformin (Somogyi et al., 1987), procainamide (Somogyi et al., 1983; Christian et al., 1984), triamterene (Muirhead et al., 1986), pilsicainide (Shiga et al., 2000), and varenicline (Feng et al., 2008). The Ki value (147 μM) of cimetidine for OCT2 as a competitive inhibitor is higher than plasma concentrations (Tsuda et al., 2009b). Therefore, it was suggested that cimetidine would inhibit MATE1 and MATE2-K activities and decrease the tubular secretion of cationic drugs. In addition, two other histamine H2 receptor antagonists, famotidine and ranitidine, also inhibit MATE more than OCT, but their clinical concentrations are lower than their IC_{50} values (Tsuda et al., 2009b). Furthermore, the antimalarial agent pyrimethamine is a potent inhibitor of MATE1 and MATE2-K (Ito et al., 2010). In mice, the coadministration of pyrimethamine and metformin increased the renal concentration of metformin at clinical doses. Cimetidine and pyrimethamine are useful probe inhibitors of MATE.

9.4 REGULATION OF *SLC47* GENE EXPRESSION

The regulatory mechanisms for *SLC47A1* gene expression were characterized, focusing on the basal transcriptional regulation. Sp1 was identified as a general transcription factor of the human *SLC47A1* gene by a deletion analysis, mutational analysis, and electrophoretic mobility assay (Kajiwara et al., 2007). In a TATA-less promoter, Sp1 binds to the GC-rich region to recruit TATA-binding proteins and to fix the transcription start site (Baumann et al., 2010). In the proximal promoter region of the human, rat, and mouse *SLC47A1* genes, two Sp1-binding consensus sequences are conserved. It was also reported that transcription of *SLC47A1* gene was regulated by AP-1 (Ha Choi et al., 2009). The binding site of AP-1 was located at −64 to −70 bp from the translational ATG site, which could be the 5′-UTR region of the *SLC47A1* gene. The ubiquitous transcription factor Sp1 is considered a constitutive activator of TATA-less genes, which are usually not highly regulated. On the other hand, AP1 activity is induced by several factors—growth factors, cytokines, neurotransmitters, polypeptide hormones, infections, and stress. Therefore, the general transcription factor Sp1 and transcriptional regulator AP-1 could coordinately play a role in the transcriptional regulation of the *SLC47A1* gene (Table 9.1). Yet, the regulatory mechanisms for *SLC47A2* gene expression have been unclear.

9.5 PHARMACOKINETICS IN MATE1 KO MICE

Experiments with Mate1 KO mice have revealed a role of MATE1 in the pharmacokinetics of several drugs. In Mate1 KO mice, plasma and renal concentrations of metformin were dramatically increased, and urinary excretion was also decreased (Tsuda et al., 2009a). The renal secretory clearance was 14% of that in wild-type mice. These results indicated that MATE plays a predominant role in the tubular secretion of metformin. In heterozygous Mate1 KO mice, the pharmacokinetics of metformin was not significantly different from that in wild-type mice, although the level of MATE1 protein was decreased (Toyama et al., 2010). It was suggested that the rate-limiting step in the excretion of metformin is not the efflux from tubular cells mediated by MATE1, but the renal plasma flow, because renal clearance is comparable to renal plasma flow. The biliary clearance of metformin was around 100 times lower than the renal clearance, although the hepatic concentration was one-third of the renal concentration (Ito et al., 2010). The hepatic concentration and Kp ratio (tissue/plasma) of metformin were significantly higher in Mate1 KO mice and in the mice coadministered with the MATE1 inhibitor pyrimethamine than in control mice (Tsuda et al., 2009a; Ito et al., 2010). The hepatic concentration of metformin was 43-fold higher in Mate1 KO mice than wild-type mice, in spite of a 7-fold increase

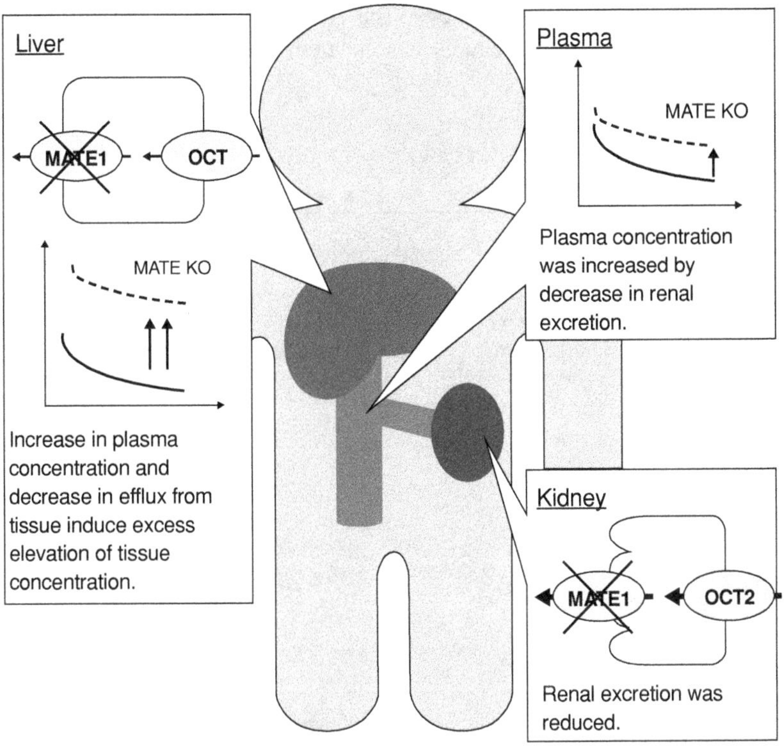

Figure 9.3 Pharmacokinetics of metformin in the dysfunction of MATE1. Metformin was excreted into urine. Disruption of MATE1 reduced the renal excretion of metformin, and the plasma concentration increased. MATE1 is also expressed in the liver. An increase in the plasma concentration and a decrease in efflux from the tissue induced an excess elevation of the hepatic concentration.

in the plasma concentration (Toyama et al., 2010). Therefore, MATE1 is also a determinant of the hepatic concentrations of cationic drugs. It is possible that MATE can control the tissue concentrations as well as renal excretion of drugs (Fig. 9.3).

Zwitterionic drug cephalexin was also transported by MATE1. In Mate1 KO mice, renal accumulation of cephalexin was much higher than in wild-type mice (Watanabe et al., 2010). Plasma concentrations were slightly increased. Tubular secretion was 35% of the wild-type level. Thus, MATE1 could contribute to the renal tubular secretion not only of cationic drugs but also of zwitterionic drugs. These results are consistent with the previous findings using rat renal brush-border membrane vesicles (Inui et al., 1985).

Furthermore, an anticancer platinum agent cisplatin is secreted into urine by tubular secretion, mediated by organic cation transport systems (Yonezawa and Inui, 2011b). In OCT1/2 double-KO mice, the renal accumulation and toxicity of cisplatin were suppressed (Filipski et al., 2009; Ciarimboli et al., 2010). In Mate1 KO mice, however, cisplatin was markedly accumulated in the kidney and was highly nephrotoxic (Nakamura et al., 2010). In addition, a MATE1 inhibitor, pyrimethamine, potentiated cisplatin-induced nephrotoxicity. The balance of uptake and efflux by OCT and MATE could determine the renal accumulation and toxicity of cisplatin.

9.6 GENETIC POLYMORPHISMS OF THE *SLC47A* GENES

In the *SLC47A1* and *SLC47A2* genes, 11 and 2 nonsynonymous single-nucleotide polymorphisms (SNPs) have been found in Japanese, Caucasians, African Americans, Asian Americans, and Mexican Americans, some of which affect function (Fig. 9.4 and Table 9.3). The mutations G64D and V480M in MATE1 and G211V

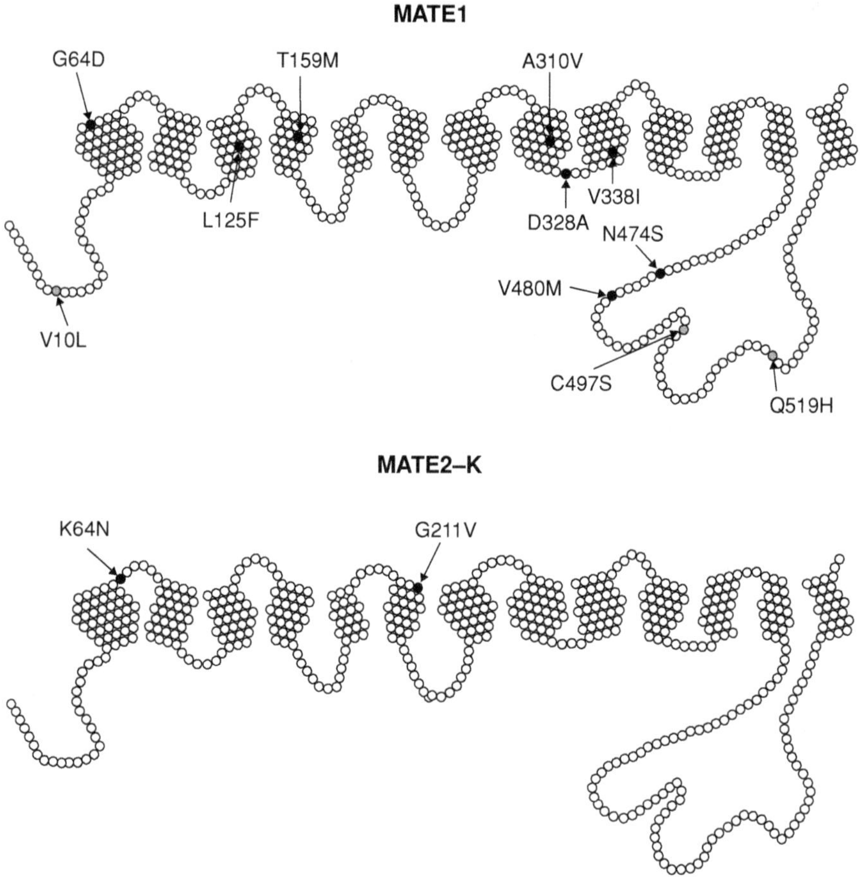

Figure 9.4 Predicted topology and polymorphisms of human MATE1 and MATE2-K. The x-ray structure of the MATE transporter from *Vibrio cholerae* was determined at 3.65 Å resolution (He et al., 2010). The topology of human MATE1 and MATE2-K was predicted. Dysfunctional mutations are shown in black. The mutations not affecting the function are shown in gray.

TABLE 9.3 SNPs of MATE1/SLC47A1 and MATE2-K/SLC47A2

Gene	Location	dbSNP	Nucleotide	Amino acid	Function	Allele frequency (%)	Ethnicity	Reference
MATE1/SLC47A1 (NM_018242.2)	Promoter	rs72466470	G>A	–	↓	1.9	JP	Kajiwara et al. (2007)
	5′UTR	rs2252281	T>C	–	↓	23.1–44.5	AS, CA, ME, AF	Ha Choi et al. (2009)
	Exon 1	rs111060521	G>T	V10L	→	2.2	JP	Kajiwara et al. (2009)
	Exon 2	rs77630697	G>A	G64D	↓↓	0.6–2.1	JP, AS	Chen et al. (2009); Kajiwara et al. (2009); Toyama et al. (2010)
	Exon 4	rs77474263	C>T	L125F	↓	0.7–5.1	JP, AS, ME	Chen et al. (2009); Toyama et al. (2010)
		rs35646404	T>C	T159M	↓↓	1	JP	Meyer zu Schwabedissen et al. (2010)
	Intron 10	rs2289669	G>A	–	N.D.	43	CA	Becker et al. (2009, 2010); Tzvetkov et al. (2009)
	Exon 11	rs111060526	C>T	A310V	↓↓	2.2	JP	Kajiwara et al. (2009)
		rs111060527	A>C	D328A	↓↓	0.6–1	JP	Kajiwara et al. (2009); Toyama et al. (2010)
		rs35790011	G>A	V338I	↓↓	0.4–10	CA, AF	Chen et al. (2009); Meyer zu Schwabedissen et al. (2010)
	Exon 16	rs111060528	A>G	N474S	↓	0.6	JP	Kajiwara et al. (2009)
		rs76645859	G>A	V480M	↓↓	0.8	AS	Chen et al. (2009)
	Exon 17	rs35395280	G>C	C497S	→	2.4	AF	Chen et al. (2009)
		rs78700676	G>C	Q519H	→	0.8	AF	Chen et al. (2009)
MATE2-K/SLC47A2 (NM_152908.3)	Exon 2	rs111060529	G>T	K64N	↓	0.6	JP	Kajiwara et al. (2009)
	Exon 8	rs111060532	GC>TT	G211V	↓↓	1.7–2.1	JP	Kajiwara et al. (2009); Toyama et al. (2010)

N.D., no data; ↓, decrease in transcription and transport activity; ↓↓, more than 50% decrease in transport activity; →, no change in transport activity; JP, Japanese; CA, Caucasian; AF, African American; AS, Asian American; ME, Mexican American.

in MATE2-K caused a complete loss of function (Chen et al., 2009; Kajiwara et al., 2009). The expression of these variants in the plasma membrane was abolished. Therefore, the variants can affect the pharmacokinetics of all substrates. The transport activity of the T159M, A310V, D328A, and V338I variants was reduced by more than 50% compared to the wild-type level. The allelic frequency of the dysfunctional mutations was less than 5%, and homozygous carriers have not been found. In addition, polymorphisms in the promoter region of MATE1 were identified. It was reported that the SNP rs72466470 decreased Sp1 binding and transcription activity (Kajiwara et al., 2007). The allelic frequency of this polymorphism was 3.7%. At the AP-1-binding site, rs2252281 SNP, causing reduced transcriptional activity, was also identified at a relatively high frequency (23.1–44.5%) (Ha Choi et al., 2009). These nonsynonymous and transcriptional polymorphisms could explain the interindividual variation in the disposition of cationic drugs.

9.7 RELATIONSHIP BETWEEN NONSYNONYMOUS MUTATIONS AND PHARMACOKINETICS OF METFORMIN IN HUMANS

Metformin (*N*,*N*-dimethylbiguanide) is widely used in the treatment of type II diabetes mellitus. It exhibits pharmacological effects in the liver, and is almost entirely excreted into urine in an unchanged form (Scheen, 1996). The renal clearance of metformin is five times higher than the glomerular filtration rate, suggesting that tubular secretion via OCT2 and MATE is the most important pathway for the excretion of metformin.

A pharmacokinetic analysis of metformin was conducted in heterozygous carriers of the MATE1 and MATE2-K variants (Toyama et al., 2010). Among 48 Japanese diabetic patients treated with metformin, two carriers of the MATE1-G64D variant, two carriers of the MATE1-L125F variant, one carrier of the MATE1-D328A variant, and two carriers of the MATE2-K-G211V variant were found. However, no patients were homozygous for the MATE variants. There was no significant difference in the plasma concentration and oral clearance of metformin with or without heterozygous MATE variants. These results corresponded with the data for KO mice. The renal clearance of metformin was reported to be in the range of 335–615 mL/min in humans (Scheen, 1996), which is comparable with renal plasma flow. These results suggested that renal blood flow is a rate-limiting factor for metformin secretion. Therefore, heterozygous patients for the MATE variants possibly show only minor interindividual variation in pharmacokinetics of metformin in clinical situations.

Kusuhara et al. (Kusuhara et al., 2011) examined the effect of a MATE inhibitor, pyrimethamine, on the renal clearance of metformin in humans in the microdose study. It was found that the oral administration of pyrimethamine (50 mg) decreased the renal clearance of metformin. Ki values of pyrimethamine for MATE1 and MATE2-K are 93 nM and 59 nM, respectively. In this clinical study, the plasma concentration of pyrimethamine was 2.3 μM. Therefore, potent inhibition of MATEs decreased the renal clearance of metformin, consistent with the findings in mice (Ito et al., 2010). It was suggested that MATE should mediate the renal clearance of metformin in humans.

9.8 ROLE OF THE INTRON SNP RS2289669 IN PHARMACODYNAMICS OF METFORMIN

The main mechanism of metformin efficacy is the suppression of gluconeogenesis in the liver. It was reported that OCT1 mediated the hepatic uptake of metformin, and polymorphisms of the *SLC22A1* gene (OCT1) affected the efficacy of metformin (Shu et al., 2007). Because MATE1 is also expressed in the liver, it could be a determinant of metformin efficacy. The relationship between the efficacy of metformin and MATE1 SNPs was recently examined in 116 diabetic patients of the Rotterdam Study (Becker et al., 2009, 2010). For this study, the authors selected nine tagging SNPs (rs894680, rs2018675, rs2440154, rs2440155, rs16960201, rs2453568, rs2244280, rs2289669, and ra1961669) in the *SLC47A1* gene and three tagging SNPs (rs2453594, rs2453589, and rs2165894) in the 10-kbp-downstream region. Among them, one SNP rs2289669 in the 10th intron of *SLC47A1* was significantly associated with a decrease in the A1C level after the start of metformin therapy (Becker et al., 2009). The decrease in the A1C level was 0.3% larger per copy of the allele. In addition, the effect of the MATE1 rs2289669 polymorphism on the glucose-lowering effect of metformin is larger in incident users with the OCT1 rs622342 minor CC genotype (Becker et al., 2010). These results suggested that polymorphisms in MATE1 could regulate the efficacy of metformin.

However, it was reported that renal clearance of metformin was not changed in the carriers of MATE1 rs2289669 (Tzvetkov et al., 2009). In addition, the allelic frequency was 43%, much higher than that of any nonsynonymous polymorphisms found to date (Table 9.3), suggesting that this intron SNP could not be linked to nonsynonymous polymorphisms. It is unclear whether rs2289669 SNP can alter the function of MATE1. Taken together, MATE1 polymorphisms would affect the pharmacodynamics of metformin, although further study is needed.

9.9 CONCLUSIONS

In this chapter, we summarized the basic information and recent findings on MATE1/SLC47A1 and MATE2-K/SLC47A2. The first mammalian MATE1/SLC47A1 was identified in 2005, based on the bacterial NorM. Human MATE2-K/SLC47A2 and rodent MATE were found thereafter. Functional characterization revealed MATE1 and MATE2-K to be H^+/organic cation antiporters, mediating the renal tubular secretion of cationic drugs in cooperation with OCT2. In addition, several nonsynonymous, promoter and intron SNPs in MATE1 and MATE2-K were found, and the relationship between these mutations and the pharmacokinetics/pharmacodynamics of metformin has been studied. Further pharmacogenomics research in MATE1 and MATE2-K should help overcome the interindividual variation of cationic drugs and establish personalized pharmacotherapy.

ACKNOWLEDGMENTS

This work is supported in part by a grant-in-aid for Scientific Research (KAKENHI) from the Ministry of Education, Culture, Sports, Science and Technology of Japan.

REFERENCES

Baumann M, Pontiller J, Ernst W (2010) Structure and Å basal transcription complex of RNA polymerase II core promoters in the mammalian genome: an overview. *Mol Biotechnol*, 45, 241–247.

Becker ML, Visser LE, van Schaik RH, Hofman A, Uitterlinden AG, Stricker BH (2009) Genetic variation in the multidrug and toxin extrusion 1 transporter protein influences the glucose-lowering effect of metformin in patients with diabetes: a preliminary study. *Diabetes*, 58, 745–749.

Becker ML, Visser LE, van Schaik RH, Hofman A, Uitterlinden AG, Stricker BH (2010) Interaction between polymorphisms in the OCT1 and MATE1 transporter and metformin response. *Pharmacogenet Genomics*, 20, 38–44.

Chen Y, Teranishi K, Li S, Yee SW, Hesselson S, Stryke D, Johns SJ, Ferrin TE, Kwok P, Giacomini KM (2009) Genetic variants in multidrug and toxic compound extrusion-1, hMATE1, alter transport function. *Pharmacogenomics J*, 9, 127–136.

Christian CD, Jr., Meredith CG, Speeg KV, Jr. (1984) Cimetidine inhibits renal procainamide clearance. *Clin Pharmacol Ther*, 36, 221–227.

Ciarimboli G, Deuster D, Knief A, Sperling M, Holtkamp M, Edemir B, Pavenstadt H, Lanvers-Kaminsky C, am Zehnhoff-Dinnesen A, Schinkel AH, Koepsell H, Jurgens H, Schlatter E (2010) Organic cation transporter 2 mediates cisplatin-induced oto- and nephrotoxicity and is a target for protective interventions. *Am J Pathol*, 176, 1169–1180.

Feng B, Obach RS, Burstein AH, Clark DJ, de Morais SM, Faessel HM (2008) Effect of human renal cationic transporter inhibition on the pharmacokinetics of varenicline, a new therapy for smoking cessation: an *in vitro–in vivo* study. *Clin Pharmacol Ther*, 83, 567–576.

Filipski KK, Mathijssen RH, Mikkelsen TS, Schinkel AH, Sparreboom A (2009) Contribution of organic cation transporter 2 (OCT2) to cisplatin-induced nephrotoxicity. *Clin Pharmacol Ther*, 86, 396–402.

Grundemann D, Gorboulev V, Gambaryan S, Veyhl M, Koepsell H (1994) Drug excretion mediated by a new prototype of polyspecific transporter. *Nature*, 372, 549–552.

Ha Choi J, Wah Yee S, Kim MJ, Nguyen L, Ho Lee J, Kang JO, Hesselson S, Castro RA, Stryke D, Johns SJ, et al. (2009) Identification and characterization of novel polymorphisms in the basal promoter of the human transporter, MATE1. *Pharmacogenet Genomics*, 19, 770–780.

He X, Szewczyk P, Karyakin A, Evin M, Hong WX, Zhang Q, Chang G (2010) Structure of a cation-bound multidrug and toxic compound extrusion transporter. *Nature*, 467, 991–914.

Hiasa M, Matsumoto T, Komatsu T, Omote H, Moriyama Y (2007) Functional characterization of testis-specific rodent multidrug and toxic compound extrusion 2, a class III MATE-type polyspecific H^+/organic cation exporter. *Am J Physiol Cell Physiol*, 293, C1437–C1444.

Inui K, Masuda S, Saito H (2000) Cellular and molecular aspects of drug transport in the kidney. *Kidney Int*, 58, 944–958.

Inui K, Okuda M (1998) Cellular and molecular mechanisms of renal tubular secretion of organic anions and cations. *Clin Exp Nephrol*, 2, 100–108.

Inui K, Takano M, Okano T, Hori R (1985) H^+ gradient-dependent transport of aminocephalosporins in rat renal brush-border membrane vesicles: role of H^+/organic cation antiport system. *J Pharmacol Exp Ther*, 233, 181–185.

Ito S, Kusuhara H, Kuroiwa Y, Wu C, Moriyama Y, Inoue K, Kondo T, Yuasa H, Nakayama H, Horita S, Sugiyama Y (2010) Potent and specific inhibition of mMate1-mediated efflux of type I organic cations in the liver and kidney by pyrimethamine. *J Pharmacol Exp Ther*, 333, 341–350.

Kajiwara M, Terada T, Asaka J, Ogasawara K, Katsura T, Ogawa O, Fukatsu A, Doi T, Inui K (2007) Critical roles of Sp1 in gene expression of human and rat H^+/organic cation antiporter MATE1. *Am J Physiol Renal Physiol*, 293, F1564–F1570.

Kajiwara M, Terada T, Ogasawara K, Iwano J, Katsura T, Fukatsu A, Doi T, Inui K (2009) Identification of multidrug and toxin extrusion (MATE1 and MATE2-K) variants with complete loss of transport activity. *J Hum Genet*, 54, 40–46.

Koepsell H, Lips K, Volk C (2007) Polyspecific organic cation transporters: structure, function, physiological roles, and biopharmaceutical implications. *Pharm Res*, 24, 1227–1251.

Kusuhara H, Ito S, Kumagai Y, Jiang M, Shiroshita T, Moriyama Y, Inoue K, Yuasa H, Sugiyama Y (2011) Effects of a MATE protein inhibitor, pyrimethamine, on the renal elimination of metformin at oral microdose and at therapeutic dose in healthy subjects. *Clin Pharmacol Ther*, 89, 837–844.

Masuda S, Terada T, Yonezawa A, Tanihara Y, Kishimoto K, Katsura T, Ogawa O, Inui K (2006) Identification and functional characterization of a new human kidney-specific H^+/organic cation antiporter, kidney-specific multidrug and toxin extrusion 2. *J Am Soc Nephrol*, 17, 2127–2135.

Matsushima S, Maeda K, Inoue K, Ohta KY, Yuasa H, Kondo T, Nakayama H, Horita S, Kusuhara H, Sugiyama Y (2009) The inhibition of human multidrug and toxin extrusion 1 is involved in the drug–drug interaction caused by cimetidine. *Drug Metab Dispos*, 37, 555–559.

Meyer zu Schwabedissen HE, Verstuyft C, Kroemer HK, Becquemont L, Kim RB (2010) Human multidrug and toxin extrusion 1 (MATE1/SLC47A1) transporter: functional characterization, interaction with OCT2 (SLC22A2), and single-nucleotide polymorphisms. *Am J Physiol Renal Physiol*, 298, F997–F1005.

Muirhead MR, Somogyi AA, Rolan PE, Bochner F (1986) Effect of cimetidine on renal and hepatic drug elimination: studies with triamterene. *Clin Pharmacol Ther*, 40, 400–407.

Nakamura T, Yonezawa A, Hashimoto S, Katsura T, Inui K (2010) Disruption of multidrug and toxin extrusion MATE1 potentiates cisplatin-induced nephrotoxicity. *Biochem Pharmacol*, 80, 1762–1767.

Ohta K, Imamura Y, Okudaira N, Atsumi R, Inoue K, Yuasa H (2009a) Functional characterization of multidrug and toxin extrusion protein 1 as a facilitative transporter for fluoroquinolones. *J Pharmacol Exp Ther*, 328, 628–634.

Ohta K, Inoue K, Hayashi Y, Yuasa H (2006) Molecular identification and functional characterization of rat multidrug and toxin extrusion type transporter 1 as an organic cation/H^+ antiporter in the kidney. *Drug Metab Dispos*, 34, 1868–1874.

Ohta K, Inoue K, Yasujima T, Ishimaru M, Yuasa H (2009b) Functional characteristics of two human MATE transporters: kinetics of cimetidine transport and profiles of inhibition by various compounds. *J Pharm Pharm Sci*, 12, 388–396.

Okuda M, Saito H, Urakami Y, Takano M, Inui K (1996) cDNA cloning and functional expression of a novel rat kidney organic cation transporter, OCT2. *Biochem Biophys Res Commun*, 224, 500–507.

Omote H, Hiasa M, Matsumoto T, Otsuka M, Moriyama Y (2006) The MATE proteins as fundamental transporters of metabolic and xenobiotic organic cations. *Trends Pharmacol Sci*, 27, 587–593.

Otsuka M, Matsumoto T, Morimoto R, Arioka S, Omote H, Moriyama Y (2005) A human transporter protein that mediates the final excretion step for toxic organic cations. *Proc Natl Acad Sci USA*, 102, 17923–17928.

Scheen AJ (1996) Clinical pharmacokinetics of metformin. *Clin Pharmacokinet*, 30, 359–371.

Shiga T, Hashiguchi M, Urae A, Kasanuki H, Rikihisa T (2000) Effect of cimetidine and probenecid on pilsicainide renal clearance in humans. *Clin Pharmacol Ther*, 67, 222–228.

Shu Y, Sheardown SA, Brown C, Owen RP, Zhang S, Castro RA, Ianculescu AG, Yue L, Lo JC, Burchard EG, et al. (2007) Effect of genetic variation in the organic cation transporter 1 (OCT1) on metformin action. *J Clin Invest*, 117, 1422–1431.

Somogyi A, McLean A, Heinzow B (1983) Cimetidine–procainamide pharmacokinetic interaction in man: evidence of competition for tubular secretion of basic drugs. *Eur J Clin Pharmacol*, 25, 339–345.

Somogyi A, Stockley C, Keal J, Rolan P, Bochner F (1987) Reduction of metformin renal tubular secretion by cimetidine in man. *Br J Clin Pharmacol*, 23, 545–551.

Takano M, Inui K, Okano T, Saito H, Hori R (1984) Carrier-mediated transport systems of tetraethylammonium in rat renal brush-border and basolateral membrane vesicles. *Biochim Biophys Acta*, 773, 113–124.

Tanihara Y, Masuda S, Sato T, Katsura T, Ogawa O, Inui K (2007) Substrate specificity of MATE1 and MATE2-K, human multidrug and toxin extrusions/H^+-organic cation antiporters. *Biochem Pharmacol*, 74, 359–371.

Terada T, Inui K (2008) Physiological and pharmacokinetic roles of H^+/organic cation antiporters (MATE/SLC47A). *Biochem Pharmacol*, 75, 1689–1696.

Terada T, Masuda S, Asaka J, Tsuda M, Katsura T, Inui K (2006) Molecular cloning, functional characterization, and tissue distribution of rat H^+/organic cation antiporter MATE1. *Pharm Res*, 23, 1696–1701.

Toyama K, Yonezawa A, Tsuda M, Masuda S, Yano I, Terada T, Osawa R, Katsura T, Hosokawa M, Fujimoto S, et al. (2010) Heterozygous variants of multidrug and toxin extrusions (MATE1 and MATE2-K) have little influence on the disposition of metformin in diabetic patients. *Pharmacogenet Genomics*, 20, 135–138.

Tsuda M, Terada T, Asaka J, Ueba M, Katsura T, Inui K (2007) Oppositely directed H^+ gradient functions as a driving force of rat H^+/organic cation antiporter MATE1. *Am J Physiol Renal Physiol*, 292, F593–F598.

Tsuda M, Terada T, Mizuno T, Katsura T, Shimakura J, Inui K (2009a) Targeted disruption of the multidrug and toxin extrusion 1 (MATE1) gene in mice reduces renal secretion of metformin. *Mol Pharmacol*, 75, 1280–1286.

Tsuda M, Terada T, Ueba M, Sato T, Masuda S, Katsura T, Inui K (2009b) Involvement of human multidrug and toxin extrusion 1 in the drug interaction between cimetidine and metformin in renal epithelial cells. *J Pharmacol Exp Ther*, 329, 185–191.

Tzvetkov MV, Vormfelde SV, Balen D, Meineke I, Schmidt T, Sehrt D, Sabolic I, Koepsell H, Brockmoller J (2009) The effects of genetic polymorphisms in the organic cation transporters OCT1, OCT2, and OCT3 on the renal clearance of metformin. *Clin Pharmacol Ther*, 86, 299–306.

Watanabe S, Tsuda M, Terada T, Katsura T, Inui K (2010) Reduced renal clearance of a zwitterionic substrate cephalexin in MATE1-deficient mice. *J Pharmacol Exp Ther*, 334, 651–656.

Yasujima T, Ohta KY, Inoue K, Ishimaru M, Yuasa H (2010) Evaluation of 4′,6-diamidino-2-phenylindole as a fluorescent probe substrate for rapid assays of the functionality of human multidrug and toxin extrusion proteins. *Drug Metab Dispos*, 38, 715–721.

Yokoo S, Yonezawa A, Masuda S, Fukatsu A, Katsura T, Inui K (2007) Differential contribution of organic cation transporters, OCT2 and MATE1, in platinum agent-induced nephrotoxicity. *Biochem Pharmacol*, 74, 477–487.

Yonezawa A, Inui K (2011a) Importance of the multidrug and toxin extrusion MATE/SLC47A family to pharmacokinetics, pharmacodynamics/toxicodynamics and pharmacogenomics. *Br J Pharmacol*, 164, 1817–1825.

Yonezawa A, Inui K (2011b) Organic cation transporter OCT/SLC22A and H^+/organic cation antiporter MATE/SLC47A are key molecules for nephrotoxicity of platinum agents. *Biochem Pharmacol*, 81, 563–568.

Yonezawa A, Masuda S, Yokoo S, Katsura T, Inui K (2006) Cisplatin and oxaliplatin, but not carboplatin and nedaplatin, are substrates for human organic cation transporters (SLC22A1-3 and multidrug and toxin extrusion family). *J Pharmacol Exp Ther*, 319, 879–886.

Zhang J, Wang C, Liu Q, Meng Q, Cang J, Sun H, Gao Y, Kaku T, Liu K (2010) Pharmacokinetic interaction between JBP485 and cephalexin in rats. *Drug Metab Dispos*, 38, 930–938.

Zhang X, Cherrington NJ, Wright SH (2007) Molecular identification and functional characterization of rabbit MATE1 and MATE2-K. *Am J Physiol Renal Physiol*, 293, F360–F370.

CHAPTER **10**

PEPT (SLC15A) FAMILY

Tomoko Sugiura
Saki Umeda
Akira Tsuji
Yukio Kato

10.1 INTRODUCTION

Orally ingested proteins are hydrolyzed to oligopeptides and amino acids inside the gastrointestinal lumen. This reduction of the molecular size may serve to minimize immunogenicity and to maximize absorption of nutrients into the systemic circulation. However, these small molecules are highly hydrophilic, and consequently it is considered that transport mechanisms are required for their transfer across plasma membrane of small intestinal epithelium.

In the 1970s, it was reported that carrier-mediated uptake is also involved in the absorption of a certain peptidomimetic therapeutic agents, including penicillins and various β-lactam antibiotics, in the small intestine (Tsuji et al., 1977). Characterization of the uptake mechanism using isolated tissues and perfusion system indicated that a proton-dependent oligopeptide transporter was involved in the absorption of several β-lactam antibiotics (Tamai et al., 1995; Tamai et al., 1997). Thus, oligopeptide transporters recognize a variety of compounds including not only nutrients, but also xenobiotic compounds as substrates.

So far, four members (SLC15A1–SLC15A4) of the oligopeptide transporter family have been identified. This chapter focuses on the genetic polymorphism and physiological function of two well-characterized members, PEPT1/SLC15A1 and PEPT2/SLC15A2.

10.2 MOLECULAR NATURE AND PHYSIOLOGICAL FUNCTION OF PEPT1 AND PEPT2

10.2.1 Identification and Characterization of PEPT1

In 1994, a cDNA clone encoding H^+/oligopeptide transporter PEPT1 (SLC15A1) was isolated from a rabbit intestinal cDNA library (Fei et al., 1994) and other mammalian

Pharmacogenomics of Human Drug Transporters: Clinical Impacts, First Edition.
Edited by Toshihisa Ishikawa, Richard B. Kim and Jörg König

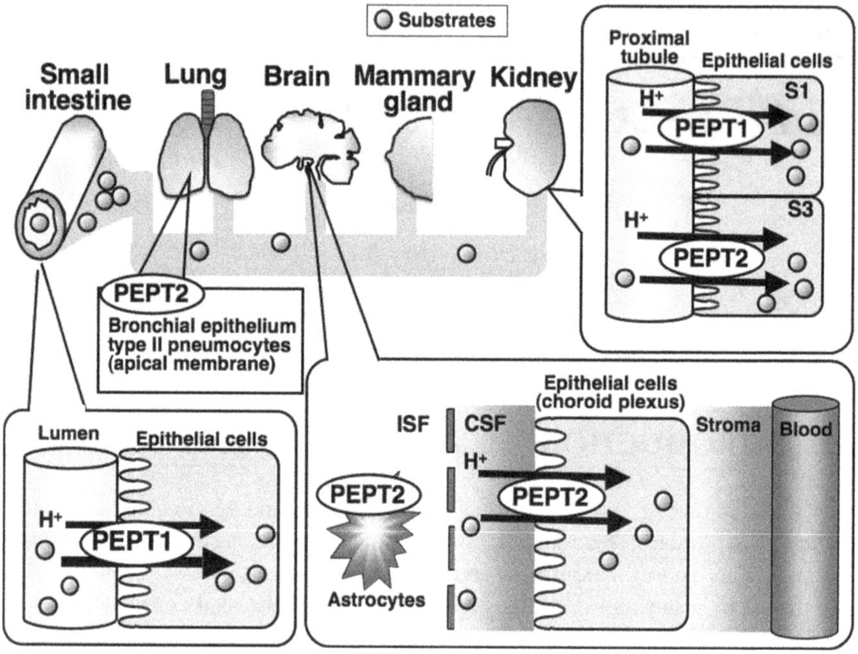

Figure 10.1 Localization of PEPT1 and PEPT2 in the body.

species (Liang et al., 1995; Saito et al., 1995). PEPT1 is localized on brush-border membranes of small intestinal cells (Sai et al., 1996; Fig. 10.1) and recognizes di- and tripeptides, but not amino acids, as substrates. The driving force for the uptake of these oligopeptides by PEPT1 is an inward H^+ gradient. PEPT1 thus works as an electrogenic symporter for proton and peptide. PEPT1 also transports peptide-like drugs, including β-lactam antibiotics. Other orally administered drugs, including bestatin, temocapril, enalapril, and oseltamivir, are also transported by PEPT1 (Fig. 10.2; Thwaites et al., 1995; Nakanishi et al., 1997; Tsuji, 2002; Ogihara et al., 2009).

Gene expression of PEPT1 in the small intestine is highest in the duodenum, followed by jejunum and ileum, and is highly induced during fasting (Naruhashi et al., 2002), implying a role in nutrient absorption. Expression of PEPT1 is also regulated by various hormones and growth factors, such as thyroid hormone, leptin, insulin, and epidermal growth factor (Adibi, 2003). The expression level of PEPT1 is also affected by development and diurnal rhythm (Adibi, 2003). For example, gene expression of PEPT1 in rat small intestine was found to be higher at night (in the dark) than in the day. Intestinal transport of glycylsarcosine (Gly-Sar), a typical substrate of PEPTs, showed a similar pattern (Pan et al., 2002). Such circadian rhythm in PEPT1 expression appears to be related to the eating habits of rats (Pan et al., 2003), although further study is needed to establish the precise physiological meaning.

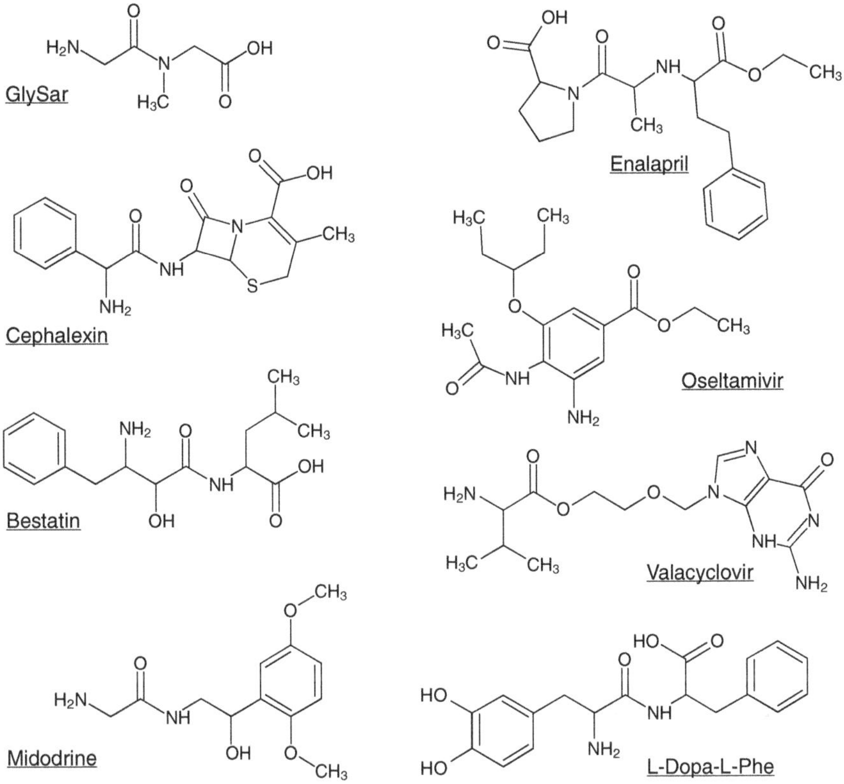

Figure 10.2 Structure of PEPT1 substrates.

Pept1 gene knockout mice (*pept1*$^{-/-}$) have recently been generated (Hu et al., 2008). Despite the possibly pivotal role of this transporter in the absorption of nutrients, *pept1*$^{-/-}$ mice grow to normal size and weight, and do not display any gross phenotypic abnormalities (Hu et al., 2008). On the other hand, oral absorption of Gly-Sar, a typical substrate of PEPTs, is dramatically reduced with a concomitant decrease in intestinal permeability of Gly-Sar in *pept1*$^{-/-}$ mice, compared with wild-type mice (Hu et al., 2008; Ma et al., 2010). Thus, PEPT1 could be involved in small intestinal absorption of oligopeptides, but does not appear to be solely responsible for the uptake of protein-derived nutrition. It seems likely that other transporter(s) compensate for the loss of PEPT1, at least in part.

PEPT1 is thus believed to be involved at least in gastrointestinal absorption of certain oligopeptides and other substrates (Fig. 10.2). Therefore, a number of therapeutic agents have been designed to be targeted to PEPT1 by incorporating peptidomimetic structure, with the aim of increasing their oral absorption. These agents include α-methyldopa-L-Phe (Tsuji et al., 1990), L-dopa-L-Phe (Tamai et al., 1998), valacyclovir (Balimane et al., 1998), and midodrine (Tsuda et al., 2006). These compounds are actually recognized by PEPT1. For example, valacyclovir is the L-valine ester prodrug of an acyclic nucleoside, acyclovir, which has no peptide bond (Balimane et al., 1998). Oral bioavailability of valacyclovir is indeed higher than that of acyclovir (Balimane et al., 1998). The driving force for PEPT1, the proton gradient, has also been considered to improve the intestinal absorption of therapeutic agents, and a proton-releasing polymer, Eudragit L100-55®, was designed to enhance PEPT1-mediated uptake of substrate drugs by supplying protons inside the intestinal lumen (Nozawa et al., 2003). Indeed, the intestinal permeability of cefixime is significantly increased by this macromolecule (Nozawa et al., 2003). Thus, PEPT1 is a suitable target for the development of orally bioavailable therapeutic agents.

10.2.2 Effect of Genetic Variants for PEPT1 on Transport Activity and Drug Absorption

Currently, about 40 coding polymorphisms of human *PEPT1* gene have been identified (Table 10.1). The most common PEPT1 SNPs are 350G>A (S117N) and 1256G>C (G419A). The allelic frequencies of these two SNPs exhibit no significant race difference. However, in the case of some rare SNPs, ethnic variations have been observed. For example, SNP 83T>A (F28Y) and 364G>A (V122M) are only present in African Americans, whereas 843G>A (T281T) is specific to Asians.

Various studies have been performed to examine the possible effect of genetic variation of PEPT1 on the expression and/or function of this transporter. Although those studies have used various types of compounds, including β-lactam antibiotics (ampicillin, amoxicillin, cephalexin, and cefadroxil), antineoplastic agents (δ-ALA and bestatin), and the neuropeptide NAAG as substrates, no significant alteration was found for most of the mutants, including the major two SNPs (S117N and G419A), compared with wild-type PEPT1 (Sala-Rabanal et al., 2006a). In addition, the effect of other nonsynonymous variants of PEPT1 was also examined, but similar pH-dependence and K_m values to those of the wild type were found (Zhang et al., 2004). The rare SNP 1758C>T (P586L), on the other hand, showed significantly reduced uptake, lower expression of immunoreactive gene product, and lower localization of gene product on plasma membranes, compared with the wild type, when it was heterologously transfected into HeLa cells (Zhang et al., 2004). Therefore, P586 may play an important role in translation, membrane insertion, and/or degradation of PEPT1. In addition, the SNP 83T>A (P28Y), which is also found with low frequency and is specific to African Americans, shows reduced uptake of cephalexin compared with the wild-type (Anderle et al., 2006). The K_m value of P28Y for Gly-Sar uptake was significantly higher than that of the wild-type, but there was no difference of V_{max} (Anderle et al., 2006). Therefore, reduced uptake by the P28Y variant may be caused by lower affinity for the substrate.

There is little information available on clinically observable effects of PEPT1 SNPs on oral drug absorption, probably because of the low frequency of PEPT1 SNPs that result in a dramatic change in transport activity or expression, at least based on *in vitro* data. Phan et al. carried out a pharmacokinetic study of valacyclovir in 16 healthy volunteers and evaluated the impact of PEPT1 genetic polymorphisms. They found that the SNPs examined do not contribute to the interindividual differences in bioavailability (Phan et al., 2003). These studies, however, do not exclude the possible

TABLE 10.1 Genetic Variants for *PEPT1* Gene

Nucleotide change	Amino acid change	dbSNP rs# cluster id	Allelic frequency	Expression	Function	Reference
61G>A	Val21Ile	rs8187818	0.000–0.005		CEX uptake (→) Inhibition profile (→)	Anderle et al. (2006)
68A>G	Glu23Gly	rs113158640	ND			
83T>A	Phe28Tyr	rs8187817	0.000–0.005	Membrane localization (→)	CEX uptake (↓) Inhibition profile (→) K_m(↑), V_{max} (→) Change in pH dependence	Anderle et al. (2006)
342C>T	Thr114Ile		ND	mRNA (→) Protein (→)	Gly-Sar uptake (→) Inhibition profile (→)	Zhang et al. (2004)
350G>A	Ser117Asn	rs2297322	0.071–0.500	mRNA (→) Protein (→)	Gly-Sar, CEX uptake (→) Inhibition profile (→) pH-dependence (→)	Anderle et al. (2006) Sala-Rabanal et al. (2006a) Zhang et al. (2004)
351C>A	Ser117Arg	rs8187821	0.000–0.005		CEX uptake (→) Inhibition profile (→)	Anderle et al. (2006)
364G>A	Val122Met	rs8187820	0.000–0.020	mRNA (→) Protein (→)	Gly-Sar, CEX uptake (→) Inhibition profile (→)	Anderle et al. (2006) Zhang et al. (2004)
483A>T	Arg161Ser	rs79092209	ND			
632T>G	Val211Gly	rs78068268	ND			
678C>A	Phe226Leu	rs78350626	ND			
679A>G	Lys227Glu	rs112948078	ND			
752G>A	Arg251Gln	rs74889332	ND			
769A>G	Lys257Glu	rs75442485	ND			
1082T>C	Met361Thr	rs112181280	ND			
1147G>A	Asp383Asn	rs1782674	ND			

(*continued*)

TABLE 10.1 (*Continued*)

Nucleotide change	Amino acid change	dbSNP rs# cluster id	Allelic frequency	Expression	Function	Reference
1248G>C	Val416Leu		ND	mRNA (→) Protein (→)	Gly-Sar uptake (→) Inhibition profile (→)	Zhang et al. (2004)
1256G>C	Gly419Ara	rs4646227	0.000–0.115	mRNA (→) Protein (→)	Gly-Sar, CEX uptake (→) Inhibition profile (→) pH-dependence (→)	Anderle et al. (2006) Sala-Rabanal et al. (2006a) Zhang et al. (2004)
1321_1322i→A	Frame shift	rs72547504	ND			
1348G>A	Val450Ile	rs2274828	0.000–0.143	mRNA (→) Protein (→)	Gly-Sar uptake (→) Inhibition profile (→)	Zhang et al. (2004)
1352C>A	Thr451Asn	rs8187838	0.000–0.020	mRNA (→) Protein (→)	Gly-Sar, CEX uptake (→) Inhibition profile (→)	Anderle et al. (2006) Zhang et al. (2004)
1375C>T	Arg459Cys	rs2274827	ND	mRNA (→) Protein (→)	Gly-Sar uptake (→) Inhibition profile (→)	Zhang et al. (2004)
1399C>T	Pro467Ser	rs62637593	ND			
1609C>T	Pro537Ser	rs8187830	0.000–0.005	Membrane localization (→)	CEX uptake (→) Inhibition profile (→)	Anderle et al. (2006)
1657G>A	Ala553Thr	rs113384238	ND			
1673T>C	Val558Ala	rs112021351	ND			
1730A>G	Asn577ser	rs113484071	ND			
1758C>T	Pro586Leu	rs56120058	ND	mRNA (→) Protein (↓) Membrane localization (→)	Gly-Sar uptake (↓) K_m(→), V_{max} (↓) Inhibition profile (→)	Zhang et al. (2004)
1883T>G	Val628Gly	rs117037854	ND			

ND, not determined; Gly-Sar, glycylsarcosine; CEX, cephalexin.

presence of unknown pharmacokinetically important SNPs, so further studies are still needed.

10.2.3 Effect of Genetic Variations of PEPT1 on Inflammatory Diseases

PEPT1 is minimally expressed in stomach, but is significantly induced in intestinal metaplasia (Terada et al., 2005). PEPT1 is poorly expressed in colon, but is highly induced in the colon during inflammatory bowel disease (IBD), such as ulcerative colitis and Crohn's disease (Rioux et al., 2007). These studies suggest a pathophysiological role of PEPT1 in the gastrointestinal tract. Certain proinflammatory bacterial peptides, including muramyl dipeptide (MDP), *N*-formylmethionyl-leucyl-phenylalanine (fMLP), and L-Ala-γ-D-Glu-meso-diaminopimelic acid (Tri-DAP), have been reported to be transported by PEPT1 (Merlin et al., 2001; Buyse et al., 2003). Among them, MDP is recognized by intercellular nucleotide-binding oligomerization domain 2 (NOD2) and induces NF-kappaB activation and subsequent inflammation (Kobayashi et al., 2005). These results may imply an association of PEPT1 with inflammatory reaction in the above diseases.

Zucchelli et al. have examined the possible association of PEPT1 polymorphisms with IBD. In Swedish and Finnish subjects, 12 PEPT1 SNPs were genotyped in IBD patients and controls. The common allele of a coding polymorphism (S117N) is associated with CD susceptibility both in Sweden and in Finland, although it is a risk allele in Sweden, but a protective one in Finland (Zucchelli et al., 2009). The best evidence of association was found in both the populations when the analysis was performed on the individuals not carrying NOD2 common risk alleles (Zucchelli et al., 2009). Thus, PEPT1 could play a role in the pathogenesis of IBD, although additional studies in other populations are needed to confirm this.

PEPT1 could also be a target to treat intestinal inflammation because of the high expression of this transporter in inflammatory tissues. For example, tripeptide KPV (Lys-Pro-Val) is a substrate of PEPT1 and possesses anti-inflammatory activity. The KPV peptide inhibits the activation of inflammatory signaling pathways involving NF-kappaB and MAP kinase, and thereby reduces the secretion of proinflammatory cytokine (Dalmasso et al., 2008); thus, it may be possible to control inflammation by using PEPT1 for the delivery of anti-inflammatory agents.

10.2.4 Background for PEPT2

In kidney, proteins and peptides are hydrolyzed into oligopeptides on apical membranes of proximal tubules after glomerular filtration. Several *in vivo* studies and kidney perfusion experiments have indicated the involvement of active transport systems in the reabsorption of the oligopeptides thus formed in the kidney (Nutzenadel and Scriver, 1976; Adibi et al., 1977; Adibi and Krzysik, 1977; Ganapathy and Leibach, 1982, 1983). The cDNA encoding oligopeptide transporter PEPT2 (SLC15A2) was then cloned as a homolog of the rabbit intestinal peptide transporter PEPT1 by plaque hybridization using a human kidney cDNA library (Liu et al., 1995). The human PEPT2 is 729 amino acids in size with 12 proposed transmembrane-spanning regions, and a large extracellular loop between transmembrane domains 9 and 10 (Fig. 10.3). PEPT2 is primarily localized on the brush-border membranes of renal epithelial cells (Liu et al., 1995). PEPT2 is capable of transporting many di- and tripeptides using the transmembrane electrochemical H^+ gradient as a driving force (Daniel and Kottra, 2004). Therefore, PEPT2 in the kidney is thought to play an important role in the reabsorption of oligopeptides, serving to prevent peptide leakage into urine and to recover amino acid nitrogen.

In rat renal tubules, both PEPT1 and PEPT2 are expressed on apical membranes (Shen et al., 1999). PEPT1 is found in the S1 segment of the proximal tubules, whereas PEPT2 is localized in both S2 and S3 segments (Shen et al., 1999). In addition, PEPT1 is a low-affinity and high-capacity transporter for Gly-Sar, whereas PEPT2 is a high-affinity and low-capacity transporter. Such localization and characterization of both the transporters could be suitable for efficient reabsorption of oligopeptides in the kidney. Like PEPT1, PEPT2 can also recognize peptidomimetic drugs, including β-lactam antibiotics, certain angiotensin-converting enzyme (ACE) inhibitors, and the antineoplastic agent bestatin (Biegel et al., 2006). Therefore, PEPT2 is thought to affect the renal disposition of peptidomimetic drugs.

Compared with PEPT1, PEPT2 is expressed relatively widely, that is, it is also expressed in the lungs, mammary gland, brain, choroid plexus, and astrocytes (Döring et al., 1998; Berger and Hediger, 1999; Shen et al., 2004) (Fig. 10.1). In rat brain, PEPT2 is localized on apical membranes of choroid plexus epithelial cells (Shen et al., 2004) and could contribute to the active efflux of oligopeptides and certain neuropeptides from cerebrospinal fluid (CSF) into the systemic circulation (Shu et al., 2002). In human, rat, and mouse lungs, PEPT2 is expressed in bronchial, bronchiolar, and alveolar type II epithelial cells (Groneberg et al., 2001, 2002). A fluorophore-conjugated dipeptide, D-Ala-L-Lys-AMCA, is taken up by isolated human lung specimens, and the uptake is inhibited by Gly-L-Glu (Groneberg et al., 2002), suggesting that PEPT2 is involved in the transport of pulmonary peptides and peptidomimetics in the lung.

10.2.5 Physiological Roles of PEPT2 Based on Studies in Gene Knockout Mice

Pept2 gene knockout mice (*pept2*$^{-/-}$) have recently been established (Shen et al., 2003; Rubio-Aliaga et al., 2003). Despite the putative key role of PEPT2 in reabsorption of nutrients, the *pept2*$^{-/-}$ mice showed no obvious phenotypic abnormalities (Rubio-Aliaga et al., 2003; Frey et al., 2006) and no compensatory upregulation in the expression of related oligopeptide transporters, such as PEPT1 and histidine/peptide transporter 1 (PHT1; SLC15A4) in the kidney (Rubio-Aliaga et al., 2003). Frey et al. used metabolomics technology to analyze amino acids and dipeptides in urine of wild-type and *pept2*$^{-/-}$ mice, with the aim of evaluating the role of PEPT2 in renal peptide transport and amino acid metabolism (Frey et al., 2007). They found that urinary excretion of peptide-bound glycine and cysteine is significantly increased in *pept2*$^{-/-}$ mice, indicating drastically increased excretion of cysteinylglycine. Moreover, proteomics analysis of the kidney showed an increase in the gene product of glutathione S-transferase Mu 1 (GSTM1). These results suggest that PEPT2 is

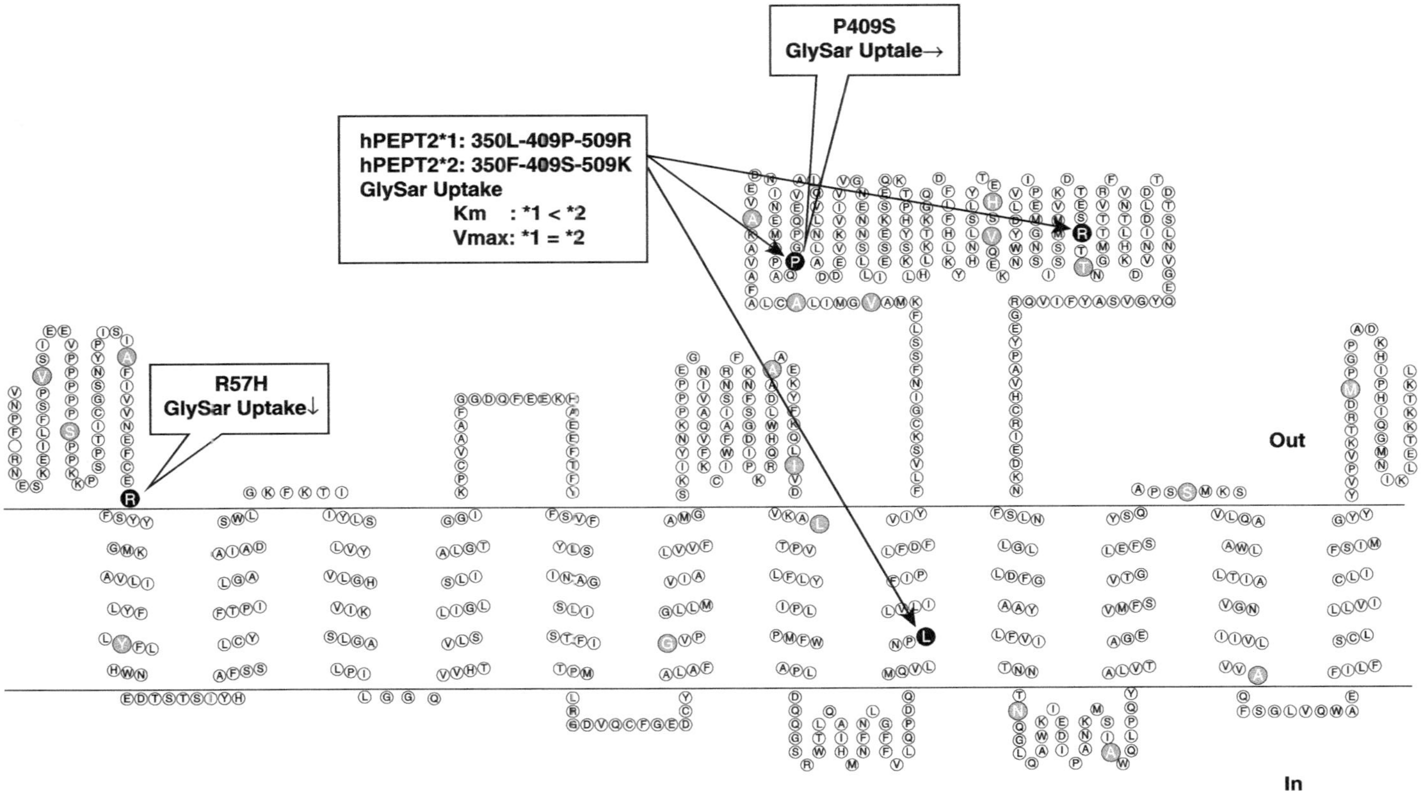

Figure 10.3 Genetic variants of PEPT2. Taken from Pinsonneault et al. (2004). Black and gray indicate amino acids involved in SNPs according to the NCBI dbSNP (http://www.ncbi.nlm.nih.gov/projects/SNP/). Black indicates amino acid variants that have been reported to show altered function.

responsible for the reabsorption of cysteinylglycine originating from glutathione breakdown (Frey et al., 2007), implying a key role of PEPT2 in GSH homeostasis.

Systemic clearance of carnosine (β-Ala-L-His), an endogenous dipeptide substrate of PEPT2, is significantly increased in *pept2*$^{-/-}$ mice, compared with wild-type mice, because of the reduction in renal reabsorption in *pept2*$^{-/-}$ mice (Kamal et al., 2009). In addition, accumulation of the fluorophore-conjugated dipeptide, D-Ala-Lys-AMCA, in the kidney after intravenous administration is significantly reduced in *pept2*$^{-/-}$ mice, compared to wild-type mice (Rubio-Aliaga et al., 2003). In *pept2*$^{-/-}$ mice, plasma disappearance of Gly-Sar after intravenous injection is more rapid, though the glomerular filtration rate (GFR) is unchanged (Ocheltree et al., 2005). On the other hand, renal clearance of Gly-Sar is increased in *pept2*$^{-/-}$ mice to a level close to the GFR, indicating that net reabsorption of Gly-Sar is almost negligible in *pept2*$^{-/-}$ mice (Ocheltree et al., 2005). All these results suggest that PEPT2 plays a predominant role in the reabsorption of oligopeptides in kidney. With regard to renal reabsorption of the β-lactam antibiotic cefadroxil, the plasma concentration in *pept2*$^{-/-}$ mice after intravenous administration at a dose of 1 nmol/g was significantly lower than that in wild-type mice (Shen et al., 2007). However, at a higher dose (100 nmol/g), the plasma level of cefadroxil is similar in wild-type and *pept2*$^{-/-}$ mice (Shen et al., 2007). These results suggest that PEPT2-mediated reabsorption of cefadroxil is mediated by PEPT2, but can be saturated at a high dose of cefadroxil. Thus, it would be important to choose an appropriate substrate concentration to avoid saturation when the role of PEPT2 in kidney is examined using *pept2*$^{-/-}$ mice *in vivo*.

At least in rodents, PEPT2 is also expressed on apical membranes of choroid plexus epithelial cells, which form the blood–CSF barrier in the brain (Shen et al., 2004). In fact, the CSF-to-plasma concentration ratio of the neuropeptide carnosine is much higher in *pept2*$^{-/-}$ mice, whereas the concentration of carnosine in the choroid plexus tissue of *pept2*$^{-/-}$ mice is lower than that in wild-type mice (Kamal et al., 2009). Similarly, the concentration in CSF of PEPT2 substrates, including glycylsarcosine, cefadroxil, and 5-aminolevulinic acid, after intracerebroventricular injection is significantly higher in *pept2*$^{-/-}$ mice than in wild-type mice (Ocheltree et al., 2005; Kamal et al., 2008; Smith et al., 2010). Thus, PEPT2 may act as an active efflux system for carnosine and other substrates from the brain. Carnosine has many physiological roles, acting as a cytosolic buffer, neurotransmitter, neuromodulator, and metabolic reservoir of histidine, which is converted to histamine during physiological stress in mammals. Therefore, changes of the intracerebral concentration of carnosine in *pept2*$^{-/-}$ mice may alter the response to physiological stress (Kamal et al., 2009). Moreover, the elimination clearance in CSF of another neuropeptide, L-kyotorphin (L-Tyr-L-Arg), after intracerebroventricular injection is significantly lower in *pept2*$^{-/-}$ mice than in wild-type mice (Jiang et al., 2009). Kyotorphin is an endogenous neuroactive dipeptide, which plays a role in pain regulation in the brain. Therefore, the lower clearance in CSF may result in greater L-kyotorphin-induced analgesia in *pept2*$^{-/-}$ mice (Jiang et al., 2009). In addition to choroid plexus, L-kyotorphin is also taken up by astrocytes via a saturable transport mechanism; this involves a high-affinity site with a $K_m$of approximately 50 μM, which is reduced in *pept2*$^{-/-}$ mice, and therefore the transporter could be PEPT2 (Xiang et al., 2010). It has been thought that disappearance of neuropeptides in the brain is mainly governed by peptidases, but the above result in *pept2*$^{-/-}$ mice may suggest that uptake into astrocytes, primarily mediated by PEPT2, also contributes to the intracellular degradation of neuropeptides. Thus, PEPT2 may be an important factor in the homeostasis of neuropeptides and in neurotoxicity.

10.2.6 Effect of Genetic Variations of PEPT2 on Transport Activity and Drug Disposition

At least 26 SNPs located in exons of the human *PEPT2* gene have already been registered in the public database (http://www.ncbi.nlm.nih.gov/projects/SNP/) (Fig. 10.3). However, most of these polymorphisms have not been characterized in detail. Terada et al. evaluated the functional consequences of rare genetic variations in human PEPT2 resulting in the amino acid changes R57H and P409S (Terada et al., 2004). The Gly-Sar transport activity of the R57H variant is almost completely abolished, though the expression level of R57H is not changed (Terada et al., 2004). On the other hand, the P409S mutation has no effect on the expression level or transport of Gly-Sar by PEPT2 (Terada et al., 2004). There are three histidine residues in PEPT2 (H87, H142, and H278) which are important for the catalytic function and are conserved among several animal species (Fei et al., 1997). Among them, H87 was found to be essential for catalytic activity, and the loss of activity was not due to a change in the expression level of PEPT2 (Fei et al., 1997). However, no polymorphism leading to change in H87 has yet been found in humans.

Two major haplotypes (*PEPT2*1* and *PEPT2*2*) have been identified in all ethnic groups (Pinsonneault et al., 2004). The *PEPT2*1* gene product contains the amino acids L350, P409, and R509, whereas hPEPT2*2 has F350, S409, and K509 (Fig. 10.3). The *hPEPT2*1* and *2 variants are almost equally distributed at 44–47% of the population (Pinsonneault et al., 2004). *In vitro* studies revealed that the K_m value of Gly-Sar uptake by the *PEPT2*2* variant (233 μM) is slightly higher than that of the *PEPT2*1* variant (83 μM), whereas the V_{max} values are similar (Pinsonneault et al., 2004). Liu et al. have reported ethnic differences in *PEPT2* haplotype distribution: The frequencies of the **1* and **2* alleles are 0.307 and 0.693 in Chinese, 0.495 and 0.505 in Malays, and 0.729 and 0.271 in Asian Indians, respectively (Liu et al., 2009). However, the pharmacokinetics of cephalexin after oral administration showed no significant difference between Chinese and Asian Indians (Liu et al., 2009), probably because the haplotypes exhibit little difference in the uptake kinetics of cephalexin. Indeed, the inhibitory effects of cephalexin on Gly-Sar uptake by the *PEPT2*1* and **2* variants are similar (Pinsonneault et al., 2004), indicating that the two variants have similar affinity for cephalexin. Recently, however, the effects of pretreatment with sodium bicarbonate and ammonium chloride on the renal clearance of cephalexin after oral administration were examined in humans and the effects were greater in homozygous *PEPT2*2/*2* carriers than in homozygous *PEPT2*1/*1* carriers (Liu et al., 2011). Pretreatment with sodium bicarbonate and ammonium chloride alters the urinary pH: Sodium bicarbonate is indicated to treat metabolic acidosis to alkalinize the urine, and ammonium chloride is conversely used in treatment of severe metabolic alkalosis. From *in vitro* studies, the two haplotypes differ in their pH sensitivity of

Gly-Sar uptake (Pinsonneault et al., 2004; Sala-Rabanal et al., 2006b): Gly-Sar uptake via *PEPT2*2* is more pH sensitive than that via PEPT2*1 (Pinsonneault et al., 2004; Sala-Rabanal et al., 2006b). Consequently, in homozygous *PETP2*2/*2* carriers renal clearance of cephalexin may be different after pretreatment with sodium bicarbonate, compared with that after ammonium chloride (Liu et al., 2011).

In human lung, PEPT2 is expressed in bronchial, bronchiolar, and alveolar type II epithelial cells (Groneberg et al., 2002). A bacterial dipeptide, γ-D-glutamyl-meso-diaminopimelic acid (γ-iE-DAP), is a substrate of PEPT2, and uptake of γ-iE-DAP in primary cultures of human airway epithelial cells was proposed to be PEPT2-mediated (Swaan et al., 2008). Exposure to γ-iE-DAP on the apical surface may induce the innate immune response. Therefore, PEPT2 could be associated with the recognition of airborne pathogens. If so, PEPT2 could be a potential target for therapeutic or biological agents to counter infection in the lung. In addition, when lung epithelial cells were isolated from human donors with homozygous or heterozygous *PEPT2*1* and *PEPT2*2*, it was found that Gly-Sar uptake by *PEPT2*1/*1* was strongly inhibited by several substrates, including dipeptides, penicillins, cephalosporins, and ACE inhibitors, compared with *PEPT**2/*2 (Bahadduri et al., 2005), implying that PEPT2 could be involved in pulmonary disposition of substrate drugs.

The geometric mean of the blood lead level in Mexican American/Hispanic children homozygous for the *PEPT2*2* polymorphism is significantly higher than in those heterozygous for *PEPT2*2* or with no polymorphism (Sobin et al., 2009). Therefore, PEPT2 could affect the disposition of lead in the body, and PEPT2*2 could be a genetic biomarker of neurotoxic risk from lead exposure (Sobin et al., 2009).

10.2.7 Possible Roles of PEPTs as a Target of PET Tracers

Both PEPT1 and PEPT2 are expressed in cancer cells. Gene expression analysis indicates that PEPT2 is more widely expressed than PEPT1 in cancer cell lines (Mitsuoka et al., 2008). Oligopeptide transport activity was originally found in human tumor cell lines such as HT1080, AsPC-1, Capan-2, MKN45, MG-63, T24, and SK-ChA-1 (Gonzalez et al., 1998; Nakanishi et al., 1997, 2000; Knütter et al., 2002). An antitumor agent, bestatin, (2S,3R)-3-amino-2-hydroxy-4-phenylbutyryl-L-leucine, which induces apoptosis in glioma, is also a substrate of PEPTs (Tomita et al., 1990).

Because of the marked gene expression of PEPT1 and PEPT2 in various human cancer cell lines, a radiolabeled tracer for positron emission tomography (PET) was recently developed with the aim of investigating whether cancer prognosis is related to PEPT activity (Mitsuoka et al., 2008). Three tumor xenografts, AsPC-1, PC-3, and MKN45, transplanted in mice were well visualized using [^{11}C]Gly-Sar (Mitsuoka et al., 2008). ^{18}F-Fluorodeoxyglucose (FDG) is in clinical use for the same purpose, taking advantage of the highly active glucose metabolism in tumor cells. However, FDG is also accumulated in normal organs, including brain, heart, kidney, bladder, and stomach. PEPTs may be advantageous in terms of the limited distribution volume of their substrates. This advantage arises at least partially because the organ distributions of PEPT1 and PEPT2 are limited, and the transporters are localized on the apical side in most cases; therefore, exposure of normal tissues to the substrate would be limited. In addition, ^{18}F-FDG is distributed to inflammatory tissues, whereas distribution of [^{11}C]Gly-Sar to inflammatory tissues is minimal (Mitsuoka et al., 2008). Thus, [^{11}C]Gly-Sar is superior to [^{18}F]FDG in terms of distinguishing between tumors and inflammatory tissues. [^{11}C]Gly-Sar was also used as a PET tracer to investigate the localization and function of PEPT2 in mouse kidney (Nabulsi et al., 2005). The renal tubular reabsorption of [^{11}C]Gly-Sar was significantly decreased in *pept2*$^{-/-}$ mice (Nabulsi et al., 2005), suggesting that renal transport of the dipeptide is mediated by PEPT2. Thus, this technology using PET tracers could be useful for chasing *in vivo* changes in the function of PEPTs over time in the same animal.

It has been speculated that the overexpression of PEPTs in cancer cells could be associated with the uptake of oligopeptides as nutrients to support the rapid cell growth, although their exact role in cancer cells is still unknown. Mitsuoka et al. recently reported that two novel synthetic dipeptides, L-phenylalanylsarcosine (Phe-Sar) and 4-(4-methoxyphenyl)-L-phenylalanylsarcosine (Bip(OMe)-Sar), inhibit the growth of human pancreatic cancer AsPC-1 cells, which highly express PEPT1. These two peptidomimetic compounds also inhibit the uptake of [^{3}H]Gly-Sar in AsPC-1 cells. Thus, oligopeptide transporter(s) could be a potential target to inhibit the growth of AsPC-1 cells (Mitsuoka et al., 2010). On the other hand, PEPT2 is expressed in primary culture of human glioblastoma, but not oligodendroglioma (Zimmermann and Stan, 2010). The expression level of PEPT2 is dependent on the grade of glial cell differentiation: The PEPT2 mRNA level is lower in tumors with a higher grade of differentiation than in those with a lower grade of differentiation (Zimmermann and Stan, 2010), implying a possible association of PEPT2 with cellular differentiation.

10.2.8 Intracellular Trafficking Signals in PEPTs

Recent observations have partially clarified the molecular mechanisms of intracellular trafficking and cell-surface stabilization of PEPTs. The C-terminal cytosolic domain, which is located after the 12th transmembrane domain, has a signal amino acid sequence that is involved in these events. The extreme C-terminal amino acids of PEPT1 and PEPT2 in most species comprise a so-called class I PDZ (PSD-95/Dlg/ZO-1) binding motif, which consists of -S/T-X-Φ (X is any amino acid and Φ is a hydrophobic amino acid; Fig. 10.4). The PDZ-binding motif in PEPT2 is involved in direct binding to PDZ domain-containing protein PDZK1 (Kato et al., 2004; Noshiro et al., 2006). In a cell line *in vitro*, cotransfection of PEPT2 with PDZK1 resulted in higher expression of PEPT2 gene product on the cell surface, leading to a higher transport activity with higher V_{max}, compared with that after transfection of PEPT2 alone (Noshiro et al., 2006). Since deletion of the C-terminal PDZ-binding motif in PEPT2 diminishes the interaction with PDZK1 and the stimulatory effect of PDZK1, the interaction with PDZK1 could be involved in cell-surface stabilization of PEPT2.

The role of PDZK1 in cell-surface expression of PEPT1 has been demonstrated in small intestinal epithelial cells using *pdzk1* gene knockout (*pdzk1*$^{-/-}$) mice. In *pdzk1*$^{-/-}$ mice, expression of PEPT1 gene product on apical membranes is reduced, and PEPT1 is localized in multivesicular bodies (Sugiura et al., 2008), implying that

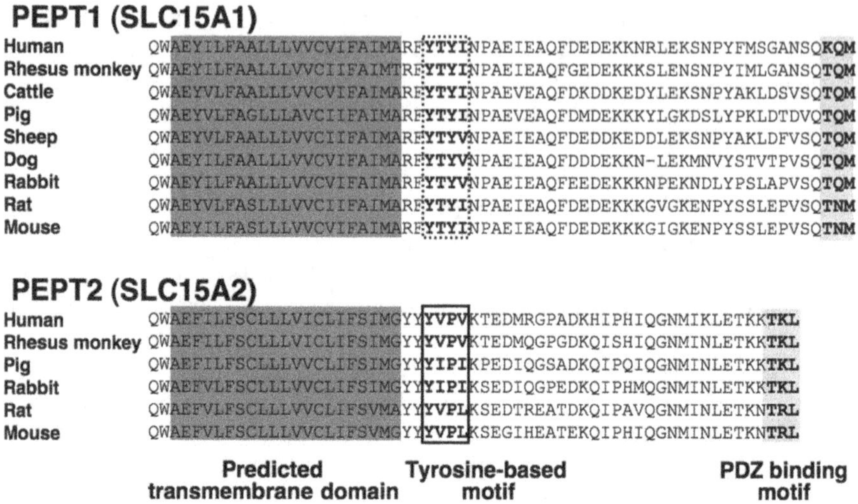

Figure 10.4 Sorting signals located in C-terminal domains of PEPTs. Putative transmembrane domains (Zhang et al., 2004 for PEPT1; Liu et al., 1995 for PEPT2) are shaded dark gray. The dotted line box (PEPT1; predicted) and solid line box (PEPT2; Klapper et al., 2006) show the tyrosine-based motif Y-X-X-Ø (X represents any amino acid and Ø represents a hydrophobic amino acid). Bold-faced sequence within the shaded light gray color indicates a PDZ-binding motif (-S/T-X-Φ).

PEPT1 might be intracellularly degraded because of unstable expression on the cell surface due to the absence of the scaffold protein.

Thus, genetic variations in the signal sequence of PEPTs and/or the PDZ proteins may affect the expression, localization, or function of the transporters. Although there has been no report on *PEPT* gene variants with altered protein–protein interaction, several genetic variants of the *PDZK1* gene have been reported. One *PDZK1* variant (E195K) causes internalization of transporters and downregulation of transport activity when it is cotransfected with PEPT2 (Sugiura et al., 2006).

In addition to the PDZ-binding motif, the C-terminal domain of PEPT2 also contains tyrosine-based motifs (Fig. 10.4), which interact with adaptor protein complex. The tyrosine-based motifs, as well as three-amino acid residue sequences including isoleucine and leucine (INL), within the C-terminal domain of PEPT2 are proposed to be involved in the steady-state apical membrane localization of PEPT2 in polarized proximal tubule cell lines (Klapper et al., 2006). A similar tyrosine-based motif is also observed in the C-terminal domain of PEPT1 (Fig. 10.4), although its biological role has not yet been clarified.

The small GTP-binding protein, Rab8, is essential for apical localization of PEPT1 (Sato et al., 2007). In *rab8* gene knockout mice, several transporters for nutrients are simultaneously internalized from apical membrane of small intestine, leading to lethality just after weaning, due to malnutrition (Sato et al., 2007). Thus, PEPT1 may form a complex with the other transporters and scaffold proteins to maintain homeostasis.

10.2.9 Other SLC15A Family Members

Peptide/histidine transporters 1 (PHT1: SLC15A4) and 2 (PHT2: SLC15A3) have been cloned from a rat brain cDNA library (Yamashita et al., 1997; Sakata et al., 2001). Both the transporters are associated with the uptake of free histidine, as well as di- and tripeptides, in a H^+-dependent manner (Yamashita et al., 1997; Sakata et al., 2001; Bhardwaj et al., 2006). PHT1 is expressed in brain, eye, lung, gastrointestinal tract, and placenta (Yamashita et al., 1997; Herrera-Ruiz et al., 2001; Bhardwaj et al., 2006), whereas PHT2 is mainly expressed in lymphatic system, lung, spleen, and thymus (Sakata et al., 2001).

At present, 15 coding polymorphisms have been identified for human PHT1. Some genetic variants of PHT1 appear to be associated with disease. In recent genome-wide association studies of systemic lupus erythematosus (SLE) in a Chinese Han population, one SNP (rs10847697) was found to have a stronger association with SLE with discoid rash (He et al., 2010). In another study, genetic susceptibility to type 2 diabetes (T2D) was evaluated in a Japanese population, and molecular variations in the *SLC15A4* gene (rs3765108) on chromosome 12q appeared to increase the risk of T2D (Takeuchi et al., 2008). However, there has been no report regarding the functional effect of these SNPs on the transporter. On the other hand, 12 coding polymorphisms have been identified for hPHT2, but the genetic variants of PHT2 have not been studied in detail, and little is known about the clinical relevance of these polymorphisms.

10.3 CONCLUSION

The proton-coupled oligopeptide transporters, PEPT1 and PEPT2, are expressed in epithelial cells of several tissues. Recent studies indicate that PEPT1 plays a role in intestinal absorption of oligopeptides and various peptidomimetic drugs, including β-lactam antibiotics, in the small intestine. On the other hand, PEPT2 functions in the reabsorption of peptides and peptidomimetics in the kidney, and in the efflux of waste products or toxins from the brain and lungs. Therefore, PEPT1 and PEPT2 should be available as targets for improving absorption and drug delivery to the above tissues. To date, there is only limited information about the genetic variation of these transporters in human. PEPT1 expression is highly induced in the colon during IBD, and its variants are associated with Crohn's disease. Polymorphisms of PEPT2 affect the disposition of lead in the body, and so influence the risk of neurotoxicity, although the pharmacokinetic effects of genetic variants of PEPT2 remain controversial. *In vitro* studies suggest that several variants of PEPTs are associated with altered expression level and transport activity. Further studies are needed to establish whether or not the expression of PEPT variants influences the absorption and disposition of substrates.

REFERENCES

Adibi SA (2003) Regulation of expression of the intestinal oligopeptide transporter (Pept-1) in health and disease. *Am J Physiol Gastrointest Liver Physiol*, 285, G779–G788.

Adibi SA, Krzysik BA (1977) Effect of nephrectomy and enterectomy on plasma clearance of intravenously administered dipeptides in rats. *Clin Sci Mol Med*, 52, 205–213.

Adibi SA, Krzysik BA, Drash AL (1977) Metabolism of intravenously administered dipeptides in rats: effects on amino acid pools, glucose concentration, and insulin and glucagon secretion. *Clin Sci Mol Med*, 52, 193–204.

Anderle P, Nielsen CU, Pinsonneault J, Krog PL, Brodin B, Sadée W (2006) Genetic variants of the human dipeptide transporter PEPT1. *J Pharmacol Exp Ther*, 316, 636–646.

Bahadduri PM, D'Souza VM, Pinsonneault JK, Sadée W, Bao S, Knoell DL, Swaan PW (2005) Functional characterization of the peptide transporter PEPT2 in primary cultures of human upper airway epithelium. *Am J Respir Cell Mol Biol*, 32, 319–325.

Balimane PV, Tamai I, Guo A, Nakanishi T, Kitada H, Leibach FH, Tsuji A, Sinko PJ (1998) Direct evidence for peptide transporter (PepT1)-mediated uptake of a nonpeptide prodrug, valacyclovir. *Biochem Biophys Res Commun*, 250, 246–251.

Berger UV, Hediger MA (1999) Distribution of peptide transporter PEPT2 mRNA in the rat nervous system. *Anat Embryol (Berl)*, 199, 439–449.

Bhardwaj RK, Herrera-Ruiz D, Eltoukhy N, Saad M, Knipp GT (2006) The functional evaluation of human peptide/histidine transporter 1 (hPHT1) in transiently transfected COS-7 cells. *Eur J Pharm Sci*, 27, 533–542.

Biegel A, Knütter I, Hartrodt B, Gebauer S, Theis S, Luckner P, Kottra G, Rastetter M, Zebisch K, Thondorf I, *et al.* (2006) The renal type H+/peptide symporter PEPT2: structure–affinity relationships. *Amino Acids*, 31, 137–156.

Buyse M, Charrier L, Sitaraman S, Gewirtz A, Merlin D (2003) Interferon-gamma increases hPepT1-mediated uptake of di-tripeptides including the bacterial tripeptide fMLP in polarized intestinal epithelia. *Am J Pathol*, 163, 1969–1977.

Dalmasso G, Charrier-Hisamuddin L, Nguyen HT, Yan Y, Sitaraman S, Merlin D (2008) PepT1-mediated tripeptide KPV uptake reduces intestinal inflammation. *Gastroenterology*, 134, 166–178.

Daniel H, Kottra G (2004) The proton oligopeptide cotransporter family SLC15 in physiology and pharmacology. *Pflugers Arch*, 447, 610–618.

Döring F, Walter J, Will J, Föcking M, Boll M, Amasheh S, Clauss W, Daniel H (1998) Delta-aminolevulinic acid transport by intestinal and renal peptide transporters and its physiological and clinical implications. *J Clin Invest*, 101, 2761–2767.

Fei YJ, Kanai Y, Nussberger S, Ganapathy V, Leibach FH, Romero MF, Singh SK, Boron WF, Hediger MA (1994) Expression cloning of a mammalian proton-coupled oligopeptide transporter. *Nature*, 368, 563–566.

Fei YJ, Liu W, Prasad PD, Kekuda R, Oblak TG, Ganapathy V, Leibach FH (1997) Identification of the histidyl residue obligatory for the catalytic activity of the human H+/peptide cotransporters PEPT1 and PEPT2. *Biochemistry*, 36, 452–460.

Frey IM, Rubio-Aliaga I, Klempt M, Wolf E, Daniel H (2006) Phenotype analysis of mice deficient in the peptide transporter PEPT2 in response to alterations in dietary protein intake. *Pflugers Arch*, 452, 300–306.

Frey IM, Rubio-Aliaga I, Siewert A, Sailer D, Drobyshev A, Beckers J, de Angelis MH, Aubert J, Bar Hen A, Fiehn O, *et al.* (2007) Profiling at mRNA, protein, and metabolite levels reveals alterations in renal amino acid handling and glutathione metabolism in kidney tissue of Pept2-/- mice. *Physiol Genomics*, 28, 301–310.

Ganapathy V, Leibach FH (1982) Peptide transport in rabbit kidney. Studies with L-carnosine. *Biochim Biophys Acta*, 691, 362–366.

Ganapathy V, Leibach FH (1983) Role of pH gradient and membrane potential in dipeptide transport in intestinal and renal brush-border membrane vesicles from the rabbit. Studies with L-carnosine and glycyl-L-proline. *J Biol Chem*, 258, 14189–14192.

Gonzalez DE, Covitz KM, Sadée W, Mrsny RJ (1998) An oligopeptide transporter is expressed at high levels in the pancreatic carcinoma cell lines AsPc-1 and Capan-2. *Cancer Res*, 58, 519–525.

Groneberg DA, Eynott PR, Döring F, Dinh QT, Oates T, Barnes PJ, Chung KF, Daniel H, Fischer A (2002) Distribution and function of the peptide transporter PEPT2 in normal and cystic fibrosis human lung. *Thorax*, 57, 55–60.

Groneberg DA, Nickolaus M, Springer J, Döring F, Daniel H, Fischer A (2001) Localization of the peptide transporter PEPT2 in the lung: implications for pulmonary oligopeptide uptake. *Am J Pathol*, 158, 707–714.

He CF, Liu YS, Cheng YL, Gao JP, Pan TM, Han JW, Quan C, Sun LD, Zheng HF, Zuo XB, *et al.* (2010) TNIP1, SLC15A4, ETS1, RasGRP3 and IKZF1 are associated with clinical features of systemic lupus erythematosus in a Chinese Han population. *Lupus*, 19, 1181–1186.

Herrera-Ruiz D, Wang Q, Gudmundsson OS, Cook TJ, Smith RL, Faria TN, Knipp GT (2001) Spatial expression patterns of peptide transporters in the human and rat gastrointestinal tracts, Caco-2 *in vitro* cell culture model, and multiple human tissues. *AAPS Pharm Sci*, 3, E9.

Hu Y, Smith DE, Ma K, Jappar D, Thomas W, Hillgren KM (2008) Targeted disruption of peptide transporter Pept1 gene in mice significantly reduces dipeptide absorption in intestine. *Mol Pharm*, 5, 1122–1130.

Jiang H, Hu Y, Keep RF, Smith DE (2009) Enhanced antinociceptive response to intracerebroventricular kyotorphin in Pept2 null mice. *J Neurochem*, 109, 1536–1563.

Kamal MA, Jiang H, Hu Y, Keep RF, Smith DE (2009) Influence of genetic knockout of Pept2 on the in vivo disposition of endogenous and exogenous carnosine in wild-type and Pept2 null mice. *Am J Physiol Regul Integr Comp Physiol*, 296, R986–R991.

Kamal MA, Keep RF, Smith DE (2008) Role and relevance of PEPT2 in drug disposition, dynamics, and toxicity. *Drug Metab Pharmacokinet*, 23, 236–242.

Kato Y, Yoshida K, Watanabe C, Sai Y, Tsuji A (2004) Screening of the interaction between xenobiotic transporters and PDZ proteins. *Pharm Res*, 21, 1886–1894.

Klapper M, Daniel H, Doring F (2006) Cytosolic COOH terminus of the peptide transporter PEPT2 is involved in apical membrane localization of the protein. *Am J Physiol*, 290, C472–C483.

Knütter I, Rubio-Aliaga I, Boll M, Hause G, Daniel H, Neubert K, Brandsch M (2002) H+-peptide cotransport in the human bile duct epithelium cell line SK-ChA-1. *Am J Physiol Gastrointest Liver Physiol*, 283, G222–G229.

Kobayashi KS, Chamaillard M, Ogura Y, Henegariu O, Inohara N, Nuñez G, Flavell RA (2005) Nod2-dependent regulation of innate and adaptive immunity in the intestinal tract. *Science*, 307, 731–734.

Liang R, Fei YJ, Prasad PD, Ramamoorthy S, Han H, Yang-Feng TL, Hediger MA, Ganapathy V, Leibach FH (1995) Human intestinal H+/peptide cotransporter. Cloning, functional expression, and chromosomal localization. *J Biol Chem*, 270, 6456–6463.

Liu R, Tang AM, Tan YL, Limenta LM, Lee EJ (2009) Interethnic differences of PEPT2 (SLC15A2) polymorphism distribution and associations with cephalexin pharmacokinetics in healthy Asian subjects. *Eur J Clin Pharmacol*, 65, 65–70

Liu R, Tang AM, Tan YL, Limenta LM, Lee EJ (2011) Effects of sodium bicarbonate and ammonium chloride pre-treatments on the PEPT2 (SLC15A2) mediated renal clearance of cephalexin in healthy subjects. *Drug Metab Pharmacokinet*, 26, 87–93.

Liu W, Liang R, Ramamoorthy S, Fei YJ, Ganapathy ME, Hediger MA, Ganapathy V, Leibach FH (1995) Molecular cloning of PEPT2, a new member of the H+/peptide cotransporter family, from human kidney. *Biochim Biophys Acta*, 1235, 461–466.

Ma K, Hu Y, Smith DE (2010) Peptide transporter 1 is responsible for intestinal uptake of the dipeptide glycylsarcosine: studies in everted jejunal rings from wild-type and pept1 null mice. *J Pharm Sci*, 100, 767–774.

Merlin D, Si-Tahar M, Sitaraman SV, Eastburn K, Williams I, Liu X, Hediger MA, Madara JL (2001) Colonic epithelial hPepT1 expression occurs in inflammatory bowel disease: transport of bacterial peptides influences expression of MHC class 1 molecules. *Gastroenterology*, 120, 1666–1679.

Mitsuoka K, Kato Y, Miyoshi S, Murakami Y, Hiraiwa M, Kubo Y, Nishimura S, Tsuji A (2010) Inhibition of oligopeptide transporter suppress growth of human pancreatic cancer cells. *Eur J Pharm Sci*, 40, 202–208.

Mitsuoka K, Miyoshi S, Kato Y, Murakami Y, Utsumi R, Kubo Y, Noda A, Nakamura Y, Nishimura S, Tsuji A (2008) Cancer detection using a PET tracer, 11C-glycylsarcosine, targeted to H+/peptide transporter. *J Nucl Med*, 49, 615–622.

Nabulsi NB, Smith DE, Kilbourn MR (2005) [11C]Glycylsarcosine: synthesis and in vivo evaluation as a PET tracer of PEPT2 transporter function in kidney of PEPT2 null and wild-type mice. *Bioorg Med Chem*, 13, 2003–3001.

Nakanishi T, Tamai I, Sai Y, Sasaki T, Tsuji A (1997) Carrier-mediated transport of oligopeptides in the human fibrosarcoma cell line HT1080. *Cancer Res*, 57, 4118–4122.

Nakanishi T, Tamai I, Takaki A, Tsuji A (2000) Cancer cell-targeted drug delivery utilizing oligopeptide transport activity. *Int J Cancer*, 88, 274–280.

Naruhashi K, Sai Y, Tamai I, Suzuki N, Tsuji A (2002) PepT1 mRNA expression is induced by starvation and its level correlates with absorptive transport of cefadroxil longitudinally in the rat intestine. *Pharm Res*, 19, 1417–1423.

Noshiro R, Anzai N, Sakata T, Miyazaki H, Terada T, Shin HJ, He X, Miura D, Inui K, Kanai Y, Endou H (2006) The PDZ domain protein PDZK1 interacts with human peptide transporter PEPT2 and enhances its transport activity. *Kidney Int*, 70, 275–282.

Nozawa T, Toyobuku H, Kobayashi D, Kuruma K, Tsuji A, Tamai I (2003) Enhanced intestinal absorption of drugs by activation of peptide transporter PEPT1 using proton-releasing polymer. *J Pharm Sci*, 92, 2208–2216.

Nutzenadel W, Scriver CR (1976) Uptake and metabolism of beta-alanine and L-carnosine by rat tissues *in vitro*: role in nutrition. *Am J Physiol*, 230, 643–651.

Ocheltree SM, Shen H, Hu Y, Keep RF, Smith DE (2005) Role and relevance of peptide transporter 2 (PEPT2) in the kidney and choroid plexus: *in vivo* studies with glycylsarcosine in wild-type and PEPT2 knockout mice. *J Pharmacol Exp Ther*, 315, 240–247.

Ogihara T, Kano T, Wagatsuma T, Wada S, Yabuuchi H, Enomoto S, Morimoto K, Shirasaka Y, Kobayashi S, Tamai I (2009) Oseltamivir (tamiflu) is a substrate of peptide transporter 1. *Drug Metab Dispos*, 37, 1676–1681.

Pan X, Terada T, Irie M, Saito H, Inui K (2002) Diurnal rhythm of H+-peptide cotransporter in rat small intestine. *Am J Physiol Gastrointest Liver Physiol*, 283, G57–G64.

Pan X, Terada T, Okuda M, Inui K (2003) Altered diurnal rhythm of intestinal peptide transporter by fasting and its effects on the pharmacokinetics of ceftibuten. *J Pharmacol Exp Ther*, 307, 626–632.

Phan DD, Chin-Hong P, Lin ET, Anderle P, Sadee W, Guglielmo BJ (2003) Intra- and interindividual variabilities of valacyclovir oral bioavailability and effect of coadministration of an hPEPT1 inhibitor. *Antimicrob Agents Chemother*, 47, 2351–2353.

Pinsonneault J, Nielsen CU, Sadée W (2004) Genetic variants of the human H+/dipeptide transporter PEPT2: analysis of haplotype functions. *J Pharmacol Exp Ther*, 311, 1088–1096.

Rioux JD, Xavier RJ, Taylor KD, Silverberg MS, Goyette P, Huett A, Green T, Kuballa P, Barmada MM, Datta LW, *et al.* (2007) Genome-wide association study identifies new susceptibility loci for Crohn's disease and implicates autophagy in disease pathogenesis. *Nat Genet*, 39, 596–604.

Rubio-Aliaga I, Frey I, Boll M, Groneberg DA, Eichinger HM, Balling R, Daniel H (2003) Targeted disruption of the peptide transporter Pept2 gene in mice defines its physiological role in the kidney. *Mol Cell Biol*, 23, 3247–3252.

Sai Y, Tamai I, Sumikawa H, Hayashi K, Nakanishi T, Amano O, Numata M, Iseki S, Tsuji A (1996) Immunolocalization and pharmacological relevance of oligopeptide transporter PEPT1 in intestinal absorption of β-lactam antibiotics. *FEBS Lett*, 392, 25–29.

Saito H, Okuda M, Terada T, Sasaki S, Inui K (1995) Cloning and characterization of a rat H+/peptide cotransporter mediating absorption of beta-lactam antibiotics in the intestine and kidney. *J Pharmacol Exp Ther*, 275, 1631–1637.

Sakata K, Yamashita T, Maeda M, Moriyama Y, Shimada S, Tohyama M (2001) Cloning of a lymphatic peptide/histidine transporter. *Biochem J*, 15, 53–60.

Sala-Rabanal M, Loo DD, Hirayama BA, Turk E, Wright EM (2006a) Molecular interactions between dipeptides, drugs, and the human intestinal H+–oligopeptide cotransporter hPEPT1. *J Physiol*, 574, 149–166.

Sala-Rabanal M, Loo DD, Hirayama BA, Wright EM (2006b) Molecular mechanism of dipeptide and drug transport by the human renal H+/oligopeptide cotransporter hPEPT2. *Am J Physiol Renal Physiol*, 294, F1422–F1432.

Sato T, Mushiake S, Kato Y, Sato K, Sato M, Takeda N, Ozono K, Miki K, Kubo Y, Tsuji A, Harada R, Harada A (2007) The Rab8 GTPase regulates apical protein localization in intestinal cells. *Nature*, 448, 366–369.

Shen H, Ocheltree SM, Hu Y, Keep RF, Smith DE (2007) Impact of genetic knockout of PEPT2 on cefadroxil pharmacokinetics, renal tubular reabsorption, and brain penetration in mice. *Drug Metab Dispos*, 35, 1209–1216.

Shen H, Smith DE, Keep RF, Brosius FC 3rd (2004) Immunolocalization of the proton-coupled oligopeptide transporter PEPT2 in developing rat brain. *Mol Pharm*, 1, 248–256.

Shen H, Smith DE, Keep RF, Xiang J, Brosius FC 3rd (2003) Targeted disruption of the PEPT2 gene markedly reduces dipeptide uptake in choroid plexus. *J Biol Chem*, 278, 4786–4791.

Shen H, Smith DE, Yang T, Huang YG, Schnermann B, and Brosius FC (1999) Localization of PEPT1 and PEPT2 proton-coupled oligopeptide transporter mRNA and protein in rat kidney. *Am J Physiol Renal Physiol*, 276, F658–F665.

Shu C, Shen H, Teuscher NS, Lorenzi PJ, Keep RF, Smith DE (2002) Role of PEPT2 in peptide/mimetic trafficking at the blood–cerebrospinal fluid barrier: studies in rat choroid plexus epithelial cells in primary culture. *J Pharmacol Exp Ther*, 301, 820–829.

Smith DE, Hu Y, Shen H, Nagaraja TN, Fenstermacher JD, Keep RF (2010) Distribution of glycylsarcosine and cefadroxil among cerebrospinal fluid, choroid plexus, and brain parenchyma after intracerebroventricular injection is markedly different between wild-type and Pept2 null mice. *J Cereb Blood Flow Metab*, 31, 250–261.

Sobin C, Gutierrez M, Alterio H (2009) Polymorphisms of delta-aminolevulinic acid dehydratase (ALAD) and peptide transporter 2 (PEPT2) genes in children with low-level lead exposure. *Neurotoxicology*, 30, 881–887.

Sugiura T, Kato Y, Kubo Y, Tsuji A (2006) Mutation in an adaptor protein PDZK1 affects transport activity of organic cation transporter OCTNs and oligopeptide transporter PEPT2. *Drug Metab Pharmacokinet*, 21, 375–383.

Sugiura T, Kato Y, Wakayama T, Silver DL, Kubo Y, Iseki S, Tsuji A (2008) PDZK1 regulates two intestinal solute carriers (SLC15A1 and SLC22A5) in mice. *Drug Metab Dispos*, 36, 1181–1188.

Swaan PW, Bensman T, Bahadduri PM, Hall MW, Sarkar A, Bao S, Khantwal CM, Ekins S, Knoell DL (2008) Bacterial peptide recognition and immune activation facilitated by human peptide transporter PEPT2. *Am J Respir Cell Mol Biol*, 39, 536–542.

Takeuchi F, Ochiai Y, Serizawa M, Yanai K, Kuzuya N, Kajio H, Honjo S, Takeda N, Kaburagi Y, Yasuda K, *et al.* (2008) Search for type 2 diabetes susceptibility genes on chromosomes 1q, 3q, and 12q. *J Hum Genet*, 53, 314–324.

Tamai I, Nakanishi T, Hayashi K, Terao T, Sai Y, Shiraga T, Miyamoto K, Takeda E, Higashida H, Tsuji A (1997) The predominant contribution of oligopeptide transporter PEPT1 to intestinal absorption of beta-lactam antibiotics in the rat small intestine. *J Pharm Pharmacol*, 49, 796–801.

Tamai I, Nakanishi T, Nakahara H, Sai Y, Ganapathy V, Leibach FH, Tsuji A (1998) Improvement of L-dopa absorption by dipeptidyl derivation, utilizing peptide transporter PEPT1. *J Pharm Sci*, 87, 1542–1546.

Tamai I, Tomizawa N, Takeuchi T, Nakayama K, Higashida H, Tsuji A (1995) Functional expression of transporter for beta-lactam antibiotics and dipeptides in *Xenopus laevis* oocytes injected with messenger RNA from human, rat and rabbit small intestines. *J Pharmacol Exp Ther*, 273, 26–31.

Terada T, Irie M, Okuda M, Inui K (2004) Genetic variant Arg57His in human H+/peptide cotransporter 2 causes a complete loss of transport function. *Biochem Biophys Res Commun*, 316, 416–420.

Terada T, Shimada Y, Pan X, Kishimoto K, Sakurai T, Doi R, Onodera H, Katsura T, Imamura M, Inui K (2005) Expression profiles of various transporters for oligopeptides, amino acids and organic ions along the human digestive tract. *Biochem Pharmacol*, 70, 1756–1763.

Thwaites DT, Cavet M, Hirst BH, Simmons NL (1995) Angiotensin-converting enzyme (ACE) inhibitor transport in human intestinal epithelial (Caco-2) cells. *Br J Pharmacol*, 114, 981–986.

Tomita Y, Katsura T, Okano T, Inui K, Hori R (1990) Transport mechanisms of bestatin in rabbit intestinal brush-border membranes: role of H+/dipeptide cotransport system. *J Pharmacol Exp Ther*, 252, 859–862.

Tsuda M, Terada T, Irie M, Katsura T, Niida A, Tomita K, Fujii N, Inui K (2006) Transport characteristics of a novel peptide transporter 1 substrate, antihypotensive drug midodrine, and its amino acid derivatives. *J Pharmacol Exp Ther*, 318, 455–460.

Tsuji A (2002) Transporter-mediated drug interactions. *Drug Metab Pharmacokinet*, 17, 253–274.

Tsuji A, Nakashima E, Kagami I, Honjo N, Yamana T (1977) Effect of dose-concentration on the absorption of amoxicillin and ampicillin from the rat intestine. *J Pharm Pharmcol*, 29, 707–708.

Tsuji A, Tamai I, Nakanishi M, Amidon GL (1990) Mechanism of absorption of the dipeptide alpha-methyldopa-phe in intestinal brush-border membrane vesicles. *Pharm Res*, 7, 308–309.

Xiang J, Jiang H, Hu Y, Smith DE, Keep RF (2010) Kyotorphin transport and metabolism in rat and mouse neonatal astrocytes. *Brain Res*, 1347, 11–18.

Yamashita T, Shimada S, Guo W, Sato K, Kohmura E, Hayakawa T, Takagi T, Tohyama M (1997) Cloning and functional expression of a brain peptide/histidine transporter. *J Biol Chem*, 272, 10205–10211.

Zhang EY, Fu DJ, Pak YA, Stewart T, Mukhopadhyay N, Wrighton SA, Hillgren KM (2004) Genetic polymorphisms in human proton-dependent dipeptide transporter PEPT1: implications for the functional role of Pro586. *J Pharmacol Exp Ther*, 310, 437–445.

Zimmermann M, Stan AC (2010) PEPT2 transporter protein expression in human neoplastic glial cells and mediation of fluorescently tagged dipeptide derivative beta-Ala-Lys-Nepsilon-7-amino-4-methyl-coumarin-3-acetic acid accumulation. *J Neurosurg*, 112, 1005–1014.

Zucchelli M, Torkvist L, Bresso F, Halfvarson J, Hellquist A, Anedda F, Assadi G, Lindgren GB, Svanfeldt M, Janson M, *et al.* (2009) PEPT1 oligopeptide transporter (SLC15A1) gene polymorphism in inflammatory bowel disease. *Inflamm Bowel Dis*, 15, 1562–1569.

CHAPTER 11

NUCLEOSIDE TRANSPORTERS (SLC28 AND SLC29) FAMILY

Míriam Molina-Arcas
Marçal Pastor-Anglada

11.1 INTRODUCTION

Nucleosides can be structurally modified to generate pharmacologically active derivatives. These drugs include nucleobase and nucleoside analogs, and phosphorylated nucleoside-derived molecules. Such compounds are currently used in antiviral and antitumoral therapy and in the treatment of autoimmune and inflammatory diseases. Both purine and pyrimidine derivatives are suitable for use in cancer therapy, although, in general, pyrimidine analogs are employed to treat solid tumors whereas purine nucleosides and nucleobases are used to treat certain lymphoproliferative disorders. Although nucleoside analogs carry structural modifications distinguishing them from natural nucleosides, they retain most of the metabolic properties of the parent compounds. To exert their pharmacological action, the drugs must initially enter into the cells, via the action of plasma membrane transporters, and must then be metabolized. As cell uptake is the first step in the pharmacological action of nucleoside derivatives, drug pharmacokinetics and cytotoxic actions will be influenced by the expression levels of drug transporters.

Several gene families encode membrane proteins involved in the uptake of nucleoside analogs. In general, structurally related compounds are transported by members of the same gene family. Proteins encoded by solute carrier (SLC) gene families SLC28 and SCL29, which translocate natural nucleosides, are also responsible for the uptake of many nucleoside-derived drugs commonly used in antiviral and anticancer therapy. However, some nucleoside analogs are not efficiently translocated by nucleoside transporters, but rather are taken up by members of other protein families, including SLC15 and SLC22. In the present chapter, we will focus on the SLC28 and SLC29 gene families, whereas SLC22- and SLC15-encoded transporters will be reviewed in Chapters 8 and 10, respectively.

Pharmacogenomics of Human Drug Transporters: Clinical Impacts, First Edition.
Edited by Toshihisa Ishikawa, Richard B. Kim and Jörg König

11.2 NUCLEOSIDE TRANSPORTER PROTEINS

Two unrelated gene families, SLC28 and SLC29, encode proteins responsible for the uptake of natural nucleosides in mammals (Table 11.1). The SLC28 family encodes concentrative nucleoside transporter (CNT) proteins mediating unidirectional sodium-coupled nucleoside transport, and has three members (CNT1, CNT2, and CNT3). The SLC29 family encodes equilibrative nucleoside transporter (ENT) proteins and has four members (ENT1, ENT2, ENT3, and ENT4), all of which mediate facilitated-diffusion, sodium-independent transport.

11.2.1 The SLC28 Gene Family: Concentrative Nucleoside Transporters

CNTs mediate the unidirectional flow of nucleosides in a sodium-dependent manner. Thus, nucleoside movement is not dependent on a nucleoside concentration gradient, and therefore has a high energy cost because transport is coupled to the transmembrane sodium gradient. The three members of the SLC28 family differ in substrate selectivity and sodium:nucleoside stoichiometry (Gray et al., 2004b) (Table 11.1).

The hCNT1 protein (encoded by *SLC28A1*) transports pyrimidine nucleosides (Ritzel et al., 1997), whereas hCNT2 (*SLC28A2*) prefers purine nucleosides, although it also transports uridine (Ritzel et al., 1998). The hCNT3 (*SLC28A3*) protein mediates uptake of both purine and pyrimidine nucleosides (Ritzel et al., 2001). Human CNTs show high affinities for their substrates, with apparent Km values in the low micromolar range (10–100 μM). No specific pharmacological inhibitor of any hCNT has yet been identified, although intense effort is currently being devoted to the generation of such inhibitors (Gupte and Buolamwini, 2009; Wang et al., 2010; Damaraju et al., 2011). The sodium:nucleoside stoichiometry of both hCNT1 and hCNT2 is 1:1 (Ritzel et al., 1998; Smith et al., 2004), whereas hCNT3 exhibits a 2:1 stoichiometry, thus translocating two sodium atoms per molecule of nucleoside (Ritzel et al., 2001). Moreover, hCNT3 can couple transport of nucleosides to that of protons and lithium, but with 1:1 stoichiometry (Hu et al., 2006). Thus, hCNT3 demonstrates a broader selectivity than either hCNT1 or hCNT2, and also has a higher concentrative capacity. In fact, phylogenetic and homology analyses suggest that hCNT3 is closer to the CNT of hagfish (which shows broad substrate selectivity) rather than to hCNT1/2 (Ritzel et al., 2001).

The homology between hCNT3 and hCNT1 and hCNT2 is only 48% and 47% respectively, whereas hCNT2 is 72% identical to hCNT1. The three transporter proteins share a putative general topology of 13 transmembrane domains, with an extracellular carboxy terminus and the amino-terminus facing the cytosol (Hamilton et al., 2001) (Fig. 11.1). This accepted topology has been revisited (Slugoski et al., 2009) and challenged by the recent crystallization of a nucleoside transporter protein from *Vibrio cholerae* (Johnson et al., 2012). Current information on the functions of particular residues and domains has been obtained using topological models, site-directed mutagenesis, and chimera construction. Truncated hCNT1, lacking the first N-terminal 174 amino acids, retained sodium coupling ability and significant transport activity (Hamilton et al., 2001). Studies on hCNT3 demonstrate that the

TABLE 11.1 Characteristics of Cloned Human Nucleoside Transporters

Concentrative Nucleoside Transporters					
Protein	Gene	Human gene locus	Protein length	Substrates	Coupling ions/Stoichiometry
hCNT1	*SLC28A1*	15q25-q26	650 aa	Pyrimidine nucleosides	$1Na^{+}$:1 Nucleoside
hCNT2	*SLC28A2*	15q15	658 aa	Purine nucleosides and uridine	$1Na^{+}$:1 Nucleoside
hCNT3	*SLC28A2*	9q22.2	691 aa	Purine and pyrimidine nucleosides	$2Na^{+}$:1 Nucleoside $1H^{+}$:1 Nucleoside $1Li^{+}$:1 Nucleoside

Equilibrative Nucleoside Transporters					
Protein	Gene	Human gene locus	Protein length	Substrates	Inhibitors
hENT1	*SLC29A1*	6p21.1-p21.2	465 aa	Purine and pyrimidine nucleosides	NBTI (0.005 μM) dipyridamole (0.05 μM) dilazep (0.02 μM)
hENT2	*SLC29A2*	11q13	465 aa	Purine pyrimidine nucleosides and nucleobases	NBTI (>1 μM) dipyridamole (5 μM) dilazep (120 μM)
hENT3	*SLC29A3*	10q22.1	475 aa	Purine and pyrimidine nucleosides and some nucleobases	NBTI (>10 μM μM) dipyridamole (>10 μM μM) dilazep (>10 μM)
hENT4	*SLC29A4*	7p22.1	530 aa	Adenosine and monoamines	NBTI (>1 μM) dipyridamole (1 μM) dilazep (1 μM)

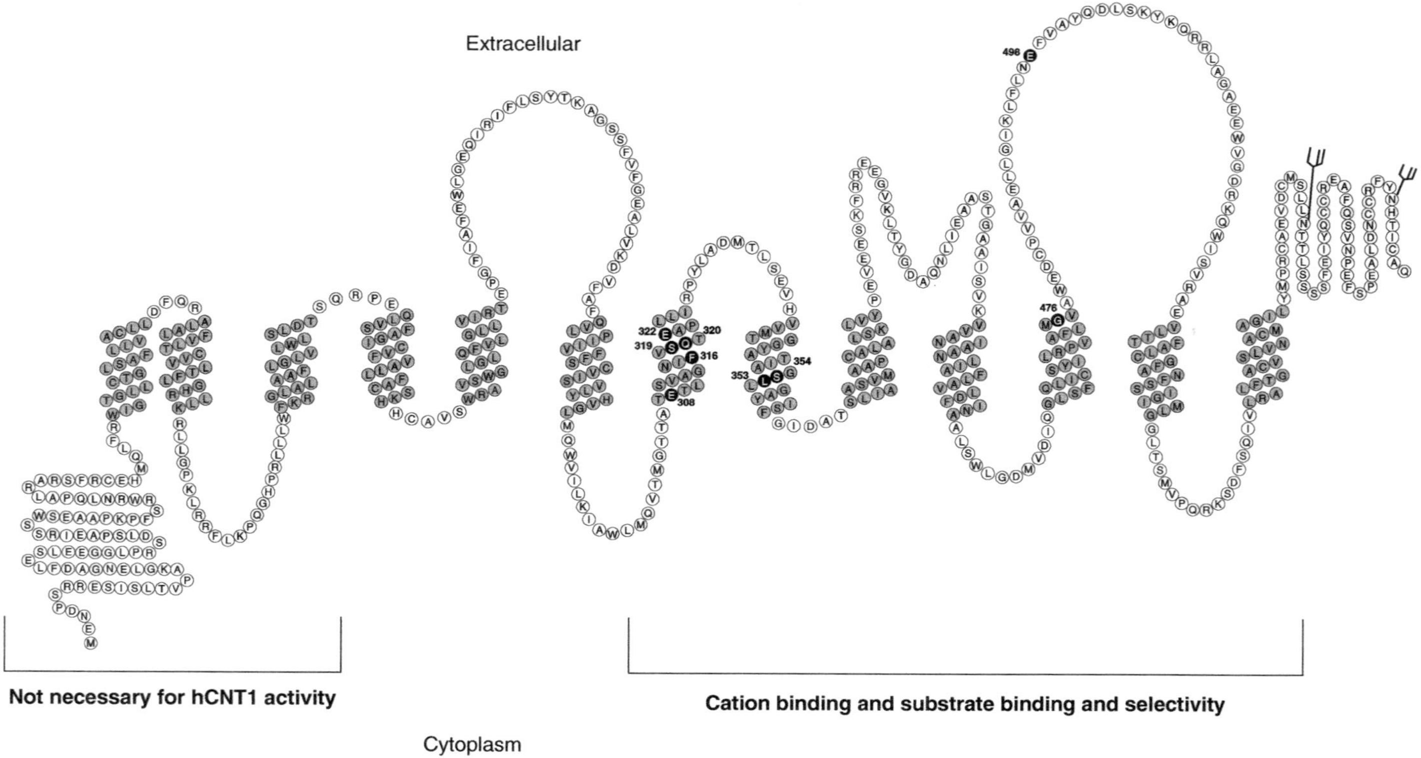

Figure 11.1 Predicted topology of concentrative nucleoside transporter 1 (hCNT1). Black circles show amino acids relevant in determining the functional characteristics of hCNT1. Regions implicated in substrate and cation binding are highlighted.

control of cation stoichiometry and binding resides in the C-terminal half of the protein (Smith et al., 2005). In fact, transmembrane domains 8 and 9 appear to contain residues involved in substrate binding and selectivity. Substitution of such domains of rCNT1 by the corresponding domains of rCNT2 changed a pyrimidine nucleoside-preferring transporter into a protein that preferentially transported purines (Wang and Giacomini, 1997). Substitution of Ser319/Gln320 of the transmembrane domain of hCNT1 by the (similarly located) Gly313/Met314 of hCNT2 transformed hCNT1 into a transporter exhibiting broad selectivity, thus similar in nature to hCNT3 (Loewen et al., 1999). Moreover, Ser353 substitution induced changes in nucleoside specificity, whereas the Leu354Val alteration increased the apparent affinities for sodium and lithium ions (Slugoski et al., 2007). These results suggest that coding-region single-nucleotide polymorphisms (SNPs) in specific domains may affect not only transporter activity but also selectivity.

11.2.2 The SLC29 Gene Family: Equilibrative Nucleoside Transporters

ENTs mediate a passive diffusion process and are thus essentially bidirectional carriers mediating both influx and efflux (Baldwin et al., 2004). The SLC29 gene family contains four members (Table 11.1). The proteins hENT1 (*SLC28A1*) and hENT2 (*SLC29A2*), the two best-characterized transporters, show broad substrate selectivity, accepting both purines and pyrimidines, but with lower affinities than concentrative transporters (the apparent Km values are in the high micromolar range) (Ward et al., 2000). hENT2 can also transport some purine and pyrimidine nucleobases (Osses et al., 1996). hENT1 and hENT2 can be distinguished on the basis of their sensitivity to inhibition by the nucleoside analog nitrobenzylthioinosine (NBTI); the IC_{50} values are 0.4 nM and 2800 nM respectively. Both transporters are also inhibited by vasodilation potentiators including dipyridamole and dilazep; however, hENT1 is more sensitive to inhibition than hENT2 (Visser et al., 2002). hENT1 and hENT2 are major determinants of equilibrative nucleoside transport across the plasma membrane, although both proteins have also been found within intracellular organelles, such as mitochondria (hENT1) and nuclei (hENT1 and hENT2) (Mani et al., 1998; Lai et al., 2004; Lee et al., 2006).

The gene-encoding hENT3 (*SLC29A3*) was cloned in 2005; before this time, its activity had not been described in any detail (Baldwin et al., 2005). The protein is only 29% identical in sequence to hENT1 and was initially thought to be a lysosomal pH-dependent transporter with an optimum pH of 5.5. hENT3 exhibits wide substrate selectivity and can also transport the nucleobase adenine. However, the protein is not sensitive to inhibition by NBTI. Recently, it has been suggested that hENT3 is also a mitochondrial transporter (Govindarajan et al., 2009), although its subcellular localization needs to be further explored. A better understanding of hENT3 substrate selectivity, as well as the subcellular, cellular, and tissue distribution of the protein, are currently major goals in the field. This is because *SLC29A3* is the only nucleoside transporter studied to date that has been associated with inherited human diseases. Several syndromes, including the H syndrome (Molho-Pessach et al., 2008) and the Pigmented Hypertrichosis with Insulin dependent Diabetes (PHID) syndrome

(Spiegel et al., 2010) have been associated with mutations in the *SLC29A3* gene. The broad spectrum of clinical defects caused by *SLC29A3* mutations may reflect the apparently complex biology of this transporter.

Finally, hENT4 (*SLC28A4*), originally described as a monoamine/organic transporter (PMAT), can also translocate adenosine under acidic pH conditions (Barnes et al., 2006), although the issue of whether the protein plays a role as a nucleoside transporter remains contentious.

hENT transporters are predicted to share a topological structure featuring 11 transmembrane domains, an intracellular amino-terminus, an extracellular carboxy-terminus, and a large cytoplasmic loop between transmembrane (TM) domains 6 and 7 (Fig. 11.2). These transporters are post-translationally modified by glycosylation. hENT1 and hENT2 are glycosylated in the extracellular loop lying between TM1 and TM2, hENT1 at Asn48, and hENT2 at Asn48 and Asn57 (Sundaram et al., 2001a; Ward et al., 2003). Glycosylation is not essential for transporter activity, but can modulate the sensitivity to inhibition by NBTI and coronary vasodilators (Vickers et al., 1999). hENT3 and hENT4 also have putative glycosylation sites in the TM1–TM2 region and the C-terminal loop, respectively. However, the glycosylation status of these proteins has not been fully characterized.

Chimera studies have helped to identify protein domains involved in substrate and inhibitor interaction (Fig. 11.2). The region lying between TM1 and TM6 in hENT2 appears to be responsible for the recognition of deoxynucleosides, whereas the TM5–TM6 region has been implicated in the ability to recognize nucleobases (Yao et al., 2001, 2002). The N-terminal half of the protein is also important in terms of inhibitor binding. The region encompassing TM3–TM6 contains residues responsible for the expression of sensitivity to the inhibitors NBTI, dipyridamole, and dilazep (Sundaram et al., 1998, 2001b). Similarly to concentrative transporters, substitution of even a single amino acid can affect substrate and inhibitor recognition. In hENT1, substitution of Gly154 resulted in decreased affinity for both adenosine and cytidine, and resulted in loss of the ability to bind NBTI (Hyde et al., 2001; SenGupta and Unadkat, 2004). The amino acid residues Pro71, Glu72, and Asn74 are important in terms of mitochondrial targeting of hENT1 (Lai et al., 2004; Lee et al., 2006).

11.3 PHARMACOLOGICAL PROFILES OF NUCLEOSIDE TRANSPORTERS

11.3.1 Anticancer Nucleoside Analogs

All currently available anticancer nucleoside analogs share similar mechanisms of activation and action. However, small structural differences may be expected to cause heterogeneity in their activity against various diseases as well as in their interaction with plasma membrane transporters. The structures of selected nucleoside-derived anticancer drugs and the principal clinical uses of these compounds are shown in Figure 11.3.

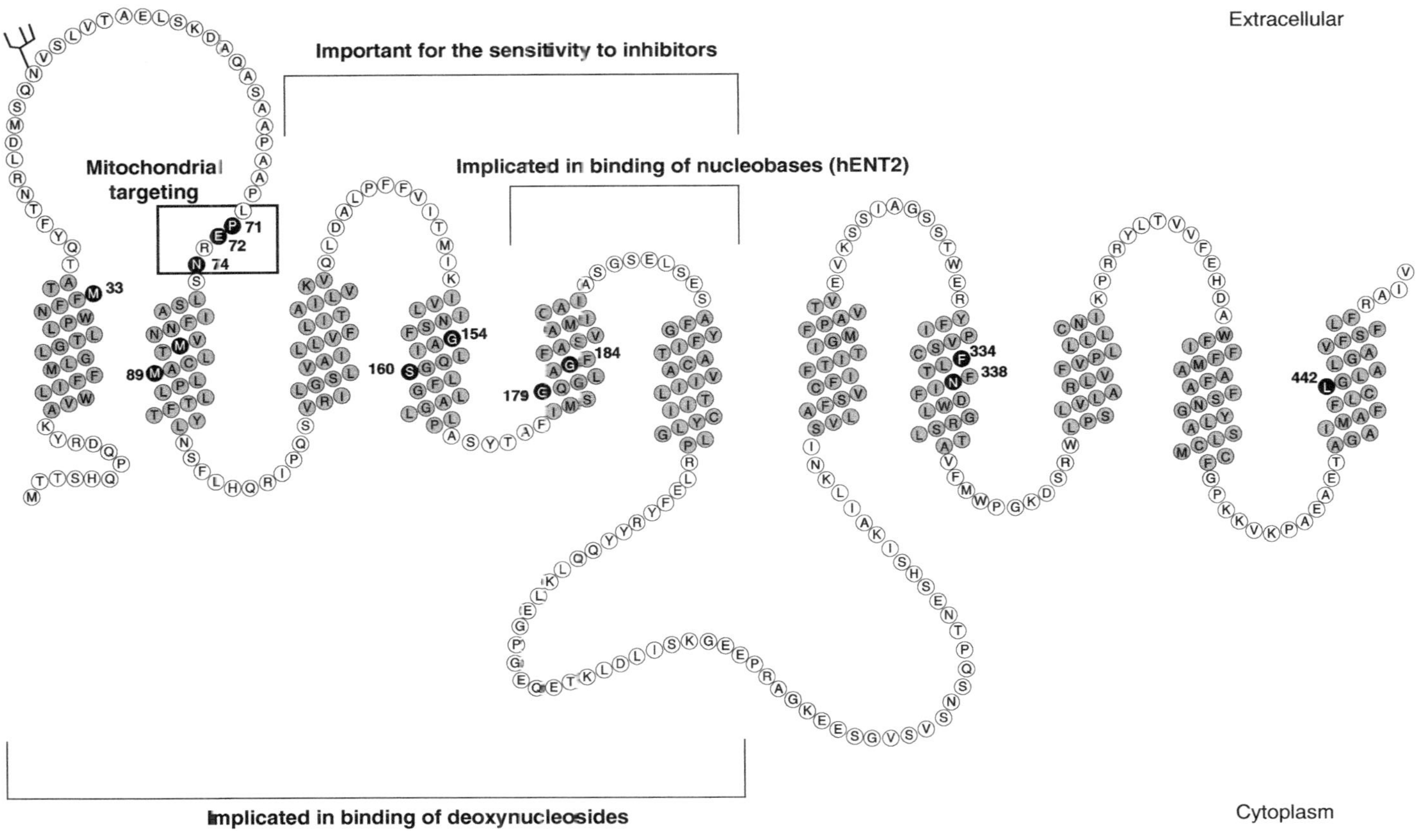

Figure 11.2 Predicted topology of equilibrative nucleoside transporter 1 (hENT1). Black circles show amino acids relevant in determining the functional characteristics of hENT1. Regions implicated in substrate and inhibitor binding are highlighted.

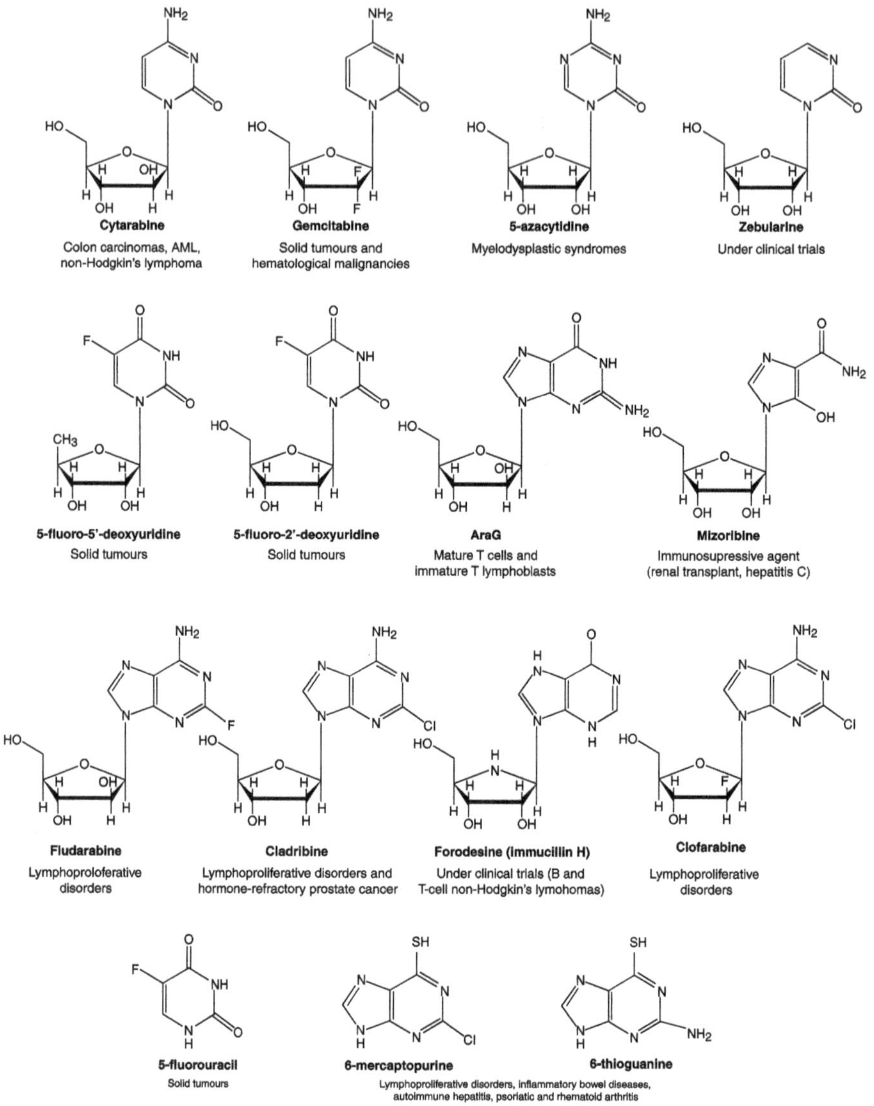

Figure 11.3 Molecular structures and current clinical uses of nucleoside-derived anticancer and immunosuppressive drugs.

Both purine and pyrimidine nucleoside analogs are currently employed as anticancer drugs. The purine derivatives cladribine (2-CdA; 2-chlorodeoxyadenosine) and fludarabine (F-ara-A; 9-β-D-arabinosyl-2-fluoroadenine) are extensively used for chemotherapy of hematological malignancies (Robak et al., 2009). More recently, clofarabine (2-chloro-9-[2′-deoxy-2′-fluoro-β-D-arabinofuranosyl]adenine) has been employed to treat lymphoproliferative disorders, as has forodesine (Immucillin-H); the latter drug is currently in clinical trials (Robak et al., 2009). In contrast, the pyrimidine analogs gemcitabine (dFdC; 2′,2′-difluorodeoxycytidine) and capecitabine are currently used for the treatment of solid tumors (Saiko et al., 2005; Wong et al., 2009). However, both gemcitabine and cytarabine (ara-C; 1-β-D-arabinosylfuranosylcytosine) are also effective when used to treat certain hematological malignancies (Johnson, 2012; Wong et al., 2009). Capecitabine is a prodrug metabolized to the nucleobase 5-fluorouracil inside tumor cells, and 5′-deoxy-5-fluorouridine (5′-DFUR) is its immediate precursor. 5-fluorouracil has been extensively used in the treatment of various solid tumors, whereas the nucleobases 6-mercaptopurine and 6-thioguanine are effective in patients with particular hematological malignancies, and in those with certain autoimmune and inflammatory diseases (Longley et al., 2003; Sahasranaman et al., 2008).

The uptake of anticancer nucleoside derivatives is mediated mostly, if not exclusively, by ENT- and CNT-type transporters. However, some transporters have been implicated in the efflux of nucleoside derivatives and/or their phosphorylated metabolites (Errasti-Murugarren and Pastor-Anglada, 2010). Pharmacological profiles of such compounds have been obtained using a combination of substrate flux measurements, cross-inhibition studies, and electrophysiology (Table 11.2). Drug substrate selectivity is similar to that observed using natural nucleosides, although, in general, the affinities for drugs are lower. hENT1 and hENT2 exhibit broad substrate selectivity, transporting both purine and pyrimidine nucleosides (Mackey et al., 1998; Molina-Arcas et al., 2003; Baldwin et al., 2004). In fact, all tested drugs were transported, albeit with variable efficacy, by both hENT1 and hENT2. hCNT3 also exhibits broad selectivity toward nucleoside derivatives (Hu et al., 2006; Errasti-Murugarren et al., 2007). hCNT3 may be the most effective nucleoside transporter identified to date because it accepts both purines and pyrimidines, with affinities higher than those of the equilibrative transporters. In contrast, hCNT2 can translocate the adenosine analog clofarabine but does not take up fludarabine, even though

TABLE 11.2 Pharmacological Profiles of Nucleoside Transporters for Antitumoral Drugs

Antitumoral drugs	CNT1	CNT2	CNT3	ENT1	ENT2	ENT3
Cytarabine	S	NT	S	S	S	
Gemcitabine	S (24 μM)	NT	S (60 μM)	S (160 μM)	S (740 μM)	S
5-azacytidine	S (63 μM)			S		
Zebularine			S			
5-fluoro-5′-deoxyuridine	S (209 μM)	S	S	S	S	
5-fluoro-2′-deoxyuridine	S	S	S			
Fludarabine	NT	NT[a]	S	S (107 μM)	S	
Cladribine	NT	S	S	S (23 μM)	S	
Clofarabine	NT	S (81 μM)	S (52 μM)	S (108 μM)	S (328 μM)	
Forodesine				S	S	
AraG				S	S	
Mizoribine	S	S[b]		S	S	
6-mercaptopurine			S		S	
6-thioguanine			S		S	

S: substrate (when it is known, apparent Km values are given). S[b]: data obtained in rat ortholog. NT: not substrate. NT[a]: not substrate but inhibitor the chemical structures of the two drugs are similar (Lang et al., 2001; King et al., 2006). Moreover, human CNT2 transports cladribine only poorly, whereas rat CNT2 is ten-fold more effective in this respect (Owen et al., 2006a). The pharmacology of hENT3, in terms of anticancer drug transport, has not been well studied. Only gemcitabine has been shown to be transported by hENT3, although some antiviral drugs are also substrates of this protein (Govindarajan et al., 2009). hENT2 and hCNT3 can translocate the nucleobase derivatives 6-mercaptopurine and 6-thioguanine (Fotoohi et al., 2006). However, 5-fluorouracil is not taken up by any of the nucleoside transporters discussed above. Whether 5-fluorouracil is transported by a carrier protein, or enters the cell via passive diffusion, remains controversial (Hsu, 1982; Nakamura et al., 2003).

11.3.2 Antiviral Nucleoside Analogs

Some classical antiviral drugs employed in HIV therapy are nucleoside-derived reverse transcriptase inhibitors (NRTIs). These include the pyrimidine analogs zidovudine (AZT; azidothymidine), zalcitabine (ddC; 2′,3′-dideoxycytidine), stavudine (d4T; dideoxythymidine), and lamivudine (3tC; 2′,3′-dideoxy-3′-thiacytidine), and the purine derivative didanosine (ddI; 2′,3′-dideoxyinosine) (Tozzi, 2010). Other nucleoside analogs including ribavirin and clevudine are employed in the treatment of various viral infections (Fig. 11.4).

The antiviral drug AZT is translocated by hCNT1, hCNT2, hENT2, and hENT3, but, surprisingly, not by hENT1 (Yao et al., 2001; Cano-Soldado et al., 2004; Errasti-Murugarren et al., 2007; Govindarajan et al., 2009). However, AZT can inhibit hENT1-mediated transport. Similarly, ddC is not taken up by hCNT1 even though it is a pyrimidine-preferring transporter; this drug can, however, inhibit hCNT1-mediated transport (Cano-Soldado et al., 2004). Thus, some drugs bind to the transporter but are not translocated. This is also the case with the natural nucleoside adenosine, which can bind to hCNT1 with high affinity but is not translocated, thus blocking the ability of hCNT1 to transport other nucleosides (Larrayoz et al., 2004).

Table 11.3 illustrates the concept that nucleoside transporters, particularly hCNT1, generally transport NRTIs less effectively than antitumor drugs. The data, which are based on selectivity and specificity profiles, demonstrate that the apparent Km values are often higher when antiviral (compared to antitumor) nucleoside analogs are examined. For example, hCNT1 cannot transport the cytidine derivatives ddC or 3tC, but efficiently takes up gemcitabine and cytarabine. When the structures of antitumor and antiviral compounds are compared, the main difference is the lack of 3′-hydroxyl group of the sugar in the antiviral drugs. Analysis of structural determinants implicated in nucleoside uptake have demonstrated the importance of this hydroxyl group in substrate recognition (Zhang et al., 2003). This indicates that slight modifications in nucleoside structure can produce dramatic changes in transportability and may even switch drug recognition from the expected transporter to other carrier proteins. In fact, proteins of the SLC22 gene family can translocate some nucleoside-derived antiviral drugs (Errasti-Murugarren and Pastor-Anglada, 2010). Both organic anion transporters (OATs) and organic cation transporters (OCTs) take

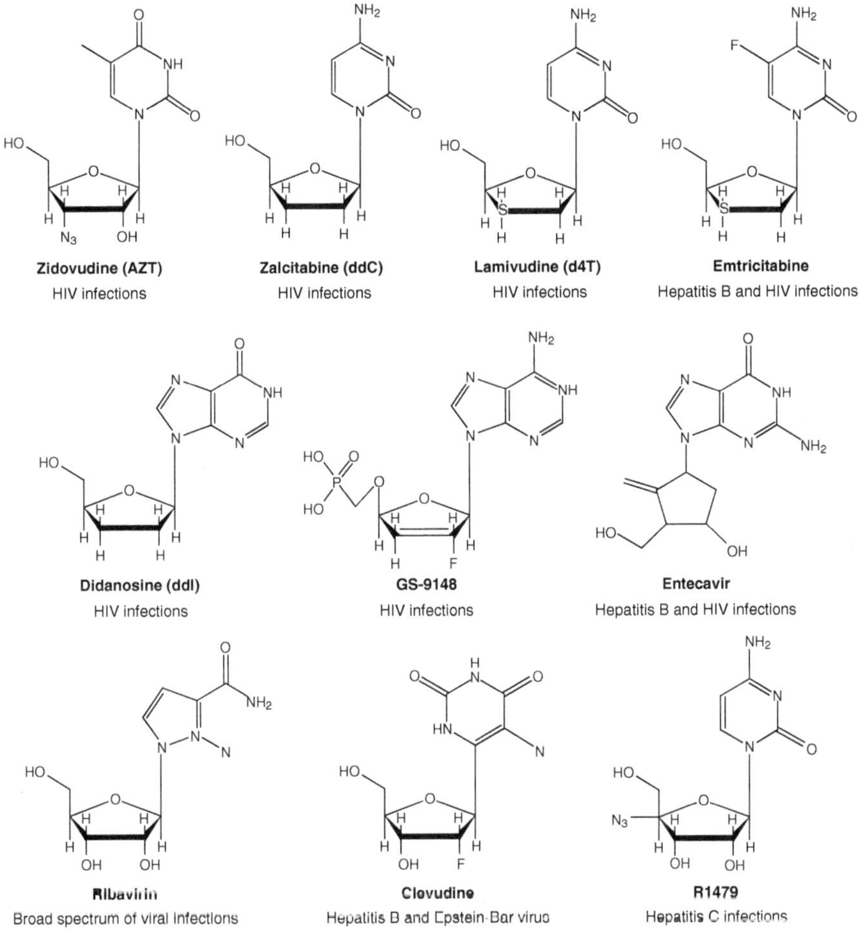

Figure 11.4 Molecular structures and current clinical uses of nucleoside-derived antiviral drugs.

TABLE 11.3 Pharmacological Profiles of Nucleoside Transporters for Antiviral Drugs

Antiviral drugs	CNT1	CNT2	CNT3	ENT1	ENT2	ENT3
Zidovudine (AZT)	S (450μM)	NT	S (310μM)	NT[a]	S	S
Zalcitabine (ddC)	NT	NT	S	S	S (>7,5 mM)	S
Didanosine (ddI)	NT	S	S	S	S (3 mM)	S
Stavudine (d4T)	S (15,6 mM)	NT	S	NT		S
Lamivudine (3tC)	NT	NT		NT		S
Ribavirin	NT	S (81μM)	S (52μM)	S (160μM)	S (328μM)	S
Clevudine				S	S	

S: substrate (when it is known, apparent Km values are given). NT: not substrate. NT[a]: not substrate but inhibitor up AZT and, in some instances, the apparent Km values are in the low micromolar range (Takeda et al., 2002). Thus, members of this transporter family are also good candidates to mediate nucleoside-derived antiviral drug uptake. The SLC22 gene family is reviewed in Chapter 9.

Other well-known non-NRTI antiviral drugs, such as ribavirin, are nucleoside derivatives. Ribavirin is a guanosine analog used to treat a broad spectrum of viral infections (Deval, 2009). The compound has the 3′-hydroxyl group and, in contrast to NRTI drugs, is taken up by most nucleoside transporters, albeit with relatively low affinity (Govindarajan et al., 2008; Yamamoto et al., 2010).

In summary, nucleoside transporters belonging to SLC28 and SLC29 gene families have been implicated in the uptake of most nucleoside-derived drugs, although transport selectivity and affinity vary upon chemical modification of the compounds. For this reason, the expression levels of the relevant transporters are important not only when chemosensitivity is considered, but also in the context of drug pharmacokinetics.

11.4 TISSUE DISTRIBUTION AND REGULATION; A CLUE TO THE UNDERSTANDING OF NUCLEOSIDE-DERIVED DRUG PHARMACOKINETICS

CNT and ENT proteins are expressed in all epithelial barriers that play major roles in drug pharmacokinetics, including the intestinal and renal resorptive epithelia, and hepatocytes, among many others (Fig. 11.5). Protein localization within the intestinal and renal epithelial barriers appears to be very asymmetrical. CNT-type proteins are found mostly, if not exclusively, in the apical membrane, whereas ENT-type proteins are largely concentrated basolaterally (Mangravite et al., 2001; Mangravite and Giacomini, 2003; Pastor-Anglada et al., 2007). This distribution in mature enterocytes and renal tubule epithelial cells might permit vectorial flux of nucleoside-derived drugs across such epithelial barriers (Lai et al., 2002; Errasti-Murugarren et al., 2007). Evidence supporting the view that differentiation promotes the expression of CNT-type proteins comes from both *in vitro* and *in vivo* studies. When the intestinal crypt-derived epithelial cell line IEC-6 was grown in the presence of glucocorticoids such as dexamethasone, which is known to promote maturation of enterocytes, CNT-related expression levels were upregulated (Aymerich et al., 2004). Also, hCNT1 and hCNT2 are present at low or even negligible levels in the crypts of human duodenum biopsy samples, but are easily detectable in mature enterocytes (Govindarajan et al., 2007). On the other hand, ENT-type proteins, particularly ENT1, may be relevant for proliferation. Pharmacological inhibition of ENT1 using NBTI inhibits MCSF-induced proliferation of murine bone marrow macrophages, an event that is associated with a significant decrease in the incorporation of extracellular tritiated thymidine into DNA (Soler et al., 2001a). Macrophages also express both CNT1 and CNT2. However, neither transporter is regulated by cell proliferation but rather by cell activation triggered by either TNF-α or LPS (Soler et al., 2001b). Also using the intestinal epithelial cell line IEC-6, it was observed that addition of either TGF-α or EGF, both of which are known to promote proliferation of immature intestinal cells,

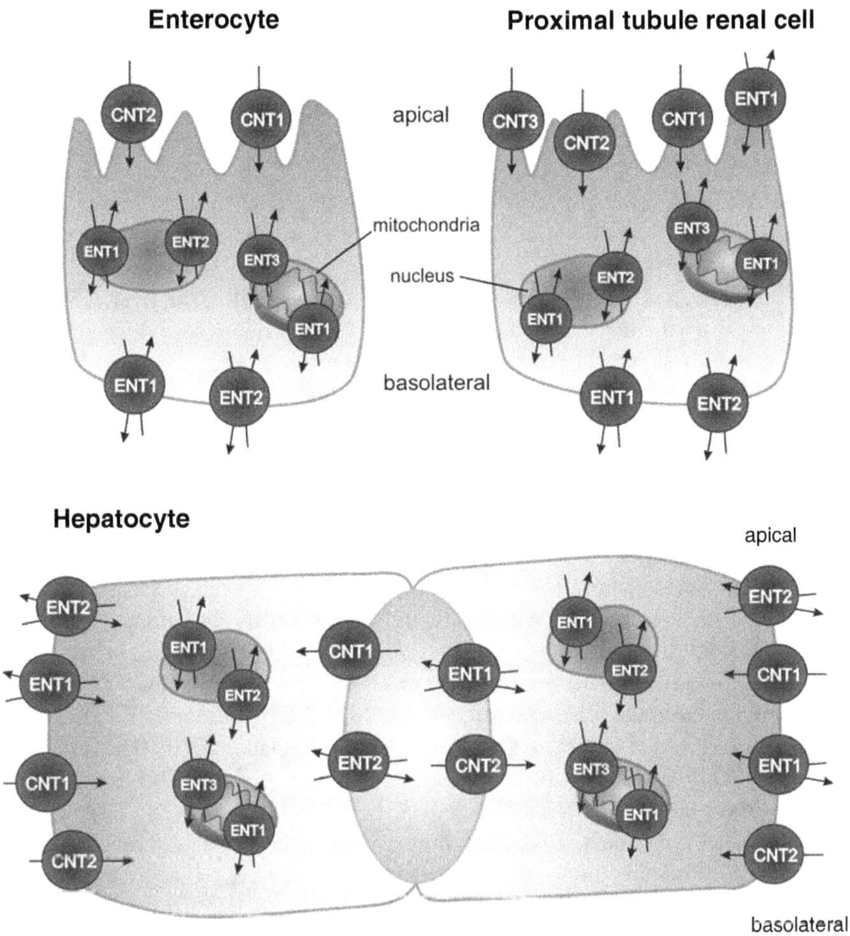

Figure 11.5 Subcellular localization of nucleoside transporters in cell types implicated in nucleoside-derived drug pharmacokinetics.

resulted in upregulation of ENT-type protein expression, particularly that of ENT1, and this was mimicked by cell wounding (Aymerich et al., 2004). Unlike the situation with the hCNTs, both hENT1 and hENT2 are much more abundant in crypt cells than in differentiated enterocytes of the human duodenum (Govindarajan et al., 2007).

Interestingly, NT distribution in apical and basolateral membrane domains may differ among epithelial barriers. Hepatocytes, which are assumed to be both consumers and producers of nucleosides, show a less asymmetrical disposition of CNT and ENT proteins in the basal (sinusoidal) and apical (canalicular) membrane domains (Fig. 11.5). This has been assessed for both rat and human hepatocytes (Duflot et al., 2002; Govindarajan et al., 2008). hCNT1, hCNT2, and hENT1 appear to be located at both poles, thus anticipating the need for nucleoside export and recovery mechanisms at the sinusoidal and canalicular faces of such cells. Moreover, the basolateral (sinusoidal) localization of hENT1 and hCNT2 evident in human hepatocytes supports the view that these membrane proteins may mediate the uptake of particular antiviral drugs used to treat patients infected with hepatitis C or B (Govindarajan et al., 2008).

When other relevant barriers such as that of placental syncytiotrophoblasts and the blood–brain barrier (BBB) were studied, CNT isoform-specific expression was also evident. CNT2 appears to be the major (if not the only) CNT-type protein expressed in the BBB (Li et al., 2001; Cansev, 2006), whereas preliminary data suggest that hCNT1 is the only hCNT-type protein expressed in the at-term syncytiotrophoblast (Errasti-Murugarren and Pastor-Anglada, 2010). Considering that the proteins under discussion differ in substrate selectivity profiles, the overall impact of variations in tissue distribution on drug bioavailability might also depend on the chemical nature of the nucleoside analog administered.

CNT proteins were initially thought to be expressed principally in epithelia, being their expression highly dependent upon differentiation, as reviewed above. This suggests that CNT downregulation in solid tumors of epithelial origin (see below) might influence nucleoside-derived drug bioavailability and action. Nevertheless, CNTs are expressed in other nonepithelial cell types, including murine bone marrow macrophages (Soler et al., 2001a, 2001b) and B cell-derived lines (Soler et al., 1998, 2000). In fact, primary chronic lymphocytic leukemia (CLL) cells express hCNT2, hCNT3, hENT1, and hENT2. However, although hENT- and hCNT2-related activities are often detected in primary CLL cells, hCNT3-related function is either very low or negligible in such cells (Molina-Arcas et al., 2003). Interestingly, hCNT3 activity can be upregulated in the CLL-derived cell line MEC1 by treatment with all-*trans*-retinoic acid (ATRA) (Fernandez-Calotti and Pastor-Anglada, 2010). This effect is mediated by TGF-β1 and involves trafficking of intracellular hCNT3-containing vesicles to the plasma membrane. Moreover, both hENTs and hCNT2 are abundant in primary T lymphocytes, whereas monocytes, monocyte-derived macrophages (MDMs), and immature monocyte-derived dendritic cells express predominantly hCNT3 (Minuesa et al., 2008). The possibility that hCNT proteins, together with the more widely expressed hENT family members, might contribute to hematotoxicity in patients undergoing genotoxic nucleoside-derived drug-based therapies needs to be further addressed. To date, no detailed information on the NT expression profiles of hematopoietic progenitor cells is available.

In summary, the tissue distribution of NT proteins is sufficiently broad to anticipate that these membrane proteins may play major roles in nucleoside-derived drug pharmacokinetics, therapeutic efficacy, and adverse drug reactions.

11.5 GENETIC VARIABILITY AND FUNCTIONAL CONSEQUENCES

Neither the SLC28 nor SLC29 gene family appears to be highly polymorphic in humans (Errasti-Murugarren and Pastor-Anglada, 2010), compared to other gene families, principally the ABC superfamily covered elsewhere in this book. Nevertheless, although particular polymorphisms may be relevant in terms of function, there still is a need for a better knowledge of their clinical relevance (Table 11.4).

TABLE 11.4 Reported SNPs for Nucleoside Transporters with Functional and/or Clinical Repercussions

Gene	Nucleoside position and change	Aminoacid change	Position	Functional consequence	Clinical relevance
SLC28A1	565 G>A	V189I	TM4	Increase Thy uptake	Higher bioavailability of mizoribine in kidney transplants
(hCNT1)	568 G>T	A190S	TM4	Increase Thy uptake	
	689 C>T	A230V	Loop 5–6	Increase Thy uptake	
	709 C>A	Q237K	Loop 5–6	Increase Thy uptake	
	1153 G>del	V385del	TM9	Nonfunctional	
	1561 G>A	D521N	Loop 11–12	Increase Thy uptake	Higher risk of hematologic toxicity in patients treated with gemcitabine
	1636 T>C	S546P	TM12	Nonfunctional	
	1903 C>G	L635V	C-terminal	Decrease Thy uptake	
SLC28A2	65 C>T	P22L	N-terminal	No effect on Guo uptake	More favorable prognosis in lung carcinoma patients treated with gemcitabine
(hCNT2)	225 C>A	S75R	N-terminal	Increase Guo uptake	
	1064 T>C	F355S	TM8	Decrease Guo uptake	
	1153 G>A	E385K	TM9	Decrease Ino, Urd and ribavirine uptake	
	1835 T>C	M612T	C-terminal	Decrease Ino and Urd uptake	
	−115 T>G	–		Decrease promoter activity	
	−146 T>A	–		Increase promoter activity	
	−564 G>A	–		Decrease promoter activity	
	−861 A>C	–		Decrease promoter activity	
	−880 T>C	–		Decrease promoter activity	
	−960 C>T	–		Decrease promoter activity	

(continued)

TABLE 11.4 (*Continued*)

Gene	Nucleoside position and change	Aminoacid change	Position	Functional consequence	Clinical relevance
SLC28A3	267 A>G	T89T	N-terminal		Association with neutropenia toxicity in patients treated with gemcitabine
(hCNT3)	1099 G>A	G367R	TM8	Decrease Thy and Ino uptake	
	1383 C>T	L461L	TM10		Overall survival in patients treated with gemcitabine[b]
	1804 T>C	C602R	TM13	Decrease uptake	
SLC29A1	647 T>C	I216T	TM6		Increased risk of alcohol withdrawal-induced seizures
(hENT1)	−1050 G>A	–		Increase promoter activity	
	0	–		Increase promoter activity	
	IVS12 −210 A>G	–			Progression-free survival in patients treated with gemcitabine[a]
	IVS2 −549T>C	–			Overall survival in patients treated with gemcitabine[b]
	IVS2 +913C>T	–			Grade 3–4 neutropenia and overall survival[b] in patients treated with gemcitabine

SLC29A2	13 G>T	D5Y	N-terminal	Decrease uptake	
(hENT2)	93 C>A	N68K	Loop 1–2	Increase Guo uptake	
	551–556 del	S184M/ deletes G185 and V186	TM5	Decrease Urd, Guo and Ino uptake	
	845–846 del	Frameshift	Truncated from loop 6–7	Nonfunctional	
SLC29A3	1045 delC	Frameshift	Truncated from TM8	Almost complete reduction in Ado uptake. Mistargeting to Golgi	
(hENT3)	1087 C>T	R363W	Loop 8–9		
	1083 G>A	R363Q	Loop 8–9		H syndrome
	1279 G>A	G427S	TM19	Almost complete reduction in Ado uptake	
	1309 G>A	G437R	TM10	Reduction in Ado uptake. Accelerated protein turnover	
	347 T>G	M116R	TM2	Almost complete reduction in Ado uptake. Mistargeting to ÉR	
	940 T>del	Frameshift	Truncated from TM7	Almost complete reduction in Ado uptake. Mistargeting to Golgi	Pigmented hypertrichosis and insulin-dependent diabetes mellitus
	1330 G>T	E444X	Loop 10–11	Reduction in Ado uptake. Accelerated protein turnover	
	1346 C>G	T449R	Loop 10–11	Reduction in Ado uptake	

[a]Combined genotype effect of CDA A-76C, RRM1 A33G, RRM1 C-27A and hENT1 A-201G.

[b]Combined genotype effect of CDA A-76C, dCK C-1205T, DCTD T-47C, hCNT3 1383 C<T, hENT1 T-549C and hENT1 C913T

References: *SLC28A1* (Gray et al., 2004a; Soo et al., 2009; Naito et al., 2010); *SLC28A2* (Owen et al., 2005; Li et al., 2007, 2009; Soo et al., 2009; Yee et al., 2009); *SLC28A3* (Badagnani et al., 2005; Errasti-Murugarren et al., 2008; Okazaki et al., 2010): *SLC29A1* (Myers et al., 2006; Okazaki et al., 2010; Tanaka et al., 2010; Kim et al., 2011); *SLC29A2* (Owen et al., 2006b); *SLC29A3* (Molho-Pessach et al., 2008; Cliffe et al., 2009)

In principle, and considering tissue distribution patterns, it might be expected that NT variants affecting transport activity are likely to influence drug resorption, thus in turn affecting drug disposition.

Nearly 60 SNPs of the *SLC28A1* gene have been identified to date, but only 15 alter the amino acid sequence (13 are nonsynonymous changes; one amino acid insertion and one single base-pair deletion have also been noted). All but two of the mutant proteins, hCNT1-Ser546Pro and hCNT1-1153del, retain functional activity (Gray et al., 2004a) (Table 11.4). The former mutation affects a highly conserved Ser residue, whereas the latter may result in the production of a truncated protein. Interestingly, some variants (hCNT1+140Val, hCNT1-Val189Ile, hCNT1-Gln237Lys, and hCNT1-Asp521Asn) had allele frequencies ≥20% in the population studied (Gray et al., 2004a). Moreover, several variants appear to show some gain-of-function, based upon the thymidine uptake values of *Xenopus laevis* oocytes expressing the relevant proteins. Presence of the hCNT1-Val189Ile is associated with increased thymidine uptake but also with a significantly lower affinity for the nucleoside analog gemcitabine (Gray et al., 2004a). This hCNT1 variant has also been linked to higher bioavailability of mizoribine in kidney transplant recipients (Naito et al., 2010) Another *SLC28A1* variant, hCNT1-Asp521Asn, has been shown to be associated with the occurrence of neutropenia and thrombocytopenia in nonsmall cell lung cancer (NSCLC) patients receiving gemcitabine therapy, thus providing the first indication that the hCNT1 protein may be involved in the development of hematotoxicity and adverse drug reactions (ADRs) (Soo et al., 2009).

About 10 nonsynonymous variants of *SLC28A2* (encoding hCNT2) have been identified. All retain function, although some are associated with either increased (hCNT2-Ser75Arg) or lowered (hCNT2-Phe355Ser, hCNT2-Glu385Lys, and hCNT2-Met612Thr) biological activity (Owen et al., 2005; Li et al., 2007) (Table 11.4). To date, no strong clinical correlation between mutation and altered function has been found for any variant. However, the hCNT2-Pro22Leu mutant, which apparently displays normal biological activity, has been associated with a more favorable prognosis of NSCLC patients undergoing gemcitabine chemotherapy (Soo et al., 2009).

Importantly, several nucleotide changes in a noncoding putative promoter region of the *SLC28A2* gene have been shown to alter basal transcriptional activity (Li et al., 2009; Yee et al., 2009). Although the clinical relevance of these variants remains to be fully assessed, it is tempting to speculate that polymorphisms affecting *SLC28A2* gene transcription would indirectly impact hCNT2 protein expression levels, thus in turn influencing drug pharmacokinetics and bioavailability. Such considerations may be of particular relevance when considering, for example, ribavirin uptake by hepatocytes, which is mediated by hCNT2.

Very few polymorphic variants of the *SLC28A3* gene (encoding hCNT3) bearing functional and/or clinical implications have been identified to date (Table 11.4). Indeed, it has even been suggested that this transporter does not tolerate nonsynonymous changes; the implication is that any change critically compromises "fitness" in humans (Badagnani et al., 2005). However, two variants, hCNT3-Gly367Arg and hCNG3-Cys602Arg have low biological activities (Badagnani et al., 2005; Errasti-Murugarren et al., 2008). The latter mutation was identified in a Spanish population, with an allele frequency of 1% and is a paradigm of functional complexity among polymorphic transporter variants. The hCNT3-Cys602Arg variant is associated with low-level transport activity, likely because one of the two Na^+ ions required for nucleoside translocation cannot properly bind to the transporter protein (Errasti-Murugarren et al., 2008). Moreover, the hCNT3-Cys602Arg protein shows missorting to nonlipid raft fractions, wherein the transporter appears to be less active than when the appropriate location is attained. This differentially affects the selectivity of nucleoside-derived drugs (Errasti-Murugarren et al., 2010). Very recently, the variant *SLC28A3* haplotype rs10868138G/rs56350726T has been associated with protection from hemolytic anemia in hepatitis C patients under ribavirin treatment (Doehring et al., 2011). However, this finding is difficult to interpret from a mechanistic viewpoint, because erythrocytes express only hENT1 and this transporter contributes significantly to ribavirin uptake by red blood cells (Endres et al., 2009).

Genetic variation within the SLC29 gene family may be relevant in terms of drug pharmacokinetics and action because such transporter proteins are widely but differentially expressed; this is particularly true of hENT1 (encoded by *SLC29A1*) in normal and tumor cells. Heterogeneity in hENT1 expression and its probable relationship with outcomes of cancer treatment is of clinical interest, and will be briefly discussed below. Although gene expression levels may be highly dependent on tumor microenvironment (i.e., hENT1 expression is sensitive to hypoxia), it is tempting to speculate that particular polymorphic variants in the *SLC29A1* promoter region (Myers et al., 2006) (Table 11.4) might contribute to inter-individual heterogeneity in basal hENT1 expression levels.

Four nonsynonymous variants of *SLC29A2*, altering hENT2-related functions, have been identified to date. All but one mutant yield active proteins, although the biological activities vary (Table 11.4). No correlation between any known *SLC29A2* variant and clinical outcome has yet been reported. Turning to *SLC29A3*, it appears that genetic changes may be very important when seeking to understand the etiology of a variety of syndromes with which such changes have recently been associated. The relevant clinical conditions include H syndrome (Molho-Pessach et al., 2008) and the PHID syndrome (Spiegel et al., 2010). *SLC29A3* variation affects hENT3-related biological functions. However, all mutations of the gene studied to date are associated with rare diseases and any general relevance of polymorphic variants in *SLC29A3* remains to be addressed.

Finally, it should be mentioned that the structural determinants of substrate (drug)-transporter interaction in nucleoside transporter proteins have been quite well described (Cano-Soldado and Pastor-Anglada, 2012, for review), but any impact of a particular genetic variation on the transporter structure remains difficult to evaluate because no NT crystal structures are available. Although the structures of ENT-type proteins have been modeled in a preliminary manner (Valdes et al., 2009), the basic topology of CNT-type proteins remains controversial (Slugoski et al., 2009). Prokaryotic CNT orthologs such as NupC lack the intracellular N-terminal tail of typical CNT-type proteins, and also the first three transmembrane domains, but retain functionality (Loewen et al., 2004), thus supporting the view that these regions are not essential for activity but might contribute to provide regulatory properties to eukaryotic (i.e., human) transporters. Interestingly, a few genetic variants located in these domains appear to modify protein function. In any case, the understanding of how certain naturally occurring variations lead to structural changes, in turn altering biological functions, remains relatively poor.

11.6 CORRELATION OF NUCLEOSIDE TRANSPORTER PROTEIN EXPRESSION WITH SENSITIVITY TO NUCLEOSIDE-DERIVED DRUGS

It is evident that NT proteins are necessary for many nucleoside analogs to enter into the cells and exert their pharmacological action. Early *in vitro* studies showed that cells lacking selected transport functions were resistant to the action of nucleoside-derived analogs (Aran and Plagemann, 1992; Mackey et al., 1998). In contrast, expression of a high-affinity hCNT increased the sensitivity of cells to nucleoside-derived drugs (Lang et al., 2001; Mata et al., 2001). Thus, NT expression levels are important in terms of drug sensitivity. The real question is whether inter-individual variability in NT levels within tumors can result in individual heterogeneity in terms of response to drugs.

Several studies have explored NT expression levels in different tumors, and found heterogeneity therein (Pennycooke et al., 2001; Mackey et al., 2002). Immunohistochemical analysis of hENT1, hENT2, and hCNT1 levels in 300 gynecologic tumors revealed great variation in expression levels; a significant number of tumors did not express hCNT1 at all, but all tumors retained equilibrative transporter expression (Farre et al., 2004). In breast cancer samples, the levels of both hCNT1 and hENT1 were lower in tumor samples than in normal background tissue (Lane et al., 2010). Although the tumor microenvironment, the state of tissue inflammation within the disease context, and the phenotypic characteristics of individual patients affect transporter expression levels and activities, we cannot exclude the possibility that genetic changes in promoter regions, as summarized above, might also contribute to the observed variability. Indeed, NT-level heterogeneity has been found in normal tissues (Lane et al., 2010; Marvi et al., 2010), although different factors can contribute to this variability.

Work linking NT expression to drug sensitivity and the clinical outcomes of cancer patients was initially performed on patients with lymphoproliferative diseases. Several studies have shown that hENT1 expression level determines the extent of responsiveness to cytarabine treatment (Gati et al., 1997; Wright et al., 2002; Hubeek et al., 2005). In mantle cell lymphoma patients, hENT1 levels correlated with better therapeutic responses to gemcitabine (Marce et al., 2006). Most studies analyzing NT expression in solid tumors have appeared in the last 8 years, demonstrating the increasing importance attached to the roles played by such transporters in drug responsiveness. In pancreatic adenocarcinoma patients, the therapeutic response to gemcitabine was better in individuals with hENT1-expressing tumors than in those with tumors with undetectable or very low levels of hENT1 (Spratlin et al., 2004; Giovannetti et al., 2006; Marechal et al., 2009).

The role played by NTs in terms of drug responsiveness is not always as clear as described above. Indeed, some studies found that high-level expression of a particular NT isoform may be associated with poor patient prognosis. In CLL patients, elevated expression of hCNT3 resulted in a poor complete response rate to fludarabine therapy (Mackey et al., 2005). However, it is likely that hCNT3 is principally localized intracellularly, and recent data reveal that upregulation of hCNT3 trafficking into the plasma membrane of primary CLL cells improves fludarabine *ex vivo* sensitivity. High hCNT1 levels in breast cancer patients were inversely correlated with disease-free survival in those undergoing cyclophosphamide–methotrexate–5FU therapy. However, these drugs are not substrates for NTs and it is probable that, under such circumstances, high hCNT1 levels promoted nucleoside salvage within tumor cells, thus counteracting drug genotoxic actions (Gloeckner-Hofmann et al., 2006). In contrast, lower hCNT1 levels were correlated with poorer clinical responses to cladribine in patients with Wandentröm's macroglobulinemia and small lymphocytic leukemia (Rabascio et al., 2010). It is tempting to hypothesize that high expression of transporters responsible for drug uptake would lead to better antitumor responses. However, the negative correlations found in some studies are difficult to comprehend from a mechanistic viewpoint. Probably, the observed effects are not related to drug transport per se but rather reflect genetic and phenotypic changes occurring within tumors.

As reviewed above, some genetic changes in NT promoter regions might contribute to variability in NT expression levels and thus, in turn, to clinical responses, but this has not yet been demonstrated. On the other hand, some polymorphisms affect nucleoside-mediated uptake, and may therefore also influence nucleoside-derived drug translocation. Any possible role for such genetic alterations cannot be determined using the methodology of the studies reviewed above, because in most cases only protein and mRNA levels were measured. In CLL patients, inter-individual variability in hENT2 protein levels correlated directly with the transport activity of and the response to fludarabine (Molina-Arcas et al., 2003), demonstrating a direct correlation between protein (but not mRNA) levels, transporter activity, and drug responsiveness. In this example, the existence of polymorphisms affecting transport activity could explain the existence of some outlier patients.

11.7 FUTURE PERSPECTIVES

Several issues in the field must be addressed in the very near future. Nucleoside-derived drugs are not the exclusive substrates of SLC28 and SLC29 transporter proteins. In fact, as mentioned above, some antiviral drugs used in the treatment of AIDS patients also interact with members of the SLC22 gene family, some of which are highly polymorphic in humans. Also, drug metabolites (usually phosphorylated nucleosides) can be extruded from cells via the action of certain ABC transporter proteins, which, again, can be polymorphic. Overall, we would suggest that the application of a systems biology approach would aid in the better understanding of drug pharmacokinetics, drug action, and risk of ADR. This requires an integrated "transportome" analysis of all putative protein players. In terms of the two gene families covered in this chapter, we believe that consistent clinical linkages between protein expression and clinical outcomes are few in number, and that those associations that have been documented require further evaluation from a more mechanistic perspective. NT proteins exhibit a narrower substrate selectivity than do the other transporter proteins discussed in this book, and they may have been underestimated in many clinical assessments. However, NTs are key players in nucleoside salvage pathways and perform major roles in terms of provision of nucleic acid precursors, especially in cells that are proliferating rapidly (i.e., tumor cells). Thus, the expression levels of such proteins should be evaluated in patients receiving genotoxic therapies, regardless of whether the relevant drugs are nucleoside-based.

ACKNOWLEDGMENTS

The authors would like to thank the past and present members of the Regulation of Transport Systems (RST) research group at the University of Barcelona for contributing to the development of the field of nucleoside transporter biology. Funding from the following agencies and foundations is gratefully acknowledged: SAF2011-23660 (Ministerio de Ciencia e Innovación, MICINN, Spain), SGR2009-624 (Talencia, Generalitat de Catalunya, Spain), Project 36621/06 (Fundación para la Investigación y la Prevención del SIDA en España, FIPSE), Fundació la Marató de TV3, and Fundación Ramón Areces. M.M-A is funded by Cancer Research UK and Marie Curie Intra-european fellowship.

REFERENCES

Aran JM, Plagemann PG (1992) Nucleoside transport-deficient mutants of PK-15 pig kidney cell line. *Biochim Biophys Acta*, 1110, 51–58.

Aymerich I, Pastor-Anglada M, Casado FJ (2004) Long term endocrine regulation of nucleoside transporters in rat intestinal epithelial cells. *J Gen Physiol*, 124, 505–512.

Badagnani I, Chan W, Castro RA, Brett CM, Huang CC, Stryke D, Kawamoto M, Johns SJ, Ferrin TE, Carlson EJ, Burchard EG, Giacomini KM (2005) Functional analysis of genetic variants in the human concentrative nucleoside transporter 3 (CNT3; SLC28A3). *Pharmacogenomics J*, 5, 157–165.

Baldwin SA, Beal PR, Yao SY, King AE, Cass CE, Young JD (2004) The equilibrative nucleoside transporter family, SLC29. *Pflugers Arch*, 447, 735–743.

Baldwin SA, Yao SY, Hyde RJ, Ng AM, Foppolo S, Barnes K, Ritzel MW, Cass CE, Young JD (2005) Functional characterization of novel human and mouse equilibrative nucleoside transporters (hENT3 and mENT3) located in intracellular membranes. *J Biol Chem*, 280, 15880–15887.

Barnes K, Dobrzynski H, Foppolo S, Beal PR, Ismat F, Scullion ER, Sun L, Tellez J, Ritzel MW, Claycomb WC, et al. (2006) Distribution and functional characterization of equilibrative nucleoside transporter-4, a novel cardiac adenosine transporter activated at acidic pH. *Circ Res*, 99, 510–519.

Cano-Soldado P, Pastor-Anglada M (2012) Transporters that translocate nucleosides and structural similar drugs: structural requirements for substrate recognition. *Med Res Rev*, 32, 428–457.

Cano-Soldado P, Lorrayoz IM, Molina-Arcas M, Casado FJ, Martinez-Picado J, Lostao MP, Pastor-Anglada M (2004) Interaction of nucleoside inhibitors of HIV-1 reverse transcriptase with the concentrative nucleoside transporter-1 (SLC28A1). *Antivir Ther*, 9, 993–1002.

Cansev M (2006) Uridine and cytidine in the brain: their transport and utilization. *Brain Res Rev*, 52, 389–397.

Cliffe ST, Kramer JM, Hussain K, Robben JH, de Jong EK, de Brouwer AP, Nibbeling E, Kamsteeg EJ, Wong M, Prendiville J, et al. (2009) SLC29A3 gene is mutated in pigmented hypertrichosis with insulin-dependent diabetes mellitus syndrome and interacts with the insulin signaling pathway. *Hum Mol Genet*, 18, 2257–2265.

Damaraju VL, Smith KM, Mowles D, Nowak I, Karpinski E, Young JD, Robins MJ, Cass CE (2011) Interaction of fused-pyrimidine nucleoside analogs with human concentrative nucleoside transporters: high-affinity inhibitors of human concentrative nucleoside transporter 1. *Biochem Pharmacol*, 81, 82–90.

Deval J (2009) Antimicrobial strategies: inhibition of viral polymerases by 3′-hydroxyl nucleosides. *Drugs*, 69, 151–166.

Doehring A, Hofmann WP, Schlecker C, Zeuzem S, Sarrazin C, Berg T, Muller T, Herrmann E, Geisslinger G, Lotsch J (2011) Role of nucleoside transporters SLC28A2/3 and SLC29A1/2 genetics in ribavirin therapy: protection against anemia in patients with chronic hepatitis C. *Pharmacogenet Genomics*, 21, 289–296.

Duflot S, Calvo M, Casado FJ, Enrich C, Pastor-Anglada M (2002) Concentrative nucleoside transporter (rCNT1) is targeted to the apical membrane through the hepatic transcytotic pathway. *Exp Cell Res*, 281, 77–85.

Endres CJ, Moss AM, Govindarajan R, Choi DS, Unadkat JD (2009) The role of nucleoside transporters in the erythrocyte disposition and oral absorption of ribavirin in the wild-type and equilibrative nucleoside transporter 1-/- mice. *J Pharmacol Exp Ther*, 331, 287–296.

Errasti-Murugarren E, Cano-Soldado P, Pastor-Anglada M, Casado FJ (2008) Functional characterization of a nucleoside-derived drug transporter variant (hCNT3C602R) showing altered sodium-binding capacity. *Mol Pharmacol*, 73, 379–386.

Errasti-Murugarren E, Molina-Arcas M, Casado FJ, Pastor-Anglada M (2010) The human concentrative nucleoside transporter-3 C602R variant shows impaired sorting to lipid rafts and altered specificity for nucleoside-derived drugs. *Mol Pharmacol*, 78, 157–165.

Errasti-Murugarren E, Pastor-Anglada M (2010) Drug transporter pharmacogenetics in nucleoside-based therapies. *Pharmacogenomics*, 11, 809–841.

Errasti-Murugarren E, Pastor-Anglada M, Casado FJ (2007) Role of Cnt3 in the transepithelial flux of nucleosides and nucleoside-derived drugs. *J Physiol*, 582(Pt 3), 1249–1260.

Farre X, Guillen-Gomez E, Sanchez L, Hardisson D, Plaza Y, Lloberas J, Casado FJ, Palacios J, Pastor-Anglada M (2004) Expression of the nucleoside-derived drug transporters hCNT1, hENT1 and hENT2 in gynecologic tumors. *Int J Cancer*, 112, 959–966.

Fernandez-Calotti P, Pastor-Anglada M (2010) All-trans-retinoic acid promotes trafficking of human concentrative nucleoside transporter-3 (hCNT3) to the plasma membrane by a TGF-beta1-mediated mechanism. *J Biol Chem*, 285, 13589–13598.

Fotoohi AK, Lindqvist M, Peterson C, Albertioni F (2006) Involvement of the concentrative nucleoside transporter 3 and equilibrative nucleoside transporter 2 in the resistance of T-lymphoblastic cell lines to thiopurines. *Biochem Biophys Res Commun*, 343, 208–215.

Gati WP, Paterson AR, Larratt LM, Turner AR, Belch AR (1997) Sensitivity of acute leukemia cells to cytarabine is a correlate of cellular es nucleoside transporter site content measured by flow cytometry with SAENTA-fluorescein. *Blood*, 90, 346–353.

Giovannetti E, Del Tacca M, Mey V, Funel N, Nannizzi S, Ricci S, Orlandini C, Boggi U, Campani D, Del Chiaro M, et al. (2006) Transcription analysis of human equilibrative nucleoside transporter-1 predicts survival in pancreas cancer patients treated with gemcitabine. *Cancer Res*, 66, 3928–3935.

Gloeckner-Hofmann K, Guillen-Gomez E, Schmidtgen C, Porstmann R, Ziegler R, Stoss O, Casado FJ, Ruschoff J, Pastor-Anglada M (2006) Expression of the high-affinity fluoropyrimidine-preferring nucleoside transporter hCNT1 correlates with decreased disease-free survival in breast cancer. *Oncology*, 70, 238–244.

Govindarajan R, Bakken AH, Hudkins KL, Lai Y, Casado FJ, Pastor-Anglada M, Tse CM, Hayashi J, Unadkat JD (2007) In situ hybridization and immunolocalization of concentrative and equilibrative nucleoside transporters in the human intestine, liver, kidneys, and placenta. *Am J Physiol Regul Integr Comp Physiol*, 293, R1809–R1822.

Govindarajan R, Endres CJ, Whittington D, LeCluyse E, Pastor-Anglada M, Tse CM, Unadkat JD (2008) Expression and hepatobiliary transport characteristics of the concentrative and equilibrative nucleoside transporters in sandwich-cultured human hepatocytes. *Am J Physiol Gastrointest Liver Physiol*, 295, G570–G580.

Govindarajan R, Leung GP, Zhou M, Tse CM, Wang J, Unadkat JD (2009) Facilitated mitochondrial import of antiviral and anticancer nucleoside drugs by human equilibrative nucleoside transporter-3. *Am J Physiol Gastrointest Liver Physiol*, 296, G910–G922.

Gray JH, Mangravite LM, Owen RP, Urban TJ, Chan W, Carlson EJ, Huang CC, Kawamoto M, Johns SJ, Stryke D, Ferrin TE, Giacomini KM (2004a) Functional and genetic diversity in the concentrative nucleoside transporter, CNT1, in human populations. *Mol Pharmacol*, 65, 512–519.

Gray JH, Owen RP, Giacomini KM (2004b) The concentrative nucleoside transporter family, SLC28. *Pflugers Arch*, 447, 728–734.

Gupte A, Buolamwini JK (2009) Synthesis and biological evaluation of phloridzin analogs as human concentrative nucleoside transporter 3 (hCNT3) inhibitors. *Bioorg Med Chem Lett*, 19, 917–921.

Hamilton SR, Yao SY, Ingram JC, Hadden DA, Ritzel MW, Gallagher MP, Henderson PJ, Cass CE, Young JD, Baldwin SA (2001) Subcellular distribution and membrane topology of the mammalian concentrative Na+-nucleoside cotransporter rCNT1. *J Biol Chem*, 276, 27981–27988.

Hsu LS (1982) The kinetics of uptake of 5-fluorouracil by rat liver. *Br J Pharmacol*, 77, 413–417.

Hu H, Endres CJ, Chang C, Umapathy NS, Lee EW, Fei YJ, Itagaki S, Swaan PW, Ganapathy V, Unadkat JD (2006) Electrophysiological characterization and modeling of the structure activity relationship of the human concentrative nucleoside transporter 3 (hCNT3). *Mol Pharmacol*, 69, 1542–1553.

Hubeek I, Stam RW, Peters GJ, Broekhuizen R, Meijerink JP, van Wering ER, Gibson BE, Creutzig U, Zwaan CM, Cloos J, Kuik DJ, Pieters R, Kaspers GJ (2005) The human equilibrative nucleoside transporter 1 mediates in vitro cytarabine sensitivity in childhood acute myeloid leukaemia. *Br J Cancer*, 93, 1388–1394.

Hyde RJ, Cass CE, Young JD, Baldwin SA (2001) The ENT family of eukaryote nucleoside and nucleobase transporters: recent advances in the investigation of structure/function relationships and the identification of novel isoforms. *Mol Membr Biol*, 18, 53–63.

Johnson SA (2001) Nucleoside analogues in the treatment of haematological malignancies. *Expert Opin Pharmacother*, 2, 929–943.

Johnson ZL, Cheong CG, Lee SY (2012) Crystal structure of a concentrative nucleoside transporter from *Vibrio cholerae* at 2.4 Å. *Nature*, 483(7390), 489–93.

Kim JH, Karpyak VM, Biernacka JM, Nam HW, Lee MR, Preuss UW, Zill P, Yoon G, Colby C, Mrazek DA, Choi DS (2011) Functional role of the polymorphic 647 T/C variant of ENT1 (SLC29A1) and its association with alcohol withdrawal seizures. *PLoS One*, 6, e16331.

King K, Damaraju V, Vickers M, Yao S, Lang T, Tackaberry T, Mowles D, Ng A, Young J, Cass C (2006) A comparison of the transportability, and its role in cytotoxicity, of clofarabine, cladribine and fludarabine by recombinant human nucleoside transporters produced in three model expression systems. *Mol Pharmacol*, 69, 346–353.

Lai Y, Bakken AH, Unadkat JD (2002) Simultaneous expression of hCNT1-CFP and hENT1-YFP in Madin-Darby canine kidney cells. Localization and vectorial transport studies. *J Biol Chem*, 277, 37711–37717.

Lai Y, Tse CM, Unadkat JD (2004) Mitochondrial expression of the human equilibrative nucleoside transporter 1 (hENT1) results in enhanced mitochondrial toxicity of antiviral drugs. *J Biol Chem*, 279, 4490–4497.

Lane J, Martin TA, McGuigan C, Mason MD, Jiang WG (2010) The differential expression of hCNT1 and hENT1 i n breast cancer and the possible impact on breast cancer therapy. *J Exp Ther Oncol*, 8, 203–210.

Lang TT, Selner M, Young JD, Cass CE (2001) Acquisition of human concentrative nucleoside transporter 2 (hcnt2) activity by gene transfer confers sensitivity to fluoropyrimidine nucleosides in drug-resistant leukemia cells. *Mol Pharmacol*, 60, 1143–1152.

Larrayoz IM, Casado FJ, Pastor-Anglada M, Lostao MP (2004) Electrophysiological characterization of the human Na(+)/nucleoside cotransporter 1 (hCNT1) and role of adenosine on hCNT1 function. *J Biol Chem*, 279, 8999–9007.

Lee EW, Lai Y, Zhang H, Unadkat JD (2006) Identification of the mitochondrial targeting signal of the human equilibrative nucleoside transporter 1 (hENT1): implications for interspecies differences in mitochondrial toxicity of fialuridine. *J Biol Chem*, 281, 16700–16706.

Li JY, Boado RJ, Pardridge WM (2001) Cloned blood-brain barrier adenosine transporter is identical to the rat concentrative Na+ nucleoside cotransporter CNT2. *J Cereb Blood Flow Metab*, 21, 929–936.

Li L, Koo SH, Hong IH, Lee EJ (2009) Identification of functional promoter haplotypes of human concentrative nucleoside transporter 2, hCNT2 (SLC28A2). *Drug Metab Pharmacokinet*, 24, 161–166.

Li L, Tan CM, Koo SH, Chong KT, Lee EJ (2007) Identification and functional analysis of variants in the human concentrative nucleoside transporter 2, hCNT2 (SLC28A2) in Chinese, Malays and Indians. *Pharmacogenet Genomics*, 17, 783–786.

Loewen SK, Ng AM, Yao SY, Cass CE, Baldwin SA, Young JD (1999) Identification of amino acid residues responsible for the pyrimidine and purine nucleoside specificities of human concentrative Na(+) nucleoside cotransporters hCNT1 and hCNT2. *J Biol Chem*, 274, 24475–24484.

Loewen SK, Yao SY, Slugoski MD, Mohabir NN, Turner RJ, Mackey JR, Weiner JH, Gallagher MP, Henderson PJ, Baldwin SA, Cass CE, Young JD (2004) Transport of physiological nucleosides and anti-viral and anti-neoplastic nucleoside drugs by recombinant *Escherichia coli* nucleoside-H(+) cotransporter (NupC) produced in *Xenopus laevis* oocytes. *Mol Membr Biol*, 21, 1–10.

Longley DB, Harkin DP, Johnston PG (2003) 5-fluorouracil: mechanisms of action and clinical strategies. *Nat Rev Cancer*, 3, 330–338.

Mackey JR, Galmarini CM, Graham KA, Joy AA, Delmer A, Dabbagh L, Glubrecht D, Jewell LD, Lai R, Lang T, Hanson J, Young JD, Merle-Béral H, Binet JL, Cass CE, Dumontet C (2005) Quantitative analysis of nucleoside transporter and metabolism gene expression in chronic lymphocytic leukemia (CLL): identification of fludarabine-sensitive and -insensitive populations. *Blood*, 105, 767–774.

Mackey JR, Jennings LL, Clarke ML, Santos CL, Dabbagh L, Vsianska M, Koski SL, Coupland RW, Baldwin SA, Young JD, Cass CE (2002) Immunohistochemical variation of human equilibrative nucleoside transporter 1 protein in primary breast cancers. *Clin Cancer Res*, 8, 110–116.

Mackey JR, Mani RS, Selner M, Mowles D, Young JD, Belt JA, Crawford CR, Cass CE (1998) Functional nucleoside transporters are required for gemcitabine influx and manifestation of toxicity in cancer cell lines. *Cancer Res*, 58, 4349–4357.

Mangravite LM, Giacomini KM (2003) Sorting of rat SPNT in renal epithelium is independent of N-glycosylation. *Pharm Res*, 20, 319–323.

Mangravite LM, Lipschutz JH, Mostov KE, Giacomini KM (2001) Localization of GFP-tagged concentrative nucleoside transporters in a renal polarized epithelial cell line. *Am J Physiol Renal Physiol*, 280, F879–F885.

Mani RS, Hammond JR, Marjan JM, Graham KA, Young JD, Baldwin SA, Cass CE (1998) Demonstration of equilibrative nucleoside transporters (hENT1 and hENT2) in nuclear envelopes of cultured human choriocarcinoma (BeWo) cells by functional reconstitution in proteoliposomes. *J Biol Chem*, 273, 30818–30825.

Marce S, Molina-Arcas M, Villamor N, Casado FJ, Campo E, Pastor-Anglada M, Colomer D (2006) Expression of human equilibrative nucleoside transporter 1 (hENT1) and its correlation with gemcitabine uptake and cytotoxicity in mantle cell lymphoma. *Haematologica*, 91, 895–902.

Marechal R, Mackey JR, Lai R, Demetter P, Peeters M, Polus M, Cass CE, Young J, Salmon I, Deviere J, Van Laethem JL (2009) Human equilibrative nucleoside transporter 1 and human concentrative nucleoside transporter 3 predict survival after adjuvant gemcitabine therapy in resected pancreatic adenocarcinoma. *Clin Cancer Res*, 15, 2913–2919.

Marvi M, Rose JB, Bang A, Moon BC, Pozeg Z, Ibrahim M, Peniston C, Coe IR (2010) Nucleoside transporter expression profiles in human cardiac tissue show striking individual variability with overall predominance of hENT1. *Eur J Pharm Sci*, 41, 685–691.

Mata JF, Garcia-Manteiga JM, Lostao MP, Fernandez-Veledo S, Guillen-Gomez E, Larrayoz IM, Lloberas J, Casado FJ, Pastor-Anglada M (2001) Role of the human concentrative nucleoside transporter (hCNT1) in the cytotoxic action of 5[Prime]-deoxy-5-fluorouridine, an active intermediate metabolite of capecitabine, a novel oral anticancer drug. *Mol Pharmacol*, 59, 1542–1548.

Minuesa G, Purcet S, Erkizia I, Molina-Arcas M, Bofill M, Izquierdo-Useros N, Casado FJ, Clotet B, Pastor-Anglada M, Martinez-Picado J (2008) Expression and functionality of anti-human immunodeficiency virus and anticancer drug uptake transporters in immune cells. *J Pharmacol Exp Ther*, 324, 558–567.

Molho-Pessach V, Lerer I, Abeliovich D, Agha Z, Abu Libdeh A, Broshtilova V, Elpeleg O, Zlotogorski A (2008) The H syndrome is caused by mutations in the nucleoside transporter hENT3. *Am J Hum Genet*, 83, 529–534.

Molina-Arcas M, Bellosillo B, Casado FJ, Montserrat E, Gil J, Colomer D, Pastor-Anglada M (2003) Fludarabine uptake mechanisms in B-cell chronic lymphocytic leukemia. *Blood*, 101, 2328–2334.

Myers SN, Goyal RK, Roy JD, Fairfull LD, Wilson JW, Ferrell RE (2006) Functional single nucleotide polymorphism haplotypes in the human equilibrative nucleoside transporter 1. *Pharmacogenet Genomics*, 16, 315–320.

Naito T, Tokashiki S, Mino Y, Otsuka A, Ozono S, Kagawa Y, Kawakami J (2010) Impact of concentrative nucleoside transporter 1 gene polymorphism on oral bioavailability of mizoribine in stable kidney transplant recipients. *Basic Clin Pharmacol Toxicol*, 106, 310–316.

Nakamura J, Horimoto T, Hirayama R, Mukai T, Nakashima M, Sasaki H, Nishida K (2003) Effect of the absorption enhancer saponin on the intrarenal distribution of 5-fluorouracil following its kidney surface application in rats. *Biol Pharm Bull*, 26, 1049–1051.

Okazaki T, Javle M, Tanaka M, Abbruzzese JL, Li D (2010) Single nucleotide polymorphisms of gemcitabine metabolic genes and pancreatic cancer survival and drug toxicity. *Clin Cancer Res*, 16, 320–329.

Osses N, Pearson JD, Yudilevich DL, Jarvis SM (1996) Hypoxanthine enters human vascular endothelial cells (ECV 304) via the nitrobenzylthioinosine-insensitive equilibrative nucleoside transporter. *Biochem J*, 317, 843–848.

Owen RP, Badagnani I, Giacomini KM (2006a) Molecular determinants of specificity for synthetic nucleoside analogs in the concentrative nucleoside transporter, CNT2. *J Biol Chem*, 281, 26675–26682.

Owen RP, Gray JH, Taylor TR, Carlson EJ, Huang CC, Kawamoto M, Johns SJ, Stryke D, Ferrin TE, Giacomini KM (2005) Genetic analysis and functional characterization of polymorphisms in the human concentrative nucleoside transporter, CNT2. *Pharmacogenet Genomics*, 15, 83–90.

Owen RP, Lagpacan LL, Taylor TR, De La Cruz M, Huang CC, Kawamoto M, Johns SJ, Stryke D, Ferrin TE, Giacomini KM (2006b) Functional characterization and haplotype analysis of polymorphisms in the human equilibrative nucleoside transporter, ENT2. *Drug Metab Dispos*, 34, 12–15.

Pastor-Anglada M, Errasti-Murugarren E, Aymerich I, Casado FJ (2007) Concentrative Nucleoside Transporters (CNTs) in epithelia: from absorption to cell signalling. *J Physiol Biochem*, 63, 97–110.

Pennycooke M, Chaudary N, Shuralyova I, Zhang Y, Coe IR (2001) Differential expression of human nucleoside transporters in normal and tumor tissue. *Biochem Biophys Res Commun*, 280, 951–959.

Rabascio C, Laszlo D, Andreola G, Saronni L, Radice D, Rigacci L, Fabbri A, Frigeri F, Calabrese L, Billio A, Bertolini F, Martinelli G (2010) Expression of the human concentrative nucleotide transporter 1 (hCNT1) gene correlates with clinical response in patients affected by Waldenstrom's Macroglobulinemia (WM) and small lymphocytic lymphoma (SLL) undergoing a combination treatment with 2-chloro-2′-deoxyadenosine (2-CdA) and Rituximab. *Leuk Res*, 34, 454–457.

Ritzel MW, Ng AM, Yao SY, Graham K, Loewen SK, Smith KM, Ritzel RG, Mowles DA, Carpenter P, Chen XZ, Karpinski E, Hyde RJ, Baldwin SA, Cass CE, Young JD (2001) Molecular identification and characterization of novel human and mouse concentrative Na^+-nucleoside cotransporter proteins (hCNT3 and mCNT3) broadly selective for purine and pyrimidine nucleosides (system cib). *J Biol Chem*, 276, 2914–2927.

Ritzel MW, Yao SY, Huang MY, Elliott JF, Cass CE, Young JD (1997) Molecular cloning and functional expression of cDNAs encoding a human Na+-nucleoside cotransporter (hCNT1). *Am J Physiol*, 272, C707–C714.

Ritzel MW, Yao SY, Ng AM, Mackey JR, Cass CE, Young JD (1998) Molecular cloning, functional expression and chromosomal localization of a cDNA encoding a human Na+/nucleoside cotransporter (hCNT2) selective for purine nucleosides and uridine. *Mol Membr Biol*, 15, 203–211.

Robak T, Korycka A, Lech-Maranda E, Robak P (2009) Current status of older and new purine nucleoside analogues in the treatment of lymphoproliferative diseases. *Molecules*, 14, 1183–1226.

Sahasranaman S, Howard D, Roy S (2008) Clinical pharmacology and pharmacogenetics of thiopurines. *Eur J Clin Pharmacol*, 64, 753–767.

Saiko P, Horvath Z, Jaeger W, Schott H, Fritzer-Szekeres M, Szekeres T (2005) Heterodinucleoside phosphates of 5-fluorodeoxyuridine and arabinofuranosylcytosine–new drugs in cancer chemotherapy? *In Vivo*, 19, 205–214.

SenGupta DJ, Unadkat JD (2004) Glycine 154 of the equilibrative nucleoside transporter, hENT1, is important for nucleoside transport and for conferring sensitivity to the inhibitors nitrobenzylthioinosine, dipyridamole, and dilazep. *Biochem Pharmacol*, 67, 453–458.

Slugoski MD, Loewen SK, Ng AM, Smith KM, Yao SY, Karpinski E, Cass CE, Baldwin SA, Young JD (2007) Specific mutations in transmembrane helix 8 of human concentrative Na+/nucleoside cotransporter hCNT1 affect permeant selectivity and cation coupling. *Biochemistry*, 46, 1684–1693.

Slugoski MD, Ng AM, Yao SY, Lin CC, Mulinta R, Cass CE, Baldwin SA, Young JD (2009) Substituted cysteine accessibility method analysis of human concentrative nucleoside transporter hCNT3 reveals a novel discontinuous region of functional importance within the CNT family motif (G/A)XKX3NEFVA(Y/M/F). *J Biol Chem*, 284, 17281–17292.

Smith KM, Ng AM, Yao SY, Labedz KA, Knaus EE, Wiebe LI, Cass CE, Baldwin SA, Chen XZ, Karpinski E, Young JD (2004) Electrophysiological characterization of a recombinant human Na+-coupled nucleoside transporter (hCNT1) produced in Xenopus oocytes. *J Physiol*, 558, 807–823.

Smith KM, Slugoski MD, Loewen SK, Ng AM, Yao SY, Chen XZ, Karpinski E, Cass CE, Baldwin SA, Young JD (2005) The broadly selective human Na+/nucleoside cotransporter (hCNT3) exhibits novel cation-coupled nucleoside transport characteristics. *J Biol Chem*, 280, 25436–25449.

Soler C, Felipe A, Casado FJ, Celada A, Pastor-Anglada M (2000) Nitric oxide regulates nucleoside transport in activated B lymphocytes. *J Leukoc Biol*, 67, 345–349.

Soler C, Felipe A, Mata JF, Casado FJ, Celada A, Pastor-Anglada M (1998) Regulation of nucleoside transport by lipopolysaccharide, phorbol esters, and tumor necrosis factor-alpha in human B-lymphocytes. *J Biol Chem*, 273, 26939–26945.

Soler C, Garcia-Manteiga J, Valdes R, Xaus J, Comalada M, Casado FJ, Pastor-Anglada M, Celada A, Felipe A (2001a) Macrophages require different nucleoside transport systems for proliferation and activation. *FASEB J*, 15, 1979–1988.

Soler C, Valdés R, Garcia-Manteiga J, Xaus J, Comalada M, Casado FJ, Modolell M, Nicholson B, MacLeod C, Felipe A, Celada A, Pastor-Anglada M (2001b) Lipopolysaccharide-induced apoptosis of macrophages determines the up-regulation of concentrative nucleoside transporters Cnt1 and Cnt2 through tumor necrosis factor-alpha-dependent and -independent mechanisms. *J Biol Chem*, 276, 30043–30049.

Soo RA, Wang LZ, Ng SS, Chong PY, Yong WP, Lee SC, Liu JJ, Choo TB, Tham LS, Lee HS, Goh BC, Soong R (2009) Distribution of gemcitabine pathway genotypes in ethnic Asians and their association with outcome in non-small cell lung cancer patients. *Lung Cancer*, 63, 121–127.

Spiegel R, Cliffe ST, Buckley MF, Crow YJ, Urquhart J, Horovitz Y, Tenenbaum-Rakover Y, Newman WG, Donnai D, Shalev SA (2010) Expanding the clinical spectrum of SLC29A3 gene defects. *Eur J Med Genet*, 53, 309–313.

Spratlin J, Sangha R, Glubrecht D, Dabbagh L, Young JD, Dumontet C, Cass C, Lai R, Mackey JR (2004) The absence of human equilibrative nucleoside transporter 1 is associated with reduced survival in patients with gemcitabine-treated pancreas adenocarcinoma. *Clin Cancer Res*, 10, 6956–6961.

Sundaram M, Yao SY, Ingram JC, Berry ZA, Abidi F, Cass CE, Baldwin SA, Young JD (2001a) Topology of a human equilibrative, nitrobenzylthioinosine (NBMPR)-sensitive nucleoside transporter (hENT1) implicated in the cellular uptake of adenosine and anti-cancer drugs. *J Biol Chem*, 276, 45270–45275.

Sundaram M, Yao SY, Ng AM, Cass CE, Baldwin SA, Young JD (2001b) Equilibrative nucleoside transporters: mapping regions of interaction for the substrate analogue nitrobenzylthioinosine (NBMPR) using rat chimeric proteins. *Biochemistry*, 40, 8146–8151.

Sundaram M, Yao SY, Ng AM, Griffiths M, Cass CE, Baldwin SA, Young JD (1998) Chimeric constructs between human and rat equilibrative nucleoside transporters (hENT1 and rENT1) reveal hENT1 structural domains interacting with coronary vasoactive drugs. *J Biol Chem*, 273, 21519–21525.

Takeda M, Khamdang S, Narikawa S, Kimura H, Kobayashi Y, Yamamoto T, Cha SH, Sekine T, Endou H (2002) Human organic anion transporters and human organic cation transporters mediate renal antiviral transport. *J Pharmacol Exp Ther*, 300, 918–924.

Tanaka M, Javle M, Dong X, Eng C, Abbruzzese JL, Li D (2010) Gemcitabine metabolic and transporter gene polymorphisms are associated with drug toxicity and efficacy in patients with locally advanced pancreatic cancer. *Cancer*, 116, 5325–5335.

Tozzi V (2010) Pharmacogenetics of antiretrovirals. *Antiviral Res*, 85, 190–200.

Valdes R, Arastu-Kapur S, Landfear SM, Shinde U (2009) An ab Initio structural model of a nucleoside permease predicts functionally important residues. *J Biol Chem*, 284, 19067–19076.

Vickers MF, Mani RS, Sundaram M, Hogue DL, Young JD, Baldwin SA, Cass CE (1999) Functional production and reconstitution of the human equilibrative nucleoside transporter (hENT1) in *Saccharomyces cerevisiae*. Interaction of inhibitors of nucleoside transport with recombinant hENT1 and a glycosylation-defective derivative (hENT1/N48Q). *Biochem J*, 339 (Pt 1), 21–32.

Visser F, Vickers MF, Ng AM, Baldwin SA, Young JD, Cass CE (2002) Mutation of residue 33 of human equilibrative nucleoside transporters 1 and 2 alters sensitivity to inhibition of transport by dilazep and dipyridamole. *J Biol Chem*, 277, 395–401.

Wang C, Pimple S, Buolamwini JK (2010) Interaction of benzopyranone derivatives and related compounds with human concentrative nucleoside transporters 1, 2 and 3 heterologously expressed in porcine PK15 nucleoside transporter deficient cells. Structure-activity relationships and determinants of transporter affinity and selectivity. *Biochem Pharmacol*, 79, 307–320.

Wang J, Giacomini KM (1997) Molecular determinants of substrate selectivity in Na^+-dependent nucleoside transporters. *J Biol Chem*, 272, 28845–28848.

Ward JL, Leung GP, Toan SV, Tse CM (2003) Functional analysis of site-directed glycosylation mutants of the human equilibrative nucleoside transporter-2. *Arch Biochem Biophys*, 411, 19–26.

Ward JL, Sherali A, Mo ZP, Tse CM (2000) Kinetic and pharmacological properties of cloned human equilibrative nucleoside transporters, ENT1 and ENT2, stably expressed in nucleoside transporter-deficient PK15 cells. Ent2 exhibits a low affinity for guanosine and cytidine but a high affinity for inosine. *J Biol Chem*, 275, 8375–8381.

Wong A, Soo RA, Yong WP, Innocenti F (2009) Clinical pharmacology and pharmacogenetics of gemcitabine. *Drug Metab Rev*, 41, 77–88.

Wright AM, Paterson AR, Sowa B, Akabutu JJ, Grundy PE, Gati WP (2002) Cytotoxicity of 2-chlorodeoxyadenosine and arabinosylcytosine in leukaemic lymphoblasts from paediatric patients: significance of cellular nucleoside transporter content. *Br J Haematol*, 116, 528–537.

Yamamoto T, Sugawara M, Kikukawa T, Miyauchi S, Yamaguchi M, Tero A, Takagi S, Nakagaki T (2010) Kinetic study of anti-viral ribavirin uptake mediated by hCNT3 and hENT1 in *Xenopus laevis* oocytes. *Biophys Chem*, 147, 59–65.

Yao SY, Ng AM, Sundaram M, Cass CE, Baldwin SA, Young JD (2001) Transport of antiviral 3′-deoxy-nucleoside drugs by recombinant human and rat equilibrative, nitrobenzylthioinosine (NBMPR)-insensitive (ENT2) nucleoside transporter proteins produced in Xenopus oocytes. *Mol Membr Biol*, 18, 161–167.

Yao SY, Ng AM, Vickers MF, Sundaram M, Cass CE, Baldwin SA, Young JD (2002) Functional and molecular characterization of nucleobase transport by recombinant human and rat equilibrative nucleoside transporters 1 and 2. Chimeric constructs reveal a role for the ENT2 helix 5–6 region in nucleobase translocation. *J Biol Chem*, 277, 24938–24948.

Yee SW, Shima JE, Hesselson S, Nguyen L, De Val S, Lafond RJ, Kawamoto M, Johns SJ, Stryke D, Kwok PY, et al. (2009) Identification and characterization of proximal promoter polymorphisms in the human concentrative nucleoside transporter 2 (SLC28A2). *J Pharmacol Exp Ther*, 328, 699–707.

Zhang J, Visser F, Vickers MF, Lang T, Robins MJ, Nielsen LP, Nowak I, Baldwin SA, Young JD, Cass CE (2003) Uridine binding motifs of human concentrative nucleoside transporters 1 and 3 produced in *Saccharomyces cerevisiae*. *Mol Pharmacol*, 64, 1512–1520.

CHAPTER **12**

P-GLYCOPROTEIN (MDR1/ABCB1)

Ingolf Cascorbi

12.1 INTRODUCTION

P-glycoprotein (P-gp) is the most prominent drug efflux transporter. It came into focus of the scientific community when it was identified as multidrug resistance gene (MDR1) in tumor cells conferring to the commonly known phenomenon of multidrug resistance against certain antineoplastic agents (Biedler and Riehm, 1970; Juranka et al., 1989). It belongs to the group of ATP-binding cassette (ABC) membrane transporters. According to the common nomenclature, it is termed ABCB1. Due to its important role as drug efflux pump, P-gp is one of the best characterized human efflux transporters with increasing understanding of its function, regulation, and impact of genetic variants.

Interestingly, absence of P-gp is comfortable with life. Within collies, a common dog breed in Europe and North America, a naturally occurring deletion polymorphism in ABCB1 leads to a frame shift with total loss of any protein activity (Mealey et al., 2001). Such dogs have a normal life until they are exposed to toxic compounds, being substrates of this export pump. A typical example is ivermectin, an antihelmintic drug potentially causing severe neurotoxic effects in homozygote carriers. The same was observed before in mice, being double knockout for the homologues *Abcb1a* and *Abcb1b* (Schinkel et al., 1997). The investigators did not find any physiological abnormalities or altered viability, but the pharmacokinetics of digoxin, a typical P-gp substrate was significantly changed. There was a decreased elimination and increased penetration into the brain, indicating the significant role of ABCB1 in transport across the blood–brain barrier. In the long run, such knockout mice can develop inflammatory bowel disease-like symptoms, similar to Crohn's disease (Panwala et al., 1998). In humans, however, no genetic variant of ABCB1 is known, being associated with absence of any activity underlining the importance of P-gp as one defense mechanism for the elimination of xenobiotics.

Pharmacogenomics of Human Drug Transporters: Clinical Impacts, First Edition.
Edited by Toshihisa Ishikawa, Richard B. Kim and Jörg König

12.2 TISSUE DISTRIBUTION AND SIGNIFICANCE FOR DRUG TRANSPORT

P-gp is expressed in a polarized manner and the ATP-driven active transport occurs at most compartment barriers. The first defense against intrusion of xenobiotics occurs already at brush-border membranes of intestinal enterocytes at the apical membrane of the entire intestine (Thorn et al., 2005; Canaparo et al., 2007). ABCB1 contributes to the lowering of bioavailability of a large number of drugs not only as a functional barrier against drug entry but also because of active excretion (Schinkel et al., 1996; Terao et al., 1996; Zhang and Benet, 2001). The pivotal role at the intestinal barrier is mirrored in a higher protein expression in the duodenum and jejunum than in the liver (von Richter et al., 2004). Within hepatocytes, P-gp is expressed at the canalicular site of hepatocytes exhibiting a large interindividual variability (Schuetz et al., 1995). Transporters also play a substantial role in tubular cells in the kidney, here ABCB1 is expressed at the luminal side of cells thereby contributing to the active secretion of lipophilic drugs and other xenobiotics (del Moral et al., 1998).

The blood–brain barrier is crucial for the bioavailability of drugs acting in the central nervous system and possibly more important as a defense mechanism against damages of the brain by exogenous toxins (Miller et al., 2008). As mentioned above, *Abcb1* knockout mice exhibited higher loperamide concentrations in the cerebrospinal fluid (CSF) than wild types (Schinkel et al., 1996). Also inhibition of P-gp by, for example, verapamil or downregulation after inflammatory stimuli led to a dramatic increment of typical substrates such as digoxin in the liquor (Goralski et al., 2003). Accordingly, the antinociceptive effects of morphin-6-glucuronide increased after inhibition of P-gp by PSC833 in a rat model (Lotsch et al., 2000) or central nervous effects of loperamide occurred after coadministration of quinidine (Skarke et al., 2003). Due to its pivotal role, alteration of P-gp function or expression may have major impact in the treatment of neurologic or psychiatric diseases. The overexpression of P-gp in epilepsy is considered as one of the major hypotheses of resistance toward anticonvulsants (Loscher et al., 2009).

However, P-gp is not only expressed at compartment barriers but also within the membranes of single cells like lymphocytes and hematopoietic stem cells. Therefore, P-gp contributes also to the phenomenon of drug resistance in the treatment of HIV (Chandler et al., 2007) or leukemia (Svirnovski et al., 2009) on lymphocyte level. The co-medication of P-gp inhibitors like ritonavir to other protease inhibitors such as lopinavir contributed significantly to an improved bioavailability and clinical outcome (Lucia et al., 2001; Janneh et al., 2007). Interestingly, P-gp occurs also intracellular in the endoplasmic reticulum and cell vesicles, thus actively contributing to the sequestration and transport of drugs outward the cells (Molinari et al., 2002).

12.3 MOLECULAR PROPERTIES AND FUNCTION

ABCB1, the gene of P-gp was firstly cloned by Riordan et al. (1985), having a size of 170 Da. The ABCB1 gene is located approximately 25 kb apart from ABCB4 on chromosome 7q21.12, spans 209.6 kb and contains 29 exons; the first two are not translated but encompass the promoter in exon1/intron1 (Bodor et al., 2005). The cDNA is 3843 bp in length (RefSeq accession NM_000927.3), coding for a protein of 1280 amino acids. The secondary structure of P-gp reveals two homologous halves to the protein, each containing six transmembrane domains and a nucleotide-binding site (Fig. 12.1).

P-gp is post-translationally modified by glycosilation and phosphorylation influencing the activity of the transporter (Lelong-Rebel and Cardarelli, 2005). The substrate-binding site is located at the intracellular part of the protein. After phosphorylation of the nucleotide-binding site, the substrate molecule is extruded by a "flip-flop" mechanism to the luminal side. Dephosphorylation leads to transformation of the protein back into the initial state.

P-gp exhibits a low specificity and transports a large variety of drugs used in the treatment of various diseases (Table 12.1). The substrate specificity overlaps with other ABC transporters like ABCC1, ABCC2, and ABCG2 (Cascorbi, 2011). These transporters partly act synergistically with cytochrome P450 enzymes in particular with CYP3A4 as defense mechanism against the intrusion of drugs and toxins. ABCB1, ABCC2, and CYP3A4 are co-regulated by the nuclear receptor PXR (von Richter et al., 2004), they are both induced by PXR/CAR ligands like rifampicin (Geick et al., 2001) or St. John's wort (Johne et al., 1999).

12.4 ABCB1 GENETIC POLYMORPHISMS

According to the PharmGKB database (www.pharmgkb.org), there are 1279 SNPs in the ABCB1 gene region, among them 41 nonsynonymous SNPs leading to amino acid exchanges, 22 synonymous SNPs, and one in the start codon. 124 SNPs have an allele frequency higher than 5%. Further information is given in the University of California, San Francisco Pharmacogenetics of Membrane Transporters Database (http://pharmacogenetics.ucsf.edu). It lists a total of 68 SNPs being genotyped in ethnical diverse populations. Seventeen SNPs are in the 5′-region (six being located on the untranslated first two exons), nineteen in exonic regions (14 missenses and 5 senses), twenty-five intronic variants, and thirteen in the 3′-UTR of the ABCB1 gene.

However, deep sequencing has not been performed so far. Table 12.2 lists the frequency of exonic ABCB1 SNPs in various populations as well as the observed functional impact. The initial observation of a functional correlation of the silent polymorphism in exon 26 (3435C>T; rs1045642) with intestinal P-gp expression levels and oral bioavailability of digoxin (Hoffmeyer et al., 2000) was later inconsistently reproduced. This SNP is in linkage disequilibrium with the triallelic SNP 2677G>T/A (rs2032582) (Cascorbi et al., 2001; Kim et al., 2001; Tanabe et al., 2001; Horinouchi et al., 2002). The frequency of the putatively most interesting 3435C>T SNP differs significantly between ethnicities (Table 12.2). Due to its assumed clinical relevance, it was discussed whether such genotypic differences may contribute to interethnic differences of drug responses in certain populations; however, later performed large meta-analyses failed to prove this assumption.

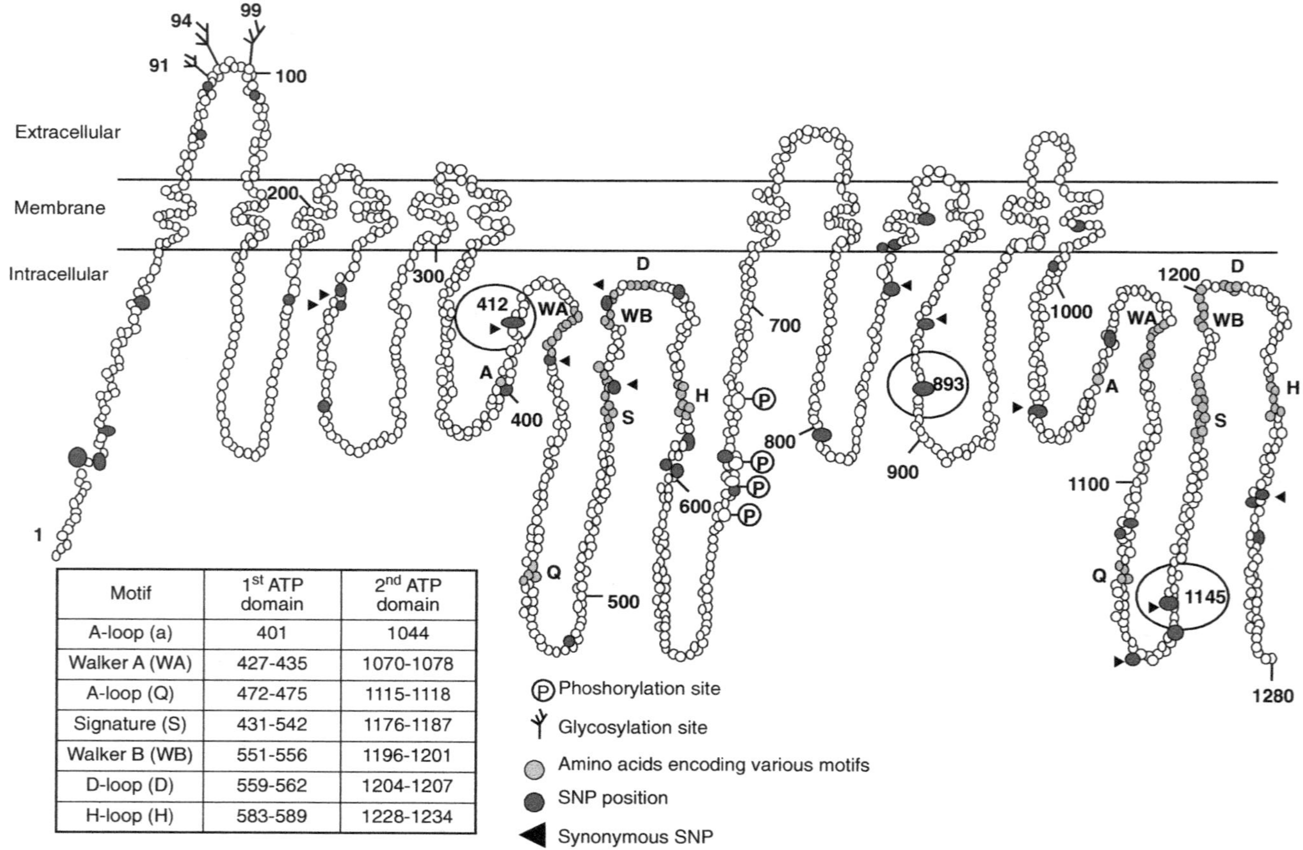

Motif	1st ATP domain	2nd ATP domain
A-loop (a)	401	1044
Walker A (WA)	427-435	1070-1078
A-loop (Q)	472-475	1115-1118
Signature (S)	431-542	1176-1187
Walker B (WB)	551-556	1196-1201
D-loop (D)	559-562	1204-1207
H-loop (H)	583-589	1228-1234

Figure 12.1 Hypothetical two-dimensional model of ABCB1 highlighting single-nucleotide polymorphsims and amino acid exchanges. Red circles represent the amino acids affected by SNPs. Amino acid residues affected by synonymous SNPs are marked with triangles. The amino acids, affected by the three major ABCB1 SNPs often related to certain diseases or drug disposition (c.1236C>T, p.412G=G; c.2677G>T/A, p.893A>S/T; and c.3435C>T, p.1145I=I) are circled. Phosphorylation and glycosylation sites are also shown. Adapted from Fung and Gottesman (2009). (*See insert for color representation of the figure.*)

TABLE 12.1 Substrates, Inducers, and Inhibitors of P-Glycoprotein (ABCB1)

Class	Drug
Anticancer drugs	Docetaxel, doxorubicin, etoposide, imatinib, paclitaxel, teniposide, topotecan, vinblastine, vincristine
Steroids	Budenoside, aldosterone, cortisol, dexamethasone, methylprednisolone
Immunosuppressants	Cyclosporine, sirolimus, tacrolimus
HIV protease inhibitors	Amprenavir, indinavir, nelfinavir, saquinavir, ritonavir
Antibiotics	Erythromycin, levofloxacin, ofloxacin, tetracycline
β-blockers:	Bunitrolol, carvedilol, celiprolol, tanilolol
Ca^{2+}-channel blockers:	Diltiazem, verapamil
Cardiac drugs	Digoxin, digitoxin, quinidine
HMG-CoA inhibitors	Artovastatin, lovastatin
Antihistamines	Fexofenadine, ranitidine, terfenadine,
Antiemetics	Ondansetron
Opioids	Loperamide, morphine,
Diverse	Amitryptiline, cimetidine, colchicine, domperideone, erythromycin, itraconazole, ivermectin, lansoprazole, losartan, phenytoin, rifampicin, vecuronium
Fluorescent dyes	Rhodamine 123
Inducers	
Anticonvulsants	Carbamazepine, phenytoin, phenobarbital, primidon
Tuberculostatics	Rifampicin
Herbals	Hyperforin (constituent of St. John's wort)
Inhibitors	
Calcium channel antagonists	Verapamil
Makrolide antibiotics	Erythromycin, clarythromycin, *not* azithromycin
HIV protease inhibitors	Ritonavir
Immunosuppressants	Cyclosporine
Antiarrhythmics	Chinidin, propafenon
Antimycotics	Ketocoanzole, itraconazole,
Designed inhibitors	Elacridar (GF120918), LY335979, valspodar (PSC 833)

Source: Cascorbi and Haenisch (2010); Fromm (2004); www.fda.gov.

The functional significance of the ABCB1 variants is still discussed controversial, though there is increasing evidence that it has only weak impact on the intestinal ABCB1 expression and systemic drug disposition (Cascorbi, 2006). The most recent hypothesis on the functional impact suggests that the timing of cotranslational folding could be affected in presence of the nucleotide codon, requiring a rare tRNA for the synonymous amino acid. As a consequence, this deceleration of the translation may alter the protein structure and substrate affinity (Kimchi-Sarfaty et al., 2007). However, as far as this hypothesis comes true, the expression and function of P-gp appears to be influenced by many other confounders, diminishing the effect of 3435C>T. From a pharmacokinetic point of view, the missense variant 2677G>T/A appears to be more effective *in vitro*. In an Sf9 insect cell expression system, the three different amino acids A893S/T exhibited altered transport properties. 893T had a higher V_{max}

TABLE 12.2 Exonic ABCB1 Genetic Variants, Position on cDNA, rs-Number, Amino Acid Effects, Frequency in Various Populations, and Observed Effects

		Peptide position					
cDNA position	rs-no.		*Caucasians*	*African Americans*	*Asians*	*Mexicans*	Association to expression, kinetics, or drug response
c.266T>C	rs35810889	p.89 M>T	0.005[a]	0.00[a]	0.00[a]	0.00[a]	Increased cytostatics resistance in transformed yeast[b]
c. 548 A>G		p.183 N>S	0.01[c]				
c. 571 G>A		p.191 G>R	0.07[d]				Reduced chemotherapy resistance[d]
c.781A>G	rs36008564	p.261 I>V	0.000[a]	0.01[a]	0.00[a]	0.00[a]	
c. 1199 G>A/T	rs2229109	p.400 S>N/I	0.05[e]/0.02[f]	0.01[a]	0.00[a]	0.00[a]	Elevated activity[f]
c. 1236 C>T	rs1128503	p. 412G=G	0.41[e]	0.21[a]	0.68[a]	0.45[a]	Increased imatinib disposition and therapy response[g]
c. 1474 C>T		p.492 R>C	0.01[c]				
c. 1985T>G	rs35657960	p.662 L>R	0.005[a]	0.00[a]	0.00[a]	0.00[a]	Increased cytostatics resistance in transformed yeast[b]
c.2005C>T	rs35023033	p.669 R>C	0.000[a]	0.01[a]	0.00[a]	0.00[a]	Increased cytostatics resistance in transformed yeast[b]
c. 2650 C>T	rs9282563	884 L=L	0.03[c]	0.005[a]	0.00[a]	0.00[a]	
c. 2677 G>T/A	rs2032582	p.893 A>S/T	0.42/0.02[e]	0.01/0.005[a]	0.45/0.07[a]	0.40/0.00[a]	Increased imatinib response in CML[g]; *in vitro* increased V_{max}[h],
c. 2956 A>G		p. 986 M>V	0.005[i]				
c. 3151C>G	rs28401798	p.1051 P>A	0.000[a]	0.005[a]	0.00[a]	0.00[a]	
c. 3320 A>C	rs55852620	p.1107 Q>P	0.002[e]				
c. 3322T>C	rs35730308	p. 1108 T>R	0.000[a]	0.005[a]	0.00[a]	0.00[a]	Decreased cytostatics resistance in transformed yeast[b]
c. 3421 T>A	rs2229107	p. 1141 S>T	0.00[j]	0.11[a]	0.00[a]	0.05[a]	

c. 3435 C>T	rs1045642	1145 I=I	0.54[e]	0.20[a]	0.40[a]	0.50[a]	Decreased mRNA and protein expression[c, k], decreased *in vitro* transport[l], no effect on expression and bioavailability of talinolol[m], no effect on *in vitro* transport[n, o], decreased digoxin bioavailability[p], increased etoposid disposition[q], no effect on AML or ALL outcome[o], better prognosis of multiple myeloma[r], better chemotherapy response in breast cancer[s], no effect in colon cancer[t], decreased opioid need in outpatients[u], various effects on drug resistance in epilepsy[v]
c.3751G>A	rs28364274	Val1251Ile	0.00[b]	0.00[b]	0.00[b]	0.05[b]	Decreased function in transfected cells[w]
c.3767C>A	rs35721439	Thr1256Lys	0.005[b]	0.00[b]	0.00[b]	0.00[b]	

[a](http://pharmacogenetics.ucsf.edu); [b](Jeong et al., 2007), [c](Hoffmeyer et al., 2000), [d](Yang et al., 2008), [e](Cascorbi et al., 2001); [f](Crouthamel et al., 2006), [g](Dulucq et al., 2008), [h](Schaefer et al., 2006); [i](Tanabe et al., 2001), [j](Kim et al., 2001), [k](Seedhouse et al., 2007), [l](Hitzl et al., 2001), [m](Siegmund et al., 2002), [n](Kimchi-Sarfaty et al., 2002), [o](van der Holt et al., 2006; Hur et al., 2008; Semsei et al., 2008); [p](Sakaeda et al., 2001a), [q](Kishi et al., 2004), [r](Buda et al., 2007), [s](Kafka et al., 2003), [t](Ho et al., 2006; Petrova et al., 2008), [u](Lotsch et al., 2009), [v](Haerian et al., 2011), [w](Gow et al., 2008)

for the anticancer drug vincristine than 893S, and V_{max} of 893S was higher than the wild-type 893A whereas K_m was higher for 893S compared to 893T or A (Schaefer et al., 2006). The functional difference could be also observed investigating the ATPase activity in a similar expression system (Ishikawa et al., 2005, 2010).

The rare missense SNP 1199 G>T (Ser400Asn) was associated with lower transport capacity *in vitro* leading to putatively higher sensitivity against cytostatics. In contrast, HEK cells transfected with ABCB1 carrying 1199G>A exhibited elevated chemoresistance indicating elevated transport of the modified P-gp (Crouthamel et al., 2006). Other very rare variants caused increased resistance to various cytostatics in transformed yeast cells (Jeong et al., 2007).

12.5 IMPACT ON DRUG BIOAVAILABILITY

Digoxin is widely accepted as probe drug of P-gp. A study in 2000 revealed a strong impact of 3435C>T on digoxin plasma distribution with elevated levels among 3435TT carriers after the intake of a single dose of 1 mg (Hoffmeyer et al., 2000). In subjects being in steady state with 0.25 mg digoxin per day, 3435T carriers had a 20% elevated area under the curve (AUC) within the first 4 h (Johne et al., 2002). Another study in patients under chronic digoxin treatment revealed a significance of the ABCB1 haplotype 1236T/2677T/3435T on digoxin trough levels (Aarnoudse et al., 2008). Interestingly, applying a nonlinear mixed-effect model, Comets et al. found out that homozygous carriers of 3435C>T had a lower apparent volume of distribution than carriers of the C allele (Comets et al., 2007). In Asians, the opposite tendency was observed in three independent studies, with lower digoxin AUC in the CC group than in subjects homozygous for TT (Sakaeda et al., 2001b).

However, the majority of studies on the influence of ABCB1 variants (in particular, 3435C>T) on the pharmacokinetics digoxin revealed no differences between the genotypes as demonstrated by a meta-analysis in 2005 (Chowbay et al., 2005). Same results were obtained from pharmacokinetic studies using the beta-blocker talinolol that is also a suitable P-gp substrate. Homozygous ABCB1 2677T/A variant carriers showed significantly higher AUCs of talinolol compared to all other genotypes ($p < .05$). However, multiple comparisons with combinations of putative functional SNPs did not confirm a significant influence of the ABCB1 genotype on talinolol disposition (Siegmund et al., 2002).

Fexofenadine, considered also as P-gp substrate revealed contradictory results with higher bioavailability in 3435C carriers (Kim et al., 2001), but no effects in another study (Drescher et al., 2002).

As reviewed earlier, various hormonal and immunological influences on ABCB1 expression (Cascorbi, 2006) or saturation of the intestinal ABCB1 transport capacity may surpass any genetic effects (Sakaeda, 2005). Finally, other uptake and efflux transporters may contribute to the disposition of digoxin in humans.

The major immunosuppressive calcineurin inhibitors, cyclosporine and tacrolimus, as well as the TOR-inhibitor sirolimus are also well established as P-gp substrates. Small differences of mean plasma levels may hamper the successful immunosuppression and bear the risk of organ rejection. In a meta-analysis from

14 papers published between 1997 and 2007, a total of 1036 individuals were included (Jiang et al., 2008). Overall, there was lack of evidence of a major influence of 3435C>T on the pharmacokinetics of cyclosporine, though the authors noticed a somewhat lower AUC within the first 12 h in subjects homozygous for 3435C and minor ethnic differences. All changes of the plasma concentrations appeared to be clinically not relevant. The results from this meta-analysis are essentially supported by further studies in renal or heart transplant recipients (Taegtmeyer et al., 2010; Staatz et al., 2010; Bouamar et al., 2011; Singh et al., 2011). However, most important are physiological factors and co-medication, that is, the body weight was identified as the most important covariate and explained 35% of the random interindividual variability in cyclosporine clearance. Co-medication of ≥20 mg per day prednisolone was associated with a 22% higher clearance of cyclosporine (Press et al., 2010).

For tacrolimus, we found lack of any impact of ABCB1 variants on any pharmacokinetic parameter, but a significant dependency of kinetics on the presence of active cytochrome P450 3A5 (CYP3A5) (Renders et al., 2007). A study investigating tacrolimus-related neurotoxicity also denied any impact of ABCB1, but again of the CYP3A5 genotype (Yanagimachi et al., 2010).

There are a number of further studies, partly showing effects of ABCB1 SNPs; however, they were not always controlled for CYP3A5. Based on our observation in sigmoidal tissue that ABCB1 mRNA expression is significantly correlated with CYP3A5 mRNA expression (Ufer et al., 2008), a combined analysis appears to be crucial. Indeed, the most recent studies revealed that among many other pharmacogenetic factors, only CYP3A5 resulted as a promising genetic factor, suitable for the prediction of tacrolimus dosages in renal transplant therapy (Kuypers et al., 2010; Wang et al., 2010).

The figure remains similar for sirolimus. In a small sample of 20 patients (Renders et al., 2007), we could not identify any impact on the pharmacokinetics which is in line with other studies in this field (Anglicheau et al., 2005; Mourad et al., 2005; Miao et al., 2008).

All the above mentioned studies described the impact of ABCB1 (and partly CYP3A5) on pharmacokinetics rather than on the clinical output. It should be taken into account that finally intracellular concentrations of lymphocytes are crucial for immunosuppression but not plasma concentrations. Interestingly, in a study determining intracellular lymphocyte concentrations from 64 stable renal, liver, or lung transplant recipients, there was a higher cyclosporine concentration among lymphocytes containing homozygously the ABCB1 3435T allele than in CC carriers (Crettol et al., 2008b). Such differences are much more pronounced than ever determined in the plasma. The expression of ABCB1 is the major determinant of intracellular levels, whereas plasma concentrations are dependent on ABCB1 levels in the intestine, liver, and kidney, and also hepatic CYP3A4 activity.

The CYP3A activity is also of major importance for systemic side effects such as tacrolimus-related neurotoxicity. It could be shown that the CYP3A5 genotype, but not of the ABCB1 genotyped modulates the risk of this adverse event (Yanagimachi et al., 2010). There are a number of further studies, partly showing effects of ABCB1 SNPs; however, they were not always controlled for CYP3A5. Based on our observation in sigmoidal tissue that ABCB1 mRNA expression is significantly correlated with CYP3A5 mRNA expression (Ufer et al., 2008), a combined analysis appears to be crucial.

Concluding, the impact of ABCB1 genotypes on immunosuppressants seems to be negligible. However, there is increasing evidence that consideration of the CYP3A5 status could be beneficial for tacrolimus dosing.

12.6 IMPACT ON INTRACELLULAR CONCENTRATIONS

As outlined above, protease inhibitors obligatorily used in the treatment of HIV infections, are clearly P-gp substrates of ABCB1. However, there is only weak evidence that ABCB1 genetic variants may have an impact on plasma concentrations of lopinavir (Winzer et al., 2003). For indinavir, only small studies have been undertaken showing results that are difficult to interpret, for example, ABCB1 CC carriers led to moderately lower absorption but CYP3A5 had an additional impact (Solas et al., 2007; Curras et al., 2009). For nelfinavir, no influence of ABC transporter variants was observed (Colombo et al., 2005), but a significant impact of CYP2C19 (Hirt et al., 2008; Saitoh et al., 2009). The more recent protease inhibitor, atazanavir, is also a substrate of a P450 enzyme, namely CYP3A5. Thus, this confounder has to be taken into account when determining the impact of ABCB1 genotypes. Interestingly, in a study in US Caucasians and Blacks, ABCB1 1236C/2677G/3435C haplotype carriers had a decreased clearance, but the study was too small to prove the ABCB1 effects in subgroups of CYP3A5 expressors and nonexpressors alone (Anderson et al., 2009). Strikingly, another study pointed into the same direction of an opposite effect of 3435C>T toward lower plasma concentrations (Rodriguez-Novoa et al., 2007).

12.7 IMPACT ON TREATMENT OF DISEASES

12.7.1 Epilepsy

Failure of drug-response is a major limitation in the treatment of epilepsy. Approximately, every second patient does not respond sufficiently to the first drug of choice and about one-third to one-quarter remain resistant to anticonvulsants. One major mechanistic hypothesis is based on the overexpression of efflux transporters in particular P-gp at the blood–brain barrier leading to reduced CSF concentrations and thus decreased local bioavailability of anticonvulsants at their target site (Hughes, 2008). P-gp overexpression may be the consequence of seizure, PXR-mediated upregulation or genetic variation. Moreover, the hypothesis of efflux transporter overexpression is based on the assumption that most anticonvulsants are P-gp substrates. First evidence for a role of ABCB1 variants to the risk of pharmacotherapy resistance resolved from a study in 115 British treatment-responsive and 200 nonresponsive patients (Siddiqui et al., 2003). The frequency of carriers being homozygous for the putatively low-active 3435T allele was significantly higher among responsive patients than in controls and responsive patients. Though, in this study no functional analyses on ABCB1 expression were performed and no data on the medication was provided. Assuming that further candidate SNPs could be causal for the ABCB1 expression, an additional attempt was made within a cohort partially overlapping with the study of Siddiqui (Siddiqui et al., 2003). Again ABCB1 34335C>T turned out to be associated with treatment responsiveness, but an SNP at IVS 26 + 80 T>C was additionally identified that was 1.5-fold less causal than 3435C>T (Soranzo et al., 2004). A further study from Austria found a result going into the same direction, although the definition of therapy response differed (Zimprich et al., 2004). A small study from Egypt was also in line with the initial study (Ebid et al., 2007). All further studies performed within Caucasian populations (nine were identified comprising 2772 cases) failed to confirm any association between genetic variants of ABCB1 and therapy responsiveness of epilepsy. Only one study from Croatia showed an impact of 3435C>T on phenobarbital CSF–plasma ratios. In this study, 3435CC carriers had a significantly lower CSF/PB concentration ratio than CT or TT carriers. Moreover, low CSF/S PB concentration ratio correlated with increased seizure frequency (Basic et al., 2008).

There are two studies from Asian population showing a positive association with ABCB1 3435C>T, one with a similar result as Siddiqui study (Hung et al., 2005) whereas a Japanese study pointed to the opposite direction with an over-representation of variant T-allele carriers among drug-resistant patients (Seo et al., 2006). All other studies revealed lack of evidence of any association (four were identified comprising 1074 cases). Also a study in a mixed cohort from Scottish, Australian, and Hong Kong origin failed to find an association (Szoeke et al., 2009).

A meta-analysis summarizing 11 studies involving a total of 3371 patients (1725 responsive and 1646 drug-resistant patients) calculated an odds ratio for drug responsiveness of T-allele carriers of 1.15 (0.78–1.70) $p = .48$, indicating lack of association. Stratification to ethnic groups revealed no further evidence (Bournissen et al., 2009). The most comprehensive meta-analysis was published very recently comprising 7067 patients (Haerian et al., 2011) (Table 12.3).

The current data allows the well-substantiated conclusion that ABCB1 variants are not associated with drug responsiveness to anticonvulsants. This discrepancy may be explained on one hand by the fact that in contrast to earlier assumptions many anticonvulsants are not substrates of human P-gp (Baltes et al., 2007a, 2007b) and on the other hand that the impact of ABCB1 variants on ABCB1 expression and function in brain tissue is modest (Mosyagin et al., 2008).

12.7.2 Depression

There is not too much literature dealing with the association of ABC drug transporters and the clinical outcome of treatment of mood disorders with antidepressants. However, Uhr et al. firstly demonstrated in a mouse knockout model that venlafaxine, a potent dual reuptake inhibitor of norepinephrine and serotonin, or citalopram, a specific serotonin reuptake inhibitor, is a substrate of P-gp. Brain concentrations of citalopram, venlafaxine, and D-venlafaxine were 3, 1.7, and 4.1 times higher in the mutant mice compared to wild types (Uhr et al., 2008). Secondly, they could show that polymorphisms in the ABCB1 gene could predict the response to antidepressant

TABLE 12.3 Association of ABCB1 3435C>T (rs1045642) to Therapy Responsiveness in the Treatment of Epilepsy

Reference	Year	Origin	Size	Responders/ nonresponders	Association of 3435C>T to nonresponse	Association of other ABCB1 variants to nonresponse
Positive association in Caucasians						
(Siddiqui et al., 2003)	2003	United Kingdom	315	115/200	C more frequent	none
(Soranzo et al., 2004)	2004	UK overlapping with (Siddiqui et al., 2003)	421	135/286	C more frequent	IVS 26 + 80
(Zimprich et al., 2004)	2004	Austria	193	44/83/66*	C more frequent	Haplotype CGC
(Ebid et al., 2007)	2007	Egypt	100	37/63	C more frequent	n.a.
Negative association in Caucasians						
(Tan et al., 2004)	2004	Australia	609	208/401	none	n.a.
(Sills et al., 2005)	2005	United Kingdom	400	230/170	none	n.a.
(Shahwan et al., 2007)	2006	Ireland	366	242/124	none	none
(Leschziner et al., 2007)	2007	United Kingdom	149	76/73	none	none
(Dericioglu et al., 2008)	2008	Turkey	189	100[b]/89	none	n.a.
(Ozgon et al., 2008)	2008	Turkey	271	174[b]/97[c]	none	n.a.
(Ufer et al., 2009)	2009	Germany	221	103/118	none	none
(Lakhan et al., 2009)	2009	North India	325	231/94	none	none
(Vahab et al., 2009)	2009	South India	242	129/113	none	n.a.

Positive association in Asians						
(Hung et al., 2005)	2005	Taiwan	331	223/108	C more frequent	Haplotype CGC. TGC, TTT
(Seo et al., 2006)	2006	Japan	210	84/126	T more frequent	Haplotype TTT
Negative association in Asians						
(Kim et al., 2006c)	2006	Korea	199	100/99	none	none
(Kim et al., 2006b)	2006	Korea	160	101/59	none	n.a.
(Kwan et al., 2007)	2007	China	501	286/215	none	none
(Chen et al., 2007)	2007	China	214	164/50	none	n.a.
Negative association in a mixed cohort						
(Szoeke et al., 2009)	2009	Australia, China, United Kingdom	542	208/334	none	n.a.
Meta-analysis						
(Bournissen et al., 2009)	2009	Various	3370	1725/1645	none	n.a.
(Haerian et al., 2011)	2011	Various	7067		none	none

Source: Adapted from Cascorbi (2011).
n.a., not applied; IVS, intervening sequence.
[a]stratification according to seizure frequency.
[b]healthy controls.
[c]carbamazepine-resistant.

treatment in those depressed patients receiving drugs that have been identified as P-gp substrates. They identified an intronic C>T SNP (rs2032583) showing the strongest differences in therapy response. It has a frequency of 0.14 in Caucasians. No further data regarding this SNP is available.

12.7.3 Pain

Opioids like morphine and some of their metabolites such as morphine 6-glucuronide are substrates of ABC transporters. Accordingly, antinociceptive effects of morphin-6-glucuronide increased after inhibition of P-gp by the PSC833 in a rat model (Lotsch et al., 2000) or central nervous effects of orally administered loperamide occurred after co-administration of the inhibitor quinidine (Skarke et al., 2003). A number of studies have been performed investigating the role of ABCB1 variants on the pharmacodynamics and pharmacokinetics of different opioids.

In a study on 98 former severe heroin-dependent subjects recruited by an Israeli methadone maintenance treatment program, the methadone requirement was determined in relation to the ABCB1 SNPs 1236C>T, 2677G>T, and 3435C>T. Compared to individuals being homozygous wild-type at all three loci, subjects being homozygote variant had a 5.5-fold chance of requiring methadone doses >150 mg than doses ≤150 mg. The difference of heterozygotes to wild-types was not significant. Interestingly, 1236C>T turned out as SNP with the strongest impact (Levran et al., 2008).

A study in 145 Italian patients undergoing morphine therapy, ABCB1 3435C>T showed a highly significant association with pain relief variability, accounting for 16% of total deviance. Homozygous 3435 T/T carriers were significantly associated with greater pain relief than homozygous wild-type CC carriers (Campa et al., 2008). A French study in patients undergoing colorectal surgery also revealed no impact of ABCB1 variants on morphine dose, but a significant association with morphine side effects and antiemetic drug requirement (Coulbault et al., 2006). Similarly, the ABCB1 wild-types, 2677G and 1236C, were significantly associated with decreased levels of drowsiness and confusion or hallucinations in 228 British cancer patients (Ross et al., 2008). Also, the analgetic effects of oxycodone appear to be affected. Zwisler et al. found that the ABCB1 2677G allele was associated with significantly reduced antinociceptive effect as measured by pain-tolerance thresholds to single electrical nerve stimulation (8% increase vs. 25% for the wild-type carriers), but carriers of the variant 3435T allele had fewer adverse drug reactions on oxycodone than the carriers of the wild-type genotype (Zwisler et al., 2010). Moreover, ABCB1 34356C>T showed a significant effect on the opioid dosing in 352 German outpatients. The opioid dose ranged from 92 mg per day among 3435TT carriers to 178 mg per day in 3435CC carriers (Lotsch et al., 2009).

The results of the study are nearly all pointing to the same direction. Carriers of the putative high-active ABCB1 variants required higher opioid doses or had fewer side effects, indicating that P-gp at the blood–brain barrier may play a crucial role for opioid-mediated central pain suppression or modulation of central side effects.

However, these intriguing results could not always be confirmed. Crettol et al. could not observe any impact of ABCB1 variants on dose requirements in 279

Swiss methadone maintenance patients (Crettol et al., 2008a), and Lotsch et al. did not find any impact on the pharmacokinetics and pharmacodynamics of methadone in 51 healthy German volunteers (Lotsch et al., 2006). Overall, the results suggest that ABCB1 variants have a stronger clinical effect on CNS-expressed P-gp function than in other tissues.

12.7.4 Leukemia

The expression of ABCB1 in lymphocyte membrane extruding drugs out of the cell may be of importance in the response to cytostatics used in the treatment of various types of leukemia. Indeed, there is an increasing number of studies suggesting an association between ABCB1 genotypes and clinical outcome (Illmer et al., 2002; Jamroziak et al., 2004; Stanulla et al., 2005; Kim et al., 2006a). In fact, the ABCB1 expression was significantly higher among the upper percentile lymphocytes from British acute myeloic leukemia (AML) patients being 3435C carriers than among T carriers (Seedhouse et al., 2007). In a study in 101 Korean AML patients, 3435CC was significantly correlated with lower functional ABCB1 function in a daunorubicin intracellular accumulation assay (Kim et al., 2006a). Surprisingly, presence of the putative high-active genotypes 3435CC and 2677GG were strongly associated with the higher probability of complete remission and 3-year event-free survival. However, no differences were noted in overall survival according to the ABCB1 SNPs. This lack of association was confirmed in a clinical study among elderly AML patients, ABCB1 variants failed to show any association to the treatment outcome or ABCB1 expression and function, as evidenced by rhodamine efflux experiments controlled with the P-gp inhibitor PSC833 (van der Holt et al., 2006).

In French chronic myeloic leukemia (CML) patients, however, the molecular response to the BRC-ABL-inhibitor imatinib was dependent on ABCB1 genotypes (Dulucq et al., 2008). Although 3435C>T failed to show a significant influence, patients with 1236T had higher imatinib plasma concentrations and had a better therapy response whereas presence of the wild-type 2677G variant worsened the clinical response. In another study, in a small sample of CML patients and gastrointestinal stromal tumors (GIST), 1236C>T, 2677G>T/A, and 3435C>T led to decreased imatinib clearance among variant carriers.

In a Korean study on acute lymphoblastic leukemia (ALL), however, there was lack of association of clinical endpoints such as complete remission rates, or relapse-free and event-free survival rates to ABCB1 variants (Hur et al., 2008). Only rare ABCB1 haplotypes of 2677G>T/A and 3435C>T differed in a large Hungarian ALL study, but overall, the genotype distribution was not statistically different (Semsei et al., 2008). In multiple myelomas treated with dexamethasone, doxorubicin, and vincristine, ABCB1 3435CT or TT carriers had a better prognosis than 3435CC carriers ($p = .02$) (Buda et al., 2007), and the pharmacokinetics of vincristine were only marginally influenced by the ABCB1 SNPs, 2677G>T/A and 3435C>T, in another study in childhood ALL (Plasschaert et al., 2004).

The attempt to find any further ABCB1 variant influencing the activity or being associated to the clinical response revealed a novel ABCB1 571G>A missense variant detected in 6.4% of leukemia patients, causing a Gly191Arg amino acid change

(Yang et al., 2008). The functional impact was investigated in a stable recombinant expression cell model. Among 571A carriers, there was a selectively reduced degree of ABCB1-mediated resistance against the anticancer drugs, such as doxorubicin, daunorubicin, vinblastine, vincristine, paclitaxel, and etoposide. In particular, the resistance on vinblastine, vincristine, paclitaxel, and etoposide was fivefold reduced, indicating lower transport capacity of the protein carrying the 191Arg-variant. It was suggested that individuals with the ABCB1 571A genotype may be more sensitive to the specific anticancer drugs that are P-gp substrates but may also exhibit a higher risk of side effects.

12.8 CONCLUSION

Numerous studies on the impact of ABCB1 variants on pharmacokinetics and pharmacodynamics give no clear picture. The current knowledge of the functional significance genetic variants of ABC membrane transporters does not allow selection of a particular SNP to predict an individual's pharmacokinetics. As far as the associations of ABCB1 variants to clinical outcome were achieved, mostly the variant 3435T was over-represented among those cases exhibiting stronger drug effects, though these observations have been made in particular among patients under opioid treatment and to some proportion in leukemia patients, thus further investigation of the role of the ABC transporters for intracellular drug bioavailability are strongly desired.

However, the effects of any ABCB1 variant on pharmacokinetics are modest and do not justify any dose recommendation based on ABCB1 genotypes. Current clinical considerations for P-gp are, therefore, related to its important role in multidrug resistance and drug–drug interactions, due to its broad substrate specificity and variable intrinsic and drug-induced expression.

REFERENCES

Aarnoudse AJ, Dieleman JP, Visser LE, Arp PP, van dH, I, Van Schaik RH, Molokhia M, Hofman A, Uitterlinden AG, Stricker BH (2008) Common ATP-binding cassette B1 variants are associated with increased digoxin serum concentration. *Pharmacogenet Genomics*, 18, 299–305.

Anderson PL, Aquilante CL, Gardner EM, Predhomme J, McDaneld P, Bushman LR, Zheng JH, Ray M, MaWhinney S (2009) Atazanavir pharmacokinetics in genetically determined CYP3A5 expressors versus non-expressors. *J Antimicrob Chemother*, 64, 1071–1079.

Anglicheau D, Le Corre D, Lechaton S, Laurent-Puig P, Kreis H, Beaune P, Legendre C, Thervet E (2005) Consequences of genetic polymorphisms for sirolimus requirements after renal transplant in patients on primary sirolimus therapy. *Am J Transplant*, 5, 595–603.

Baltes S, Fedrowitz M, Tortos CL, Potschka H, Loscher W (2007a) Valproic acid is not a substrate for P-glycoprotein or multidrug resistance proteins 1 and 2 in a number of *in vitro* and *in vivo* transport assays. *J Pharmacol Exp Ther*, 320, 331–343.

Baltes S, Gastens AM, Fedrowitz M, Potschka H, Kaever V, Loscher W (2007b) Differences in the transport of the antiepileptic drugs phenytoin, levetiracetam, and carbamazepine by human and mouse P-glycoprotein. *Neuropharmacology*, 52, 333–346.

Basic S, Hajnsek S, Bozina N, Filipcic I, Sporis D, Mislov D, Posavec A (2008) The influence of C3435T polymorphism of ABCB1 gene on penetration of phenobarbital across the blood–brain barrier in patients with generalized epilepsy. *Seizure*, 17, 524–530.

Biedler JL, Riehm H (1970) Cellular resistance to actinomycin D in Chinese hamster cells *in vitro*: cross-resistance, radioautographic, and cytogenetic studies. *Cancer Res*, 30, 1174–1184.

Bodor M, Kelly EJ, Ho RJ (2005) Characterization of the human MDR1 gene. *AAPS J*, 7, E1–E5.

Bouamar R, Hesselink DA, Van Schaik RH, Weimar W, Macphee IA, de Fijter JW, Gelder TV (2011) Polymorphisms in CYP3A5, CYP3A4, and ABCB1 are not associated with cyclosporine pharmacokinetics nor with cyclosporine clinical end points after renal transplantation. *Ther Drug Monit*, 33, 178–184.

Bournissen FG, Moretti ME, Juurlink DN, Koren G, Walker M, Finkelstein Y (2009) Polymorphism of the MDR1/ABCB1 C3435T drug-transporter and resistance to anticonvulsant drugs: a meta-analysis. *Epilepsia*, 50, 898–903.

Buda G, Maggini V, Galimberti S, Martino A, Giuliani N, Morabito F, Genestreti G, Iacopino P, Rizzoli V, Barale R, et al. (2007) MDR1 polymorphism influences the outcome of multiple myeloma patients. *Br J Haematol*, 137, 454–456.

Campa D, Gioia A, Tomei A, Poli P, Barale R (2008) Association of ABCB1/MDR1 and OPRM1 gene polymorphisms with morphine pain relief. *Clin Pharmacol Ther*, 83, 559–566.

Canaparo R, Finnstrom N, Serpe L, Nordmark A, Muntoni E, Eandi M, Rane A, Zara GP (2007) Expression of CYP3A isoforms and P-glycoprotein in human stomach, jejunum, and ileum. *Clin Exp Pharmacol Physiol*, 34, 1138–1144.

Cascorbi I (2006) Role of pharmacogenetics of ATP-binding cassette transporters in the pharmacokinetics of drugs. *Pharmacol Ther*, 112, 457–473.

Cascorbi I (2011) P-glycoprotein: tissue distribution, substrates, and functional consequences of genetic variations. In: *Handbook of Experimental Pharmacology*, Vol. 201. Heidelberg, Germany: Springer. pp 261–283.

Cascorbi I, Gerloff T, Johne A, Meisel C, Hoffmeyer S, Schwab M, Schaeffeler E, Eichelbaum M, Brinkmann U, Roots I (2001) Frequency of single-nucleotide polymorphisms in the P-glycoprotein drug transporter MDR1 gene in white subjects. *Clin Pharmacol Ther*, 69, 169–174.

Cascorbi I, Haenisch S (2010) Pharmacogenetics of ATP-binding cassette transporters and clinical implications. *Methods Mol Biol*, 596, 95–121.

Chandler B, Detsika M, Khoo SH, Williams J, Back DJ, Owen A (2007) Factors impacting the expression of membrane-bound proteins in lymphocytes from HIV-positive subjects. *J Antimicrob Chemother*, 60, 685–689.

Chen L, Liu CQ, Hu Y, Xiao ZT, Chen Y, Liao JX (2007) [Association of a polymorphism in MDR1 C3435T with response to antiepileptic drug treatment in ethic Han Chinese children with epilepsy]. *Zhongguo Dang. Dai Er. Ke. Za Zhi.*, 9, 11–14.

Chowbay B, Li H, David M, Cheung YB, Lee EJ (2005) Meta-analysis of the influence of MDR1 C3435T polymorphism on digoxin pharmacokinetics and MDR1 gene expression. *Br J Clin Pharmacol*, 60, 159–171.

Colombo S, Soranzo N, Rotger M, Sprenger R, Bleiber G, Furrer H, Buclin T, Goldstein D, Decosterd L, Telenti A (2005) Influence of ABCB1, ABCC1, ABCC2, and ABCG2 haplotypes on the cellular exposure of nelfinavir *in vivo*. *Pharmacogenet Genomics*, 15, 599–608.

Comets E, Verstuyft C, Lavielle M, Jaillon P, Becquemont L, Mentre F (2007) Modelling the influence of MDR1 polymorphism on digoxin pharmacokinetic parameters. *Eur J Clin Pharmacol*, 63, 437–449.

Coulbault L, Beaussier M, Verstuyft C, Weickmans H, Dubert L, Tregouet D, Descot C, Parc Y, Lienhart A, Jaillon P, Becquemont L (2006) Environmental and genetic factors associated with morphine response in the postoperative period. *Clin Pharmacol Ther*, 79, 316–324.

Crettol S, Deglon JJ, Besson J, Croquette-Krokar M, Hammig R, Gothuey I, Monnat M, Eap CB (2008a) No influence of ABCB1 haplotypes on methadone dosage requirement. *Clin Pharmacol Ther*, 83, 668–669.

Crettol S, Venetz JP, Fontana M, Aubert JD, Ansermot N, Fathi M, Pascual M, Eap CB (2008b) Influence of ABCB1 genetic polymorphisms on cyclosporine intracellular concentration in transplant recipients. *Pharmacogenet Genomics*, 18, 307–315.

Crouthamel MH, Wu D, Yang Z, Ho RJ (2006) A novel MDR1 G1199T variant alters drug resistance and efflux transport activity of P-glycoprotein in recombinant Hek cells. *J Pharm Sci*, 95, 2767–2777.

Curras V, Hocht C, Mangano A, Niselman V, Marino HE, Caceres GP, Mecikovsky D, Bellusci C, Bologna R, Sen L, Rubio MC, Bramuglia GF (2009) Pharmacokinetic study of the variability of indinavir drug levels when boosted with ritonavir in HIV-infected children. *Pharmacology*, 83, 59–66.

del Moral RG, Olmo A, Aguilar M, O'Valle F (1998) P glycoprotein: a new mechanism to control drug-induced nephrotoxicity. *Exp Nephrol*, 6, 89–97.

Dericioglu N, Babaoglu MO, Yasar U, Bal IB, Bozkurt A, Saygi S (2008) Multidrug resistance in patients undergoing resective epilepsy surgery is not associated with C3435T polymorphism in the ABCB1 (MDR1) gene. *Epilepsy Res*, 80, 42–46.

Drescher S, Schaeffeler E, Hitzl M, Hofmann U, Schwab M, Brinkmann U, Eichelbaum M, Fromm MF (2002) MDR1 gene polymorphisms and disposition of the P-glycoprotein substrate fexofenadine. *Br J Clin Pharmacol*, 53, 526–534.

Dulucq S, Bouchet S, Turcq B, Lippert E, Etienne G, Reiffers J, Molimard M, Krajinovic M, Mahon FX (2008) Multidrug resistance gene (MDR1) polymorphisms are associated with major molecular responses to standard-dose imatinib in chronic myeloid leukemia. *Blood*, 112, 2024–2027.

Ebid AH, Ahmed MM, Mohammed SA (2007) Therapeutic drug monitoring and clinical outcomes in epileptic Egyptian patients: a gene polymorphism perspective study. *Ther Drug Monit*, 29, 305–312.

Fromm MF (2004) Importance of P-glycoprotein at blood–tissue barriers. *Trends Pharmacol Sci*, 25, 423–429.

Fung KL, Gottesman MM (2009) A synonymous polymorphism in a common MDR1 (ABCB1) haplotype shapes protein function. *Biochim Biophys Acta*, 1794, 860–871.

Geick A, Eichelbaum M, Burk O (2001) Nuclear receptor response elements mediate induction of intestinal MDR1 by rifampin. *J Biol Chem*, 276, 14581–14587.

Goralski KB, Hartmann G, Piquette-Miller M, Renton KW (2003) Downregulation of MDR1A expression in the brain and liver during CNS inflammation alters the *in vivo* disposition of digoxin. *Br J Pharmacol*, 139, 35–48.

Gow JM, Hodges LM, Chinn LW, Kroetz DL (2008) Substrate-dependent effects of human ABCB1 coding polymorphisms. *J Pharmacol Exp Ther*, 325, 435–442.

Haerian BS, Lim KS, Tan CT, Raymond AA, Mohamed Z (2011) Association of ABCB1 gene polymorphisms and their haplotypes with response to antiepileptic drugs: a systematic review and meta-analysis. *Pharmacogenomics*, 12, 713–725.

Hirt D, Mentre F, Tran A, Rey E, Auleley S, Salmon D, Duval X, Treluyer JM (2008) Effect of CYP2C19 polymorphism on nelfinavir to M8 biotransformation in HIV patients. *Br J Clin Pharmacol*, 65, 548–557.

Hitzl M, Drescher S, van der KH, Schaffeler E, Fischer J, Schwab M, Eichelbaum M, Fromm MF (2001) The C3435T mutation in the human MDR1 gene is associated with altered efflux of the P-glycoprotein substrate rhodamine 123 from CD56+ natural killer cells. *Pharmacogenetics*, 11, 293–298.

Ho GT, Soranzo N, Nimmo ER, Tenesa A, Goldstein DB, Satsangi J (2006) ABCB1/MDR1 gene determines susceptibility and phenotype in ulcerative colitis: discrimination of critical variants using a gene-wide haplotype tagging approach. *Hum Mol Genet*, 15, 797–805.

Hoffmeyer S, Burk O, von Richter O, Arnold HP, Brockmoller J, Johne A, Cascorbi I, Gerloff T, Roots I, Eichelbaum M, Brinkmann U (2000) Functional polymorphisms of the human multidrug-resistance gene: multiple sequence variations and correlation of one allele with P-glycoprotein expression and activity *in vivo*. *Proc Natl Acad Sci U S A*, 97, 3473–3478.

Horinouchi M, Sakaeda T, Nakamura T, Morita Y, Tamura T, Aoyama N, Kasuga M, Okumura K (2002) Significant genetic linkage of MDR1 polymorphisms at positions 3435 and 2677: functional relevance to pharmacokinetics of digoxin. *Pharm Res*, 19, 1581–1585.

Hughes JR (2008) One of the hottest topics in epileptology: ABC proteins. Their inhibition may be the future for patients with intractable seizures. *Neurol Res*, 30, 920–925.

Hung CC, Tai JJ, Lin CJ, Lee MJ, Liou HH (2005) Complex haplotypic effects of the ABCB1 gene on epilepsy treatment response. *Pharmacogenomics*, 6, 411–417.

Hur EH, Lee JH, Lee MJ, Choi SJ, Lee JH, Kang MJ, Seol M, Jang YE, Lee HJ, Kang IS, et al. (2008) C3435T polymorphism of the MDR1 gene is not associated with P-glycoprotein function of leukemic blasts and clinical outcome in patients with acute myeloid leukemia. *Leuk Res*, 32, 1601–1604.

Illmer T, Schuler US, Thiede C, Schwarz UI, Kim RB, Gotthard S, Freund D, Schakel U, Ehninger G, Schaich M (2002) MDR1 gene polymorphisms affect therapy outcome in acute myeloid leukemia patients. *Cancer Res*, 62, 4955–4962.

Ishikawa T, Sakurai A, Kanamori Y, Nagakura M, Hirano H, Takarada Y, Yamada K, Fukushima K, Kitajima M (2005) High-speed screening of human ATP-binding cassette transporter function and genetic polymorphisms: new strategies in pharmacogenomics. *Methods Enzymol*, 400, 485–510.

Ishikawa T, Sakurai A, Hirano H, Lezhava A, Sakurai M, Hayashizaki Y (2010) Emerging new technologies in pharmacogenomics: rapid SNP detection, molecular dynamic simulation, and QSAR analysis methods to validate clinically important genetic variants of human ABC transporter ABCB1 (P-gp/MDR1). *Pharmacol Ther*, 126, 69–81.

Jamroziak K, Mlynarski W, Balcerczak E, Mistygacz M, Trelinska J, Mirowski M, Bodalski J, Robak T (2004) Functional C3435T polymorphism of MDR1 gene: an impact on genetic susceptibility and clinical outcome of childhood acute lymphoblastic leukemia. *Eur J Haematol*, 72, 314–321.

Janneh O, Jones E, Chandler B, Owen A, Khoo SH (2007) Inhibition of P-glycoprotein and multidrug resistance-associated proteins modulates the intracellular concentration of lopinavir in cultured CD4 T cells and primary human lymphocytes. *J Antimicrob Chemother*, 60, 987–993.

Jeong H, Herskowitz I, Kroetz DL, Rine J (2007) Function-altering SNPs in the human multidrug transporter gene ABCB1 identified using a Saccharomyces-based assay. *PLoS Genet*, 3, e39.

Jiang ZP, Wang YR, Xu P, Liu RR, Zhao XL, Chen FP (2008) Meta-analysis of the effect of MDR1 C3435T polymorphism on cyclosporine pharmacokinetics. *Basic Clin Pharmacol Toxicol*, 103, 433–444.

Johne A, Brockmoller J, Bauer S, Maurer A, Langheinrich M, Roots I (1999) Pharmacokinetic interaction of digoxin with an herbal extract from St John's wort (Hypericum perforatum). *Clin Pharmacol Ther*, 66, 338–345.

Johne A, Kopke K, Gerloff T, Mai I, Rietbrock S, Meisel C, Hoffmeyer S, Kerb R, Fromm MF, Brinkmann U, et al. (2002) Modulation of steady-state kinetics of digoxin by haplotypes of the P-glycoprotein MDR1 gene. *Clin Pharmacol Ther*, 72, 584–594.

Juranka PF, Zastawny RL, Ling V (1989) P-glycoprotein: multidrug-resistance and a superfamily of membrane-associated transport proteins. *FASEB J*, 3, 2583–2592.

Kafka A, Sauer G, Jaeger C, Grundmann R, Kreienberg R, Zeillinger R, Deissler H (2003) Polymorphism C3435T of the MDR-1 gene predicts response to preoperative chemotherapy in locally advanced breast cancer. *Int J Oncol*, 22, 1117–1121.

Kim DH, Park JY, Sohn SK, Lee NY, Baek JH, Jeon SB, Kim JG, Suh JS, Do YR, Lee KB (2006a) Multidrug resistance-1 gene polymorphisms associated with treatment outcomes in de novo acute myeloid leukemia. *Int J Cancer*, 118, 2195–2201.

Kim DW, Kim M, Lee SK, Kang R, Lee SY (2006b) Lack of association between C3435T nucleotide MDR1 genetic polymorphism and multidrug-resistant epilepsy. *Seizure*, 15, 344–347.

Kim RB, Leake BF, Choo EF, Dresser GK, Kubba SV, Schwarz UI, Taylor A, Xie HG, McKinsey J, Zhou S, et al. (2001) Identification of functionally variant MDR1 alleles among European Americans and African Americans. *Clin Pharmacol Ther*, 70, 189–199.

Kim YO, Kim MK, Woo YJ, Lee MC, Kim JH, Park KW, Kim EY, Roh YI, Kim CJ (2006c) Single nucleotide polymorphisms in the multidrug resistance 1 gene in Korean epileptics. *Seizure*, 15, 67–72.

Kimchi-Sarfaty C, Gribar JJ, Gottesman MM (2002) Functional characterization of coding polymorphisms in the human MDR1 gene using a vaccinia virus expression system. *Mol Pharmacol*, 62, 1–6.

Kimchi-Sarfaty C, Oh JM, Kim IW, Sauna ZE, Calcagno AM, Ambudkar SV, Gottesman MM (2007) A "silent" polymorphism in the MDR1 gene changes substrate specificity. *Science*, 315, 525–528.

Kishi S, Yang W, Boureau B, Morand S, Das S, Chen P, Cook EH, Rosner GL, Schuetz E, Pui CH, Relling MV (2004) Effects of prednisone and genetic polymorphisms on etoposide disposition in children with acute lymphoblastic leukemia. *Blood*, 103, 67–72.

Kuypers DR, de Jonge H, Naesens M, Vanrenterghem Y (2010) A prospective, open-label, observational clinical cohort study of the association between delayed renal allograft function, tacrolimus exposure, and CYP3A5 genotype in adult recipients. *Clin Ther*, 32, 2012–2023.

Kwan P, Baum L, Wong V, Ng PW, Lui CH, Sin NC, Hui AC, Yu E, Wong LK (2007) Association between ABCB1 C3435T polymorphism and drug-resistant epilepsy in Han Chinese. *Epilepsy Behav*, 11, 112–117.

Lakhan R, Misra UK, Kalita J, Pradhan S, Gogtay NJ, Singh MK, Mittal B (2009) No association of ABCB1 polymorphisms with drug-refractory epilepsy in a North Indian population. *Epilepsy Behav*, 14, 78–82.

Lelong-Rebel IH, Cardarelli CO (2005) Differential phosphorylation patterns of P-glycoprotein reconstituted into a proteoliposome system: insight into additional unconventional phosphorylation sites. *Anticancer Res*, 25, 3925–3935.

Leschziner GD, Andrew T, Leach JP, Chadwick D, Coffey AJ, Balding DJ, Bentley DR, Pirmohamed M, Johnson MR (2007) Common ABCB1 polymorphisms are not associated with multidrug resistance in epilepsy using a gene-wide tagging approach. *Pharmacogenet Genomics*, 17, 217–220.

Levran O, O'Hara K, Peles E, Li D, Barral S, Ray B, Borg L, Ott J, Adelson M, Kreek MJ (2008) ABCB1 (MDR1) genetic variants are associated with methadone doses required for effective treatment of heroin dependence. *Hum Mol Genet*, 17, 2219–2227.

Loscher W, Klotz U, Zimprich F, Schmidt D (2009) The clinical impact of pharmacogenetics on the treatment of epilepsy. *Epilepsia*, 50, 1–23.

Lotsch J, Skarke C, Wieting J, Oertel BG, Schmidt H, Brockmoller J, Geisslinger G (2006) Modulation of the central nervous effects of levomethadone by genetic polymorphisms potentially affecting its metabolism, distribution, and drug action. *Clin Pharmacol Ther*, 79, 72–89.

Lotsch J, Tegeder I, Angst MS, Geisslinger G (2000) Antinociceptive effects of morphine-6-glucuronide in homozygous MDR1a P-glycoprotein knockout and in wild-type mice in the hotplate test. *Life Sci*, 66, 2393–2403.

Lotsch J, von Hentig N, Freynhagen R, Griessinger N, Zimmermann M, Doehring A, Rohrbacher M, Sittl R, Geisslinger G (2009) Cross-sectional analysis of the influence of currently known pharmacogenetic modulators on opioid therapy in outpatient pain centers. *Pharmacogenet Genomics*, 19, 429–436.

Lucia MB, Rutella S, Leone G, Vella S, Cauda R (2001) HIV-protease inhibitors contribute to P-glycoprotein efflux function defect in peripheral blood lymphocytes from HIV-positive patients receiving HAART. *J Acquir Immune Defic Syndr*, 27, 321–330.

Mealey KL, Bentjen SA, Gay JM, Cantor GH (2001) Ivermectin sensitivity in collies is associated with a deletion mutation of the MDR1 gene. *Pharmacogenetics*, 11, 727–733.

Miao LY, Huang CR, Hou JQ, Qian MY (2008) Association study of ABCB1 and CYP3A5 gene polymorphisms with sirolimus trough concentration and dose requirements in Chinese renal transplant recipients. *Biopharm Drug Dispos*, 29, 1–5.

Miller DS, Bauer B, Hartz AM (2008) Modulation of P-glycoprotein at the blood–brain barrier: opportunities to improve central nervous system pharmacotherapy. *Pharmacol Rev*, 60, 196–209.

Molinari A, Calcabrini A, Meschini S, Stringaro A, Crateri P, Toccacieli L, Marra M, Colone M, Cianfriglia M, Arancia G (2002) Subcellular detection and localization of the drug transporter P-glycoprotein in cultured tumor cells. *Curr Protein Pept Sci*, 3, 653–670.

Mosyagin I, Runge U, Schroeder HW, Dazert E, Vogelgesang S, Siegmund W, Warzok RW, Cascorbi I (2008) Association of ABCB1 genetic variants 3435C>T and 2677G>T to ABCB1 mRNA and protein expression in brain tissue from refractory epilepsy patients. *Epilepsia*, 49, 1555–1561.

Mourad M, Mourad G, Wallemacq P, Garrigue V, Van Bellingen C, Van Kerckhove V, De Meyer M, Malaise J, Eddour DC, Lison D, et al. (2005) Sirolimus and tacrolimus trough concentrations and dose requirements after kidney transplantation in relation to CYP3A5 and MDR1 polymorphisms and steroids. *Transplantation*, 80, 977–984.

Ozgon GO, Bebek N, Gul G, Cine N (2008) Association of MDR1 (C3435T) polymorphism and resistance to carbamazepine in epileptic patients from Turkey. *Eur Neurol*, 59, 67–70.

Panwala CM, Jones JC, Viney JL (1998) A novel model of inflammatory bowel disease: mice deficient for the multiple drug resistance gene, MDR1A, spontaneously develop colitis. *J Immunol*, 161, 5733–5744.

Petrova DT, Nedeva P, Maslyankov S, Toshev S, Yaramov N, Atanasova S, Toncheva D, Oellerich M, von Ahsen N (2008) No association between MDR1 (ABCB1) 2677G>T and 3435C>T polymorphism and sporadic colorectal cancer among Bulgarian patients. *J Cancer Res Clin Oncol*, 134, 317–322.

Plasschaert SL, Groninger E, Boezen M, Kema I, de Vries EG, Uges D, Veerman AJ, Kamps WA, Vellenga E, de Graaf SS, de Bont ES (2004) Influence of functional polymorphisms of the MDR1 gene on vincristine pharmacokinetics in childhood acute lymphoblastic leukemia. *Clin Pharmacol Ther*, 76, 220–229.

Press RR, Ploeger BA, den Hartigh J, van der ST, van Pelt H, Danhof M, de Fijter H, Guchelaar HJ (2010) Explaining variability in ciclosporin exposure in adult kidney transplant recipients. *Eur J Clin Pharmacol*, 66, 579–590.

Renders L, Frisman M, Ufer M, Mosyagin I, Haenisch S, Ott U, Caliebe A, Dechant M, Braun F, Kunzendorf U, Cascorbi I (2007) CYP3A5 genotype markedly influences the pharmacokinetics of tacrolimus and sirolimus in kidney transplant recipients. *Clin Pharmacol Ther*, 81, 228–234.

Riordan JR, Deuchars K, Kartner N, Alon N, Trent J, Ling V (1985) Amplification of P-glycoprotein genes in multidrug-resistant mammalian cell lines. *Nature*, 316, 817–819.

Rodriguez-Novoa S, Martin-Carbonero L, Barreiro P, Gonzalez-Pardo G, Jimenez-Nacher I, Gonzalez-Lahoz J, Soriano V (2007) Genetic factors influencing atazanavir plasma concentrations and the risk of severe hyperbilirubinemia. *AIDS*, 21, 41–46.

Ross JR, Riley J, Taegetmeyer AB, Sato H, Gretton S, du Bois RM, Welsh KI (2008) Genetic variation and response to morphine in cancer patients: catechol-O-methyltransferase and multidrug resistance-1 gene polymorphisms are associated with central side effects. *Cancer*, 112, 1390–1403.

Saitoh A, Capparelli E, Aweeka F, Sarles E, Singh KK, Kovacs A, Burchett SK, Wiznia A, Nachman S, Fenton T, Spector SA (2009) CYP2C19 genetic variants affect nelfinavir pharmacokinetics and virologic response in HIV-1-infected children receiving highly active antiretroviral therapy. *J Acquir Immune Defic Syndr*, 54, 285–289.

Sakaeda T (2005) MDR1 genotype-related pharmacokinetics: fact or fiction? *Drug Metab Pharmacokinet*, 20, 391–414.

Sakaeda T, Nakamura T, Horinouchi M, Kakumoto M, Ohmoto N, Sakai T, Morita Y, Tamura T, Aoyama N, Hirai M, et al. (2001a) MDR1 genotype-related pharmacokinetics of digoxin after single oral administration in healthy Japanese subjects. *Pharm Res*, 18, 1400–1404.

Sakaeda T, Okamura N, Nagata S, Yagami T, Horinouchi M, Okumura K, Yamashita F, Hashida M (2001b) Molecular and pharmacokinetic properties of 222 commercially available oral drugs in humans. *Biol Pharm Bull*, 24, 935–940.

Schaefer M, Roots I, Gerloff T (2006) *In vitro* transport characteristics discriminate wild-type ABCB1 (MDR1) from ALA893SER and ALA893THR polymorphisms. *Pharmacogenet Genomics*, 16, 855–861.

Schinkel AH, Mayer U, Wagenaar E, Mol CA, van Deemter L, Smit JJ, van der Valk MA, Voordouw AC, Spits H, van Tellingen O, et al. (1997) Normal viability and altered pharmacokinetics in mice lacking MDR1-type (drug-transporting) P-glycoproteins. *Proc Natl Acad Sci U S A*, 94, 4028–4033.

Schinkel AH, Wagenaar E, Mol CA, van Deemter L (1996) P-glycoprotein in the blood–brain barrier of mice influences the brain penetration and pharmacological activity of many drugs. *J Clin Invest*, 97, 2517–2524.

Schuetz EG, Furuya KN, Schuetz JD (1995) Interindividual variation in expression of P-glycoprotein in normal human liver and secondary hepatic neoplasms. *J Pharmacol Exp Ther*, 275, 1011–1018.

Seedhouse CH, Grundy M, White P, Li Y, Fisher J, Yakunina D, Moorman AV, Hoy T, Russell N, Burnett A, Pallis M (2007) Sequential influences of leukemia specific and genetic factors on P-glycoprotein expression in blasts from 817 patients entered into the National Cancer Research Network acute myeloid leukemia 14 and 15 trials. *Clin Cancer Res*, 13, 7059–7066.

Semsei AF, Erdelyi DJ, Ungvari I, Kamory E, Csokay B, Andrikovics H, Tordai A, Csagoly E, Falus A, Kovacs GT, Szalai C (2008) Association of some rare haplotypes and genotype combinations in the MDR1 gene with childhood acute lymphoblastic leukaemia. *Leuk Res*, 32, 1214–1220.

Seo T, Ishitsu T, Ueda N, Nakada N, Yurube K, Ueda K, Nakagawa K (2006) ABCB1 polymorphisms influence the response to antiepileptic drugs in Japanese epilepsy patients. *Pharmacogenomics*, 7, 551–561.

Shahwan A, Murphy K, Doherty C, Cavalleri GL, Muckian C, Dicker P, McCarthy M, Kinirons P, Goldstein D, Delanty N (2007) The controversial association of ABCB1 polymorphisms in refractory epilepsy: an analysis of multiple SNPs in an Irish population. *Epilepsy Res*, 73, 192–198.

Siddiqui A, Kerb R, Weale ME, Brinkmann U, Smith A, Goldstein DB, Wood NW, Sisodiya SM (2003) Association of multidrug resistance in epilepsy with a polymorphism in the drug-transporter gene ABCB1. *N Engl J Med*, 348, 1442–1448.

Siegmund W, Ludwig K, Giessmann T, Dazert P, Schroeder E, Sperker B, Warzok R, Kroemer HK, Cascorbi I (2002) The effects of the human MDR1 genotype on the expression of duodenal P-glycoprotein and disposition of the probe drug talinolol. *Clin Pharmacol Ther*, 72, 572–583.

Sills GJ, Mohanraj R, Butler E, McCrindle S, Collier L, Wilson EA, Brodie MJ (2005) Lack of association between the C3435T polymorphism in the human multidrug resistance (MDR1) gene and response to antiepileptic drug treatment. *Epilepsia*, 46, 643–647.

Singh R, Srivastava A, Kapoor R, Mittal RD (2011) Do drug transporter (ABCB1) SNPs influence cyclosporine and tacrolimus dose requirements and renal allograft outcome in the posttransplantation period? *J Clin Pharmacol*, 51, 603–615.

Skarke C, Jarrar M, Schmidt H, Kauert G, Langer M, Geisslinger G, Lotsch J (2003) Effects of ABCB1 (multidrug resistance transporter) gene mutations on disposition and central nervous effects of loperamide in healthy volunteers. *Pharmacogenetics*, 13, 651–660.

Solas C, Simon N, Drogoul MP, Quaranta S, Frixon-Marin V, Bourgarel-Rey V, Brunet C, Gastaut JA, Durand A, Lacarelle B, Poizot-Martin I (2007) Minimal effect of MDR1 and CYP3A5 genetic polymorphisms on the pharmacokinetics of indinavir in HIV-infected patients. *Br J Clin Pharmacol*, 64, 353–362.

Soranzo N, Cavalleri GL, Weale ME, Wood NW, Depondt C, Marguerie R, Sisodiya SM, Goldstein DB (2004) Identifying candidate causal variants responsible for altered activity of the ABCB1 multidrug resistance gene. *Genome Res*, 14, 1333–1344.

Staatz CE, Goodman LK, Tett SE (2010) Effect of CYP3A and ABCB1 single nucleotide polymorphisms on the pharmacokinetics and pharmacodynamics of calcineurin inhibitors: part I. *Clin Pharmacokinet*, 49, 141–175.

Stanulla M, Schaffeler E, Arens S, Rathmann A, Schrauder A, Welte K, Eichelbaum M, Zanger UM, Schrappe M, Schwab M (2005) GSTP1 and MDR1 genotypes and central nervous system relapse in childhood acute lymphoblastic leukemia. *Int J Hematol*, 81, 39–44.

Svirnovski AI, Shman TV, Serhiyenka TF, Savitski VP, Smolnikova VV, Fedasenka UU (2009) ABCB1 and ABCG2 proteins, their functional activity and gene expression in concert with drug sensitivity of leukemia cells. *Hematology*, 14, 204–212.

Szoeke C, Sills GJ, Kwan P, Petrovski S, Newton M, Hitiris N, Baum L, Berkovic SF, Brodie MJ, Sheffield LJ, O'Brien TJ (2009) Multidrug-resistant genotype (ABCB1) and seizure recurrence in newly treated epilepsy: data from International Pharmacogenetic Cohorts. *Epilepsia*, 50, 1689–1696.

Taegtmeyer AB, Breen JB, Smith J, Burke M, Leaver N, Pantelidis P, Lyster H, Yacoub MH, Barton PJ, Banner NR (2010) ATP-binding cassette subfamily B member 1 polymorphisms do not determine cyclosporin exposure, acute rejection or nephrotoxicity after heart transplantation. *Transplantation*, 89, 75–82.

Tan NC, Heron SE, Scheffer IE, Pelekanos JT, McMahon JM, Vears DF, Mulley JC, Berkovic SF (2004) Failure to confirm association of a polymorphism in ABCB1 with multidrug-resistant epilepsy. *Neurology*, 63, 1090–1092.

Tanabe M, Ieiri I, Nagata N, Inoue K, Ito S, Kanamori Y, Takahashi M, Kurata Y, Kigawa J, Higuchi S, et al. (2001) Expression of P-glycoprotein in human placenta: relation to genetic polymorphism of the multidrug resistance (MDR)-1 gene. *J Pharmacol Exp Ther*, 297, 1137–1143.

Terao T, Hisanaga E, Sai Y, Tamai I, Tsuji A (1996) Active secretion of drugs from the small intestinal epithelium in rats by P-glycoprotein functioning as an absorption barrier. *J Pharm Pharmacol*, 48, 1083–1089.

Thorn M, Finnstrom N, Lundgren S, Rane A, Loof L (2005) Cytochromes P450 and MDR1 mRNA expression along the human gastrointestinal tract. *Br J Clin Pharmacol*, 60, 54–60.

Ufer M, Dilger K, Leschhorn L, Daufresne L, Mosyagin I, Rosenstiel P, Haesler R, Kuehbacher T, Nikolaus S, Schreiber S, Cascorbi I (2008) Influence of CYP3A4, CYP3A5, and ABCB1 genotype and expression on budesonide pharmacokinetics: a possible role of intestinal CYP3A4 expression. *Clin Pharmacol Ther*, 84, 43–46.

Ufer M, Mosyagin I, Muhle H, Jacobsen T, Haenisch S, Hasler R, Faltraco F, Remmler C, von Spiczak S, Kroemer HK, et al. (2009) Non-response to antiepileptic pharmacotherapy is associated with the ABCC2 -24C>T polymorphism in young and adult patients with epilepsy. *Pharmacogenet Genomics*, 19, 353–362.

Uhr M, Tontsch A, Namendorf C, Ripke S, Lucae S, Ising M, Dose T, Ebinger M, Rosenhagen M, Kohli M, et al. (2008) Polymorphisms in the drug transporter gene ABCB1 predict antidepressant treatment response in depression. *Neuron*, 57, 203–209.

Vahab SA, Sen S, Ravindran N, Mony S, Mathew A, Vijayan N, Nayak G, Bhaskaranand N, Banerjee M, Satyamoorthy K (2009) Analysis of genotype and haplotype effects of ABCB1 (MDR1) polymorphisms in the risk of medically refractory epilepsy in an Indian population. *Drug Metab Pharmacokinet*, 24, 255–260.

van der Holt B, Van den Heuvel-Eibrink MM, Van Schaik RH, van dH, I, Wiemer EA, Vossebeld PJ, Lowenberg B, Sonneveld P (2006) ABCB1 gene polymorphisms are not associated with treatment outcome in elderly acute myeloid leukemia patients. *Clin Pharmacol Ther*, 80, 427–439.

von Richter O, Burk O, Fromm MF, Thon KP, Eichelbaum M, Kivisto KT (2004) Cytochrome P450 3A4 and P-glycoprotein expression in human small intestinal enterocytes and hepatocytes: a comparative analysis in paired tissue specimens. *Clin Pharmacol Ther*, 75, 172–183.

Wang P, Mao Y, Razo J, Zhou X, Wong ST, Patel S, Elliott E, Shea E, Wu AH, Gaber AO (2010) Using genetic and clinical factors to predict tacrolimus dose in renal transplant recipients. *Pharmacogenomics*, 11, 1389–1402.

Winzer R, Langmann P, Zilly M, Tollmann F, Schubert J, Klinker H, Weissbrich B (2003) No influence of the P-glycoprotein genotype (MDR1 C3435T) on plasma levels of lopinavir and efavirenz during antiretroviral treatment. *Eur J Med Res*, 8, 531–534.

Yanagimachi M, Naruto T, Tanoshima R, Kato H, Yokosuka T, Kajiwara R, Fujii H, Tanaka F, Goto H, Yagihashi T, et al. (2010) Influence of CYP3A5 and ABCB1 gene polymorphisms on calcineurin inhibitor-related neurotoxicity after hematopoietic stem cell transplantation. *Clin Transplant*, 24, 855–861.

Yang Z, Wu D, Bui T, Ho RJ (2008) A novel human multidrug resistance gene MDR1 variant G571A (G191R) modulates cancer drug resistance and efflux transport. *J Pharmacol Exp Ther*, 327, 474–481.

Zhang Y, Benet LZ (2001) The gut as a barrier to drug absorption: combined role of cytochrome P450 3A and P-glycoprotein. *Clin Pharmacokinet*, 40, 159–168.

Zimprich F, Sunder-Plassmann R, Stogmann E, Gleiss A, Dal Bianco A, Zimprich A, Plumer S, Baumgartner C, Mannhalter C (2004) Association of an ABCB1 gene haplotype with pharmacoresistance in temporal lobe epilepsy. *Neurology*, 63, 1087–1089.

Zwisler ST, Enggaard TP, Noehr-Jensen L, Mikkelsen S, Verstuyft C, Becquemont L, Sindrup SH, Brosen K (2010) The antinociceptive effect and adverse drug reactions of oxycodone in human experimental pain in relation to genetic variations in the OPRM1 and ABCB1 genes. *Fundam Clin Pharmacol*, 24, 517–524.

CHAPTER 13

BSEP (ABCB11)

Bruno Stieger
Gerd A. Kullak-Ublick

13.1 INTRODUCTION

Bile formation is one of the key functions of the liver. Bile is delivered from the liver to the duodenum. There, it is needed for the digestion of fat and the absorption of lipids from our diet. Fat digestion takes advantage of the detergent properties of bile salts (Hofmann and Hagey, 2008; Hofmann, 2009a). In the small intestine, bile salts are almost quantitatively reabsorbed and moved back to the liver via the portal circulation. About 500 mg of bile salts are lost daily and excreted via feces (Dawson, 2011). In the liver, bile salts are actively removed from the blood plasma and secreted again into the bile for delivery to the intestine. This entire process is called enterohepatic circulation of bile salts (Dawson et al., 2009; Hofmann, 2009b).

In hepatocytes, which are the most abundant cells in liver and which drive bile formation, bile acids are synthesized in a multistep reaction cascade from cholesterol (Russell, 2009) and conjugated to glycine or taurine (Trottier et al., 2006). Consequently, bile acid biosynthesis is one way of excess cholesterol elimination from the body. As bile salts are potentially dangerous to hepatocytes and cholesterol homeostasis needs to be tightly controlled, both bile acid biosynthesis and bile salt handling of hepatocytes are tightly controlled at the transcriptional and post-transcriptional level. (Chiang, 2009; Klaassen and Aleksunes, 2010). Bile formation starts at the canaliculus of hepatocytes. Principal components of the bile fluid are bile salts, phospholipids, organic ions, small ions, and proteins (Esteller, 2008). The formation of bile is an iso-osmotic process (Forker, 1977) implying that water follows passively the osmotic activity of the solutes secreted into the canaliculi. Thereby, water reaches the canaliculi by a paracellular route via tight junctions between hepatocytes and by a transcellular route across hepatocytes mediated by aquaporins (Masyuk and LaRusso, 2006). As a consequence of this iso-osmotic process, with increasing canalicular secretion of bile constituents, bile flow increases. Animal experiments and measurements in patients with T tubes have revealed that bile flow has a bile salt-dependent and a bile salt-independent component, the relative proportion of which is species dependent (Hofmann, 1990). Any impairment of bile flow is a pathophysiologic process and is called cholestasis. If cholestasis persists, bile salts may accumulate in hepatocytes,

Pharmacogenomics of Human Drug Transporters: Clinical Impacts, First Edition.
Edited by Toshihisa Ishikawa, Richard B. Kim and Jörg König

lead to cellular damage, and ultimately to cholestatic liver disease (Trauner et al., 1998; Pauli-Magnus et al., 2005).

13.2 HEPATOCELLULAR BILE SALT TRANSPORT

Bile salts are extracted from the portal blood plasma and taken up largely in a sodium-dependent manner and to a minor portion by (a) sodium-independent process(es) (Meier, 1988). The sodium-dependent transport into hepatocytes is mediated by the so-called sodium-taurocholate cotransporting polypeptide (NTCP, *SLC10A1*) (Meier and Stieger, 2002; Stieger, 2011). NTCP has a preference for bile salts amidated with glycine or taurine over unconjugated bile acids (Stieger, 2011). The sodium-independent uptake of bile salts is mediated by the so-called organic anion transporting polypeptides (OATPs) (Meier and Stieger, 2002; Hagenbuch and Dawson, 2004). In human hepatocytes OATP1B1, OATP1B3, and OATP2B1 *(SLCO1B1, SLCO1B3,* and *SLCO2B1)* are expressed. They favor unconjugated bile acids over conjugated bile salts as substrates (Meier and Stieger, 2002; Hagenbuch and Meier, 2004). These four bile salt transporters are specifically expressed in the basolateral plasma membrane of hepatocytes (Kullak-Ublick et al., 1997, 2001). After having crossed the basolateral membrane, bile salts reach the canalicular membrane by diffusing across the hepatocytes. This process is not worked out in molecular details but most likely involves binding proteins (Agellon and Torchia, 2000). Export across the canalicular membrane occurs against a steep concentration gradient and is mediated by the ATP-binding cassette transporter, bile salt export pump (BSEP, *ABCB11*) (Stieger et al., 2007). The rate-limiting step of transcellular bile salt transport across hepatocytes is located to the canalicular cell pole (Reichen and Paumgartner, 1976). Bile salts are amphipathic molecules and are potentially cytotoxic to hepatocytes (Krahenbuhl et al., 1994; Sokol et al., 2006). Therefore, their uptake as well as their export is tightly regulated at the transcriptional and post-transcriptional level.

The intracellular level of bile salts is monitored in hepatocytes by a bile salt sensor, which is the nuclear transcription factor, farnesoid X receptor (FXR, *NR1H4*) (Eloranta and Kullak-Ublick, 2008). In order to regulate gene transcription, FXR requires the interaction with its heterodimeric partner, retinoid X receptor (RXR, *NR2B1*) (Eloranta and Kullak-Ublick, 2008; Lo Sasso et al., 2008). The activation of FXR is stronger with hydrophobic bile acids compared to hydrophilic bile acids. Increasing intracellular bile salts will activate FXR, which in turn upregulates the transcription of a number of hepatocellular transport proteins, including BSEP, MRP2, and MDR3 at the canalicular membrane, and OATP1B3 and the heterodimeric organic solute transporter OSTα/OSTβ (Landrier et al., 2006) at the basolateral hepatocytes membrane. FXR also induces the transcriptional repressor small heterodimer partner 1 (SHP-1, *NR0B2*) (Eloranta et al., 2006). This protein represses *SLC10A1* in a species-dependent manner and consequently leads to lower NTCP levels in hepatocytes (Eloranta et al., 2006). At the post-transcriptional level, NTCP is a phosphoprotein and its phosphorylation state determines the partitioning between the basolateral plasma membrane and an intracellular endosomal compartment of hepatocytes (Anwer, 2004). Thereby, bile salts activate phosphoinositide-3-kinase, which in turn activates Rab4 and consequently the insertion of NTCP into the basolateral hepatocyte membrane (Schonhoff et al., 2008). *SLCO1B1* transcription is indirectly regulated by FXR, while FXR is a direct positive regulator of *SLCO1B3* (Eloranta and Kullak-Ublick, 2008; Klaassen and Aleksunes, 2010; Tirona, 2011). In the case of *SLCO1B1*, FXR induces SHP, which in turn by interacting with hepatocyte nuclear factor (HNF) 4α inhibits HNF1α activation of *SLCO1B1*. HNF1α is the key activator of *SLCO1B1*. In addition, bile salts interact directly with HNF4α leading to decreased nuclear binding of this transcription factor (Jung and Kullak-Ublick, 2003). The concept of indirect downregulation of *SLCO1B1* through an effect of FXR on HNF1α may in part explain the decrease of OATP1B1 mRNA levels observed in liver biopsies of patients with cholestatic liver disease (Zollner et al., 2001). In normal liver, FXR may also exert an activating effect on OATP1B1 expression. Thus, in patients carrying an *FXR *1b* allele, which shows a reduced transactivation activity compared to wild-type FXR in an *in vitro* assay, OATP1B1 mRNA was significantly downregulated in liver biopsies (Marzolini et al., 2007). Two patients receiving the pregnane X receptor (PXR) ligand, carbamazepine, showed elevated OATP1B1 mRNA (Oscarson et al., 2006) compared to seven control liver samples suggesting that PXR may in some way regulate *SLCO1B1*, an interpretation, which is in concordance with induction studies in primary cultured human hepatocytes (Jigorel et al., 2006). Experiments with primary cultured hepatocytes demonstrate a downregulation of OATP1B1 and OATP2B1 by the aryl hydrocarbon receptor (AhR) and by the constitutive androstane receptor (CAR) (Jigorel et al., 2006). Taken together, the transcriptional regulation of hepatocellular OATPs is complex. Phosphorylation of OATPs leads in the case of OATP1B1 to increased plasma membrane expression (Sun et al., 2008) and to a downregulation of OATP2B1 at the plasma membrane (Kock et al., 2010).

In vitro experiments have demonstrated that *ABCB11* is under strong positive control of FXR (Ananthanarayanan et al., 2001; Plass et al., 2002). Interestingly, however, in patients carrying a *FXR*1b* allele, BSEP mRNA is not significantly reduced but only shows a tendency towards lower expression (Marzolini et al., 2007), which may be due to a nonsignificant reduction of FXR in the same samples. In primary cultured human hepatocytes, activation of AhR, CAR, and PXR leads to a downregulation of BSEP mRNA (Jigorel et al., 2006), while in livers from patients under carbamazepine treatment, BSEP was found to be upregulated (Oscarson et al., 2006).

13.3 BSEP IN NORMAL PHYSIOLOGY

BSEP is the main driving force of the enterohepatic circulation and as such critically important for many functions of the liver and the intestine. These various aspects have been reviewed extensively lately (Stieger et al., 2007; Alrefai and Gill, 2007; Kosters and Karpen, 2008; Dawson et al., 2009; Lam et al., 2010; Pauli-Magnus et al., 2010; Stieger, 2010, 2011; Stieger and Beuers, 2011; Stieger and Geier, 2011). ATP-dependent bile salt transport was identified in canalicular plasma membrane vesicles from rat in the mid 1990s (Adachi et al., 1991; Muller et al., 1991; Nishida et al., 1991; Stieger et al., 1992). These studies delineated the functional properties of the canalicular transporter(s) involved in bile salt secretion: ATP dependence and ATP hydrolysis. In the same decade, rat Bsep was cloned and identified as an ATP-dependent BSEP (Gerloff et al., 1998). This ABC transporter is member 11 of the ABCB family of ABC transporters and requires ATP hydrolysis for its transport activity. Bsep/BSEP has subsequently been cloned from multiple species including mouse (Green et al., 2000; Lecureur et al., 2000; Noe et al., 2001), skate (Cai et al., 2001), human (Byrne et al., 2002; Noe et al., 2002), dog (Yabuuchi et al., 2008), or rainbow trout (Zaja et al., 2008). BSEP is exclusively expressed at the canalicular cell pole of hepatocytes and transports almost only conjugated bile salts (Stieger, 2011). For rat Bsep, unconjugated bile acids have been shown to be very poor substrates (Gerloff et al., 1998) and in humans, patients with a defect in bile acid conjugation have practically no bile acids in their bile (Carlton et al., 2003). Comparison of the substrate pattern of rat Bsep and Mrp2 has suggested that Bsep is the only canalicular export system for bile salts (Stieger et al., 2000). The secondary bile salt lithocholate is a substrate neither for rat nor for human BSEP (Mita et al., 2005, 2006). While the sulfated dianionic bile salt taurolithocholate-3 sulfate is not (Stieger et al., 2000; Akita et al., 2001) a rat Bsep substrate in Sf9 cells or only minimally transported by rat Bsep expressed in HEK cells (Hayashi et al., 2005), human BSEP transports this bile salt metabolite (Hayashi et al., 2005). The fluorescent bile salt derivative cholylglycylamido-fluorescein is transported by BSEP (Mita et al., 2006), however, the different derivative cholyl-L-lysyl-fluorescein is not a substrate of BSEP (de Waart et al., 2010). This demonstrates that the bile salt side chain is an important determinant for potential BSEP substrates. Comparing different expression systems for BSEP/Bsep, it becomes clear that its transport function is more or less independent of the expression system (Stieger, 2011). However, there is one limitation to this: the transport activity of Bsep positively depends on the cholesterol content of its membrane environment (Kis et al., 2009; Paulusma et al., 2009). In this context, the partitioning of Bsep into cholesterol-enriched microdomains of the canalicular membrane offers the possibility that the cholesterol content in the direct vicinity of Bsep may constitute a regulatory element (Ismair et al., 2009). Comparison of the intrinsic clearance of BSEP from human, rat, and mouse showed the same rank order for all three species (Noe et al., 2002). This demonstrates that the substrate pattern of BSEP, with the notable exception of taurolithocholate-3 sulfate is conserved over different species. This suggests that findings obtained in animal models, such as rats, can be extrapolated to humans with the necessary caution. To date, only pravastatin has been identified as a drug substrate for human BSEP, which displays a much higher transport capacity specifically for this statin in comparison to the rat orthologue (Hirano et al., 2005).

13.4 INHERITED BSEP IMPAIRMENT

There are several forms of progressive familial intrahepatic cholestasis (PFIC) (Davit-Spraul et al., 2009). Patients with PFIC type 2 (PFIC2) develop severe progressive liver disease, which may lead to liver failure. Patients with PFIC2 have mutations in the *ABCB11* gene leading to the absence of functional BSEP in the canalicular membrane (Strautnieks et al., 1998; Jansen et al., 1999). This clinical observation supports the concept that no backup system exists for BSEP-mediating canalicular bile salt export in humans. Some mutations in ABCB11 lead to a milder form of disease with recurrent episodes of cholestasis, called benign recurrent intrahepatic cholestasis type 2 (BRIC2). Together, mutations in *ABCB11* cause BSEP deficiency syndrome (Pauli-Magnus et al., 2005). Mutations in *ABCB11* present with a continuum of liver disease from mild-to-severe intrahepatic cholestasis, which may be progressive from BRIC2 to PFIC2 (Pauli-Magnus et al., 2005; Lam et al., 2006; Takahashi et al., 2007). Frequently occurring BSEP mutations include missense mutations, nonsense mutations, deletions, insertions, and splice site mutations (Strautnieks et al., 2008). Molecular analysis of 82 mutations in BSEP revealed that they cluster in the two nucleotide-binding domains, which are highly conserved (Strautnieks et al., 2008; Noe et al., 2001). Immunohistochemical analysis of many patients with PFIC2 revealed in the majority of patients the absence of BSEP staining, while some patients also display aberrant and seldom normal BSEP staining (Strautnieks et al., 2008). Currently, more than 100 mutations of BSEP are known (Lam et al., 2010), 63 of which were characterized at the mRNA level (Byrne et al., 2009). Twenty of the investigated mutations resulted in reduced splicing activity and subsequent reduced BSEP mRNA levels compared to the wild type. Many of the mutations expressed *in vitro* lead to accumulation in the endoplasmic reticulum with subsequent degradation (Byrne et al., 2009). This finding is a very likely explanation for the lack of BSEP immunostaining on liver biopsies from PFIC2 patients.

13.5 ACQUIRED BSEP IMPAIRMENT

Functional BSEP impairment may not only be a consequence of mutations in the *ABCB11* gene, but may also result from reduced expression or retrieval from the canalicular membrane domain in pathophysiologic situations or from inhibition by endogenous or exogenous substances.

Percutaneous transhepatic biliary drainage in patients with obstructive cholestasis showed that BSEP is downregulated in patients with poor drainage in comparison to well-drained or controlled individuals (Shoda et al., 2001). Furthermore, BSEP staining of biopsies from cholestatic patients was fuzzy, indicating partial retrieval of BSEP from the canalicular membrane. In children with biliary atresia, BSEP mRNA is at the early stages of the disease downregulated, while in later stages it is comparable to controls (Chen et al., 2008). Contrasting the data for mRNA, immunofluorescence staining of BSEP protein in these patients showed in all stages of biliary atresia neither a change in subcellular location of BSEP protein nor in signal intensity (Chen et al., 2008). Patients with inflammatory liver disease show a divergent pattern: patients with inflammatory cholestasis are reported to have downregulated BSEP in contrast to patients with primary biliary cirrhosis stage I and II or with primary sclerosing cholangitis (Zollner et al., 2001, 2003; Kojima et al., 2003), where BSEP remains unchanged. However, upregulation of BSEP in primary biliary cirrhosis was also reported (Ros et al., 2003), leaving this issue controversial. In support of findings from patients, incubation of human liver slices with lipopolysaccharide or incubation of primary cultured human hepatocytes with interleukin-1β or with interferon-γ, which are models for inflammatory liver disease, leads to a decrease of BSEP expression (Elferink et al., 2004; Le Vee et al., 2008, 2011). Ursodeoxycholate is frequently used to treat cholestatic liver disease (Beuers, 2006; Stieger and Beuers, 2011). In healthy gallstone patients receiving ursodeoxycholate, BSEP was found to be upregulated at the protein, but not at the mRNA level (Marschall et al., 2005). The same study found no effect on BSEP expression after treatment with rifampicin, which is at variance with a later study reporting an upregulation of BSEP in patients receiving a PXR activator (Oscarson et al., 2006). After liver transplantation, BSEP mRNA is upregulated, which is paralleled by an increased bile salt secretion (Geuken et al., 2004). The expression of BSEP in hepatocellular carcinoma is highly variable, which may in part reflect the variable degree of cell differentiation in the tumors. While there is consequently no consistent pattern of BSEP expression, a trend towards reduced expression was reported (Zollner et al., 2005; Van der Borght et al., 2007).

Inhibition of BSEP leads to impaired bile salt secretion and cholestasis. The immunosuppressant cyclosporine is associated with cholestasis in patients and was found to be a competitive inhibitor of rat Bsep (Stieger et al., 2000). The same study also identified glibenclamide, rifamycin, and rifampicin as competitive inhibitors of rat Bsep. These drugs are all associated with cholestatic episodes in patients (Stieger et al., 2000). After cloning of human BSEP, all four drugs were found to inhibit BSEP competitively or in a concentration-dependent manner (Byrne et al., 2002; Noe et al., 2002). The clinical relevance of bland cholestasis initiated by BSEP inhibition is highlighted by the example of bosentan, which together with its main metabolite is a dual endothelin receptor antagonist. Patients receiving bosentan developed cholestasis, whereby the percentage of susceptible individuals was positively correlated with the administered dose of bosentan (Fattinger et al., 2001). This effect was even more pronounced when patients who were on glibenclamide received bosentan in addition. After cessation of bosentan, serum liver markers returned rapidly to normal, suggesting a reversible interaction of bosentan with BSEP. This is explained by the observation that bosentan is a competitive inhibitor of BSEP (Noe et al., 2002). A follow-up study in rats revealed that the cholestatic effect of bosentan may, in addition, be due to a stimulation of Mrp2 transport activity with concomitant lowering of intracanalicular bile salt concentration (Fouassier et al., 2002; Meier, 2002). In the meantime, several more drugs were identified as BSEP inhibitors (Stieger and Geier, 2011) and a recent study investigating over 100 drugs for their inhibitory potential towards BSEP demonstrated that drugs exhibiting a low IC_{50} value as BSEP inhibitors were all associated with liver problems observed in patients (Morgan et al., 2010).

Susceptible women develop intrahepatic cholestasis of pregnancy or cholestasis due to oral contraceptives (Pauli-Magnus et al., 2010). Intrahepatic cholestasis of pregnancy has been associated with increased serum levels of estrogen and progesterone and their metabolites (Pauli-Magnus et al., 2010). In animal models, both estradiol-17β-glucuronide and progesterone sulfate application lead to acute cholestasis (Meyers et al., 1980; Vallejo et al., 2006). Studies with Bsep expressed in Sf9 cells interestingly showed no inhibition of its transport activity by estradiol-17β-glucuronide. Only, if in the same vesicles Mrp2 was coexpressed, a time- and concentration-dependent inhibition of the transport activity of Bsep was observed

(Stieger et al., 2000), demonstrating indirect or transinhibition of Bsep. This observation parallels findings in rats, where the acute cholestatic action of estradiol-17β-glucuronide requires functional expression of Mrp2 in the canalicular membrane (Huang et al., 2000). Indirect inhibition of Bsep was confirmed and extended to progesterone sulfate (Akita et al., 2001; Vallejo et al., 2006). In rats, acute administration of cholestatic estrogens leads to retrieval of Bsep from the canalicular membrane (Crocenzi et al., 2003). However, so far no evidence from human liver for this observation has been presented. Indirect inhibition of Bsep has also been demonstrated for the HER1/HER2 inhibitor PKI166 (Takada et al., 2004).

Drug-induced cholestasis due to BSEP inhibition is a rare clinical entity. It is hence evident that such patients harbor susceptibility factors. To date, two polymorphisms in the *ABCB11* gene have been identified in addition to many mutations and variants: c.1331C>T (p.V444A) and c.2029A>G (p.M677V) have consistently been identified in unrelated cohorts with frequencies of more than 0.5% (Saito et al., 2002; Pauli-Magnus et al., 2004; Lang et al., 2006). A study analyzing the relative expression of four canalicular ABC transporters (BSEP, MDR1, MDR3, MRP2) found a considerable interindividual variability in the expression of these transporters at the protein level (Meier et al., 2006). Of note, individuals with the p.444A variant of BSEP tended to have lower BSEP expression levels. Low BSEP expression could constitute a susceptibility factor for drug-induced cholestasis. In a cohort with drug-induced cholestasis and in cohorts with intrahepatic cholestasis of pregnancy, the c.1331T>C allele was observed significantly more frequent than in controls (Lang et al., 2007; Meier et al., 2008; Dixon et al., 2009). Kinetic characterization of the p.444A and of the p444V variant revealed no differences (Lang et al., 2007). Hence, the c.1331C>T variant of *ABCB11* may be a susceptibility factor for acquired forms of cholestasis. As the frequency of this variant is around 50% in different populations, it seems evident that the additional susceptibility factors for acquired cholestasis are awaiting their discovery. Taken together, inhibition of BSEP by drugs or drug metabolites as well as by endogenous substance is now well established as a pathophysiologic mechanism for acquired cholestasis.

13.6 PATHOPHYSIOLOGIC CONSEQUENCES OF BSEP IMPAIRMENT

Inhibition of BSEP leads to an intracellular accumulation of bile salts, which due to their amphipathic properties can act as detergents (Hofmann, 2009b). As such, they have numerous effects and can, for example, damage mitochondria (Krahenbuhl et al., 1994; Palmeira and Rolo, 2004; Perez and Briz, 2009). This may lead to the generation of reactive oxygen species and oxidative stress and ultimately to the activation of apoptosis and cell death (Sokol et al., 2006; Perez and Briz, 2009). Toxicity of bile salts to mitochondria may be aggravated if a drug or a metabolite of a drug is directly toxic to mitochondria. Troglitazone is an antidiabetic drug, which was withdrawn from the market after it displayed severe hepatotoxicity. While the exact toxic action remains to some extent enigmatic at the molecular level, it is toxic to mitochondria (Masubuchi, 2006; Julie et al., 2008). Importantly, the troglitazone metabolite, troglitazone sulfate, as well as the parent compound are inhibitors of BSEP of different species (Funk et al., 2001; Marion et al., 2007; Yabuuchi et al., 2008). Hence, troglitazone can synergistically damage mitochondria by combining its direct mitochondrial toxicity with the elevation of intracellular bile salts following BSEP inhibition. Patients with severe forms of BSEP deficiency syndrome are at a considerable risk of developing hepatocellular carcinoma (Knisely et al., 2006; Strautnieks et al., 2008). The detailed mechanism is not known so far, but a role of bile salts in carcinogenesis is well established experimentally (Debruyne et al., 2001). Bile acids have been shown to interfere with cell cycle control (Atherford and Jankowski, 2006), with homeobox genes (Souza et al., 2008), and with DNA repair mechanisms (Burnat et al., 2010). Bile acids can also reversibly induce polarization of rat hepatoma cells highlighting their pleiotropic action potential (Ng et al., 2000). Recently, patients with inherited BSEP deficiency syndrome were identified, which had developed an autoimmune reaction against BSEP after liver transplantation (Jara et al., 2009; Keitel et al., 2009; Maggiore et al., 2010). It is currently not clear, whether this autoimmune reactions occurred prior to or after liver transplantation.

Elevated bile salts not only inflict damage to hepatocytes, but they also activate protective systems. For example, sulfation leads to detoxification of bile salts (Alnouti, 2009). Furthermore, via activation of nuclear transcription factors, predominantly FXR, bile salt biosynthesis is repressed (Eloranta and Kullak-Ublick, 2005; 2008; Trauner and Halilbasic, 2011; Wagner et al., 2011). FXR induces SHP-1, which in turn represses CYP7A1 and consequently bile acid biosynthesis and SHP-1 also represses the basolateral bile salt uptake systems, NTCP and OATP1B1. At the canalicular side, FXR activates bile salt (ABCB11) and lipid secretion (MDR3). Finally, in the basolateral plasma membrane MRP3, MRP4 as well as OSTα/OSTβ are positively regulated by FXR and these transporters start, in turn, working as salvage systems.

13.7 CONCLUSION

BSEP is the main driving force for enterohepatic circulation of bile salts. As such, it is a key transporter in hepatocytes in health and in disease and subject to extensive regulation at the transcriptional and post-transcriptional levels. At the transcription level, FXR is a pivotal transcription factor. A recent fascinating observation, however, identified adenosine monophosphate-activated protein kinase as a novel important upstream element of the transcriptional regulation of BSEP (Chopra et al., 2011), linking the energy state to the rate of biliary bile salt output. This opens a fascinating new area of research putting BSEP on center stage in energy homeostasis. While we now know a lot about the regulation of BSEP expression in various disease states, we still lack an insight in additional factors making patients susceptible to acquired cholestatic liver disease due to impaired BSEP function. To progress in this area, clearly large cohorts need to be investigated in genomewide association studies. The role of putative new susceptibility factors will then needed to be worked out at the molecular level. Finally, the structure of BSEP at high resolution has not yet been solved. Elucidation of the BSEP structure will certainly help during drug development to early detect molecules, which could interfere with BSEP function. In addition, this structural information could even lead to the identification of activators of BSEP function, which could be used to treat patients with chronic cholestatic liver diseases.

REFERENCES

Adachi Y, Kobayashi H, Kurumi Y, Shouji M, Kitano M, Yamamoto T (1991) ATP-dependent taurocholate transport by rat liver canalicular membrane vesicles. *Hepatology*, 14, 655–659.

Agellon LB, Torchia EC (2000) Intracellular transport of bile acids. *Biochim Biophys Acta*, 1486, 198–209.

Akita H, Suzuki H, Ito K, Kinoshita S, Sato N, Takikawa H, Sugiyama Y (2001) Characterization of bile acid transport mediated by multidrug resistance associated protein 2 and bile salt export pump. *Biochim Biophys Acta*, 1511, 7–16.

Alnouti Y (2009) Bile Acid sulfation: a pathway of bile acid elimination and detoxification. *Toxicol Sci*, 108, 225–246.

Alrefai WA, Gill RK (2007) Bile acid transporters: structure, function, regulation and pathophysiological implications. *Pharm Res*, 24, 1803–1823.

Ananthanarayanan M, Balasubramanian N, Makishima M, Mangelsdorf DJ, Suchy FJ (2001) Human bile salt export pump promoter is transactivated by the farnesoid X receptor/bile acid receptor. *J Biol Chem*, 276, 28857–28865.

Anwer MS (2004) Cellular regulation of hepatic bile acid transport in health and cholestasis. *Hepatology*, 39, 581–590.

Atherford PA, Jankowski JA (2006) Molecular biology of Barrett's cancer. *Best Pract Res Clin Gastroenterol*, 20, 813–827.

Beuers U (2006) Drug insight: mechanisms and sites of action of ursodeoxycholic acid in cholestasis. *Nat Clin Pract Gastroenterol Hepatol*, 3, 318–328.

Burnat G, Majka J, Konturek PC (2010) Bile acids are multifunctional modulators of the Barrett's carcinogenesis. *J Physiol Pharmacol*, 61, 185–192.

Byrne JA, Strautnieks SS, Ihrke G, Pagani F, Knisely AS, Linton KJ, Mieli-Vergani G, Thompson RJ (2009) Missense mutations and single nucleotide polymorphisms in *ABCB11* impair bile salt export pump processing and function or disrupt pre-messenger RNA splicing. *Hepatology*, 49, 553–567.

Byrne JA, Strautnieks SS, Mieli-Vergani G, Higgins CF, Linton KJ, Thompson RJ (2002) The human bile salt export pump: characterization of substrate specificity and identification of inhibitors. *Gastroenterology*, 123, 1649–1658.

Cai SY, Wang L, Ballatori N, Boyer JL (2001) Bile salt export pump is highly conserved during vertebrate evolution and its expression is inhibited by PFIC type II mutations. *Am J Physiol Gastrointest Liver Physiol*, 281, G316–G322.

Carlton VE, Harris BZ, Puffenberger EG, Batta AK, Knisely AS, Robinson DL, Strauss KA, Shneider BL, Lim WA, Salen G, et al. (2003) Complex inheritance of familial hypercholanemia with associated mutations in TJP2 and BAAT. *Nat Genet*, 34, 91–96.

Chen HL, Liu YJ, Wu SH, Ni YH, Ho MC, Lai HS, Hsu WM, Hsu HY, Tseng HC, Jeng YM, Chang MH (2008) Expression of hepatocyte transporters and nuclear receptors in children with early and late-stage biliary atresia. *Pediatr Res*, 63, 667–673.

Chiang JY (2009) Bile acids: regulation of synthesis. *J Lipid Res*, 50, 1955–1966.

Chopra AR, Kommagani R, Saha P, Louet JF, Salazar C, Song J, Jeong J, Finegold M, Viollet B, DeMayo F, et al. (2011) Cellular energy depletion resets whole-body energy by promoting coactivator-mediated dietary fuel absorption. *Cell Metab*, 13, 35–43.

Crocenzi FA, Mottino AD, Cao J, Veggi LM, Pozzi EJ, Vore M, Coleman R, Roma MG (2003) Estradiol-17β-D-glucuronide induces endocytic internalization of Bsep in rats. *Am J Physiol Gastrointest Liver Physiol*, 285, G449–G459.

Davit-Spraul A, Gonzales E, Baussan C, Jacquemin E (2009) Progressive familial intrahepatic cholestasis. *Orphanet J Rare Dis*, 4, 1.

Dawson PA (2011) Role of the intestinal bile acid transporters in bile acid and drug disposition. In: *Handbook of Experimental Pharmacology*, Vol. 201. Heidelberg, Germany: Springer. pp 169–203.

Dawson PA, Lan T, Rao A (2009) Bile acid transporters. *J Lipid Res*, 50, 2340–2357.

de Waart DR, Hausler S, Vlaming ML, Kunne C, Hanggi E, Gruss HJ, Oude Elferink RP, Stieger B (2010) Hepatic transport mechanisms of cholyl-L-lysyl-fluorescein. *J Pharmacol Exp Ther*, 334, 78–86.

Debruyne PR, Bruyneel EA, Li X, Zimber A, Gespach C, Mareel MM (2001) The role of bile acids in carcinogenesis. *Mutat Res*, 480–481, 359–369.

Dixon PH, van Mil SW, Chambers J, Strautnieks S, Thompson RJ, Lammert F, Kubitz R, Keitel V, Glantz A, Mattsson LA, et al. (2009) Contribution of variant alleles of *ABCB11* to susceptibility to intrahepatic cholestasis of pregnancy. *Gut*, 58, 537–544.

Elferink MG, Olinga P, Draaisma AL, Merema MT, Faber KN, Slooff MJ, Meijer DK, Groothuis GM (2004) LPS-induced downregulation of MRP2 and BSEP in human liver is due to a posttranscriptional process. *Am J Physiol Gastrointest Liver Physiol*, 287, G1008–G1016.

Eloranta JJ, Jung D, Kullak-Ublick GA (2006) The human Na+-taurocholate cotransporting polypeptide gene is activated by glucocorticoid receptor and peroxisome proliferator-activated receptor-gamma coactivator-1α, and suppressed by bile acids via a small heterodimer partner-dependent mechanism. *Mol Endocrinol*, 20, 65–79.

Eloranta JJ, Kullak-Ublick GA (2005) Coordinate transcriptional regulation of bile acid homeostasis and drug metabolism. *Arch Biochem Biophys*, 433, 397–412.

Eloranta JJ, Kullak-Ublick GA (2008) The role of FXR in disorders of bile acid homeostasis. *Physiology (Bethesda)*, 23, 286–295.

Esteller A (2008) Physiology of bile secretion. *World J Gastroenterol*, 14, 5641–5649.

Fattinger K, Funk C, Pantze M, Weber C, Reichen J, Stieger B, Meier PJ (2001) The endothelin antagonist bosentan inhibits the canalicular bile salt export pump: a potential mechanism for hepatic adverse reactions. *Clin Pharmacol Ther*, 69, 223–231.

Forker EL (1977) Mechanisms of hepatic bile formation. *Annu Rev Physiol*, 39, 323–347.

Fouassier L, Kinnman N, Lefevre G, Lasnier E, Rey C, Poupon R, Elferink RP, Housset C (2002) Contribution of mrp2 in alterations of canalicular bile formation by the endothelin antagonist bosentan. *J Hepatol*, 37, 184–191.

Funk C, Ponelle C, Scheuermann G, Pantze M (2001) Cholestatic potential of troglitazone as a possible factor contributing to troglitazone-induced hepatotoxicity: in vivo and in vitro interaction at the canalicular bile salt export pump (Bsep) in the rat. *Mol Pharmacol*, 59, 627–635.

Gerloff T, Stieger B, Hagenbuch B, Madon J, Landmann L, Roth J, Hofmann AF, Meier PJ (1998) The sister of P-glycoprotein represents the canalicular bile salt export pump of mammalian liver. *J Biol Chem*, 273, 10046–10050.

Geuken E, Visser D, Kuipers F, Blokzijl H, Leuvenink HG, de Jong KP, Peeters PM, Jansen PL, Slooff MJ, Gouw AS, Porte RJ (2004) Rapid increase of bile salt secretion is associated with bile duct injury after human liver transplantation. *J Hepatol*, 41, 1017–1025.

Green RM, Hoda F, Ward KL (2000) Molecular cloning and characterization of the murine bile salt export pump. *Gene*, 241, 117–123.

Hagenbuch B, Dawson P (2004) The sodium bile salt cotransport family SLC10. *Pflugers Arch*, 447, 566–570.

Hagenbuch B, Meier PJ (2004) Organic anion transporting polypeptides of the OATP/*SLC21* family: Phylogenetic classification as OATP/*SLCO* superfamily, new nomenclature and molecular/functional properties. *Pflugers Arch*, 447, 653–665.

Hayashi H, Takada T, Suzuki H, Onuki R, Hofmann AF, Sugiyama Y (2005) Transport by vesicles of glycine- and taurine-conjugated bile salts and taurolithocholate 3-sulfate: a comparison of human BSEP with rat Bsep. *Biochim Biophys Acta*, 1738, 54–62.

Hirano M, Maeda K, Hayashi H, Kusuhara H, Sugiyama Y (2005) Bile salt export pump (BSEP/ABCB11) can transport a nonbile acid substrate, pravastatin. *J Pharmacol Exp Ther*, 314, 876–882.

Hofmann AF (1990) Bile acid secretion, bile flow and biliary lipid secretion in humans. *Hepatology*, 12, 17S–22S.

Hofmann AF (2009a) Bile acids: trying to understand their chemistry and biology with the hope of helping patients. *Hepatology*, 49, 1403–1418.

Hofmann AF (2009b) The enterohepatic circulation of bile acids in mammals: form and functions. *Front Biosci*, 14, 2584–2598.

Hofmann AF, Hagey LR (2008) Bile acids: chemistry, pathochemistry, biology, pathobiology, and therapeutics. *Cell Mol Life Sci*, 65, 2461–2483.

Huang L, Smit JW, Meijer DK, Vore M (2000) Mrp2 is essential for estradiol-17β(β-D-glucuronide)-induced cholestasis in rats. *Hepatology*, 32, 66–72.

Ismair MG, Hausler S, Stuermer CA, Guyot C, Meier PJ, Roth J, Stieger B (2009) ABC-transporters are localized in caveolin-1-positive and reggie-1-negative and reggie-2-negative microdomains of the canalicular membrane in rat hepatocytes. *Hepatology*, 49, 1673–1682.

Jansen PLM, Strautnieks SS, Jacquemin E, Hadchouel M, Sokal EM, Hooiveld GJ, Koning JH, De Jager Krikken A, Kuipers F, et al. (1999) Hepatocanalicular bile salt export pump deficiency in patients with progressive familial intrahepatic cholestasis. *Gastroenterology*, 117, 1370–1379.

Jara P, Hierro L, Martinez-Fernandez P, Alvarez-Doforno R, Yanez F, Diaz MC, Camarena C, De la Vega A, Frauca E, Munoz-Bartolo G, et al. (2009) Recurrence of bile salt export pump deficiency after liver transplantation. *N Engl J Med*, 361, 1359–1367.

Jigorel E, Le Vee M, Boursier-Neyret C, Parmentier Y, Fardel O (2006) Differential regulation of sinusoidal and canalicular hepatic drug transporter expression by xenobiotics activating drug-sensing receptors in primary human hepatocytes. *Drug Metab Dispos*, 34, 1756–1763.

Julie NL, Julie IM, Kende AI, Wilson GL (2008) Mitochondrial dysfunction and delayed hepatotoxicity: another lesson from troglitazone. *Diabetologia*, 51, 2108–2116.

Jung D, Kullak-Ublick GA (2003) Hepatocyte nuclear factor 1α: a key mediator of the effect of bile acids on gene expression. *Hepatology*, 37, 622–631.

Keitel V, Burdelski M, Vojnisek Z, Schmitt L, Haussinger D, Kubitz R (2009) De novo bile salt transporter antibodies as a possible cause of recurrent graft failure after liver transplantation: a novel mechanism of cholestasis. *Hepatology*, 50, 510–517.

Kis E, Ioja E, Nagy T, Szente L, Heredi-Szabo K, Krajcsi P (2009) Effect of membrane cholesterol on BSEP/Bsep activity: species specificity studies for substrates and inhibitors. *Drug Metab Dispos*, 37, 1878–1886.

Klaassen CD, Aleksunes LM (2010) Xenobiotic, bile acid, and cholesterol transporters: function and regulation. *Pharmacol Rev*, 62, 1–96.

Knisely AS, Strautnieks SS, Meier Y, Stieger B, Byrne JA, Portmann BC, Bull LN, Pawlikowska L, Bilezikci B, Ozcay F, et al. (2006) Hepatocellular carcinoma in ten children under five years of age with bile salt export pump deficiency. *Hepatology*, 44, 478–486.

Kock K, Koenen A, Giese B, Fraunholz M, May K, Siegmund W, Hammer E, Volker U, Jedlitschky G, Kroemer HK, Grube M (2010) Rapid modulation of the organic anion transporting polypeptide 2B1 (OATP2B1, SLCO2B1) function by protein kinase C-mediated internalization. *J Biol Chem*, 285, 11336–11347.

Kojima H, Nies AT, Konig J, Hagmann W, Spring H, Uemura M, Fukui H, Keppler D (2003) Changes in the expression and localization of hepatocellular transporters and radixin in primary biliary cirrhosis. *J Hepatol*, 39, 693–702.

Kosters A, Karpen SJ (2008) Bile acid transporters in health and disease. *Xenobiotica*, 38, 1043–1071.

Krahenbuhl S, Talos C, Fischer S, Reichen J (1994) Toxicity of bile acids on the electron transport chain of isolated rat liver mitochondria. *Hepatology*, 19, 471–479.

Kullak-Ublick GA, Glasa J, Boker C, Oswald M, Grutzner U, Hagenbuch B, Stieger B, Meier PJ, Beuers U, Kramer W, Wess G, Paumgartner G (1997) Chlorambucil-taurocholate is transported by bile acid carriers expressed in human hepatocellular carcinomas. *Gastroenterology*, 113, 1295–1305.

Kullak-Ublick GA, Ismair MG, Stieger B, Landmann L, Huber R, Pizzagalli F, Fattinger K, Meier PJ, Hagenbuch B (2001) Organic anion-transporting polypeptide B (OATP-B) and its functional comparison with three other OATPs of human liver. *Gastroenterology*, 120, 525–533.

Lam CW, Cheung KM, Tsui MS, Yan MS, Lee CY, Tong SF (2006) A patient with novel ABCB11 gene mutations with phenotypic transition between BRIC2 and PFIC2. *J Hepatol*, 44, 240–242.

Lam P, Soroka CJ, Boyer JL (2010) The bile salt export pump: clinical and experimental aspects of genetic and acquired cholestatic liver disease. *Semin Liver Dis*, 30, 125–133.

Landrier JF, Eloranta JJ, Vavricka SR, Kullak-Ublick GA (2006) The nuclear receptor for bile acids, FXR, transactivates human organic solute transporter-alpha and -beta genes. *Am J Physiol Gastrointest Liver Physiol*, 290, G476–G485.

Lang C, Meier Y, Stieger B, Beuers U, Lang T, Kerb R, Kullak-Ublick GA, Meier PJ, Pauli-Magnus C (2007) Mutations and polymorphisms in the bile salt export pump and the multidrug resistance protein 3 associated with drug-induced liver injury. *Pharmacogenet Genomics*, 17, 47–60.

Lang T, Haberl M, Jung D, Drescher A, Schlagenhaufer R, Keil A, Mornhinweg E, Stieger B, Kullak-Ublick GA, Kerb R (2006) Genetic variability, haplotype structures, and ethnic diversity of hepatic transporters MDR3 (*ABCB4*) and bile salt export pump (*ABCB11*). *Drug Metab Dispos*, 34, 1582–1599.

Le Vee M, Gripon P, Stieger B, Fardel O (2008) Down-regulation of organic anion transporter expression in human hepatocytes exposed to the proinflammatory cytokine interleukin 1β. *Drug Metab Dispos*, 36, 217–222.

Le Vee M, Jouan E, Moreau A, Fardel O (2011) Regulation of drug transporter mRNA expression by interferon-γ in primary human hepatocytes. *Fundam Clin Pharmacol*, 25, 99–103.

Lecureur V, Sun D, Hargrove P, Schuetz EG, Kim RB, Lan LB, Schuetz JD (2000) Cloning and expression of murine sister of P-glycoprotein reveals a more discriminating transporter than MDR1/P-glycoprotein. *Mol Pharmacol*, 57, 24–35.

Lo Sasso G, Petruzzelli M, Moschetta A (2008) A translational view on the biliary lipid secretory network. *Biochim Biophys Acta*, 1781, 79–96.

Maggiore G, Gonzales E, Sciveres M, Redon MJ, Grosse B, Stieger B, Davit-Spraul A, Fabre M, Jacquemin E (2010) Relapsing features of bile salt export pump deficiency after liver transplantation in two patients with progressive familial intrahepatic cholestasis type 2. *J Hepatol*, 53, 981–986.

Marion TL, Leslie EM, Brouwer KL (2007) Use of sandwich-cultured hepatocytes to evaluate impaired bile acid transport as a mechanism of drug-induced hepatotoxicity. *Mol Pharm*, 4, 911–918.

Marschall HU, Wagner M, Zollner G, Fickert P, Diczfalusy U, Gumhold J, Silbert D, Fuchsbichler A, Benthin L, Grundstrom R, et al. (2005) Complementary stimulation of hepatobiliary transport and detoxification systems by rifampicin and ursodeoxycholic acid in humans. *Gastroenterology*, 129, 476–485.

Marzolini C, Tirona RG, Gervasini G, Poonkuzhali B, Assem M, Lee W, Leake BF, Schuetz JD, Schuetz EG, Kim RB (2007) A common polymorphism in the bile acid receptor farnesoid X receptor is associated with decreased hepatic target gene expression. *Mol Endocrinol*, 21, 1769–1780.

Masubuchi Y (2006) Metabolic and non-metabolic factors determining troglitazone hepatotoxicity: a review. *Drug Metab Pharmacokinet*, 21, 347–356.

Masyuk AI, LaRusso NF (2006) Aquaporins in the hepatobiliary system. *Hepatology*, 43, S75–S81.

Meier PJ (1988) Transport polarity of hepatocytes. *Semin Liver Dis*, 8, 293–307.

Meier PJ (2002) Canalicular bile formation: beyond single transporter functions. *J Hepatol*, 37, 272–273.

Meier PJ, Stieger B (2002) Bile salt transporters. *Annu Rev Physiol*, 64, 635–661.

Meier Y, Pauli-Magnus C, Zanger UM, Klein K, Schaeffeler E, Nussler AK, Nussler N, Eichelbaum M, Meier PJ, Stieger B (2006) Interindividual variability of canalicular ATP-binding-cassette (ABC)-transporter expression in human liver. *Hepatology*, 44, 62–74.

Meier Y, Zodan T, Lang C, Zimmermann R, Kullak-Ublick GA, Meier PJ, Stieger B, Pauli-Magnus C (2008) Increased susceptibility for intrahepatic cholestasis of pregnancy and contraceptive-induced cholestasis in carriers of the 1331T>C polymorphism in the bile salt export pump. *World J Gastroenterol*, 14, 38–45.

Meyers M, Slikker W, Pascoe G, Vore M (1980) Characterization of cholestasis induced by estradiol-17 β-D-glucuronide in the rat. *J Pharmacol Exp Ther*, 214, 87–93.

Mita S, Suzuki H, Akita H, Hayashi H, Onuki R, Hofmann AF, Sugiyama Y (2006) Inhibition of bile acid transport across Na+/taurocholate cotransporting polypeptide (SLC10A1) and bile salt export pump (ABCB 11)-coexpressing LLC-PK1 cells by cholestasis-inducing drugs. *Drug Metab Dispos*, 34, 1575–1581.

Mita S, Suzuki H, Akita H, Stieger B, Meier PJ, Hofmann AF, Sugiyama Y (2005) Vectorial transport of bile salts across MDCK cells expressing both rat Na+-taurocholate cotransporting polypeptide and rat bile salt export pump. *Am J Physiol Gastrointest Liver Physiol*, 288, G159–G167.

Morgan RE, Trauner M, van Staden CJ, Lee PH, Ramachandran B, Eschenberg M, Afshari C, Qualls CWJr, Lightfoot-Dunn R, Hamadeh HK (2010) Interference with bile salt export pump function is a susceptibility factor for human liver injury in drug development. *Toxicol Sci*, 118, 485–500.

Muller M, Ishikawa T, Berger U, Klunemann C, Lucka L, Schreyer A, Kannicht C, Reutter W, Kurz G, Keppler D (1991) ATP-dependent transport of taurocholate across the hepatocyte canalicular membrane mediated by a 110-kDa glycoprotein binding ATP and bile salt. *J Biol Chem*, 266, 18920–18926.

Ng KH, Le Goascogne C, Amborade E, Stieger B, Deschatrette J (2000) Reversible induction of rat hepatoma cell polarity with bile acids. *J Cell Sci*, 113, 4241–4251.

Nishida T, Gatmaitan Z, Che M, Arias IM (1991) Rat liver canalicular membrane vesicles contain an ATP-dependent bile acid transport system. *Proc Natl Acad Sci U S A*, 88, 6590–6594.

Noe J, Hagenbuch B, Meier PJ, St-Pierre MV (2001) Characterization of the mouse bile salt export pump overexpressed in the baculovirus system. *Hepatology*, 33, 1223–1231.

Noe J, Stieger B, Meier PJ (2002) Functional expression of the canalicular bile salt export pump of human liver. *Gastroenterology*, 123, 1659–1666.

Oscarson M, Zanger UM, Rifki OF, Klein K, Eichelbaum M, Meyer UA (2006) Transcriptional profiling of genes induced in the livers of patients treated with carbamazepine. *Clin Pharmacol Ther*, 80, 440–456.

Palmeira CM, Rolo AP (2004) Mitochondrially-mediated toxicity of bile acids. *Toxicology*, 203, 1–15.

Pauli-Magnus C, Kerb R, Fattinger K, Lang T, Anwald B, Kullak-Ublick GA, Beuers U, Meier PJ (2004) BSEP and MDR3 haplotype structure in healthy Caucasians, primary biliary cirrhosis and primary sclerosing cholangitis. *Hepatology*, 39, 779–791.

Pauli-Magnus C, Meier PJ, Stieger B (2010) Genetic determinants of drug-induced cholestasis and intrahepatic cholestasis of pregnancy. *Semin Liver Dis*, 30, 147–159.

Pauli-Magnus C, Stieger B, Meier Y, Kullak Ublick GA, Meier PJ (2005) Enterohepatic transport of bile salts and genetics of cholestasis. *J Hepatol*, 43, 342–357.

Paulusma CC, de Waart DR, Kunne C, Mok KS, Elferink RP (2009) Activity of the bile salt export pump (ABCB11) is critically dependent on canalicular membrane cholesterol content. *J Biol Chem*, 284, 9947–9954.

Perez MJ, Briz O (2009) Bile-acid-induced cell injury and protection. *World J Gastroenterol*, 15, 1677–1689.

Plass JR, Mol O, Heegsma J, Geuken M, Faber KN, Jansen PL, Muller M (2002) Farnesoid X receptor and bile salts are involved in transcriptional regulation of the gene encoding the human bile salt export pump. *Hepatology*, 35, 589–596.

Reichen J, Paumgartner G (1976) Uptake of bile acids by perfused rat liver. *Am J Physiol*, 231, 734–742.

Ros JE, Libbrecht L, Geuken M, Jansen PL, Roskams TA (2003) High expression of MDR1, MRP1, and MRP3 in the hepatic progenitor cell compartment and hepatocytes in severe human liver disease. *J Pathol*, 200, 553–560.

Russell DW (2009) Fifty years of advances in bile acid synthesis and metabolism. *J Lipid Res*, 50(Suppl), S120–S125.

Saito S, Iida A, Sekine A, Miura Y, Ogawa C, Kawauchi S, Higuchi S, Nakamura Y (2002) Three hundred twenty-six genetic variations in genes encoding nine members of ATP-binding cassette, subfamily B (ABCB/MDR/TAP), in the Japanese population. *J Hum Genet*, 47, 38–50.

Schonhoff CM, Thankey K, Webster CR, Wakabayashi Y, Wolkoff AW, Anwer MS (2008) Rab4 facilitates cyclic adenosine monophosphate-stimulated bile acid uptake and Na(+)-taurocholate cotransporting polypeptide translocation. *Hepatology*, 48, 1665–1670.

Shoda J, Kano M, Oda K, Kamiya J, Nimura Y, Suzuki H, Sugiyama Y, Miyazaki H, Todoroki T, Stengelin S, Kramer W, Matsuzaki Y, Tanaka N (2001) The expression levels of plasma membrane transporters in the cholestatic liver of patients undergoing biliary drainage and their association with the impairment of biliary secretory function. *Am J Gastroenterol*, 96, 3368–3378.

Sokol RJ, Devereaux M, Dahl R, Gumpricht E (2006) "Let there be bile"–understanding hepatic injury in cholestasis. *J Pediatr Gastroenterol Nutr*, 43(Suppl 1), S4–S9.

Souza RF, Krishnan K, Spechler SJ (2008) Acid, bile, and CDX: the ABCs of making Barrett's metaplasia. *Am J Physiol Gastrointest Liver Physiol*, 295, G211–G218.

Stieger B (2010) Role of the bile salt export pump, BSEP, in acquired forms of cholestasis. *Drug Metab Rev*, 42, 437–445.

Stieger B (2011) The role of the sodium-taurocholate cotransporting polypeptide (NTCP) and of the bile salt export pump (BSEP) in physiology and pathophysiology of bile formation. In: *Handbook of Experimental Pharmacology*, Vol. 201. Heidelberg, Germany: Springer. pp 205–259.

Stieger B, Beuers U (2011) The canalicular bile salt export pump BSEP (ABCB11) as a potential therapeutic target. *Curr Drug Targets*, 12, 661–670.

Stieger B, Fattinger K, Madon J, Kullak Ublick GA, Meier PJ (2000) Drug- and estrogen-induced cholestasis through inhibition of the hepatocellular bile salt export pump (Bsep) of rat liver. *Gastroenterology*, 118, 422–430.

Stieger B, Geier A (2011) Genetic variations of bile salt transporters as predisposing factors for drug-induced cholestasis, intrahepatic cholestasis of pregnancy and therapeutic response of viral hepatitis. *Expert Opin Drug Metab Toxicol*, 7, 411–425.

Stieger B, Meier Y, Meier PJ (2007) The bile salt export pump. *Pflugers Arch*, 453, 611–620.

Stieger B, O'Neill B, Meier PJ (1992) ATP-dependent bile-salt transport in canalicular rat liver plasma-membrane vesicles. *Biochem J*, 284, 67–74.

Strautnieks SS, Bull LN, Knisely AS, Kocoshis SA, Dahl N, Arnell H, Sokal E, Dahan K, Childs S, Ling V, et al. (1998) A gene encoding a liver-specific ABC transporter is mutated in progressive familial intrahepatic cholestasis. *Nat Genet*, 20, 233–238.

Strautnieks SS, Byrne JA, Pawlikowska L, Cebecauerova D, Rayner A, Dutton L, Meier Y, Antoniou A, Stieger B, Arnell H, et al. (2008) Severe bile salt export pump deficiency: 82 different *ABCB11* mutations in 109 families. *Gastroenterology*, 134, 1203–1214.

Sun AQ, Ponamgi VM, Boyer JL, Suchy FJ (2008) Membrane trafficking of the human organic anion-transporting polypeptide C (hOATPC). *Pharm Res*, 25, 463–474.

Takada T, Weiss HM, Kretz O, Gross G, Sugiyama Y (2004) Hepatic transport of PKI166, an epidermal growth factor receptor kinase inhibitor of the pyrrolo-pyrimidine class, and its main metabolite, ACU154. *Drug Metab Dispos*, 32, 1272–1278.

Takahashi A, Hasegawa M, Sumazaki R, Suzuki M, Toki F, Suehiro T, Onigata K, Tomomasa T, Suzuki T, Matsui A, et al. (2007) Gradual improvement of liver function after administration of ursodeoxycholic acid in an infant with a novel *ABCB11* gene mutation with phenotypic continuum between BRIC2 and PFIC2. *Eur J Gastroenterol Hepatol*, 19, 942–946.

Tirona RG (2011) Molecular mechanisms of drug transporter regulation. In: *Handbook of Experimental Pharmacology*, Vol. 201. Heidelberg, Germany: Springer. pp 373–402.

Trauner M, Halilbasic E (2011) Nuclear receptors as new perspective for the management of liver diseases. *Gastroenterology*, 140, 1120–1125.

Trauner M, Meier PJ, Boyer JL (1998) Molecular pathogenesis of cholestasis. *N Engl J Med*, 339, 1217–1227.

Trottier J, Milkiewicz P, Kaeding J, Verreault M, Barbier O (2006) Coordinate regulation of hepatic bile acid oxidation and conjugation by nuclear receptors. *Mol Pharm*, 3, 212–222.

Vallejo M, Briz O, Serrano MA, Monte MJ, Marin JJ (2006) Potential role of trans-inhibition of the bile salt export pump by progesterone metabolites in the etiopathogenesis of intrahepatic cholestasis of pregnancy. *J Hepatol*, 44, 1150–1157.

Van der Borght S, Libbrecht L, Katoonizadeh A, Aerts R, Nevens F, Verslype C, Roskams TA (2007) Nuclear beta-catenin staining and absence of steatosis are indicators of hepatocellular adenomas with an increased risk of malignancy. *Histopathology*, 51, 855–856.

Wagner M, Zollner G, Trauner M (2011) Nuclear receptors in liver disease. *Hepatology*, 53, 1023–1034.

Yabuuchi H, Tanaka K, Maeda M, Takemura M, Oka M, Ohashi R, Tamai I (2008) Cloning of the dog bile salt export pump (BSEP; ABCB11) and functional comparison with the human and rat proteins. *Biopharm Drug Dispos*, 29, 441–448.

Zaja R, Munic V, Klobucar RS, Ambriovic-Ristov A, Smital T (2008) Cloning and molecular characterization of apical efflux transporters (ABCB1, ABCB11 and ABCC2) in rainbow trout (*Oncorhynchus mykiss*) hepatocytes. *Aquat Toxicol*, 90, 322–332.

Zollner G, Fickert P, Silbert D, Fuchsbichler A, Marschall HU, Zatloukal K, Denk H, Trauner M (2003) Adaptive changes in hepatobiliary transporter expression in primary biliary cirrhosis. *J Hepatol*, 38, 717–727.

Zollner G, Fickert P, Zenz R, Fuchsbichler A, Stumptner C, Kenner L, Ferenci P, Stauber RE, Krejs GJ, Denk H, et al. (2001) Hepatobiliary transporter expression in percutaneous liver biopsies of patients with cholestatic liver diseases. *Hepatology*, 33, 633–646.

Zollner G, Wagner M, Fickert P, Silbert D, Fuchsbichler A, Zatloukal K, Denk H, Trauner M (2005) Hepatobiliary transporter expression in human hepatocellular carcinoma. *Liver Int*, 25, 367–379.

CHAPTER **14**

BCRP (ABCG2)

Agnes Basseville
Susan E. Bates
William D. Figg
Alex Sparreboom

14.1 INTRODUCTION

ABCG2 was the third ABC transporter found to be involved in chemoresistance, after ABCB1 in 1987 and ABCC1 in 1991 (Ueda et al., 1987; Cole et al., 1992). Several laboratories reported its discovery almost simultaneously in the late 1990s (Allikmets et al., 1998; Doyle et al., 1998; Miyake et al., 1999). Cell lines resistant to mitoxantrone were found to not overexpress ABCB1 or ABCC1 (MRP1), leading researchers to continue the search for another transporter protein, which was then named ABCP, BCRP, and MXR, respectively, by the three groups, and has since been assigned as ABCG2.

The role of ABCG2 in drug resistance was quickly extended in *in vitro* studies, but correlative clinical studies obtained mixed results. It was observed that ABCG2 was expressed in acute myeloid leukemia and in several solid tumors, but more evidence is needed to link it to clinical drug resistance. ABCG2 was also shown to be expressed in normal tissues where it plays a part in protection against xenobiotics, for example, in limiting intestinal drug absorption or transit across the blood–brain barrier.

ABCG2 gene sequencing has been performed in multiple clinical samples and, to date, 180 natural variants have been identified. *In vitro* experimentation has yielded important information about ABCG2 substrates and the functional consequences of polymorphisms in this gene. Increasingly, retrospective studies have aimed to correlate ABCG2 genotype with drug absorption, disposition, treatment outcome, or toxicity. Association studies were initially limited to the C421A polymorphism, but interesting results have recently been obtained with newly discovered SNPs, including several that are localized to the promoter region, resulting in an increasing interest in this field and its potential practical applications.

Pharmacogenomics of Human Drug Transporters: Clinical Impacts, First Edition.
Edited by Toshihisa Ishikawa, Richard B. Kim and Jörg König

14.2 ABCG2 GENE AND PROTEIN

The ABCG2 gene spans over 66 kb and comprises 16 exons and 15 introns. It is located on chromosome 4q22. The translational start of ABCG2 is in exon 2, and the majority of the 5′ untranslated region (UTR) is contained within exon 1. The promoter does not contain a TATA box but a CCAAT box (Bailey-Dell et al., 2001). Furthermore, it appears as though alternative promoters of ABCG2 may exist, following the discovery of three different novel forms of exon 1 with differential expression (Nakanishi et al., 2006). There has been significant interest recently in the promoter region of ABCG2, with the discovery of different consensus sequences such as estrogen, hypoxic, peroxisome proliferator-activated receptor (PPAR) response elements, and a xenobiotic site mediating AhR signaling. Sequence analysis also indicated that the promoter region has a potential CpG island and putative binding sites for SP1, AP1, and AP2 (Ee et al., 2004; Krishnamurthy et al., 2004; Tompkins et al., 2010; To et al., 2011).

The ABCG2 gene encodes a 655 amino acid half transporter including one nucleotide-binding domain and one transmembrane region, often referred to as an NBD-TM (Allikmets et al., 1998; Doyle et al., 1998; Miyake et al., 1999; Bailey-Dell et al., 2001). Unlike most functional ABC transporter genes, which typically encode a pair of NBD-TM, the ABCG2 gene encodes only a half transporter. It was originally believed that ABCG2 functioned as a homodimer, although more recently, it was shown that human ABCG2 probably exists as a homotetramer with possibly a higher form of oligomerization (Xu et al., 2004; Lepper et al., 2005). It has also recently been suggested that ABCG2 may contain more than one substrate-binding site, based on enzyme kinetics experiments. Based on K_m values, there appears to be a low-affinity binding site, in addition to the high-affinity site (Huang et al., 2006).

14.3 ABCG2 FUNCTION

14.3.1 Physiological Functions

Comprehensive screening of normal human tissues by immunohistochemistry and RT-PCR to assess ABCG2 protein levels and mRNA, respectively, found high expression in the small intestine and colon, the bile canaliculi of the liver, the lobules and ducts in the mammary gland, the placenta, and in the venous and capillary endothelium (Maliepaard et al., 2001; Meissner et al., 2006). Expression was also found in alveolar pneumocytes; cervical, prostate, and bladder epithelium; interstitial cells of testes; and in the pancreas and adrenal gland (Fetsch et al., 2006). Finally, expression in the CNS has been documented in human brain endothelial cells and confirmed by functional analysis in murine models, suggesting that ABCG2 is an important component of the blood–brain barrier (Cooray et al., 2002; Eisenblatter et al., 2003; Cisternino et al., 2004)ABCG2 in the placenta is localized to the trophoblast cells, and most likely prevents transfer of substrates to the fetus (Jonker et al., 2000; Ceckova et al., 2006; Evseenko et al., 2007). Intestinal expression restricts the exposure to ingested carcinogens and potentially toxic xenobiotics by decreasing uptake from the gut lumen (Jonker et al., 2000; Allen and Schinkel, 2002; Ejendal and Hrycyna, 2002; van Herwaarden et al., 2003; Han and Zhang, 2004). For example, several heterocyclic amine carcinogens present in overcooked meat are transported by ABCG2 and have been linked to breast and prostate cancers, providing evidence of the transporter's protective role (van Herwaarden et al., 2003). ABCG2 is also highly expressed in side population stem cells, but the normal physiological function in these cells remains unknown (Zhou et al., 2001). During late pregnancy and lactation, expression of ABCG2 has been shown to increase substantially in the mammary glands of mice, cows, and humans (Jonker et al., 2005). Expression is localized primarily in the apical membrane of the alveolar epithelial cells, where the protein transports substrates into the milk, but not in the main ducts. The evolutionary role of ABCG2 in milk ducts is unknown, since it appears to be in contradiction to the purported protective role of ABCG2. It has been suggested that one physiological role of ABCG2 in the mammary gland is the secretion of vitamin B(2) (riboflavin) and possibly other vitamins (e.g., biotin and vitamin K) into breast milk (van Herwaarden et al., 2007). Recently, it was also shown that ABCG2 is a transporter involved in urate elimination across the apical membrane of proximal tubular cells of kidneys (Woodward et al., 2009).

14.3.2 Role in Multidrug Resistance

While mutations in target proteins in cancer cells can result in primary resistance to substrate drugs, ABC transporters are more often thought to be involved in the development of secondary, or acquired, resistance. This development of resistance can occur by a number of different mechanisms, including decreased uptake, increased detoxification, alteration of target proteins, or increased excretion (Litman et al., 2001). For ABCG2, which is expressed on the cell membrane and functions as an efflux transporter, drug treatment has been shown in several cases to lead to an upregulation of protein expression, resulting in increased cell detoxification. This can result in multidrug resistance (MDR) in which the cell becomes resistant to other substrates in addition to the original treatment. For example, MCF-7 AdVp3000 cells, selected from parental MCF-7 cells by treatment with doxorubicin and verapamil, overexpress ABCG2 and exhibit MDR against numerous agents including mitoxantrone and topotecan (Litman et al., 2000). Despite the apparent overexpression of ABCG2 leading to MDR in numerous tumor cell lines *in vitro*, the role of ABCG2 in human tumors is, however, currently unclear. Several studies have linked ABCG2 expression in acute myeloid leukemia with poor prognosis (Benderra et al., 2004, 2005; Wilson et al., 2006). Additional studies are needed to understand the role of ABCG2 in MDR in cancer cells.

14.3.3 Pharmacological Functions

Interest in ABCG2 is rapidly increasing as more substrates are identified. ABCG2 transports a wide variety of molecules, including cancer therapeutics, endogenous compounds, and xenobiotics (Table 14.1). Numerous inhibitors of ABCG2 have been identified, but the majority of these appears to be nonspecific; for example, ritonavir also inhibits CYP3A4 and tariquidar also inhibits ABCB1 (Gupta et al.,

TABLE 14.1 Select Drug Substrates of ABCG2

Chemotherapy Agent		
Alkaloids:	Etoposide	Volk et al. (2000); Robey et al. (2003)
Anthracenes:	BBR3390	Hazlehurst et al. (1999)
	Bisantrene	Robey et al. (2003)
	Mitoxantrone	Doyle et al. (1998); Robey et al. (2003)
Anthracyclines:	Doxorubicin*	Robey et al. (2003)
	Daunorubicin	Doyle et al. (1998)
	Epirubicin	Burger et al. (2003)
Antifolates:	GW-1843	Shafran et al. (2005)
	Methotrexate*	Chen et al. (2003); Volk and Schneider (2003)
	Tomudex	Shafran et al. (2005)
Camptothecin derivatives:	9-aminocamptothecin*	Rajendra et al. (2003); Schellens et al. (2000)
	Diflomotecan*	Bates et al. (2004)
	Exatecan	Ishii et al. (2000)
	Homocamptothecin	Bates et al. (2004)
	SN-38 (irinotecan derivative)*	Kawabata et al. (2001)
	Topotecan*	Maliepaard (1999)
Flavonoides:	Flavopiridol*	Robey et al. (2001)
Indolocarbazoles:	J-107088	Komatani et al. (2001)
	NB-506	Komatani et al. (2001)
	UCN-01	Robey et al. (2004)
Nucleoside analogues:	Cladribine	de Wolf et al. (2008)
	Clofarabine	de Wolf et al. (2008)
	Mercaptopurine	de Wolf et al. (2008)
Platinum salts:	Oxaliplatin	Ceckova (2008)
Taxanes:	Docetaxel*	Shen et al. (2011)
Tyrosine kinase inhibitors:	Cl-1033	Erlichman et al. (2001)
	Erlotinib*	Li et al. (2007)
	Gefitinib*	Li et al. (2007)
	Imatinib*	Houghton et al. (2004)
	Nilotinib	Brendel et al. (2007)
	Sunitinib*	Tang et al. (2012)
HMG-coa Reductase Inhibitors		
Statins:	Atorvastatin*	Li et al. (2011)
	Cerivastatin	Matsushima et al. (2005)
	Fluvastatin*	Hirano et al. (2005)
	Pitavastatin*	Hirano et al. (2005)
	Pravastatin*	Matsushima et al. (2005)
	Rosuvastatin*	Hirano et al. (2005); Huang et al. (2006)

TABLE 14.1 (*Continued*)

HIV Treatment		
Nucleoside analogues:	Lamivudine*	Wang et al. (2003)
	Zidovudine	Wang et al. (2003, 2004), 2004
Antibiotic		
Nitrofurans:	Nitrofurantoin*	Merino et al. (2005)
Fluoroquinolones:	Ciprofloxacin	Merino et al. (2006)
	Norfloxacin	Merino et al. (2006)
	Ofloxacin	Merino et al. (2006)
Anti-inflammatory		
Sulfa-drugs:	Sulfasalazine*	van der Heijden et al. (2004)

*Indicates the drugs tested in pharmacogenetic studies.

2004; Robey et al., 2004). Effective inhibitors of ABCG2, including GF120918, Ko143, and (analogues of) fumitremorgin C have been described (Rabindran et al., 1998; de Bruin et al., 1999; Rabindran et al., 2000; Allen and Schinkel, 2002; Allen et al., 2002; Gottesman et al., 2002), but have never been systematically tested in patients for their ability to increase drug accumulation in sanctuary sites such as the CNS, or to alter resistance to anticancer substrate drugs.

14.4 GENETIC VARIANTS OF THE ABCG2 GENE

In the past several years, there has been rising interest in the identification of naturally occurring mutations in the ABCG2 gene. To date, ABCG2 has been systematically screened for genetic variations in 16 different ethnic groups or subpopulations, identifying 180 genetic variants (Honjo et al., 2002; Iida et al., 2002; Imai et al., 2002; Backstrom et al., 2003; Itoda et al., 2003; Zamber et al., 2003; de Jong et al., 2004; Kondo et al., 2004; Mizuarai et al., 2004; Bosch et al., 2005; Colombo et al., 2005; Kobayashi et al., 2005; Maekawa et al., 2006; Lee et al., 2007; Yoshioka et al., 2007; Poonkuzhali et al., 2008). Among these, 144 are single nucleotide polymorphisms (SNPs) in at least one of the studied populations, while the other 36 polymorphisms show an allelic frequency between 0.1% and 0.8%. Among the 180 variants, 99 are localized in introns, 47 have been found in the promoter, and 5 in the 3′ UTR. Twenty-nine polymorphisms are located in exons, including 19 nonsynonymous variants. Among them, V12M, Q126X, Q141K, I206L, N590Y, and D620N are SNPs, whereas S13L, G51C, R160Q, Q166E, F208S, S248P, P269S, G354R, F431L, S441N, F489L, the frame shift (F506S, F507L, V508L, M509X), and R575X have frequency lower than 1%. Data describing the 180 polymorphisms are found in Table 14.2.

Some of these natural genetic variants have been extensively studied for their relationship with protein expression and function *in vitro*, particularly the nonsynonymous variants. Several variants localized in the coding region are now known to modify ABCG2 stability, trafficking, or function. The majority of polymorphisms in introns and in the 3′ UTR region seem to not affect protein expression or function, with

TABLE 14.2 Summary of Genetic Variants in the ABCG2 Gene

				Tested in		Allelic frequency (%)					
Region	Position* to ATG	Change	aa substit.	*Vitro*	Clinic	Cauc.	Asian	Afr.	multi	dbSNP NCBI	Reference
5′ flank region	33 other SNPs									10	
	–49556*	–delATAA			Yes						10
	–49394*	C/A			Yes					rs2127861	10
	–34911*	C/T			Yes					rs7699188	10
	–34763*	A/C			Yes						10
	–34539*	C/T			Yes						10
	–20445*	C/T					1				9
	–20296*	A/G					11–13				9,15
	–19855*	C/T					0.5				15
	–19781*	A/G					0.3–0.5				9,16
	–19605*	A/G					0.5				15
	(–19572)–	CTCAdel			Yes	54	23.5–23.7			rs4148162	1,9,10,
	(–19569)*									rs3219191	11,13,16
	–19202*	G/C				5				rs2231134	13
	–19167*	T/C					0.3				16
	–19031*	C/T		Yes			0.6–1.6				15,16
Exon 1	–18845*	T/C								rs2231135	13
	–18631*	C/T					2.2			rs2231136	15
	–18604*	delA								rs3219192	13
	–18483*	C/T							0.55		4
	–18388*	A/G				2		0	1.1		3,4
Intron 1	37	G/T					0.3				16
	152	G/A			Yes	50		50	57	rs2622605	1,10
	614	G/A		Yes	Yes	33		15	15	rs2622604	1,10
	921	A/C				17		0	4	rs62310595	10
	6341	T/C							9	rs9999111	10
	6730	G/T							26		10
										rs1116205887	
	6918	C/A							8	rs72554037	10
	8964	G/A							1		10
	9134	T/G				0		0	6		10
	9376	A/G				1					10
	9400	A/C				1					10
	10002	T/A									1
	10123	G/A									1
	10768	G/A									1
	10791	T/C									1
	10792	G/A									1
	11028	T/C							54	rs2725246	10
	11071	A/C			Yes				44	rs17731799	10
	11751	T/C			Yes				6	rs27433332	10
	12106	delC							33	rs11287117	10
	12142	delT							24	rs72554038	10
	12538	delTCAGC							37	rs10632140	10
	13457	G/A			Yes	32		53	29	rs6857600	10
	14183	T/G								rs2622627	1
	14934	A/G			Yes	27		71	28	rs3109823	1,10
	14955	G/T				63		74	49	rs3114018	1,10
	15548	T/C				32		21	28	rs2725250	10
	15682	T/G				57		85	52	rs2622620	10

TABLE 14.2 (*Continued*)

Region	Position* to ATG	Change	aa substit.	Tested in *Vitro*	Tested in Clinic	Allelic frequency (%) Cauc.	Asian	Afr.	multi	dbSNP NCBI	Reference
	16170	C/T			Yes	59		0	10	rs2046134	10
	16179	C/T							4	rs72554039	10
	16289	delT			Yes	3		0	24	rs5860120	10
	17251	A/C								rs4148149	1
	17347	G/A								rs4148150	1
	17626	T/G								rs2725252	1
	18242	A/G or /T					0.3				16
	18271	G/A					35.9			rs1564481	1,16
	18369	G/T								rs4148151	1
Exon 2	34	G/A	V12M	Yes	Yes	2–10.3	15–18	4	12.2	rs2231137	1,3,4,7,9, 12,13,15,16
	38	C/T	S13L				0.3				16
	114	T/C	S38S			1	0	0		rs12721640	3
	151	G/T	G51C	Yes			0.1				14
Intron 2	36	A/G				3	15–19.2	5	12.2	rs4148152	1,3,4,9, 10,12,15,16
	4230	C/T									1
	4815	A/C									1
	5233	G/C							1		10
	5367	C/T							1		10
	5460	C/T							5	rs2725255	10
	5566	C/T			Yes				12	rs17731538	10
	5929	C/T							2		10
	6278	T/C				10		0	9	rs4148155	1,10
	6459	G/A				0		12	11	rs17013859	10
	7126	T/C					2.5–2.8				5,16
Intron 3	10	A/G				3.5–9.7	6.8–9.8		6	rs2231138	1,7,9,10, 15,11,16
	72	insT					0.5–0.8			rs5860119	5,9,16
	682	G/A								rs3219193	13
Exon 4	369	C/T	Y123Y	Yes		0–0.3	0.3	7		rs2231139	3,7,16
	376	C/T	Q126X	Yes	Yes		1–2.8			rs72552713	2,9,15,16
Intron 4	437	G/C				0.5					12
Exon 5	421	C/A	Q141K	yes	yes	11–14	26.5–35	0–5	6.1	rs2231142	1,2,3,6,7,8, 9,13,15,16
	474	C/T	N158N			0.3				rs150450599	7
	479	G/A	R160Q			0.5	0.3				12,16
	496	C/G	Q166E	Yes			0.3			rs1061017	14
Intron 5	9253	A/G				2.5	0.3–0.5		5–10	rs2231144	3,9,12,16
Exon 6	616	A/C	I206L	Yes		10				rs12721643	3
	623	T/C	F208S	Yes			0.3			rs1061018	14
Intron 6	3158	A/G					24.6			rs1871744	1,16
	3171	C/T					3.1–3.3			rs138944374	5,16
	3203	A/G					12.7			rs2231145	1,16
	3287	A/G					12.7			rs2231146	1,16
	3355	T/A							0.55		4
Exon 7	742	T/C	S248P	Yes			0.5			rs3116448	14
	805	C/T	P269S	yes			0.2			rs34678167	15

(continued)

TABLE 14.2 (*Continued*)

Region	Position* to ATG	Change	aa substit.	Tested in *Vitro*	Clinic	Allelic frequency (%) Cauc.	Asian	Afr.	multi	dbSNP NCBI	Reference
Intron 7	18	A/G				0.5			3.9	rs2231147	4,12
	179	T/C			Yes	0		7	13	rs1481012	1,10
Exon 9	1060	G/A	G354R				0.3			rs138606116	16
	1098	G/A	E366E				0.8–1.4		2.2	rs35622453	4,5,9,15,16
Intron 9	5200	G/T			Yes	10		21	19	rs2622628	10
	5677	T/C							14	rs2054576	1,10
	5908	T/G							4	rs41282399	10
	5974	A/T			Yes	36	16.9		24	rs2231148	1,10,12,16
Intron 10	67	T/G							1	rs2231150	10
	95	T/A					0.6–1.5				9,16
	100	A/C							2	rs35518113	10
	112	T/C							1	rs2231151	10
	496	T/G							20	rs76811519	10
	1908	G/T								rs2725263	1
	2094	G/A								rs4148156	1
	5860	A/G				0.5					12
Exon 11	1291	T/C	F431L	Yes			0.3–0.8				5,14,16
	1322	G/A	S441N	Yes			0.3–0.5				8,9,16
Intron 11	20	A/G				4.5–18	16.1–24		22.8	rs2231153	1,4,7,9, 12,15,16
	1448	C/T								rs4148157	1
	1622	insA							0.5		15
	1647	G/A or /T					1.7–22			rs80169899	5,15,16
Exon 12	1423,24 ou25	A/G	L475L						1.1		
	1465	T/C	F489L		Yes		0.5–0.8				5,9,16
Intron 12	38	G/A				0.5				rs2231155	12
	47	A/G				0.5					12
	49	G/T or /C				9.5	26.3–33.5		13.3	rs2231156	1,4,9,12,16
	1566	C/T									1
Exon 13	1515	C/del	Frameshift F506S-F507L-V508L-M509X		0.3–0.5				8,9,16		
Intron 13	40	C/T				53.5	60.2–67.4			rs2231157	1,12,15,16
	65	T/G					0.8–2.2			rs34262876	5,15,16
	405	T/del								rs4148163	1
	692–702	(T)9-11								rs4148164	1
	1659	T/C					0.3				16
	1801	A/T					0.5				9
	1823	C/T				4–13.3	16.1–23.4		17.75–22	rs2231162	1,4,7,9,10, 12,15,16
Exon 14	1723	C/T	R575X		Yes	0.5	0.3				12,16
Intron 14	118	C/T							1	rs72554044	10
	497	T/C				10		74	28	rs2725267	1,10
	815	A/G			Yes	6–16	42.9–50	68	43	rs2231164	1,9,10,12, 13,15,16
Exon 15	1768	A/T	N590Y	Yes		0.3–1		0		rs34264773	3,7

TABLE 14.2 (*Continued*)

Region	Position* to ATG	Change	aa substit.	Tested in *Vitro*	Tested in Clinic	Allelic frequency (%) Cauc.	Asian	Afr.	multi	dbSNP NCBI	Reference
Intron 15	110	C/T					12.1–20		7	rs2231165	1,10,15,16
	142	T/G					0.6				16
	548	C/T							39	rs72554045	10
	566	T/A							4	rs4148159	1,10
	639	G/A							1	rs4148160	1,10
	645–650	A(7-8)								rs4148165	1
	943	A/G							12	rs72554046	10
	1197	C/T									1
Exon 16	1858	G/A	D620N	Yes		0.5			1.1	rs34783571	4,12
3′ UTR	2597	G/A							1.1		4
	2332	A/TA					7				9
	2364	A/C					0.5				9
	2512	C/T					0.5				9
3′region	520	T/C								rs1448784	1

References: 1-Iida et al. (2002); 2-Imai et al. (2002); 3-Zamber et al. (2003); 4-Honjo et al. (2002); 5-Itoda et al. (2003); 6-de Jong et al. (2004); 7-Mizuarai et al. (2004); 8-Kondo et al. (2004); 9-Kobayashi et al. (2005); 10-Poonkuzhali et al. (2008); 11- Colombo (2005) ; 12-Bosch et al. (2005); 13-Backstrom et al. (2003); 14-Yoshioka et al. (2007); 15-Lee et al. (2007); 16- Maekawa et al. (2006).

the exception of the intron 1 C16170T (rs2046134) and G614A (rs2622604) SNPs which were reported to be associated with high expression of ABCG2 in liver and low expression in intestine, respectively (Poonkuzhali et al., 2008). Various polymorphisms in the promoter have also been linked to modifications of protein transcription (Poonkuzhali et al., 2008). The associations of several ABCG2 genetic variants with drug absorption and disposition have also been assessed in clinical studies for various ABCG2 substrates and are presented below. Table 14.3 summarizes the data obtained with the chemotherapy drugs.

14.5 LINKAGE AND HAPLOTYPE STRUCTURES

Kobayashi et al. (2005) genotyped placentas from 97 Japanese patients to assess ABCG2 haplotypes and their association with protein expression. Placentas homozygous for the C421A variant allele were found to have significantly lower expression of ABCG2 than wild-type placentas. However, no additional correlation was found with the analysis of haplotypes. Interestingly, the C421A polymorphism did not coexist on a single allele with either the G34A (V12M) or C376T (Q126X) variants. C421A has also been shown to be in significant linkage disequilibrium with a CTCA deletion at -19572 to ATG by Backstrom et al. (2003), who observed a frequency of the deletion of 0.54 in a Swedish population. In a Dutch population, haplotype analysis showed that the most frequent variant haplotype (29%) included the intron 9 A5974T and intron 13 C40T SNPs, whereas the frequency of other haplotypes was less than 5%. In another study, the ABCG2 haplotype distribution in Japanese individuals was found to be different from those reported in Caucasians (Maekawa et al., 2006).

TABLE 14.3 Impact of ABCG2 Polymorphisms on Drug Pharmacokinetics, Outcome or Toxicity of Anticancer Drugs

SNP/haplotype	Drug	Population	Effects compared to WT (statistically significant)	Reference
C421A (Q141K)	Diflopotecan	22 Caucasian patients	Plasma levels 3-fold higher for heterozygotes	1
	Topotecan	12 patients	Oral bioavibility 1.34-fold higher for heterozygotes	2
	Irinotecan	84 Caucasian patients	No significant change in PK	3
		29 Asian patients	No significant change in PK	4
		107 Korean patients	No significant change in PK	5, 6
		55 Japanese patients	No significant change in PK alone, but association with grade 3/4 neutropenia when part of a haplotype with 3 intronic SNPs	7
	9-NC/9-AC	55 patients	Higher exposure to 9-AC lactone form	8
	Imatinib	82 Caucasian patients	No significant change in PK	9
		112 patients	No significant change in PK	10
		34 Japanese patients	No significant change in PK	13
		229 patients	Associated with lower response to treatment in univariate but not in multivariate analyses	14
		67 patients	22% reduction in apparent clearance for heterozygotes	11
		67 Japanese patients	1.36 fold higher dose-adjusted trough plasma concentration, but no significant association with response to treatment	12
	Gefitinib	124 patients	Associated with grade 1-2 diarrhea occurrence	15
		27 patients	Higher drug accumulation for heterozygotes	16
		75 Japanese patients	No significant change in adverse effect	17
		94 patients	No significant change in toxicity and clinical outcome	18
	Erlotinib	42 patients	Associated with a 24% decrease in clearance.	20
		79 patients	Associated with clearance in univariate but not in multivariate analyses	21
		80 patients	No significant change in PK or toxicity	19
	Sunitinib	219 patients	Weak trend in increased thrombocytopenia, no significant change in response	22
		166 patients	No significant change in PFS or OS	23
		5 patients	Increase in exposure and severe toxicity for homozygote	24
	Docetaxel	87 patients	Increase in survival beyond 15 months (67% vs 22% for WT)	25
		54 Asian patients	Trend in a decrease in Cmax	27
	Doxorubucin	62 Asian patients	No significant change in PK	28
	Flavopiridol	51 patients	Trend in increase in AUC of the glucuronide metabolite of flavopiridol, no change in response to treatment	32

TABLE 14.3 (*Continued*)

SNP/haplotype	Drug	Population	Effects compared to WT (statistically significant)	Reference
	Combination chemotherapy	914 patients	(Platinum plus taxane) no significant change in toxicity or clinical outcome	26
		145 Korean patients	(Rituximab plus cyclophosphamide/doxorubicin/ vincristine /prednisone) increase in fever, infection, diarrhea	29
		349 Caucasian patients	(first line chemotherapy) worse OS	30
G34A (V12M)	Irinotecan	29 Asian patients	Trend in higher exposure	4
		107 Korean patients	No significant change in PK or toxicity alone but higher incidence of grade 3 diarrhea in multivariate analysis	5,6
		55 Japanese patients	Haplotype with del TCAT and 6 SNPs on intron decrease trend in grade 3/4 neutropenia for irinotecan/cisplatin therapy	7
	Erlotinib	80 patients	No significant change in PK or toxicity	19
	Imatinib	229 patients	Associated with higher response to treatment in univariate but not in multivariate analyses	14
	Sunitinib	219 patients	No significant change in toxicity, trend in increase in OS and PFS	22
		166 patients	No significant change in PFS or OS	23
	Combination chemotherapy	145 Korean patients	(rituximab plus cyclophosphamide/doxorubicin/ vincristine /prednisone) no significant change in PK, toxicity and response	29
		349 Caucasian patients	(first line chemotherapy) no significant change in PFS or OS	30
C376T (Q126X)	Irinotecan	55 Japanese patients	Related to grade 4 neutropenia (statistics could not be applied)	7
	Gefitinib	75 Japanese patients	No significant change in adverse effect	17
T1465C (F489L)	Irinotecan	55 Japanese patients	Related to grade 4 neutropenia (statistics could not be applied)	7
C1723T (R575X)	Irinotecan	55 Japanese patients	Related to grade 4 neutropenia (statistics could not be applied)	7
-19572 delTCAC	Irinotecan	29 Asian patients	Decreased conversion	4
		55 Japanese patients	Haplotype with G34A and 6 intronic SNPs decrease trends in grade 3/4 neutropenia for irinotecan/cisplatin therapy	7
C-34911T (G-15994A)	Erlotinib	80 patients	No significant change in PK or toxicity	19
C-34539T (C-15622T)	Gefitinib	94 patients	Increase in grade 2/3 diarrhea, trend in a decrease in time to progression (TTP)	18
	Erlotinib	80 patients	No change in AUC, trend in increase in C_{max}, trends to increase in diarrhea	19
	Sunitinib	219 patients	No significant change in toxicity or response	22
		166 patients	No significant change in PFS or OS	23

(*continued*)

TABLE 14.3 (*Continued*)

SNP/haplotype	Drug	Population	Effects compared to WT (statistically significant)	Reference
Intron 1 G152A	Flavopiridol	51 patients	No significant change in PKOR response to treatment	32
Intron 1 G614A	Gefitinib	94 patients	No significant change in toxicity an clinical outcome	18
(C1143T)	Erlotinib	80 patients	Increase in C_{max} and trend in increase in AUC	19
	Sunitinib	219 patients	No significant change in toxicity or response	22
		166 patients	No significant change in PFS or OS	23
	Methotrexate	374 patients	No significant change in toxicity or response to treatment	31
Intron 1 A10130G	Flavopiridol	51 patients	No significant change in PK or response to treatment	32
Intron 1 A11071C	Methotrexate	374 patients	No significant change in toxicity or response to treatment	31
Intron 1 G11520T	Methotrexate	374 patients	No significant change in toxicity or response to treatment	31
Intron 1 G12812T	Methotrexate	374 patients	No significant change in toxicity or response to treatment	31
Intron 1 G14952T	Methotrexate	374 patients	No significant change in toxicity or response to treatment	31
	Flavopiridol	51 patients	No significant change in PK or response to treatment	32
Intron 1 C16170T (G16702A)	Erlotinib	80 patients	Decrease in grade 2 > skin rash	19
Intron 1 G18271A	Methotrexate	374 patients	No significant change in toxicity or response to treatment	31
	Flavopiridol	51 patients	Increase in the AUC of the glucuronide metabolite, improved response to treatment, trend in increase in toxicity	32
Intron 2 C5566T	Methotrexate	374 patients	Better response to treatment	31
Intron 7 T179C	Methotrexate	374 patients	No significant change in toxicity or response to treatment	31
Intron 9 A928G	Methotrexate	374 patients	Better response to treatment	31
Intron 9 G3531C	Methotrexate	374 patients	No significant change in toxicity or response to treatment	31
Intron 9 A3610C	Methotrexate	374 patients	No significant change in toxicity or response to treatment	31
Intron 14 A815G	Methotrexate	374 patients	No significant change in toxicity or response to treatment	31
Haplotype	Gefitinib	94 patients	Increase in grade 2/3 diarrhea, trend in decrease in TTP	18
C-15622T C1143T	Erlotinib	80 patients	Increase in AUC and C_{max}	19
	Sunitinib	219 patients	Increase in any toxicity higher than grade 2, trend in thrombocytopenia, no significant change in response	22

References: 1-Sparreboom et al. (2004); 2- Sparreboom et al. (2005); 3-de Jong et al. (2004); 4-Zhou et al. (2005); 5-Han et al. (2007); 6-Han et al. (2009); 7-Sai et al. (2010); 8-Zamboni et al. (2006); 9-Gardner et al. (2006); 10-Raymond et al. (2008); 11-Petain et al. (2008); 12-Takahashi et al. (2010); 13-Yamakawa et al. (2011); 14-Kim et al. (2009); 15-Cusatis et al. (2006); 16-Li et al. (2007); 17-Akasaka et al. (2010); 18-Lemos et al. (2011); 19-Rudin et al. (2008); 20-Thomas et al. (2009); 21-White-Koning et al. (2011); 22-van Erp et al. (2009); 23-van der Veldt et al. (2011); 24 -Mizuno et al. (2010); 25-Hahn et al. (2006) 26-Marsh et al. (2007); 27-Chew et al. (2011); 28 Lal et al. (2008); 29-Kim et al. (2008); 30-Muller et al. (2009); 31 Warren et al. (2008); 32-Ni et al. (2010).

Three haplotypes were relatively well conserved in the Japanese. The first one includes the C421A SNP (Q141K) combined with three SNPs in introns 12, 13, and 14 (G49T, C40T, and A815G, respectively) and was observed with a frequency of 24.6%. The second one, observed in 19.2% of samples, corresponds to the intron 1 G18271A SNP. The last one has a frequency of 8.8% and includes the CTCA deletion localized in the promoter (-19572 to ATG), the nonsynonymous V12M SNP and 6 SNPs in introns (intron 2 A26G, intron 6 A3158G, A3203G, A3287G, intron 9 A5974T, and intron 13 T40C). Perfect linkage equilibrium has been reported between G34A (V12M) and the intron 2 A36G SNP, as well as between the intron 6 A3203G and A3287G, and between the intron 11 A20G and the intron 13 C1823T.

14.6 FUNCTIONAL VALIDATION OF ABCG2 POLYMORPHISMS

To clarify the possible clinical relevance of the ABCG2 polymorphisms, studies have been performed to characterize function or expression for some of the variants, mainly for the nonsynonymous ones. Numerous experiments focused on the two most frequent SNPs: G34A (V12M) and C421A (Q141K). The V12M affects the N-terminal intracellular region of the protein, and both V and M amino acids have uncharged, hydrophobic side chains. Q141K is located between the Walker A motif and the signature region and affects the ATP-binding domain. C421A leads to the replacement of the negatively charged glutamine residue with a positively charged lysine residue. Contradictory data were obtained for expression and localization for both ABCG2 variants, probably due to the differences in cell context, transfection conditions (transient vs. stable expression), copy number of cDNA incorporated in genomic DNA, or other cellular determinants. Nevertheless, a majority of the studies seem to point toward an impaired protein expression, an incomplete trafficking to the plasma membrane, and a decreased function for Q141K. In the case of V12M, data regarding an impact on function are contradictory and the question is not settled (cf. Table 14.4)

G51C, Q126X, Q166E, I206L, F208S, S248P, P269S, F431L, S441N, N590Y, and D620N were also investigated for their impact on protein function in different cell models. A number of groups described a decreased expression and impaired localization for the S441N variant, and a complete lack of function for the Q126X variant, which encodes a stop codon and generates a truncated protein. I206L, F208S, N590Y, and D620N were also linked to a lower function but more studies are required to confirm these results. Moreover, several SNPs in the ABCG2 promoter were also associated with either an increase or decrease in mRNA expression. A summary of functional assays performed on ABCG2 variants is presented in Table 14.4.

14.7 IMPACT OF ABCG2 VARIANTS ON CAMPTOTHECINS

Camptothecins are topoisomerase I inhibitors used in anticancer therapy. To date, the associations of ABCG2 variants with camptothecin pharmacokinetics have been

TABLE 14.4 Functional Assays Performed on ABCG2 Variants

Name (as in Table 14.2)	Other name or aa change	Expression change	Function change	Localization change
Variants in Noncoding Sequence				
-49556 delATAA	-43639 delATAA	Decreased mRNA (1)	nd	nd
C-49394G	C-30477G	Decreased mRNA (1)	nd	nd
C-34911T	C-15994T	Increased mRNA (1)	nd	nd
A-347363C	A-15846C	Increased mRNA (1)	nd	nd
C-34539T	C-15622T	Decreased mRNA (1)	nd	nd
C-19031T	C-19031T	No change (7)	nd	nd
intron1 G614A	A1143G	Decreased mRNA (1)	nd	nd
intron1 T11751C	T12283C	Increased mRNA (1)	nd	nd
intron1 C16170T	C16702T	Increased mRNA (1)	nd	nd
intron1 16289 delT	16823 delT	Decreased mRNA (1)	nd	nd
Nonsynonymous variants				
G34A	V12M	No change (2,3,4,5,8,9,11)	No change (2,4,5,9,10) Decreased function (3)	No change (5,9,10) Partly intracellular (3)
G151T	G51C	No change (8)	No change (8)	No change (8)
C376T	Q126X	No protein (2,4)	No function (4)	No protein (2,4)
C421A	Q141K	Decreased protein (2,4,5,8,9,11,12) No change (3)	No change (5) Decreased function (3,4) Decreased cell resistance to drug (2,8,9,10,12)	No change (3,5,9) Decreased (8, 12) Partly intracellular (10)
C496G	Q166E	No change (5) Increased amount (8)	No change (5,8)	No change (5) Slightly higher (8)
A616C	I206L	Decreased protein (6) No change (8)	Increased function (6) No change (8)	No change (6)
T623C	F208S	No/very low protein expression (8,9)	Decreased cell resistance to drug (8,9)	Not expressed (8,9)
T742C	S248P	No change (8,9)	No change (8) Decreased function (9)	No change (9)
C805T	P269S	No change in expression (5,7)	Decreased activity (7) No change (5)	No change (5,7)
T1291C	F431L	No change (8,9)	Decreased function (8,9)	No change (9)
G1322A	S441N	Decreased protein (5,8,9)	No function (4) Decreased function (5) Decreased cell resistance to drug (8,9)	Intracellular localization (5,9)
T1465C	F489L	No change (9)	Decreased function (9)	No change (9)
A1768T	N590Y	Increased expression (6) No change (8)	Decreased function (6) No change (8)	No change (6)
delC1515	F506SfsX4	nd	No function (4)	nd
G1858A	D620N	Increased expression (6) Decreased expression (8)	Decreased function (6,8) No change (10)	No change (6,10) Decreased (8)

References: 1-Poonkuzhali et al. (2008), human tissues, 2-Imai et al. (2002), PA317 cells, 3-Mizuarai et al. (2004), LLC-PK1 cells, 4-Matsuo et al. (2009), 5-Kondo et al. (2004), LLC-PK1 cells, 6-Vethanayagam et al. (2005), HEK293 cells, 7-Lee et al. (2007), SF9 insect cells, 8-Yoshioka et al. (2007), PA317 cells, 9-Tamura et al. (2006), Flp-in-293 cells, 10-Morisaki et al. (2005), HEK293 cells, 11-Mizuarai et al. (2004), LLC-PK1 cells, 12-Furukawa et al. (2009).
nd : not determined.

assessed for topotecan, diflomotecan, 9-nitrocamptothecin, and irinotecan. Of these camptothecins, the first three are ABCG2 substrates. Whereas irinotecan is not a transported substrate for ABCG2, its active metabolite SN-38 is, as is its conjugated derivative SN-38 glucuronide.

14.7.1 Diflomotecan

The first pharmacogenetic association study of ABCG2 was conducted with diflomotecan, an investigational synthetic derivative of camptothecin (Sparreboom et al., 2004). The authors observed a link between the ABCG2 C421A polymorphism and an altered drug exposure. In particular, plasma levels of diflomotecan after intravenous administration were significantly higher (300%) in patients carrying the 421A allele compared with patients with two copies of the 421C allele. However, after oral administration, the pharmacokinetic parameters of diflomotecan did not differ between the studied ABCG2 genotypes, despite the presence of ABCG2 in the intestine. Unexpectedly, the overall oral bioavailability was significantly lower ($p = 0.046$) in patients carrying the 421A allele.

14.7.2 Topotecan

Similar findings were found in a pilot study of the ABCG2 C421A polymorphism in 12 cancer patients undergoing treatment with the structurally related agent topotecan, which is a substrate for both ABCB1 and ABCG2. Heterozygotes had a 1.34-fold increase in oral bioavailability (31.4% (wild-type) vs. 42.0% (heterozygous); $p =$ 0.037), as compared to wild-type patients (Sparreboom et al., 2005). Intestinal biopsies also noted that the lowest combined mRNA expression levels of ABCG2 and ABCB1 were observed in one of the patients who was heterozygous variant for ABCG2 C421A. These results are preliminary and still need to be confirmed in a larger population.

14.7.3 Irinotecan

Contrasting the positive correlations found for diflomotecan and topotecan described above, circulating concentrations of the ABCG2 substrates SN-38 and SN-38 glucuronide (SN-38G) following the administration of irinotecan to Caucasian cancer patients were not dependent on ABCG2 C421A genotype (de Jong et al., 2004). Despite the lack of significant changes in pharmacokinetics between the wildtype (WT) and C421A carriers, one of the two homozygous variant allele carriers in this study showed extensive accumulation of SN-38 and SN-38 glucuronide.

A subsequent study was performed studying irinotecan pharmacogenomics in 29 Asian patients (Zhou et al., 2005). Of the nine SNPs analyzed for the ABCG2 gene, variant genotypes were only found for three of them: the promoter -19572 delCTCA variant, the nonsynonymous G34A (V12M) variant, and the C421A (Q141K) variant. No link was found between the C421A SNP and irinotecan pharmacokinetics, confirming the previous finding. Nevertheless, these authors did find a significant correlation between the area under curve ratio of SN-38 and irinotecan with the -19572 delCTCA polymorphism in ABCG2. Patients with the deletion had decreased relative extent of conversion (REC) of irinotecan to SN-38 as compared to wild-type patients (REC: 0.043 ± 0.020 vs. 0.064 ± 0.023). Moreover, patients who were wild-type for the G34A SNP showed a trend toward lower systemic exposures to irinotecan (i.e., C_{max}, $p = 0.056$) compared with patients with one or two variant alleles.

A Korean team took interest in the multiparametric pharmacokinetics and pharmacodynamics (PK/PD) of irinotecan. They weighed the fact that irinotecan pharmacology is complex and may be dependent on the interplay of metabolizing enzymes and transporters. They first conducted irinotecan pharmacogenetic analyses using several polymorphisms in the involved genes, including ABCG2, and evaluated their significance separately in patients with non-small-cell lung cancer. They showed that ABCG2 V12M and Q141K SNPs did not influence disposition and tumor response to irinotecan (Han et al., 2007). They then considered that the altered function caused by single gene variation could be obscured by the compensatory activity of other enzymes and transporters. They therefore conducted a multivariate analysis using 15 polymorphisms within 7 genes with putative influence on irinotecan pharmacokinetics and toxicity (Han et al., 2009). In this way, they observed an association of ABCG2 V12M SNP with a higher incidence of grade 3 diarrhea in a combination therapy of irinotecan and cisplatin (OR = 5.1).

Another study in Japanese patients focused on the possible additive effects of various transporter polymorphisms and UGT1A1 genotypes on irinotecan pharmacokinetics and pharmacodynamics (Sai et al., 2010). Using previously defined ABCG2 haplotypes (Maekawa et al., 2006), it was found that while an association of C421A along with irinotecan pharmacokinetics did not reach statistical significance, the ABCG2 #IIB haplotype (consisting of C421A combined with three SNPS in introns 12, 13, and 14) showed an association with grade 3/4 neutropenia ($p = 0.049$). No influence of ABCG2 #IIIC haplotype (including G34A, the CTCA deletion localized on the promoter, and six SNPS on introns 2, 6, 9, and 13) was detected on neutropenia in the irinotecan monotherapy, but the haplotype was associated with a decreasing trend in grade 3/4 neutropenia in the combination therapy with cisplatin. These authors also showed that the minor genetic variations C376T (Q126X), T1465C (F489L), C1723T (R575X) were possibly related to grade 4 neutropenia, but statistical analysis could not be applied due to the few variant carriers.

14.7.4 9-nitrocamptothecin

Zamboni et al. (2006) have studied the relation between the ABCG2 C421A polymorphism and disposition of 9-nitrocamptothecin (9-NC), along with its active metabolite, 9-aminocamptothecin (9-AC), in an effort to better understand the interindividual pharmacokinetics variability observed with this agent. It has previously been reported that 9-AC is a substrate for ABCG2 but not 9-NC (Rajendra et al., 2003). As expected, this polymorphism had no effect on the pharmacokinetics of the parent drug. However, patients heterozygous for the ABCG2 C421A polymorphism had higher exposure to the lactone form of 9-AC ($p = 0.032$), although there was no significant difference in total 9-AC (comprised of the lactone and hydroxy acid forms) disposition with genotype.

14.8 IMPACT OF ABCG2 VARIANTS ON SIGNAL TRANSDUCTION INHIBITORS

Tyrosine kinase inhibitors (TKIs) are commonly used in the treatment of cancer. Among them, gefitinib (EGFR tyrosine kinase inhibitor), imatinib (c-kit, bcr-abl inhibitor), sunitinib (PDGFR, VEGFR, c-kit inhibitor), and erlotinib (EGFR tyrosine kinase inhibitor) have been studied in the context of exploratory ABCG2 genotype association studies. All of these agents are known substrates of ABCG2 transport at low concentrations, and can inhibit ABCG2 activity at high concentrations (Dohse et al., 2010). While these TKIs are clearly transported by ABCG2, it is conceivable that the clinical importance of ABCG2 variants for some of them, like imatinib, may be overshadowed by the existence of multiple compensatory mechanisms of drug elimination.

14.8.1 Imatinib

In contrast to the positive *in vitro* findings, the ABCG2 C421A polymorphism was not found to correlate with imatinib pharmacokinetics in adult Caucasian patients with gastrointestinal stromal tumors (Gardner et al., 2006) or recurrent gliomas (Raymond et al., 2008), and contributed only minimally to explaining interindividual pharmacokinetic variability in a mixed group of children and adults (Petain et al., 2008). Interestingly, in populations where the C421A variant is more prevalent, such as the Japanese and Han Chinese, statistically significant associations of this SNP with imatinib trough concentrations have been reported in some (Takahashi et al., 2010) but not all studies (Yamakawa et al., 2011). Regardless, the overall rather limited contribution of ABCG2 variants to the variability in the pharmacokinetic profile of imatinib is consistent with murine studies indicating that even the complete absence of ABCG2 in knockout animals does not lead to substantially altered absorption and disposition properties of imatinib (Oostendorp et al., 2009). It is possible that inherited genetic variation in ABCG2 could impact the efficacy of imatinib treatment via direct mechanisms unrelated to an influence of SNPs on absorption or disposition. An example of this is the finding that the GG genotype in ABCG2 G34A (V12M) was significantly associated with poorer response to imatinib compared to the variant genotype in patients with chronic myeloid leukemia, especially for major or complete cytogenetic response (Kim et al., 2009).

14.8.2 Gefitinib

The first pharmacogenetics study linking ABCG2 and gefitinib-related side effects found a statistically significant association between the C421A SNP and the development of grade 1 or 2 diarrhea in 124 patients ($p = 0.0046$) (Cusatis et al., 2006), although there was no association observed with skin toxicity. The one patient with the homozygous variant genotype had no noticeable toxicity. An association between the C421A SNP and gefitinib pharmacokinetics was described later by the same investigators (Li et al., 2007). In particular, patients with the heterozygous genotype had a greater accumulation of the drug at steady state during 28 days of treatment compared to patients with the wild-type genotype (median, 5.07 vs. 3.60, $p = 0.004$). Nevertheless, in a Japanese study, no association was found between the ABCG2 C421A polymorphism and susceptibility to gefitinib-induced adverse effects (Akasaka et al., 2010). At the same time, the authors also screened the C376T SNP, resulting in the truncated, nonfunctional ABCG2 Q126X, but in the same way, no correlation was observed.

Gefitinib pharmacogenomic studies were then broadened to include other ABCG2 SNPs that have been associated with different TKIs. In this way, Lemos et al. (2011) investigated not only the C421A SNP but also the C-15622T and C1143T SNPs. They observed that the ABCG2 C-15622T polymorphism and C1143T/C-15622T haplotype resulted in a grade 2/3 gefitinib-dependent diarrhea ($p < 0.01$) suggesting that these pharmacogenetic markers should be considered to optimize NSCLC treatment. Despite no association noticed with ABCG2 SNPs and time to progression (TTP) or OS, a trend was observed between the haplotype and TTP, as well as between C-15622T SNP and TTP ($p = 0.06$ and $p = 0.07$, respectively).

14.8.3 Erlotinib

A prospective clinical study of 80 patients was performed to assess the pharmacogenomic and pharmacokinetic determinants of erlotinib toxicity (Rudin et al., 2008). C421A, G34A, G-15994A, C-15622T, G16702A, and C1143T SNPs in the ABCG2 gene were genotyped, the first two because they are common nonsynonymous variants, the last four because they cause ABCG2 expression modulation. The last four SNPs are named C-34911T, C-15622T, intron 1 C16170T, and intron 1 G614A in Table 14.2. The authors identified a novel haplotype in the ABCG2 promoter involving C-15622T and C1143T. The presence of this haplotype conferred lower ABCG2 levels associated with higher erlotinib pharmacokinetic parameters, including area under the curve ($p = 0.019$) and maximum concentration ($p = 0.006$). Separately, the C-15622T SNP showed no association with AUC and a marginal association with C_{max} ($p = 0.065$) while the C1143T SNP tends to be associated with higher erlotinib AUC and C_{max} ($p = 0.072$ and $p = 0.047$, respectively). Moreover, the authors observed a significant association between the G16702A SNP and grade $\geq$2 skin rash, where WT genotype exhibit a higher toxicity compared to variant carriers ($p = 0.027$). This association was confirmed in the multivariable analysis ($p = 0.05$), by using a regression logistic model, which is fit to examine the effect of polymorphisms on toxicity while controlling for PK. A marginally significant association was also detected between C-15622T and any grade diarrhea ($p = 0.066$), with a higher risk to develop diarrhea for the variant carriers. This association was also confirmed in the multivariant analysis ($p = 0.057$). No association was found with the C421A, G34A, and G-15994A SNPs.

Recently, the effect of C421A SNP (Q141K) in ABCG2 on erlotinib pharmacokinetic parameters was also evaluated in adults with head and neck cancer (Thomas et al., 2009) and an expanded cohort that also included pediatric cancer patients (White-Koning et al., 2011). In these studies, the presence of the ABCG2 C421A SNP was significantly associated with the apparent oral clearance of erlotinib. However, in the expanded cohort, this SNP did not remain statistically significantly associated with erlotinib pharmacokinetics when other covariables were concurrently considered. These findings collectively support the contention that ABCG2 by itself is unlikely to be a critical determinant of erlotinib absorption and/or disposition. This idea is consistent with preclinical data indicating that even the complete lack of ABCG2, in mice with a deletion of the *Abcg2* gene, does not substantially affect the plasma pharmacokinetic profile of erlotinib (Elmeliegy et al., 2011). Indeed, it has been argued that a noticeable impact of ABCG2-mediated transport of erlotinib on systemic concentrations does not become apparent unless ABCB1-mediated transport of erlotinib is simultaneously impaired (Elmeliegy et al., 2011).

14.8.4 Sunitinib

Sunitinib was first identified as an ABCG2 inhibitor, by interacting directly with the transporter substrate-binding pocket (Shukla et al., 2009), and has been found to be a transported substrate in some (Tang et al., 2012) but not all studies (Hu et al., 2009). A multicenter pharmacogenetic association study was performed in 219 patients treated with sunitinib (van Erp et al., 2009). Nineteen polymorphisms in seven genes of putative relevance to the pharmacokinetics of sunitinib were tested for their relationship with treatment-related toxicity. Among them, four ABCG2 SNPs were selected for consideration based on their frequency in the Caucasian population, their known clinical relevance, or because they encode nonsynonymous variants (C-15622T and C1143T, G34A and C421A). By correlating variant genotypes with toxicity, the investigators found that the ABCG2 variant haplotype C-15622T/C1143T was associated with an increase in the prevalence of any grade ≥ 2 toxicity (OR = 2.63; $p = 0.016$). They also observed that the C421A SNP and the C-15622T/C1143T haplotype both showed a trend toward increased thrombocytopenia (OR = 2.9, p = 0.096 and OR = 1.93, $p = 0.065$, respectively).

The same investigators also studied the progression-free survival (PFS) and overall survival (OS) in sunitinib-treated patients in relation to the same four SNPs (van der Veldt et al., 2011). Among them, only G34A SNP (V12M) was associated with PFS and OS. A potentially prolonged PFS was found in univariate analysis of patients with the A-allele (median PFS: 19.4 vs. 9 months for WT, $p = 0.077$). Likewise, the A-allele was probably associated with prolonged OS (median 39.9 vs. 15.4 months for G-allele, $p = 0.072$). Multivariate analysis confirmed that there was a trend toward an improved OS for the G34A carriers (HR: 0.416, $p = 0.069$). Since the antitumor efficacy of sunitinib is linked with circulating drug concentrations (Faivre et al., 2006), it is possible that the mechanism underlying the reported associations of variant ABCG2 genotypes with PFS and OS is due to altered systemic exposure to sunitinib in individuals carrying reduced function SNPs. This hypothesis is supported by a recent case report indicating that homozygosity for the C421A SNP (Q141K) was associated with increased sunitinib exposure and severe toxicity (Mizuno et al., 2010).

14.9 OTHER ANTICANCER DRUGS

14.9.1 Docetaxel

In line with the notion that docetaxel appears to be at best a weak substrate for ABCG2 (Baker et al., 2009; Shen et al., 2011), some studies (Marsh et al., 2007; Chew et al., 2011) but not all (Hahn et al., 2006) have found no correlation between ABCG2 genotype and the pharmacokinetic and pharmacodynamic profiles of docetaxel in cancer patients.

14.9.2 Doxorubicin

Doxorubicin was first discovered as an ABCG2 substrate in a cell line expressing a gain of function ABCG2 mutant, which has so far not been described in a clinical setting. Later, doxorubicin was found to be at best a weak substrate for the wild-type transporter (Robey et al., 2003). The impact of ABCG2 C421A polymorphism on the disposition of doxorubicin was investigated in Asian breast cancer patients receiving adjuvant chemotherapy. No significant influences on doxorubicin pharmacokinetic parameters were observed in relation to the ABCG2 C421A polymorphism (Lal et al., 2008).

14.9.3 Methotrexate

Methotrexate's relationship with ABCG2 SNPs was studied in the framework of psoriasis treatment (Warren et al., 2008), but the drug is also widely used in anticancer therapies. The authors selected 12 ABCG2 SNPs among the haplotype tagging ones: intron 1 G614A (rs2622604), intron 1 A11071C, (rs17731799), intron 1 G11520T (rs12503227), intron 1 G12812T (rs2622626), intron 1 G14952T (rs3114018), intron 1 G18271A (rs1564481), intron 2 C5566T (rs17731538), intron 7 T179C (rs1481012), intron 9 A928G (rs13120400), intron 9 G3531C (rs2622621), intron 9 A3610C (rs12505410), and intron 14 A815G (rs2231164). None of these correlated with toxicity, but two ABCG2 SNPs were associated with a good response to treatment, with $\chi 2$ for trend test of $p = 0.007$ for the intron 2 C5566T SNP, and $p = 0.03$ for the intron 9 A928G SNP.

14.9.4 Flavopiridol

Flavopiridol is a serine threonine kinase in phase II clinical development for cancer treatment. *In vitro* assays demonstrated that the drug is a substrate of ABCG2 (Robey et al., 2001). Association between ABCG2 and flavopiridol pharmacokinetics was performed on 51 patients in the USA (Ni et al., 2010). Five SNPs were selected: Q141K (rs2231142) and four located in intron 1: G152A (rs2622605), A10130G (rs2622624), G14952T (rs3114018), and G18271A (rs1564481). Among them, two ABCG2 SNPs were found to correlate with flavopiridol pharmacokinetics and treatment outcome in univariate analysis. The intron 1 G18271A and Q141K were linked to a significant increase in the AUC of the glucuronide metabolite of flavopiridol ($p =$ 0.05) and a trend of increased flavopiridol AUC ($p = 0.08$), respectively. Moreover, the intron 1 G18271A was also linked to an improved response to treatment ($p =$ 0.037), and showed a trend of associating with toxicity ($p = 0.074$).

14.9.5 Combination Chemotherapy

Therapeutic outcome and toxicity were studied in 145 Korean patients with diffuse large B-cell lymphoma treated with rituximab plus cyclophosphamide/doxorubicin /vincristine/prednisone, in relation to C421A and G34A SNPs (Kim et al., 2008). No associations were observed on patient characteristics, disease characteristics, response, survival, or hematology toxicity profiles. Nevertheless, fever and infection (grade 3–4) were statistically significant according to the C421A polymorphism ($p = 0.037$ and 0.046, respectively). On multivariate analysis, grade 1–4 diarrhea was the statistically significant factor according to the ABCG2 C421A SNP (HR 2.835; $p = 0.003$). It is not clear which substrate is causing this toxicity. Doxorubicin is at best a weak substrate; this may also be true for vinblastine (Robey et al., 2003; Shen et al., 2011).

In another association analysis of combination chemotherapy, a team observed a synergistic interaction between ABCG2 and MDR1 (ABCB1) pharmacogenomics in children with acute lymphoblastic leukemia (Erdilyi et al., 2008). Toxicity during chemotherapy was analyzed with regard to three ABCB1 SNPs and two ABCG2 SNPs (C421A and 34G4A). When the C421A SNP was analyzed separately, it was not shown to be associated with the occurrence of encephalopathy, although they observed that patients carrying the ABCG2 C421A allele tended (OR = 2.0; $p = 0.25$) to have more complications than wild-type homozygotes. However, significantly more children suffered encephalopathy in the group with ABCB1 3435T and ABCG2 C421A genotypes (OR = 12.3; $p = 0.005$). These results need to be repeated. They are consistent with findings in knockout murine models. In a series of studies, increased CNS uptake of substrates by deletion of both ABCG2 and ABCB1 was not observed extensively unless both transporters were deleted. This included topotecan, mitoxantrone, flavopiridol, dasatinib, erlotinib, imatinib, lapatinib. sunitinib, and sorafenib (de Vries et al., 2007; Polli et al., 2009; Zhou et al., 2009; Lagas et al., 2009, 2010; Kodaira et al., 2010; Tang et al., 2012).

14.10 STATINS

Statins are inhibitors of the enzyme HMG-CoA reductase, commonly used to lower cholesterol levels. Among them, atorvastatin, cerivastatin, fluvastatin, pitavastatin, pravastatin, and rosuvastatin are ABCG2 substrates (see Table 14.1), whereas simvastatin seems not to be (Li et al., 2011). They were all tested in clinical studies in their relationship with the ABCG2 C421A SNP, except cerivastatin, which was withdrawn from the market in 2001 due to grave secondary effects.

14.10.1 Rosuvastatin

The C421A polymorphism was shown to affect rosuvastatin PK in different studies. A first genetic polymorphism study conducted in healthy Chinese males showed an association between the statin pharmacokinetics and the C421A SNP after the exclusion of impact of SLCO1B1 and CYP2C9 (Zhang et al., 2006). The AUC and C_{max} were about 2-fold higher in variant carriers compared to WT subjects (with $p = 0.0018$ and $p = 0.048$, respectively) while the oral clearance (CL/F) was 1.75 higher in wild-type group compared to C421A carriers. These data were confirmed in a Finnish study (Keskitalo et al., 2009b). The AUC and C_{max} in C421A homozygote subjects were 2- and 2.08-fold greater than in the C421A heterozygotes ($p =$ 0.004 and = 0.01), and were 2.44- and 2.31-fold higher than in the homozygote WT subjects ($p < 0.001$ and = 0.003, respectively). They also observed the amount excreted in urine within 24 h and observed that it was 2-fold ($p = 0.004$) and 2.56-fold ($p < 0.001$) higher in variant homozygote carriers than in heterozygote or WT homozygote subjects. However, the renal clearance of rosuvastatin was unaffected by the ABCG2 genotype. The authors concluded that the decreased activity of the ABCG2 variant likely enhanced the absorption of rosuvastatin in the small intestine, leading to an increase in its bioavailability. They also deduced that ABCG2 may play a limited role in renal excretion of the drug, as illustrated by the lack of correlation of the SNP with renal clearance. The increase in the rosuvastatin amount excreted in the urine observed in the variant carriers was explained by the higher plasma AUC.

To confirm the C421A SNP association with rosuvastatin, two groups examined whether this polymorphism influenced the efficacy of the drug on cholesterol levels in a Caucasian and in a Chinese population (Bailey et al., 2010; Tomlinson et al., 2010). They both observed that the C421A SNP was significantly associated with greater reduction in low-density lipoprotein cholesterol (LDL-C) level, with a decrease of 6.2% in C421A (homo- and heterozygote) compared to the WT carriers ($p = 0.010$) in one study, and a decrease of 6.9% in homozygote variant carriers compared to WT ($p = 0.0006$) in the second. This decrease in cholesterol level is described to correspond to the effect obtained by more than a double exposure to rosuvastatin.

14.10.2 Other Statins with Association to C421A SNP

C421A SNP was also described to be associated with atorvastatin, fluvastatin, and simvastatin pharmacokinetics, since the polymorphism increased about twice the systemic exposure to these drugs (Keskitalo et al., 2009a, 2009b). More precisely, the homozygote variant carriers showed a 1.72-fold increase in AUC from 0 h to infinity compared to the WT carriers for atorvastatin and fluvastatin ($p = 0.049$ and $p = 0.009$, respectively) and a 2.11-fold increase ($p = 0.005$) was observed with simvastatin lactone form. Nevertheless, no significant association was found with the simvastatin acid form. Another study showed no significant difference in LDL-C level after simvastatin treatment in heterozygote and homozygote variant patients compared to the WT patients (Bailey et al., 2010).

14.10.3 Statins with Lack of Association with the C421A SNP: Pivastatin and Pravastatin

ABCG2 is involved in the biliary excretion of pitavastatin in mouse and most probably in human (Hirano et al., 2005). However, the C421A ABCG2 SNP did not appear to be associated with the altered pharmacokinetics of pitavastatin or pitavastatin lactone (Ieiri et al., 2007). In the same way, the C421A SNP seemed to have no significant effect on pravastatin pharmacokinetics (Keskitalo et al., 2009a). It has been suggested that multiple transporters are involved in the biliary excretion of pitavastatin and pravastatin in humans (Hirano et al., 2005; Matsushima et al., 2005), and could compensate for ABCG2 dysfunction.

14.11 OTHER DRUGS

14.11.1 HIV Treatment

Lamivudine is a retroviral inhibitor used in the treatment of HIV infection. Whereas *in vitro* data demonstrated ABCG2 involvement in drug transport, a study performed on 22 healthy volunteers showed no significant difference in AUC between the carriers of the G34A (V12M), C421A (Q141K) and C376T (Q126X) SNPs, and the WT subjects (Kim et al., 2007). The authors considered that disposition of the drug is probably handled by several other transporters like SLC22A6 or ABCC4.

Nelfinavir, a HIV protease inhibitor, is known to be an ABCG2 inhibitor (Weiss et al., 2007). Colombo et al. (2005) took an interest in the influence of several ABC transporter haplotypes on drug exposure. The ABCG2 gene was genotyped in a Caucasian population and five SNPs were found: -19571 delTCAT, A-19541G, G34A (V12M), intron 2 A35G, and C421A (Q141K). No significant association between AUC of the drug and ABCG2 SNPs or haplotypes was observed.

14.11.2 Nitrofurantoin (Antibiotic)

Nitrofurantoin is used to treat urinary tract infections and has been shown *in vitro* and *in vivo* (in mouse) to be an ABCG2 substrate. Nevertheless, it was reported that the C421A ABCG2 SNP had no effect on nitrofurantoin plasma and urine pharmacokinetic parameters in 32 healthy Chinese male subjects (Adkison et al., 2008).

14.11.3 Mizoribine (Immunosuppressant)

Mizoribine is an imidazole nucleoside used as an immunosuppressive agent. Knowing that ABCG2 substrates include other nucleosides like lamivudine or cladribine (Wang et al., 2004; de Wolf et al., 2008), a team studied the involvement of ABCG2 in mizoribine pharmacogenomics (Fukao et al., 2011). Bioavailability of mizoribine was calculated from the estimated total urinary excretion from time zero to infinity in 30 Japanese subjects and no correlation was found between the drug PK and the C421A SNPs.

14.11.4 Sulfasalazine—Anti-Inflammatory Agent

Sulfasalazine is an anti-inflammatory agent used in the treatment of inflammatory bowel diseases and rheumatoid diseases and known to be an ABCG2 substrate (van der Heijden et al., 2004). Contradictory studies were performed in association with the C421A SNP and sulfasalazine disposition. Two studies found a significant association between the SNP and drug disposition. In one, performed in a Japanese population, the AUC (0–48) and C_{max} showed significant differences ($p < 0.05$) among all three genotype groups, with a 1.9- and 3.46-fold higher AUC and a 1.7- and 2.6-fold higher C_{max} in heterozygote and homozygote variant carriers, respectively, compared to the WT carriers (Yamasaki et al., 2008). The second study indicated that in subjects with a C421A genotype, sulfasalazine area under the concentration–time curve was 2.4-fold greater ($p < 0.05$) compared with C421C subjects (Urquhart et al., 2008). In contrast, a third study observed high intersubject variability in a Chinese population, yet no significant difference in sulfasalazine or metabolite pharmacokinetics in variant homozygotes or heterozygotes compared to WT subjects was found (Adkison et al., 2010). It has been proposed that the different findings in these three studies are due to differences in drug formulation or dose or both.

14.12 CONCLUSION

The ABC transporter ABCG2 has an established role in the pharmacokinetic profile of many substrate drugs, including a large number of clinically important anticancer agents, and in the occurrence of MDR in malignant cells. Several polymorphic variants of ABCG2 genes have been described, of which some may alter protein expression and function in humans. The effects of genetic variants in ABCG2 in relation to its phenotypical consequences are still debatable, as contradictory results have been reported for several studied substrate drugs. Most studies published to date suffer from small sample sizes in relation to the allele and genotype frequency of the studied variants, as well as from a host of potentially confounding factors that may influence their outcome. Most important among these are environmental and physiological factors that may affect expression of ABCG2, and links to other genes or variants of putative relevance for drug absorption and disposition pathways. The use of haplotype profiles as opposed to testing unphased SNPs to predict certain phenotypes has been proposed for ABCG2 and may have clinical importance, but this remains to be clarified for most drugs. In addition, more detailed investigations into the influence of ethnicity on ABCG2 function and expression as well as on the possible dose, administration route and schedule dependence of genotype–phenotype associations are urgently needed.

It should also be emphasized that ABCG2 genotypes will typically have little, if any, effects on the pharmacokinetics of drugs unless transport is the crucial process in oral absorption or disposition for any given substrate (for e.g., in the case of topotecan). Nevertheless, the number of studies that has correlated ABCG2 genotype with toxicity or treatment outcome is increasing, demonstrating the importance of the transporter. Further, the increase in number of recognized SNPs, with the addition of the SNPs located in the promoter as well as the study of haplotypes, has provided interesting new insights. It is important to note that among the 180 existing polymorphisms in the ABCG2 gene, only 10 of them have been selected for pharmacogenomic studies. In the field of drug resistance, the pharmacogenetic aspects of ABCG2 are still largely unexplored, but may eventually play a role in allowing patient selection on the basis of tumor genotyping and in refining drug design and development.

ACKNOWLEDGMENTS

This research was supported by the Intramural Research Program of the NIH, National Cancer Institute, Center for Cancer Research (http://ccr.nci.nih.gov/).

REFERENCES

Adkison KK, Vaidya SS, Lee DY, Koo SH, Li L, Mehta AA, Gross AS, Polli JW, Humphreys JE, Lou Y, Lee EJ (2010) Oral sulfasalazine as a clinical BCRP probe substrate: pharmacokinetic effects of genetic variation (C421A) and pantoprazole coadministration. *J Pharm Sci*, 99, 1046–1062.

Adkison KK, Vaidya SS, Lee DY, Koo SH, Li L, Mehta AA, Gross AS, Polli JW, Lou Y, Lee EJ (2008) The ABCG2 C421A polymorphism does not affect oral nitrofurantoin pharmacokinetics in healthy Chinese male subjects. *Br J Clin Pharmacol*, 66, 233–239.

Akasaka K, Kaburagi T, Yasuda S, Ohmori K, Abe K, Sagara H, Ueda Y, Nagao K, Imura J, Imai Y (2010) Impact of functional ABCG2 polymorphisms on the adverse effects of gefitinib in Japanese patients with non-small-cell lung cancer. *Cancer Chemother Pharmacol*, 66, 691–698.

Allen JD, Schinkel AH (2002) Multidrug resistance and pharmacological protection mediated by the breast cancer resistance protein (BCRP/ABCG2). *Mol Cancer Ther*, 1, 427–434.

Allen JD, van Loevezijn A, Lakhai JM, van der Valk M, van Tellingen O, Reid G, Schellens JH, Koomen GJ, Schinkel AH (2002) Potent and specific inhibition of the breast cancer resistance protein multidrug transporter in vitro and in mouse intestine by a novel analogue of fumitremorgin C. *Mol Cancer Ther*, 1, 417–425.

Allikmets R, Schriml LM, Hutchinson A, Romano-Spica V, Dean M (1998) A human placenta-specific ATP-binding cassette gene (ABCP) on chromosome 4q22 that is involved in multidrug resistance. *Cancer Res*, 58, 5337–5339.

Backstrom G, Taipalensuu J, Melhus H, Brandstrom H, Svensson AC, Artursson P, Kindmark A (2003) Genetic variation in the ATP-binding cassette transporter gene ABCG2 (BCRP) in a Swedish population. *Eur J Pharm Sci*, 18, 359–364.

Bailey KM, Romaine SP, Jackson BM, Farrin AJ, Efthymiou M, Barth JH, Copeland J, McCormack T, Whitehead A, Flather MD, Samani NJ, Nixon J, Hall AS, Balmforth AJ (2010) Hepatic metabolism and transporter gene variants enhance response to rosuvastatin in patients with acute myocardial infarction: the GEOSTAT-1 Study. *Circ Cardiovasc Genet*, 3, 276–285.

Bailey-Dell KJ, Hassel B, Doyle LA, Ross DD (2001) Promoter characterization and genomic organization of the human breast cancer resistance protein (ATP-binding cassette transporter G2) gene. *Biochim Biophys Acta*, 1520, 234–241.

Baker SD, Verweij J, Cusatis GA, van Schaik RH, Marsh S, Orwick SJ, Franke RM, Hu S, Schuetz EG, Lamba V, Messersmith WA, Wolff AC, Carducci MA, Sparreboom A (2009) Pharmacogenetic pathway analysis of docetaxel elimination. *Clin Pharmacol Ther*, 85, 155–163.

Bates SE, Medina-Perez WY, Kohlhagen G, Antony S, Nadjem T, Robey RW, Pommier Y (2004) ABCG2 mediates differential resistance to SN-38 and homocamptothecins. *J Pharmacol Exp Ther*, 310, 836–842.

Benderra Z, Faussat AM, Sayada L, Perrot JY, Chaoui D, Marie JP, Legrand O (2004) Breast cancer resistance protein and P-glycoprotein in 149 adult acute myeloid leukemias. *Clin Cancer Res*, 10, 7896–7902.

Benderra Z, Faussat AM, Sayada L, Perrot JY, Tang R, Chaoui D, Morjani H, Marzac C, Marie JP, Legrand O (2005) MRP3, BCRP, and P-glycoprotein activities are prognostic factors in adult acute myeloid leukemia. *Clin Cancer Res*, 11, 7764–7772.

Bosch TM, Kjellberg LM, Bouwers A, Koeleman BP, Schellens JH, Beijnen JH, Smits PH, Meijerman I (2005) Detection of single nucleotide polymorphisms in the ABCG2 gene in a Dutch population. *Am J Pharmacogenomics*, 5, 123–131.

Brendel C, Scharenberg C, Dohse M, Robey RW, Bates SE, Shukla S, Ambudkar SV, Wang Y, Wennemuth G, Burchert A, Boudriot U, Neubauer A (2007) Imatinib mesylate and nilotinib (AMN107) exhibit high-affinity interaction with ABCG2 on primitive hematopoietic stem cells. *Leukemia*, 21, 1267–1275.

Burger H, Foekens JA, Look MP, Meijer-van Gelder ME, Klijn JG, Wiemer EA, Stoter G, Nooter K (2003) RNA expression of breast cancer resistance protein, lung resistance-related protein, multidrug resistance-associated proteins 1 and 2, and multidrug resistance gene 1 in breast cancer: correlation with chemotherapeutic response. *Clin Cancer Res*, 9, 827–836.

Ceckova M, Libra A, Pavek P, Nachtigal P, Brabec M, Fuchs R, Staud F (2006) Expression and functional activity of breast cancer resistance protein (BCRP, ABCG2) transporter in the human choriocarcinoma cell line BeWo. *Clin Exp Pharmacol Physiol*, 33, 58–65.

Ceckova M, Vackova Z, Radilova H, Libra A, Buncek M, Staud F (2008) Effect of ABCG2 on cytotoxicity of platinum drugs: interference of EGFP. *Toxicol In Vitro*, 22, 1846–1852.

Chen ZS, Robey RW, Belinsky MG, Shchaveleva I, Ren XQ, Sugimoto Y, Ross DD, Bates SE, Kruh GD (2003) Transport of methotrexate, methotrexate polyglutamates, and 17beta-estradiol 17-(beta-D-glucuronide) by ABCG2: effects of acquired mutations at R482 on methotrexate transport. *Cancer Res*, 63, 4048–4054.

Chew SC, Singh O, Chen X, Ramasamy RD, Kulkarni T, Lee EJ, Tan EH, Lim WT, Chowbay B (2011) The effects of CYP3A4, CYP3A5, ABCB1, ABCC2, ABCG2 and SLCO1B3 single nucleotide polymorphisms on the pharmacokinetics and pharmacodynamics of docetaxel in nasopharyngeal carcinoma patients. *Cancer Chemother Pharmacol*, 67, 1471–1478.

Cisternino S, Mercier C, Bourasset F, Roux F, Scherrmann JM (2004) Expression, up-regulation, and transport activity of the multidrug-resistance protein Abcg2 at the mouse blood-brain barrier. *Cancer Res*, 64, 3296–3301.

Cole SP, Bhardwaj G, Gerlach JH, Mackie JE, Grant CE, Almquist KC, Stewart AJ, Kurz EU, Duncan AM, Deeley RG (1992) Overexpression of a transporter gene in a multidrug-resistant human lung cancer cell line. *Science*, 258, 1650–1654.

Colombo S, Soranzo N, Rotger M, Sprenger R, Bleiber G, Furrer H, Buclin T, Goldstein D, Decosterd L, Telenti A (2005) Influence of ABCB1, ABCC1, ABCC2, and ABCG2 haplotypes on the cellular exposure of nelfinavir in vivo. *Pharmacogenet Genomics*, 15, 599–608.

Cooray HC, Blackmore CG, Maskell L, Barrand MA (2002) Localisation of breast cancer resistance protein in microvessel endothelium of human brain. *Neuroreport*, 13, 2059–2063.

Cusatis G, Gregorc V, Li J, Spreafico A, Ingersoll RG, Verweij J, Ludovini V, Villa E, Hidalgo M, Sparreboom A, Baker SD (2006) Pharmacogenetics of ABCG2 and adverse reactions to gefitinib. *J Natl Cancer Inst*, 15, 1739–1742.

de Bruin M, Miyake K, Litman T, Robey R, Bates SE (1999) Reversal of resistance by GF120918 in cell lines expressing the ABC half-transporter, MXR. *Cancer Lett*, 146, 117–126.

de Jong FA, Marsh S, Mathijssen RH, King C, Verweij J, Sparreboom A, McLeod HL (2004) ABCG2 pharmacogenetics: ethnic differences in allele frequency and assessment of influence on irinotecan disposition. *Clin Cancer Res*, 10, 5889–5894.

de Vries NA, Zhao J, Kroon E, Buckle T, Beijnen JH, van Tellingen O (2007) P-glycoprotein and breast cancer resistance protein: two dominant transporters working together in limiting the brain penetration of topotecan. *Clin Cancer Res*, 13, 6440–6449.

de Wolf C, Jansen R, Yamaguchi H, de Haas M, van de Wetering K, Wijnholds J, Beijnen J, Borst P (2008) Contribution of the drug transporter ABCG2 (breast cancer resistance protein) to resistance against anticancer nucleosides. *Mol Cancer Ther*, 7, 3092–3102.

Dohse M, Scharenberg C, Shukla S, Robey RW, Volkmann T, Deeken JF, Brendel C, Ambudkar SV, Neubauer A, Bates SE (2010) Comparison of ATP-binding cassette transporter interactions with the tyrosine kinase inhibitors imatinib, nilotinib, and dasatinib. *Drug Metab Dispos*, 38, 1371–1380.

Doyle LA, Yang W, Abruzzo LV, Krogmann T, Gao Y, Rishi AK, Ross DD (1998) A multidrug resistance transporter from human MCF-7 breast cancer cells. *Proc Natl Acad Sci U S A*, 95, 15665–15670.

Ee PL, Kamalakaran S, Tonetti D, He X, Ross DD, Beck WT (2004) Identification of a novel estrogen response element in the breast cancer resistance protein (ABCG2) gene. *Cancer Res*, 64, 1247–1251.

Eisenblatter T, Huwel S, Galla HJ (2003) Characterisation of the brain multidrug resistance protein (BMDP/ABCG2/BCRP) expressed at the blood-brain barrier. *Brain Res*, 971, 221–231.

Ejendal KF, Hrycyna CA (2002) Multidrug resistance and cancer: the role of the human ABC transporter ABCG2. *Curr Protein Pept Sci*, 3, 503–511.

Elmeliegy MA, Carcaboso AM, Tagen M, Bai F, Stewart CF (2011) Role of ATP-binding cassette and solute carrier transporters in erlotinib CNS penetration and intracellular accumulation. *Clin Cancer Res*, 17, 89–99.

Erdilyi DJ, Kamory E, Csokay B, Andrikovics H, Tordai A, Kiss C, Filni-Semsei A, Janszky I, Zalka A, Fekete G, Falus A, Kovacs GT, Szalai C (2008) Synergistic interaction of ABCB1 and ABCG2 polymorphisms predicts the prevalence of toxic encephalopathy during anticancer chemotherapy. *Pharmacogenomics J*, 8, 321–327.

Erlichman C, Boerner SA, Hallgren CG, Spieker R, Wang XY, James CD, Scheffer GL, Maliepaard M, Ross DD, Bible KC, Kaufmann SH (2001) The HER tyrosine kinase inhibitor CI1033 enhances cytotoxicity of 7-ethyl-10-hydroxycamptothecin and topotecan by inhibiting breast cancer resistance protein-mediated drug efflux. *Cancer Res*, 61, 739–748.

Evseenko DA, Murthi P, Paxton JW, Reid G, Emerald BS, Mohankumar KM, Lobie PE, Brennecke SP, Kalionis B, Keelan JA (2007) The ABC transporter BCRP/ABCG2 is a placental survival factor, and its expression is reduced in idiopathic human fetal growth restriction. *FASEB J*, 21, 3592–3605.

Faivre S, Delbaldo C, Vera K, Robert C, Lozahic S, Lassau N, Bello C, Deprimo S, Brega N, Massimini G, Armand JP, Scigalla P, Raymond E (2006) Safety, pharmacokinetic, and antitumor activity of SU11248, a novel oral multitarget tyrosine kinase inhibitor, in patients with cancer. *J Clin Oncol*, 24, 25–35.

Fetsch P, Abati A, Litman T, Morisaki K, Honjo Y, Mittal K, Bates S (2006) Localization of the ABCG2 mitoxantrone resistance-associated protein in normal tissues. *Cancer Lett*, 235, 84–92.

Fukao M, Ishida K, Sakamoto T, Taguchi M, Matsukura H, Miyawaki T, Hashimoto Y (2011) Effect of genetic polymorphisms of SLC28A1, ABCG2, and ABCC4 on bioavailability of mizoribine in healthy Japanese males. *Drug Metab Pharmacokinet*, 26, 538–543.

Furukawa T, Wakabayashi K, Tamura A, Nakagawa H, Morishima Y, Osawa Y, Ishikawa T (2009) Major SNP (Q141K) variant of human ABC transporter ABCG2 undergoes lysosomal and proteasomal degradations. *Pharm Res*, 26, 469–479.

Gardner ER, Burger H, van Schaik RH, van Oosterom A, de Bruijn EA, Guetens G, Prenen H, de Jong FA, Baker SD, Bates S, Figg WD, Verweij J, Sparreboom A, Nooter K (2006) Association of enzyme and transporter genotypes with the pharmacokinetics of imatinib. *Clin Pharmacol Ther*, 80, 192–201.

Gottesman MM, Fojo T, Bates SE (2002) Multidrug resistance in cancer: role of ATP-dependent transporters. *Nat Rev Cancer*, 2, 48–58.

Gupta A, Zhang Y, Unadkat JD, Mao Q (2004) HIV protease inhibitors are inhibitors but not substrates of the human breast cancer resistance protein (BCRP/ABCG2). *J Pharmacol Exp Ther*, 310, 334–341.

Hahn NM, Marsh S, Fisher W, Langdon R, Zon R, Browning M, Johnson CS, Scott-Horton TJ, Li L, McLeod HL, Sweeney CJ (2006) Hoosier Oncology Group randomized phase II study of docetaxel, vinorelbine, and estramustine in combination in hormone-refractory prostate cancer with pharmacogenetic survival analysis. *Clin Cancer Res*, 12, 6094–6099.

Han B, Zhang JT (2004) Multidrug resistance in cancer chemotherapy and xenobiotic protection mediated by the half ATP-binding cassette transporter ABCG2. *Curr Med Chem Anticancer Agents*, 4, 31–42.

Han JY, Lim HS, Park YH, Lee SY, Lee JS (2009) Integrated pharmacogenetic prediction of irinotecan pharmacokinetics and toxicity in patients with advanced non-small cell lung cancer. *Lung Cancer*, 63, 115–120.

Han JY, Lim HS, Yoo YK, Shin ES, Park YH, Lee SY, Lee JE, Lee DH, Kim HT, Lee JS (2007) Associations of ABCB1, ABCC2, and ABCG2 polymorphisms with irinotecan-pharmacokinetics and clinical outcome in patients with advanced non-small cell lung cancer. *Cancer*, 110, 138–147.

Hazlehurst LA, Foley NE, Gleason-Guzman MC, Hacker MP, Cress AE, Greenberger LW, De Jong MC, Dalton WS (1999) Multiple mechanisms confer drug resistance to mitoxantrone in the human 8226 myeloma cell line. *Cancer Res*, 59, 1021–1028.

Hirano M, Maeda K, Matsushima S, Nozaki Y, Kusuhara H, Sugiyama Y (2005) Involvement of BCRP (ABCG2) in the biliary excretion of pitavastatin. *Mol Pharmacol*, 68, 800–807.

Honjo Y, Morisaki K, Huff LM, Robey RW, Hung J, Dean M, Bates SE (2002) Single-nucleotide polymorphism (SNP) analysis in the ABC half-transporter ABCG2 (MXR/BCRP/ABCP1). *Cancer Biol Ther*, 1, 696–702.

Houghton PJ, Germain GS, Harwood FC, Schuetz JD, Stewart CF, Buchdunger E, Traxler P (2004) Imatinib mesylate is a potent inhibitor of the ABCG2 (BCRP) transporter and reverses resistance to topotecan and SN-38 in vitro. *Cancer Res*, 64, 2333–2337.

Hu S, Chen Z, Franke R, Orwick S, Zhao M, Rudek MA, Sparreboom A, Baker SD (2009) Interaction of the multikinase inhibitors sorafenib and sunitinib with solute carriers and ATP-binding cassette transporters. *Clin Cancer Res*, 15, 6062–6069.

Huang L, Wang Y, Grimm S (2006) ATP-dependent transport of rosuvastatin in membrane vesicles expressing breast cancer resistance protein. *Drug Metab Dispos*, 34, 738–742.

Ieiri I, Suwannakul S, Maeda K, Uchimaru H, Hashimoto K, Kimura M, Fujino H, Hirano M, Kusuhara H, Irie S, Higuchi S, Sugiyama Y (2007) SLCO1B1 (OATP1B1, an uptake transporter) and ABCG2 (BCRP, an efflux transporter) variant alleles and pharmacokinetics of pitavastatin in healthy volunteers. *Clin Pharmacol Ther*, 82, 541–547.

Iida A, Saito S, Sekine A, Mishima C, Kitamura Y, Kondo K, Harigae S, Osawa S, Nakamura Y (2002) Catalog of 605 single-nucleotide polymorphisms (SNPs) among 13 genes encoding human ATP-binding cassette transporters: ABCA4, ABCA7, ABCA8, ABCD1, ABCD3, ABCD4, ABCE1, ABCF1, ABCG1, ABCG2, ABCG4, ABCG5, and ABCG8. *J Hum Genet*, 47, 285–310.

Imai Y, Nakane M, Kage K, Tsukahara S, Ishikawa E, Tsuruo T, Miki Y, Sugimoto Y (2002) C421A polymorphism in the human breast cancer resistance protein gene is associated with low expression of Q141K protein and low-level drug resistance. *Mol Cancer Ther*, 1, 611–616.

Ishii M, Iwahana M, Mitsui I, Minami M, Imagawa S, Tohgo A, Ejima A (2000) Growth inhibitory effect of a new camptothecin analog, DX-8951f, on various drug-resistant sublines including BCRP-mediated camptothecin derivative-resistant variants derived from the human lung cancer cell line PC-6. *Anticancer Drugs*, 11, 353–362.

Itoda M, Saito Y, Shirao K, Minami H, Ohtsu A, Yoshida T, Saijo N, Suzuki H, Sugiyama Y, Ozawa S, Sawada J (2003) Eight novel single nucleotide polymorphisms in ABCG2/BCRP in Japanese cancer patients administered irinotacan. *Drug Metab Pharmacokinet*, 18, 212–217.

Jonker JW, Merino G, Musters S, van Herwaarden AE, Bolscher E, Wagenaar E, Mesman E, Dale TC, Schinkel AH (2005) The breast cancer resistance protein BCRP (ABCG2) concentrates drugs and carcinogenic xenotoxins into milk. *Nat Med*, 11, 127–129.

Jonker JW, Smit JW, Brinkhuis RF, Maliepaard M, Beijnen JH, Schellens JH, Schinkel AH (2000) Role of breast cancer resistance protein in the bioavailability and fetal penetration of topotecan. *J Natl Cancer Inst*, 92, 1651–1656.

Kawabata S, Oka M, Shiozawa K, Tsukamoto K, Nakatomi K, Soda H, Fukuda M, Ikegami Y, Sugahara K, Yamada Y, Kamihira S, Doyle LA, Ross DD, Kohno S (2001) Breast cancer resistance protein directly confers SN-38 resistance of lung cancer cells. *Biochem Biophys Res Commun*, 280, 1216–1223.

Keskitalo JE, Pasanen MK, Neuvonen PJ, Niemi M (2009a) Different effects of the ABCG2 c.421C>A SNP on the pharmacokinetics of fluvastatin, pravastatin and simvastatin. *Pharmacogenomics*, 10, 1617–1624.

Keskitalo JE, Zolk O, Fromm MF, Kurkinen KJ, Neuvonen PJ, Niemi M (2009b) ABCG2 polymorphism markedly affects the pharmacokinetics of atorvastatin and rosuvastatin. *Clin Pharmacol Ther*, 86, 197–203.

Kim DH, Sriharsha L, Xu W, Kamel-Reid S, Liu X, Siminovitch K, Messner HA, Lipton JH (2009) Clinical relevance of a pharmacogenetic approach using multiple candidate genes to predict response and resistance to imatinib therapy in chronic myeloid leukemia. *Clin Cancer Res*, 15, 4750–4758.

Kim HS, Sunwoo YE, Ryu JY, Kang HJ, Jung HE, Song IS, Kim EY, Shim JC, Shon JH, Shin JG (2007) The effect of ABCG2 V12M, Q141K and Q126X, known functional variants in vitro, on the disposition of lamivudine. *Br J Clin Pharmacol*, 64, 645–654.

Kim IS, Kim HG, Kim DC, Eom HS, Kong SY, Shin HJ, Hwang SH, Lee EY, Lee GW (2008) ABCG2 Q141K polymorphism is associated with chemotherapy-induced diarrhea in patients with diffuse large B-cell lymphoma who received frontline rituximab plus cyclophosphamide/doxorubicin/vincristine/prednisone chemotherapy. *Cancer Sci*, 99, 2496–2501.

Kobayashi D, Ieiri I, Hirota T, Takane H, Maegawa S, Kigawa J, Suzuki H, Nanba E, Oshimura M, Terakawa N, Otsubo K, Mine K, Sugiyama Y (2005) Functional assessment of ABCG2 (BCRP) gene polymorphisms to protein expression in human placenta. *Drug Metab Dispos* 33, 94–101.

Kodaira H, Kusuhara H, Ushiki J, Fuse E, Sugiyama Y (2010) Kinetic analysis of the cooperation of P-glycoprotein (P-gp/Abcb1) and breast cancer resistance protein (Bcrp/Abcg2) in limiting the brain and testis penetration of erlotinib, flavopiridol, and mitoxantrone. *J Pharmacol Exp Ther*, 333, 788–796.

Komatani H, Kotani H, Hara Y, Nakagawa R, Matsumoto M, Arakawa H, Nishimura S (2001) Identification of breast cancer resistant protein/mitoxantrone resistance/placenta-specific, ATP-binding cassette transporter as a transporter of NB-506 and J-107088, topoisomerase I inhibitors with an indolocarbazole structure. *Cancer Res*, 61, 2827–2832.

Kondo C, Suzuki H, Itoda M, Ozawa S, Sawada J, Kobayashi D, Ieiri I, Mine K, Ohtsubo K, Sugiyama Y (2004) Functional analysis of SNPs variants of BCRP/ABCG2. *Pharm Res*, 21, 1895–1903.

Krishnamurthy P, Ross DD, Nakanishi T, Bailey-Dell K, Zhou S, Mercer KE, Sarkadi B, Sorrentino BP, Schuetz JD (2004) The stem cell marker Bcrp/ABCG2 enhances hypoxic cell survival through interactions with heme. *J Biol Chem*, 279, 24218–24225.

Lagas JS, van Waterschoot RA, Sparidans RW, Wagenaar E, Beijnen JH, Schinkel AH (2010) Breast cancer resistance protein and P-glycoprotein limit sorafenib brain accumulation. *Mol Cancer Ther*, 9, 319–326.

Lagas JS, van Waterschoot RA, van Tilburg VA, Hillebrand MJ, Lankheet N, Rosing H, Beijnen JH, Schinkel AH (2009) Brain accumulation of dasatinib is restricted by P-glycoprotein (ABCB1) and breast cancer resistance protein (ABCG2) and can be enhanced by elacridar treatment. *Clin Cancer Res*, 15, 2344–2351.

Lal S, Wong ZW, Sandanaraj E, Xiang X, Ang PC, Lee EJ, Chowbay B (2008) Influence of ABCB1 and ABCG2 polymorphisms on doxorubicin disposition in Asian breast cancer patients. *Cancer Sci*, 99, 816–823.

Lee SS, Jeong HE, Yi JM, Jung HJ, Jang JE, Kim EY, Lee SJ, Shin JG (2007) Identification and functional assessment of BCRP polymorphisms in a Korean population. *Drug Metab Dispos*, 35, 623–632.

Lemos C, Giovannetti E, Zucali PA, Assaraf YG, Scheffer GL, van der Straaten T, D'Incecco A, Falcone A, Guchelaar HJ, Danesi R, Santoro A, Giaccone G, Tibaldi C, Peters GJ (2011) Impact of ABCG2 polymorphisms on the clinical outcome and toxicity of gefitinib in non-small-cell lung cancer patients. *Pharmacogenomics*, 12, 159–170.

Lepper ER, Nooter K, Verweij J, Acharya MR, Figg WD, Sparreboom A (2005) Mechanisms of resistance to anticancer drugs: the role of polymorphic ABC transporters ABCB1 and ABCG2. *Pharmacogenomics*, 6, 115–138.

Li J, Cusatis G, Brahmer J, Sparreboom A, Robey R, Bates S, Hidalgo M, Baker S (2007) Association of variant ABCG2 and the pharmacokinetics of epidermal growth factor receptor tyrosine kinase inhibitors in cancer patients. *Cancer Biol Ther*, 6, 432–438.

Li J, Volpe DA, Wang Y, Zhang W, Bode C, Owen A, Hidalgo IJ (2011) Use of transporter knockdown Caco-2 cells to investigate the in vitro efflux of statin drugs. *Drug Metab Dispos*, 39, 1196–1202.

Litman T, Brangi M, Hudson E, Fetsch P, Abati A, Ross DD, Miyake K, Resau JH, Bates SE (2000) The multidrug-resistant phenotype associated with overexpression of the new ABC half-transporter, MXR (ABCG2). *J Cell Sci*, 113, 2011–2021.

Litman T, Druley TE, Stein WD, Bates SE (2001) From MDR to MXR: new understanding of multidrug resistance systems, their properties and clinical significance. *Cell Mol Life Sci*, 58, 931–959.

Maekawa K, Itoda M, Sai K, Saito Y, Kaniwa N, Shirao K, Hamaguchi T, Kunitoh H, Yamamoto N, Tamura T, Minami H, Kubota K, Ohtsu A, Yoshida T, Saijo N, Kamatani N, Ozawa S, Sawada J (2006) Genetic variation and haplotype structure of the ABC transporter gene ABCG2 in a Japanese population. *Drug Metab Pharmacokinet*, 21, 109–121.

Maliepaard M, Scheffer GL, Faneyte IF, van Gastelen MA, Pijnenborg AC, Schinkel AH, van De Vijver MJ, Scheper RJ, Schellens JH (2001) Subcellular localization and distribution of the breast cancer resistance protein transporter in normal human tissues. *Cancer Res*, 61, 3458–3464.

Maliepaard M, van Gastelen MA, de Jong LA, Pluim D, van Waardenburg RC, Ruevekamp-Helmers MC, Floot BG, Schellens JH (1999) Overexpression of the BCRP/MXR/ABCP gene in a topotecan-selected ovarian tumor cell line. *Cancer Res*, 59, 4559–63.

Marsh S, Paul J, King CR, Gifford G, McLeod HL, Brown R (2007) Pharmacogenetic assessment of toxicity and outcome after platinum plus taxane chemotherapy in ovarian cancer: the Scottish Randomised Trial in Ovarian Cancer. *J Clin Oncol*, 25, 4528–4535.

Matsuo H, Takada T, Ichida K, Nakamura T, Nakayama A, Ikebuchi Y, Ito K, Kusanagi Y, Chiba T, Tadokoro S, Takada Y, Oikawa Y, Inoue H, Suzuki K, Okada R, Nishiyama J, Domoto H, Watanabe S, Fujita M, Morimoto Y, Naito M, Nishio K, Hishida A, Wakai K, Asai Y, Niwa K, Kamakura K, Nonoyama S, Sakurai Y, Hosoya T, Kanai Y, Suzuki H, Hamajima N, Shinomiya N (2009) Common defects of ABCG2, a high-capacity urate exporter, cause gout: a function-based genetic analysis in a Japanese population. *Sci Transl Med*, 1, 5ra11.

Matsushima S, Maeda K, Kondo C, Hirano M, Sasaki M, Suzuki H, Sugiyama Y (2005) Identification of the hepatic efflux transporters of organic anions using double-transfected Madin-Darby canine kidney II cells expressing human organic anion-transporting polypeptide 1B1 (OATP1B1)/multidrug resistance-associated protein 2, OATP1B1/multidrug resistance 1, and OATP1B1/breast cancer resistance protein. *J Pharmacol Exp Ther*, 314 1059–1067.

Meissner K, Heydrich B, Jedlitschky G, Meyer Zu Schwabedissen H, Mosyagin I, Dazert P, Eckel L, Vogelgesang S, Warzok RW, Bohm M, Lehmann C, Wendt M, Cascorbi I, Kroemer HK (2006) The ATP-binding cassette transporter ABCG2 (BCRP), a marker for side population stem cells, is expressed in human heart. *J Histochem Cytochem*, 54, 215–221.

Merino G, Alvarez AI, Pulido MM, Molina AJ, Schinkel AH, Prieto JG (2006) Breast cancer resistance protein (bcrp/abcg2) transports fluoroquinolone antibiotics and affects their oral availability, pharmacokinetics, and milk secretion. *Drug Metab Dispos*, 34, 690–695.

Merino G, Jonker JW, Wagenaar E, van Herwaarden AE, Schinkel AH (2005) The breast cancer resistance protein (BCRP/ABCG2) affects pharmacokinetics, hepatobiliary excretion, and milk secretion of the antibiotic nitrofurantoin. *Mol Pharmacol*, 67, 1758–1764.

Miyake K, Mickley L, Litman T, Zhan Z, Robey R, Cristensen B, Brangi M, Greenberger L, Dean M, Fojo T, Bates SE (1999) Molecular cloning of cDNAs which are highly overexpressed in mitoxantrone-resistant cells: demonstration of homology to ABC transport genes. *Cancer Res*, 59, 8–13.

Mizuarai S, Aozasa N, Kotani H (2004) Single nucleotide polymorphisms result in impaired membrane localization and reduced atpase activity in multidrug transporter ABCG2. *Int J Cancer*, 109, 238–246.

Mizuno T, Terada T, Kamba T, Fukudo M, Katsura T, Nakamura E, Ogawa O, Inui K (2010) ABCG2 421C>A polymorphism and high exposure of sunitinib in a patient with renal cell carcinoma. *Ann Oncol*, 21, 1382–1383.

Morisaki K, Robey RW, Ozvegy-Laczka C, Honjo Y, Polgar O, Steadman K, Sarkadi B, Bates SE (2005) Single nucleotide polymorphisms modify the transporter activity of ABCG2. *Cancer Chemother Pharmacol*, 56, 161–172.

Muller PJ, Dally H, Klappenecker CN, Edler L, Jager B, Gerst M, Spiegelhalder B, Tuengerthal S, Fischer JR, Drings P, Bartsch H, Risch A (2009) Polymorphisms in ABCG2, ABCC3 and CNT1 genes and their possible impact on chemotherapy outcome of lung cancer patients. *Int J Cancer*, 124, 1669–1674.

Nakanishi T, Bailey-Dell KJ, Hassel BA, Shiozawa K, Sullivan DM, Turner J, Ross DD (2006) Novel 5′ untranslated region variants of BCRP mRNA are differentially expressed in drug-selected cancer cells and in normal human tissues: implications for drug resistance, tissue-specific expression, and alternative promoter usage. *Cancer Res*, 66, 5007–5011.

Ni W, Ji J, Dai Z, Papp A, Johnson AJ, Ahn S, Farley KL, Lin TS, Dalton JT, Li X, Jarjoura D, Byrd JC, Sadee W, Grever MR, Phelps MA (2010) Flavopiridol pharmacogenetics: clinical and functional evidence for the role of SLCO1B1/OATP1B1 in flavopiridol disposition. *PLoS One*, 5, e13792.

Oostendorp RL, Buckle T, Beijnen JH, van Tellingen O, Schellens JH (2009) The effect of P-gp (Mdr1a/1b), BCRP (Bcrp1) and P-gp/BCRP inhibitors on the in vivo absorption, distribution, metabolism and excretion of imatinib. *Invest New Drugs*, 27, 31–40.

Petain A, Kattygnarath D, Azard J, Chatelut E, Delbaldo C, Geoerger B, Barrois M, Seronie-Vivien S, LeCesne A, Vassal G (2008) Population pharmacokinetics and pharmacogenetics of imatinib in children and adults. *Clin Cancer Res*, 14, 7102–7109.

Polli JW, Olson KL, Chism JP, John-Williams LS, Yeager RL, Woodard SM, Otto V, Castellino S, Demby VE (2009) An unexpected synergist role of P-glycoprotein and breast cancer resistance protein on the central nervous system penetration of the tyrosine kinase inhibitor lapatinib (N-{3-chloro-4-[(3-fluorobenzyl)oxy]phenyl}-6-[5-({[2-(methylsulfonyl)ethyl]amino}methyl)-2-furyl]-4-quinazolinamine; GW572016). *Drug Metab Dispos*, 37, 439–442.

Poonkuzhali B, Lamba J, Strom S, Sparreboom A, Thummel K, Watkins P, Schuetz E (2008) Association of breast cancer resistance protein/ABCG2 phenotypes and novel promoter and intron 1 single nucleotide polymorphisms. *Drug Metab Dispos*, 36, 780–795.

Rabindran SK, He H, Singh M, Brown E, Collins KI, Annable T, Greenberger LM (1998) Reversal of a novel multidrug resistance mechanism in human colon carcinoma cells by fumitremorgin C. *Cancer Res*, 58, 5850–5858.

Rabindran SK, Ross DD, Doyle LA, Yang W, Greenberger LM (2000) Fumitremorgin C reverses multidrug resistance in cells transfected with the breast cancer resistance protein. *Cancer Res*, 60, 47–50.

Rajendra R, Gounder MK, Saleem A, Schellens JH, Ross DD, Bates SE, Sinko P, Rubin EH (2003) Differential effects of the breast cancer resistance protein on the cellular accumulation and cytotoxicity of 9-aminocamptothecin and 9-nitrocamptothecin. *Cancer Res*, 63, 3228–3233.

Raymond E, Brandes AA, Dittrich C, Fumoleau P, Coudert B, Clement PM, Frenay M, Rampling R, Stupp R, Kros JM, Heinrich MC, Gorlia T, Lacombe D, van den Bent MJ (2008) Phase II study of imatinib in patients with recurrent gliomas of various histologies: a European Organisation for Research and Treatment of Cancer Brain Tumor Group Study. *J Clin Oncol*, 26, 4659–4665.

Robey RW, Honjo Y, Morisaki K, Nadjem TA, Runge S, Risbood M, Poruchynsky MS, Bates SE (2003) Mutations at amino-acid 482 in the ABCG2 gene affect substrate and antagonist specificity. *Br J Cancer*, 89, 1971–1978.

Robey RW, Medina-Perez WY, Nishiyama K, Lahusen T, Miyake K, Litman T, Senderowicz AM, Ross DD, Bates SE (2001) Overexpression of the ATP-binding cassette half-transporter, ABCG2 (Mxr/BCrp/ABCP1), in flavopiridol-resistant human breast cancer cells. *Clin Cancer Res*, 7, 145–152.

Robey RW, Steadman K, Polgar O, Morisaki K, Blayney M, Mistry P, Bates SE (2004) Pheophorbide a is a specific probe for ABCG2 function and inhibition. *Cancer Res*, 64, 1242–1246.

Rudin CM, Liu W, Desai A, Karrison T, Jiang X, Janisch L, Das S, Ramirez J, Poonkuzhali B, Schuetz E, Fackenthal DL, Chen P, Armstrong DK, Brahmer JR, Fleming GF, Vokes EE, Carducci MA, Ratain MJ (2008) Pharmacogenomic and pharmacokinetic determinants of erlotinib toxicity. *J Clin Oncol*, 26, 1119–1127.

Sai K, Saito Y, Maekawa K, Kim SR, Kaniwa N, Nishimaki-Mogami T, Sawada J, Shirao K, Hamaguchi T, Yamamoto N, Kunitoh H, Ohe Y, Yamada Y, Tamura T, Yoshida T, Matsumura Y, Ohtsu A, Saijo N, Minami H (2010) Additive effects of drug transporter genetic polymorphisms on irinotecan pharmacokinetics/pharmacodynamics in Japanese cancer patients. *Cancer Chemother Pharmacol*, 66, 95–105.

Schellens JH, Maliepaard M, Scheper RJ, Scheffer GL, Jonker JW, Smit JW, Beijnen JH, Schinkel AH (2000) Transport of topoisomerase I inhibitors by the breast cancer resistance protein. Potential clinical implications. *Ann N Y Acad Sci*, 922, 188–194.

Shafran A, Ifergan I, Bram E, Jansen G, Kathmann I, Peters GJ, Robey RW, Bates SE, Assaraf YG (2005) ABCG2 harboring the Gly482 mutation confers high-level resistance to various hydrophilic antifolates. *Cancer Res*, 65, 8414–8422.

Shen H, Lee FY, Gan J (2011) Ixabepilone, a novel microtubule-targeting agent for breast cancer, is a substrate for P-glycoprotein (P-gp/MDR1/ABCB1) but not breast cancer resistance protein (BCRP/ABCG2). *J Pharmacol Exp Ther*, 337, 423–432.

Shukla S, Robey RW, Bates SE, Ambudkar SV (2009) Sunitinib (Sutent, SU11248), a small-molecule receptor tyrosine kinase inhibitor, blocks function of the ATP-binding cassette (ABC) transporters P-glycoprotein (ABCB1) and ABCG2. *Drug Metab Dispos*. 37 359–365.

Sparreboom A, Gelderblom H, Marsh S, Ahluwalia R, Obach R, Principe P, Twelves C, Verweij J, McLeod HL (2004) Diflomotecan pharmacokinetics in relation to ABCG2 421C>A genotype. *Clin Pharmacol Ther*, 76, 38–44.

Sparreboom A, Loos WJ, Burger H, Sissung TM, Verweij J, Figg WD, Nooter K, Gelderblom H (2005) Effect of ABCG2 genotype on the oral bioavailability of topotecan. *Cancer Biol Ther*, 4, 650–658.

Takahashi N, Miura M, Scott SA, Kagaya H, Kameoka Y, Tagawa H, Saitoh H, Fujishima N, Yoshioka T, Hirokawa M, Sawada K (2010) Influence of CYP3A5 and drug transporter polymorphisms on imatinib trough concentration and clinical response among patients with chronic phase chronic myeloid leukemia. *J Hum Genet*, 55, 731–737.

Tamura A, Watanabe M, Saito H, Nakagawa H, Kamachi T, Okura I, Ishikawa T (2006) Functional validation of the genetic polymorphisms of human ABC transporter ABCG2: Identification of alleles that are defective in porphyrin transport. *Mol Pharmacol*, 70, 287–296.

Tang SC, Lagas JS, Lankheet NA, Poller B, Hillebrand MJ, Rosing H, Beijnen JH, Schinkel AH (2012) Brain accumulation of sunitinib is restricted by P-glycoprotein (ABCB1) and breast cancer resistance protein (ABCG2) and can be enhanced by oral elacridar and sunitinib coadministration. *Int J Cancer*, 130, 223–233.

Thomas F, Rochaix P, White-Koning M, Hennebelle I, Sarini J, Benlyazid A, Malard L, Lefebvre JL, Chatelut E, Delord JP (2009) Population pharmacokinetics of erlotinib and its pharmacokinetic/pharmacodynamic relationships in head and neck squamous cell carcinoma. *Eur J Cancer*, 45, 2316–2323.

To KK, Yu L, Liu S, Fu J, Cho CH (2011) Constitutive AhR activation leads to concomitant ABCG2-mediated multidrug resistance in cisplatin-resistant esophageal carcinoma cells. *Mol Carcinog* [Online] Available at: http://onlinelibrary.wiley.com/doi/10.1002/mc.20810/full Accessed on 15 June 2011.

Tomlinson B, Hu M, Lee VW, Lui SS, Chu TT, Poon EW, Ko GT, Baum L, Tam LS, Li EK (2010) ABCG2 polymorphism is associated with the low-density lipoprotein cholesterol response to rosuvastatin. *Clin Pharmacol Ther*, 87, 558–562.

Tompkins LM, Li H, Li L, Lynch C, Xie Y, Nakanishi T, Ross DD, Wang H (2010) A novel xenobiotic responsive element regulated by aryl hydrocarbon receptor is involved in the induction of BCRP/ABCG2 in LS174T cells. *Biochem Pharmacol*, 80, 1754–1761.

Ueda K, Cardarelli C, Gottesman MM, Pastan I (1987) Expression of a full-length cDNA for the human "MDR1" gene confers resistance to colchicine, doxorubicin, and vinblastine. *Proc Natl Acad Sci U S A*, 84, 3004–3008.

Urquhart BL, Ware JA, Tirona RG, Ho RH, Leake BF, Schwarz UI, Zaher H, Palandra J, Gregor JC, Dresser GK, Kim RB (2008) Breast cancer resistance protein (ABCG2) and drug disposition: intestinal expression, polymorphisms and sulfasalazine as an in vivo probe. *Pharmacogenet Genomics*, 18, 439–448.

van der Heijden J, de Jong MC, Dijkmans BA, Lems WF, Oerlemans R, Kathmann I, Schalkwijk CG, Scheffer GL, Scheper RJ, Jansen G (2004) Development of sulfasalazine resistance in human T cells induces expression of the multidrug resistance transporter ABCG2 (BCRP) and augmented production of TNFalpha. *Ann Rheum Dis*, 63, 138–143.

van der Veldt AA, Eechoute K, Gelderblom H, Gietema J, Guchelaar HJ, van Erp NP, van den Eertwegh AJ, Haanen JB, Mathijssen RH, Wessels JA (2011) Genetic polymorphisms associated with a prolonged progression-free survival in patients with metastatic renal cell cancer treated with sunitinib. *Clin Cancer Res*, 17, 620–629.

van Erp NP, Eechoute K, van der Veldt AA, Haanen JB, Reyners AK, Mathijssen RH, Boven E, van der Straaten T, Baak-Pablo RF, Wessels JA, Guchelaar HJ, Gelderblom H (2009) Pharmacogenetic pathway analysis for determination of sunitinib-induced toxicity. *J Clin Oncol*, 27, 4406–4412.

van Herwaarden AE, Jonker JW, Wagenaar E, Brinkhuis RF, Schellens JH, Beijnen JH, Schinkel AH (2003) The breast cancer resistance protein (Bcrp1/Abcg2) restricts exposure to the dietary carcinogen 2-amino-1-methyl-6-phenylimidazo[4,5-b]pyridine. *Cancer Res*, 63, 6447–6452.

van Herwaarden AE, Wagenaar E, Merino G, Jonker JW, Rosing H, Beijnen JH, Schinkel AH (2007) Multidrug transporter ABCG2/breast cancer resistance protein secretes riboflavin (vitamin B2) into milk. *Mol Cell Biol*, 27, 1247–1253.

Vethanayagam RR, Wang H, Gupta A, Zhang Y, Lewis F, Unadkat JD, Mao Q (2005) Functional analysis of the human variants of breast cancer resistance protein: I206L, N590Y, and D620N. *Drug Metab Dispos*, 33, 697–705.

Volk EL, Rohde K, Rhee M, McGuire JJ, Doyle LA, Ross DD, Schneider E (2000) Methotrexate cross-resistance in a mitoxantrone-selected multidrug-resistant MCF7 breast cancer cell line is attributable to enhanced energy-dependent drug efflux. *Cancer Res*, 60, 3514–3521.

Volk EL, Schneider E (2003) Wild-type breast cancer resistance protein (BCRP/ABCG2) is a methotrexate polyglutamate transporter. *Cancer Res*, 63, 5538–5543.

Wang X, Furukawa T, Nitanda T, Okamoto M, Sugimoto Y, Akiyama S, Baba M (2003) Breast cancer resistance protein (BCRP/ABCG2) induces cellular resistance to HIV-1 nucleoside reverse transcriptase inhibitors. *Mol Pharmacol*, 63, 65–72.

Wang X, Nitanda T, Shi M, Okamoto M, Furukawa T, Sugimoto Y, Akiyama S, Baba M (2004) Induction of cellular resistance to nucleoside reverse transcriptase inhibitors by the wild-type breast cancer resistance protein. *Biochem Pharmacol*, 68, 1363–1370.

Warren RB, Smith RL, Campalani E, Eyre S, Smith CH, Barker JN, Worthington J, Griffiths CE (2008) Genetic variation in efflux transporters influences outcome to methotrexate therapy in patients with psoriasis. *J Invest Dermatol*, 128, 1925–1929.

Weiss J, Rose J, Storch CH, Ketabi-Kiyanvash N, Sauer A, Haefeli WE, Efferth T (2007) Modulation of human BCRP (ABCG2) activity by anti-HIV drugs. *J Antimicrob Chemother*, 59, 238–245.

White-Koning M, Civade E, Geoerger B, Thomas F, Le Deley MC, Hennebelle I, Delord JP, Chatelut E, Vassal G (2011) Population analysis of erlotinib in adults and children reveals pharmacokinetic characteristics as the main factor explaining tolerance particularities in children. *Clin Cancer Res*, 17, 4862–4871.

Wilson CS, Davidson GS, Martin SB, Andries E, Potter J, Harvey R, Ar K, Xu Y, Kopecky KJ, Ankerst DP, Gundacker H, Slovak ML, Mosquera-Caro M, Chen IM, Stirewalt DL, Murphy M, Schultz FA, Kang H, Wang X, Radich JP, Appelbaum FR, Atlas SR, Godwin J, Willman CL (2006) Gene expression profiling of adult acute myeloid leukemia identifies novel biologic clusters for risk classification and outcome prediction. *Blood*, 108, 685–696.

Woodward OM, Kottgen A, Coresh J, Boerwinkle E, Guggino WB, Kottgen M (2009) Identification of a urate transporter, ABCG2, with a common functional polymorphism causing gout. *Proc Natl Acad Sci U S A*, 106, 10338–10342.

Xu J, Liu Y, Yang Y, Bates S, Zhang JT (2004) Characterization of oligomeric human half-ABC transporter ATP-binding cassette G2. *J Biol Chem*, 279, 19781–19789.

Yamakawa Y, Hamada A, Nakashima R, Yuki M, Hirayama C, Kawaguchi T, Saito H (2011) Association of genetic polymorphisms in the influx transporter SLCO1B3 and the efflux transporter ABCB1 with imatinib pharmacokinetics in patients with chronic myeloid leukemia. *Ther Drug Monit*, 33, 244–250.

Yamasaki Y, Ieiri I, Kusuhara H, Sasaki T, Kimura M, Tabuchi H, Ando Y, Irie S, Ware J, Nakai Y, Higuchi S, Sugiyama Y (2008) Pharmacogenetic characterization of sulfasalazine disposition based on NAT2 and ABCG2 (BCRP) gene polymorphisms in humans. *Clin Pharmacol Ther*, 84, 95–103.

Yoshioka S, Katayama K, Okawa C, Takahashi S, Tsukahara S, Mitsuhashi J, Sugimoto Y (2007) The identification of two germ-line mutations in the human breast cancer resistance protein gene that result in the expression of a low/non-functional protein. *Pharm Res*, 24, 1108–1117.

Zamber CP, Lamba JK, Yasuda K, Farnum J, Thummel K, Schuetz JD, Schuetz EG (2003) Natural allelic variants of breast cancer resistance protein (BCRP) and their relationship to BCRP expression in human intestine. *Pharmacogenetics*, 13, 19–28.

Zamboni WC, Ramanathan RK, McLeod HL, Mani S, Potter DM, Strychor S, Maruca LJ, King CR, Jung LL, Parise RA, Egorin MJ, Davis TA, Marsh S (2006) Disposition of 9-nitrocamptothecin and its 9-aminocamptothecin metabolite in relation to ABC transporter genotypes. *Invest New Drugs*, 24, 393–401.

Zhang W, Yu BN, He YJ, Fan L, Li Q, Liu ZQ, Wang A, Liu YL, Tan ZR, Fen-Jiang, Huang YF, Zhou HH (2006) Role of BCRP 421C>A polymorphism on rosuvastatin pharmacokinetics in healthy Chinese males. *Clin Chim Acta*, 373, 99–103.

Zhou L, Schmidt K, Nelson FR, Zelesky V, Troutman MD, Feng B (2009) The effect of breast cancer resistance protein and P-glycoprotein on the brain penetration of flavopiridol, imatinib mesylate (Gleevec), prazosin, and 2-methoxy-3-(4-(2-(5-methyl-2-phenyloxazol-4-yl)ethoxy)phenyl)propanoic acid (PF-407288) in mice. *Drug Metab Dispos*, 37, 946–955.

Zhou Q, Sparreboom A, Tan EH, Cheung YB, Lee A, Poon D, Lee EJ, Chowbay B (2005) Pharmacogenetic profiling across the irinotecan pathway in Asian patients with cancer. *Br J Clin Pharmacol*, 59, 415–424.

Zhou S, Schuetz JD, Bunting KD, Colapietro AM, Sampath J, Morris JJ, Lagutina I, Grosveld GC, Osawa M, Nakauchi H, Sorrentino BP (2001) The ABC transporter Bcrp1/ABCG2 is expressed in a wide variety of stem cells and is a molecular determinant of the side-population phenotype. *Nat Med*, 7, 1028–1034.

CHAPTER 15

MRP2 (ABCC2) AND MRP3 (ABCC3)

Anne T. Nies

15.1 INTRODUCTION

The human genome comprises 48 ATP-binding cassette (ABC) transporter genes. The 48 ABC proteins are grouped into seven subfamilies on the basis of amino acid sequence and phylogenetic analyses (Moitra and Dean, 2011). Together with 10 other members, multidrug resistance-associated proteins, MRP2 (ABCC2) and MRP3 (ABCC3), constitute the C branch of ABC transporters (Table 15.1). ABCC2 and ABCC3 are integral plasma membrane proteins mediating the unidirectional ATP-dependent efflux of anionic conjugates. They primarily transport physiological substrates including glutathione conjugates such as leukotriene C_4 (LTC_4) and glucuronosides such as bilirubin glucuronosides (reviewed in Deeley et al., 2006; Jedlitschky et al., 2006; Borst et al., 2007; Nies and Keppler, 2007; Keppler, 2011). In addition, they efflux natural and synthetic toxins and drugs, thereby conferring drug resistance to cells (Haimeur et al., 2004; Szakacs et al., 2006). Moreover, the ABCC2- and ABCC3-mediated efflux of conjugates, which are often generated in phase II reactions of drug metabolism, is the final step in the pathway of detoxification of many xenobiotics and several endogenous substances (Keppler, 2011). The predominant localization of ABCC2 and ABCC3 in intestinal and renal epithelia, in hepatocytes, and in blood–tissue barriers demonstrates their important role in drug absorption, distribution, and elimination, as well as in protecting tissues against the entry of xenobiotic toxins.

A number of sequence variants leading to amino acid substitutions in ABCC2 and ABCC3 have been identified. They are of considerable clinical interest because they may cause interindividual variability in drug response. Sequence variations in the *ABCC2* gene resulting in the loss of a functional ABCC2 protein are the molecular basis of the Dubin–Johnson syndrome (DJS) in humans. This rare hereditary disorder is characterized by conjugated hyperbilirubinemia (Dubin and Johnson, 1954; Sprinz and Nelson, 1954), because the lack of ABCC2-mediated efflux of bilirubin glucuronosides into bile is compensated by their efflux into blood via ABCC3.

Pharmacogenomics of Human Drug Transporters: Clinical Impacts, First Edition.
Edited by Toshihisa Ishikawa, Richard B. Kim and Jörg König

TABLE 15.1 Nomenclature and Molecular Characteristics of the Human ABCC Subfamily Members

Gene symbol	Alternative name	Entrez Gene-id	Genomic location	Protein accession number	Amino acids	Amino acid identity (%)
ABCC1	MRP1	4363	16p13.1	NP_004987	1531	48
ABCC2	MRP2	1244	10q24	NP_000383	1545	100
ABCC3	MRP3	8714	17q22	NP_003777	1527	46
ABCC4	MRP4	10257	13q32	NP_005836	1325	37
ABCC5	MRP5	10057	3q27	NP_005679	1437	35
ABCC6	MRP6	368	16p13.1	NP_001162	1503	38
CFTR	ABCC7	1080	7q31.2	NP_000483	1480	26
ABCC8	SUR1	6833	11p15.1	NP_000343	1581	33
ABCC9	SUR2A	10060	12p12.1	NP_005682	1549	32
	SUR2B	10060		NP_064693	1549	33
ABCC10	MRP7	89845	6p21.1	NP_258261	1464	34
ABCC11	MRP8	85320	16q12.1	NP_149163	1382	32
ABCC12	MRP9	94160	16q12.1	NP_150229	1359	33

Data compiled from "Gene" database at http://www.ncbi.nlm.nih.gov/sites/entrez?db=gene; regular updates available.
Amino acid identity was calculated with the SIM program for local alignment analysis (http://genome.cs.mtu.edu/align/align.html).

This chapter provides a brief overview on the basic characteristics of ABCC2 and ABCC3 (molecular structure, tissue expression, and substrates), outlines the use of knockout animals to study both the transporters, and finally discusses the functional consequences of genetic variations in the *ABCC2* and *ABCC3* genes.

15.2 BASIC CHARACTERISTICS OF HUMAN ABCC2 AND ABCC3

15.2.1 Genomic Organization

The *ABCC2* gene lies on chromosome 10q24 and contains 32 exons in a genomic region spanning 69.46 kb (Fig. 15.1). *ABCC2* encodes a 5051 bp mRNA (NM_000392.3), which gives rise to a protein that is 1545 amino acids in length (NP_000383.1). To date, no transcript variants have been described.

The *ABCC3* gene spans 56.93 kb of genomic DNA and contains 31 exons. The 5183 bp transcript variant 1 encodes the functionally active full-length ABCC3 protein (NP_003777.2). Transcript variant 2 (NM_001144070.1) lacks multiple 3' exons so that the resulting isoform 2 (NP_001137542.1) is much shorter than full-length ABCC3 and has a different C-terminus.

15.2.2 Protein Structure

Structural data of MRP transporters have so far only been obtained for ABCC1 (MRP1) (Rosenberg et al., 2010). Based on sequence similarities with ABCC1 and on hydropathy analyses, three polytopic membrane-spanning domains (MSDs) and two nucleotide-binding domains are predicted for ABCC2 and ABCC3 (Fig. 15.2). MSD0 may be required for proper ABCC2 function and localization (Fernandez et al., 2002). The extracellular location of the *N*-terminus of ABCC2 was first predicted for the rat ortholog by computational analysis (Büchler et al., 1996) and subsequently proven experimentally by immunofluorescence microscopy (Cui et al., 1999). When using the Phobius program (http://phobius.sbc.su.se) or the TMHMM program (http://www.cbs.dtu.dk/services/TMHMM), with which the probability of an amino acid sequence to be a transmembrane-spanning helix in a polytopic membrane protein is calculated, four and six transmembrane-spanning helices are predicted to constitute MSD2 in ABCC2 and ABCC3, respectively (Fig. 15.2). However, UniProt (http://www.uniprot.org) predicts six helices in MSD2 of ABCC2 and ABCC3. Experimental elucidation of the three-dimensional structure of ABCC2 and ABCC3 is, therefore, required to confirm either predictions. This is of particular interest for a better in silico prediction of functional consequences of nonsynonymous variants in ABCC2 and ABCC3.

Several *N*-glycosylation sites are predicted in the ABCC2 and ABCC3 proteins (Fig. 15.2). *N*-Glycosylation is apparently not necessary for proper localization of human ABCC2 (Keitel et al., 2000). Moreover, ABCC2 is phosphorylated at different residues (Hegedus et al., 2003; Daub et al., 2008; Dephoure et al., 2008). C-terminal phosphorylation of ABCC2 appears to be important for an increased interaction of ABCC2 with interacting proteins (Hegedus et al., 2003).

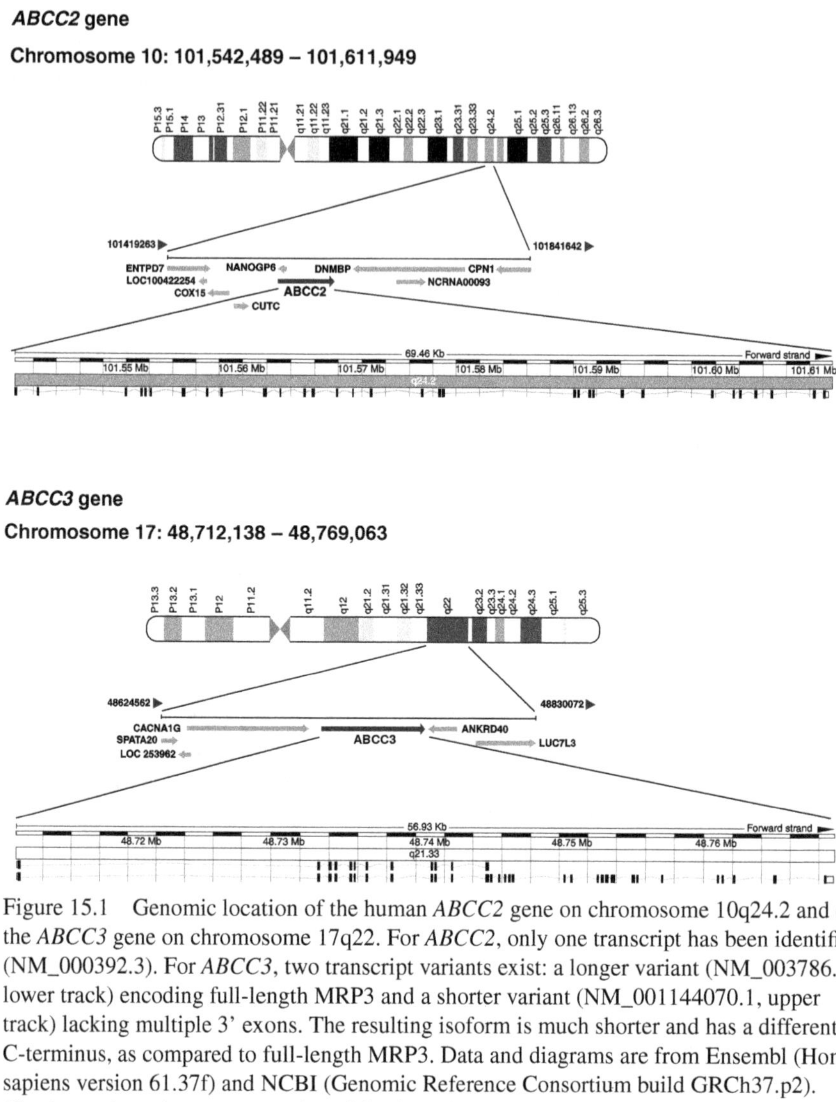

Figure 15.1 Genomic location of the human *ABCC2* gene on chromosome 10q24.2 and of the *ABCC3* gene on chromosome 17q22. For *ABCC2*, only one transcript has been identified (NM_000392.3). For *ABCC3*, two transcript variants exist: a longer variant (NM_003786.3, lower track) encoding full-length MRP3 and a shorter variant (NM_001144070.1, upper track) lacking multiple 3' exons. The resulting isoform is much shorter and has a different C-terminus, as compared to full-length MRP3. Data and diagrams are from Ensembl (Homo sapiens version 61.37f) and NCBI (Genomic Reference Consortium build GRCh37.p2). (*See insert for color representation of the figure.*)

15.2.3 Substrates and Function

The cloning of the human *ABCC2* cDNA, its expression in mammalian cell lines, and isolation of membrane vesicles from these cells (Evers et al., 1998; Cui et al., 1999; König et al., 1999a) led to the identification of various organic anions as ABCC2 substrates (Fig. 15.3, Table 15.2). Moreover, double- and multiple-transfected polarized cells stably expressing an uptake transporter of the OATP family in the basolateral membrane and human ABCC2 in the apical membrane have been used to identify ABCC2 substrates, although kinetic constants cannot be determined with this experimental setup (Cui et al., 2001; Sasaki et al., 2002; Kopplow et al., 2005). Because

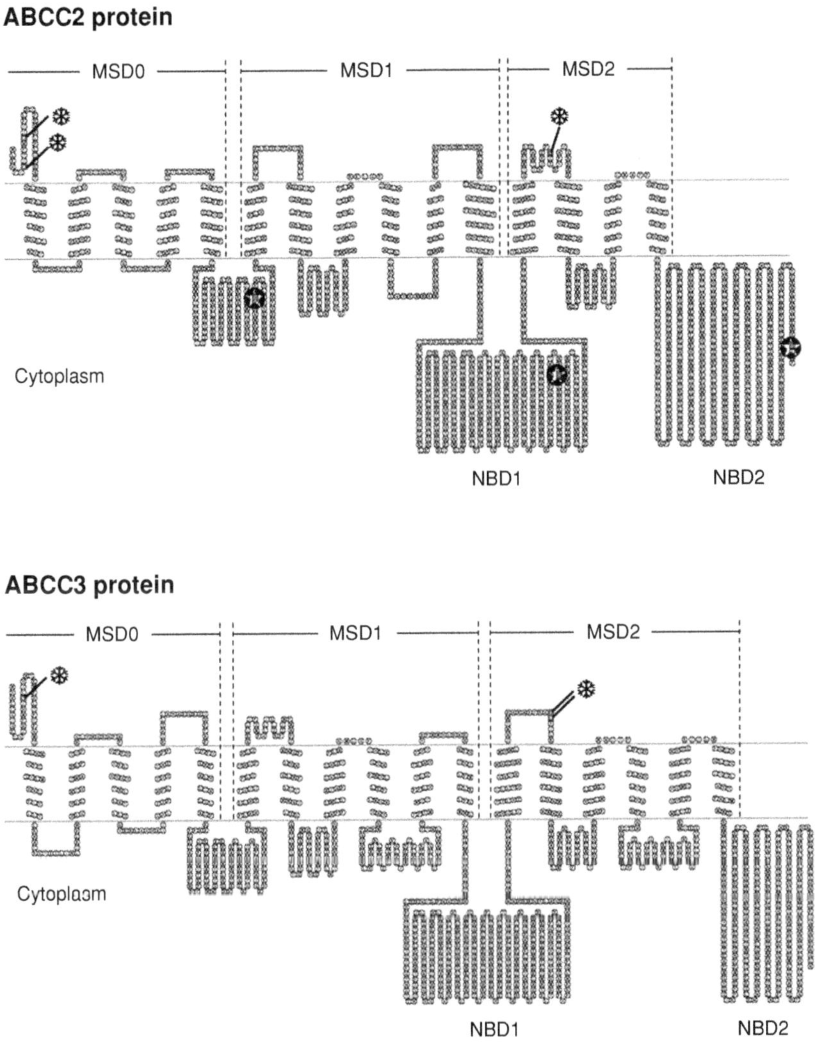

Figure 15.2 Topology models of the human ABCC2 and ABCC3 proteins. Model prediction was performed with Phobius (http://phobius.sbc.su.se) with the constraints that both nucleotide-binding domains are cytoplasmic. Topology models were drawn using TOPO2 (http://www.sacs.ucsf.edu/TOPO-run/wtopo.pl). MSD, membrane-spanning domain; (✱, potential *N*-glycosylation site; ✪, phosphorylation site. (*See insert for color representation of the figure.*)

ABCC2 confers drug resistance to cisplatin, etoposide, vincristine, and doxorubicin (Chen et al., 1999; Cui et al., 1999), these antineoplastic drugs are most likely substrates as well.

Human ABCC3 also transports a broad range of endogenous and xenobiotic, mostly conjugated organic anions, and has a preference for glucuronidated compounds (Table 15.2). ABCC3 confers resistance to several drugs such as etoposide,

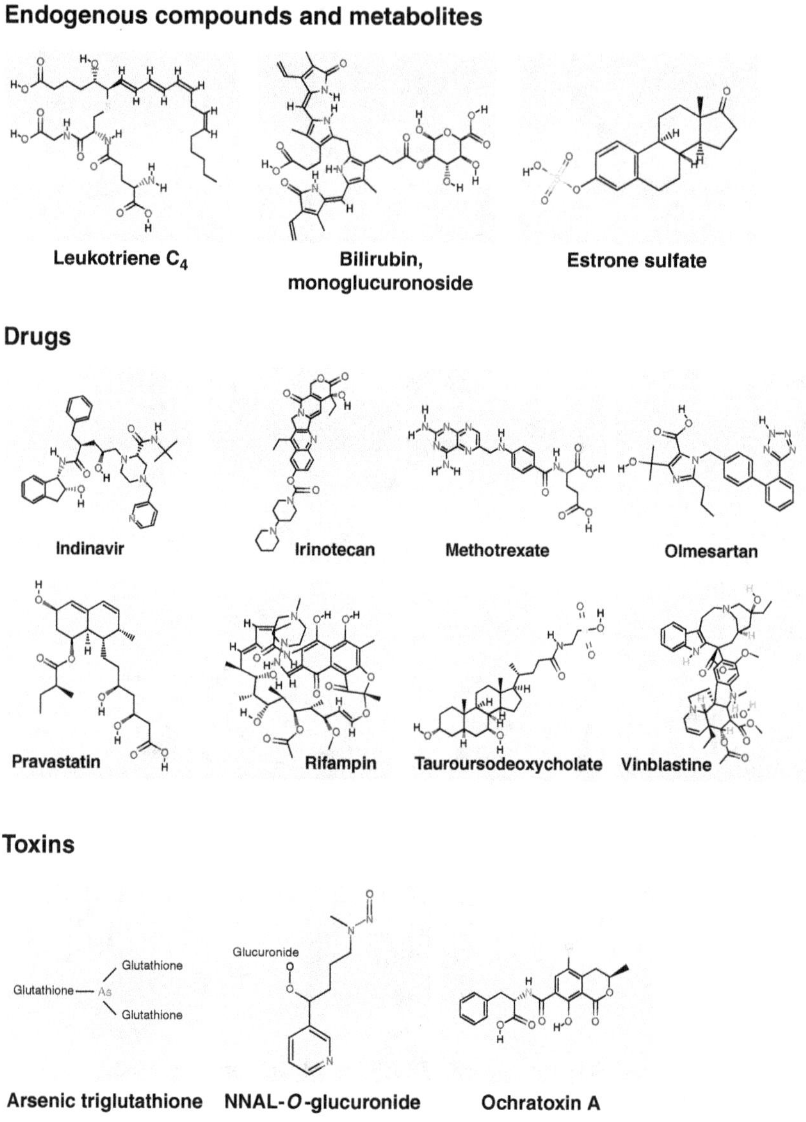

Figure 15.3 Molecular structures of selected ABCC2 substrates. Structures are from the publicly available PubChem Compound database (http://www.ncbi.nlm.nih.gov/pccompound). NNAL, 4-(methylnitrosoamino)-1-(3-pyridyl)-1-butanol. (*See insert for color representation of the figure.*)

teniposide, and methotrexate (Kool et al., 1999; Zeng et al., 1999; Zelcer et al., 2001). ABCC3 may contribute to the detoxification of glucuronidated drug conjugates and may represent an important alternative route for excretion, even under normal physiological conditions. Especially when glucuronidated drug conjugates reach high levels in hepatocytes, ABCC3-mediated transport into blood may limit their enterohepatic circulation (Borst et al., 2007).

TABLE 15.2 Selected ABCC2 and ABCC3 Substrates

ABCC2 substrates	
Physiological compounds	Bilirubin glucuronosides
	Conjugated bile salts
	Conjugated steroid hormones
	Glutathione
	Leukotriene C_4, leukotriene D_4, leukotriene E_4
Drugs and drug metabolites	Chlorambucil, monoglutathionyl
	Enalapril
	Ethinyl estradiol, glucuronide
	Fexofenadine
	HIV protease inhibitors
	Methotrexate
	Morphine-3-glucuronide
	Olmesartan
	Rifampin
	Statins
	Tauroursodeoxycholate
	Vinblastine
Toxins	Arsenic triglutathione
	NNAL-*O*-glucuronide
	Ochratoxin A
ABCC3	
Physiological compounds	Bilirubin glucuronosides
	Conjugated bile salts
	Conjugated steroid hormones
	Leukotriene C_4
Drugs and drug metabolites	Ethinyl estradiol, glucuronide
	Fexofenadine
	Etoposide glucuronide
	Methotrexate
	Morphine-3-glucuronide

Data compiled from Borst et al. (2007); Nies and Keppler (2007); and Nies et al. (2008).

15.2.4 Tissue Distribution and Localization

The ABCC2 protein is exclusively localized in the apical membrane of polarized cells, where it mediates the excretion and detoxification of many phase II conjugation products of endogenous substances and drugs into bile, urine, or intestine. Subsequent to its initial demonstration in the apical (canalicular) membrane of rat and human hepatocytes (Büchler et al., 1996; Keppler and Kartenbeck, 1996), the ABCC2 protein has been identified in many other human cell types and tissues, including proximal tubule epithelial cells of the kidney (Schaub et al., 1999), human small intestine (Fromm et al., 2000), colon (Sandusky et al., 2002), gall bladder (Rost et al., 2001), lung epithelial cells (Sandusky et al., 2002; König et al., 2003), and syncytiotrophoblast cells of placenta (St-Pierre et al., 2000). Otherwise, the ABCC2 protein is not present or below the detection limit in skin, female reproductive system, lymphatic system, cardiovascular system, connective tissue (Sandusky et al., 2002), exocrine pancreas (Sandusky et al., 2002; König et al., 2005), and some blood–tissue barriers such as the blood–testis barrier (Sandusky et al., 2002; Bart et al., 2004) and the blood–brain barrier in human cerebral cortex (Sandusky et al., 2002; Zhang et al., 2004; Nies et al., 2006). However, epilepsy in humans and rats apparently leads to ABCC2 upregulation in brain capillaries of the hippocampus, which may be a reason for the frequently observed resistance to antiepileptic drug treatment (Aronica et al., 2004; Hoffmann et al., 2006).

The ABCC3 protein is localized in the basolateral membrane of many polarized cells, including hepatocytes (Kool et al., 1999; König et al., 1999b), cholangiocytes, and gall bladder epithelial cells (Rost et al., 2001). Other tissues and cells with detectable ABCC3 protein include pancreas (König et al., 2005), kidney distal tubules, enterocytes in the ileum and the colon, spleen, adrenal cortex, and lung (Paulusma et al., 1996; Scheffer et al., 2002a, 2002b).

15.2.5 Regulation

Regulation of ABCC2 and ABCC3 expressions may occur on the transcriptional and post-transcriptional levels in response to many endogenous and xenobiotic substances, particularly by their binding to nuclear receptors, in response to different stimuli such as cell hydration, or under different pathophysiological conditions. These processes are described in detail by Eloranta et al. (2005); Geier et al. (2007); Kubitz and Häussinger (2007); Nies and Keppler (2007); and Klaassen and Aleksunes (2010). Of interest, there is evidence that ABCC2 can be post-transcriptionally regulated by SUMOylation as well (Minami et al., 2009).

15.3 KNOCKOUT ANIMALS

Two different mutant hyperbilirubinemic rat strains exist that have been very valuable in determining the substrate specificity of ABCC2. The GY/TR$^-$ rats (Jansen et al., 1985) and the Eisai hyperbilirubinemic rats (EHBR) (Takikawa et al., 1991; Hosokawa et al., 1992) have a hereditary defect in the biliary secretion of anionic conjugates. The molecular basis of this defect is the lack of the ABCC2 protein, which is due to variants in the rat *ABCC2* gene leading to premature stop codons (Büchler et al., 1996; Paulusma et al., 1996; Ito et al., 1997). Impaired ABCC2 function in the canalicular membrane is compensated by upregulation of ABCC3 in the sinusoidal membrane of hepatocytes (Donner and Keppler, 2001). ABCC3-mediated efflux into blood may protect the hepatocytes from intracellular accumulation of toxic substances, including drugs, which are finally secreted into urine (Keppler and Kartenbeck, 1996; Donner and Keppler, 2001). The two rat strains are considered as animal models for the human DJS (see paragraph 4).

Because the mutant rat strains had been readily available, it lasted several years until ABCC2 knockout mice were generated by several groups (Chu et al., 2006; Nezasa et al., 2006; Vlaming et al., 2006). Similar to the ABCC2-deficient rats, these mice have an impaired biliary secretion of organic anions and a mild conjugated hyperbilirubinemia. Otherwise, ABCC3 knockout mice show an impaired sinusoidal secretion of glucuronides (Belinsky et al., 2005; Zamek-Gliszczynski et al., 2006). These knockout mouse models, alone or as combination knockouts, are powerful tools for assessing the role of ABCC2 and ABCC3 for the pharmacokinetics of several clinically relevant drugs (Lagas et al., 2009).

15.4 GENETIC VARIATION CAUSING ABSENCE OF A FUNCTIONAL ABCC2 PROTEIN (DJS)

Certain rare genetic variants in the *ABCC2* gene may lead to either absence of the ABCC2 protein or expression of an inactive ABCC2 protein (Table 15.3), both causing the DJS in humans (Kartenbeck et al., 1996; Keppler and Kartenbeck, 1996; Paulusma et al., 1997). This benign disorder is inherited in an autosomal, recessive mode with an incidence ranging from 1:1300 among Iranian Jews (Shani et al., 1970) and 1:300,000 in the Japanese population (Kajihara et al., 1998; Wada et al., 1998). Individuals affected with the DJS have a mild conjugated hyperbilirubinemia and deposition of dark pigments in hepatocytes, but do not suffer from fatal liver insufficiency (Dubin and Johnson, 1954; Sprinz and Nelson, 1954). Since biliary efflux via ABCC2 is abolished, bilirubin conjugates are excreted into blood via ABCC3 (Lee et al., 2004) thus causing the observed conjugated hyperbilirubinemia. ABCC2 deficiency is apparently compensated by upregulation of basolateral ABCC3 as suggested by immunofluorescence microscopy in hepatocytes from individuals affected with the DJS (König et al., 1999b).

Table 15.3 summarizes variations in the *ABCC2* gene that have been detected in individuals affected with the DJS. Variations include splice site variants resulting in exon loss and subsequent premature stop codons, small and large deletion variants, deletion/in-frame insertion variants, nonsense variants leading to premature stop codons, and missense variants. When investigated *in vitro*, some DJS-associated variants lead to deficient maturation and impaired sorting of the ABCC2 protein or to an apically localized, yet inactive ABCC2 protein (Keitel et al., 2000; Mor-Cohen et al., 2001; Hashimoto et al., 2002; Keitel et al., 2003).

Rotor's syndrome is another rare inherited benign disorder that is characterized by conjugated hyperbilirubinemia but that is distinct from the DJS (Wolkoff et al., 1976). However, Rotor's syndrome is not caused by genetic variation of the *ABCC2* gene (Hrebicek et al., 2007). As recently discovered, the molecular basis of Rotor's syndrome is the inherited complete functional co-deficiency of the hepatic organic anion uptake transporters OATP1B1 and OATP1B3 (van de Steeg et al., 2012).

15.5 PHENOTYPIC CONSEQUENCES OF *ABCC2* GENETIC VARIANTS NOT ASSOCIATED WITH DJS

A large number of regulatory and coding variants in the *ABCC2* gene, which are not associated with the DJS, are known and can be found at the publicly available dbSNP database (http://www.ncbi.nlm.nih.gov/snp). From these variants, only three are common nonsynonymous variants with frequencies clearly more than 1%: V417I (rs2273697, c.1249G>A, ~22% in Sub-Saharan Africans, ~9% in Asians, and ~24%

TABLE 15.3 Variations in the Human *ABCC2* Gene Identified in Patients with the Dubin–Johnson Syndrome

Type of variant	Nucleotide change[a]	Predicted consequence[a]	Affected region in ABCC2 protein[b]	First reported ref.
Nonsense	c.298 C>T[c]	p.R100X	Truncation in MSD0	Shoda et al., 2003
Nonsense	c.974 C>G	p.S325X	Truncation in MSD1	Corpechot et al., 2006
Large-scale deletion	Genomic 1008 bp deletion	Deletion of exon 7	Intracellular loop between MSD0 and MSD1	Kanda et al., 2009
Splicing	c.1031+4A>G	Complex splicing	Truncation in MSD1	Mor-Cohen et al., 2005
Missense	c.1177 C>T	p.R393W	MSD1	Machida et al., 2005
In-frame indel	c.1256_1272del insCT[c]	p.E419_M424del insA	Disruption of MSD1	Cebecauerova et al., 2005
Missense	c.1321 C>A[c]	p.L441M	MSD1	Lee et al., 2006
Splicing	c.1815+2T>A	Exon 13 skipping	Disruption of MSD1	Wada et al., 1998
Splicing	c.1967+2T>C	p.K635X	Truncation in NBD1	Kajihara et al., 1998
Missense	c.2026 G>C[c]	p.G676R	NBD1	Wakusawa et al., 2003
Missense	c.2125 T>C	p.W709R	NBD1	Machida et al., 2005
Missense	c.2302 C>T	p.R768W	NBD1	Wada et al., 1998
Splicing	c.2439+2T>C	Exon 18 skipping	Disruption of NBD1	Toh et al., 1999
Deletion	c.2748_2838del[c]	p.S916fsX	Truncation in MSD2	Lee et al., 2006
Nonsense	c.3196 C>T	p.R1066X	Truncation in MSD2	Paulusma et al., 1997

Deletion	c.3399_3400del[c]	p.Y1134CfsX	Truncation in MSD2	Lee et al., 2006
Missense	c.3449 G>A	p.R1150H	MSD2	Mor-Cohen et al., 2001
Missense	c.3517 A>T	p.I1173F	MSD2	Mor-Cohen et al., 2001
Deletion	c.3615_3843del[c]	p.W1206fsX	Truncation in MSD2	Kanai et al., 1995
Nonsense	c.3825 C>G[c]	p.Y1275X	Truncation in MSD2	Kanai et al., 1995
Nonsense	c.3928 C>T[c]	p.R1310X	Truncation in NBD2	Tate et al., 2002
Missense	c.4054 G>C[c]	p.E1352Q	NBD2	Kanai et al., 1995
Missense	c.4145 A>G[c]	p.Q1382R	NBD2	Toh et al., 1999
Deletion	c.4175_4180del	p.R1392_M1393del	NBD2	Tsujii et al., 1999
Deletion	c.4292_4293del[c]	p.T1431RfsX	Truncation in NBD2	Cebecauerova et al., 2005

[a] As recommended by the Human Genome Variation Society (http://www.hgvs.org/mutnomen) and by den Dunnen and Antonarakis (2000),

- nucleotide position +1 is the A of the ATG of the translation initiation codon in the ABCC2 cDNA sequence;
- "c." describes a nucleotide change in relation to the ABCC2 cDNA sequence (NM_000392.3);
- "p." describes a change in relation to the deduced ABCC2 protein sequence (NP_000383.1);
- "X" denotes a premature stop codon;
- "fs" denotes a frame shift;
- "indel" denotes insertion/deletion.

[b] See Figure 15.2 for regions.
[c] Heterozygous variant.

in Caucasians), V1188E (rs8187694, c.3563T>A, ~7% in Caucasians), and C1515Y (rs8187710, c.4544G>A, ~5% in Caucasians). In addition, the promoter variant c.-24C>T (rs717620) is common in different populations as well (~20% in Caucasians, Sub-Saharan Africans, and Asians). Because of their high allele frequencies, these four variants have been studied extensively.

Identification of genetic variants and systematic genotyping of the *ABCC2* gene has been reported, for example, for the Japanese population (Ito et al., 2001; Itoda et al., 2002; Saito et al., 2002; Sai et al., 2008), for Koreans (Choi et al., 2007), Caucasians (Bosch et al., 2006; Leschziner et al., 2006; Meier et al., 2006; Niemi et al., 2006; Haenisch et al., 2007), as well as Chinese, Malay, and Indian populations of Singapore (Ho et al., 2008).

In vitro functional characterization showed that the most frequently observed V417I variant probably does not affect *in vivo* function of ABCC2, whereas S789F and A1450T, which occur with a much lower frequency, may be associated with a reduced *in vivo* function of ABCC2 (Hirouchi et al., 2004). Regarding the promoter variant c.-24C>T, this variant resulted in a reduced reporter gene activity in one study (Haenisch et al., 2007), but not in another one (Zhang et al., 2010) and its *in vitro* effect remains to be solved.

The effect of common ABCC2 variants on ABCC2 expression was analyzed in different tissues. The promoter variant allele -24T was associated with lower *ABCC2* mRNA levels in the normal human kidney (Haenisch et al., 2007), but not in the intestine (Moriya et al., 2002; Haenisch et al., 2008). The variant allele 1249A was associated with reduced *ABCC2* mRNA expression in human preterm placenta (Meyer zu Schwabedissen et al., 2005). Otherwise, variants 1515Y and 1188E resulted in a higher ABCC2 protein expression in the human liver (Meier et al., 2006).

The effect of ABCC2 variants on the pharmacokinetics of several drugs has been extensively studied and recently reviewed (Cascorbi and Haenisch, 2010; Haufroid, 2010). Pharmacokinetics, for example, of the anticancer drugs methotrexate and irinotecan (Rau et al., 2006; Innocenti et al., 2009), of the antiretroviral drug lopinavir (Elens et al., 2009), of immunosuppressive drugs such as mycophenolic acid (Naesens et al., 2006; Baldelli et al., 2007; Miura et al., 2007; Levesque et al., 2008; Zhang et al., 2008), of talinolol (Haenisch et al., 2008), telmisartan (Miura et al., 2009), and fexofenadine (Akamine et al., 2010) have been associated with certain ABCC2 variants. Interestingly, -24T allele carriers had a higher risk to develop a hepatotoxic reaction after the intake of drugs or herbal remedies (Choi et al., 2007; Daly et al., 2007).

15.6 PHENOTYPIC CONSEQUENCES OF VARIATIONS IN THE *ABCC3* GENE

The first systematic screening for ABCC3 variants was carried out in a Japanese population identifying a number of intronic and exonic variants (Saito et al., 2002). Fukushima-Uesaka et al. identified several novel heterozygous variants in the Japanese population (Fukushima-Uesaka et al., 2007). A systematic sequencing approach in 103 Caucasians also identified multiple genetic variants (Lang et al.,

2004). Common nonsynonymous variants in Caucasians with a frequency of clearly more than 1% are H68Y (rs34926034, 2%) and R1297H (rs11568591, 5%).

In vitro functional characterization of several nonsynonymous ABCC3 variants showed that variants may either be retained within the endoplasmic reticulum (R1381S), or may lack function (S346F, S607N), or may not have any effect at all (G11D, R99Q, V765L, P920S, R923Q, R1286G, R1297H, R1348C, Q1365R) (Lee et al., 2004; Kobayashi et al., 2008). Since the variants R1381S, S346F, or S607N have only limited ability to transport toxic compounds across the sinusoidal membrane, it was suggested that the presence of these ABCC3 variants in individuals poses a risk factor for the acquisition of hepatotoxicity (Kobayashi et al., 2008).

Of interest, individuals with the -211C>T promoter variant (rs4793665) have significantly lower hepatic *ABCC3* mRNA levels compared to wild-type individuals (Lang et al., 2004), while in patients with acute leukemia, no influence of the -211C>T variant on the ABCC3 expression level was found (Doerfel et al., 2006). Binding of nuclear factors to the *ABCC3* promoter appears to be affected by the variant although their nature needs to be elucidated (Lang et al., 2004; Gradhand et al., 2007).

The -211C>T promoter variant was also associated with an adverse prognostic significance in patients with acute myeloid leukemia (Müller et al., 2008) and suggested as a possible predictor of the clinical outcome of chemotherapy in lung cancer patients (Müller et al., 2009). However, it was not associated with colorectal cancer risk (Campa et al., 2008). The amplification of the *ABCC3* gene in HER2-amplified and luminal primary breast tumors may be a predictor for taxane resistance in breast cancer (O'Brien et al., 2008).

15.7 CONCLUSION

ABCC2 and ABCC3 are efflux transporters for organic anions. Because of their localization in the apical and basolateral membrane, respectively, of polarized cells, for example, in liver, kidney, and intestine, they determine the absorption, distribution, and excretion of many clinically relevant drugs. Complete loss-of-function genetic variants in the *ABCC2* gene are the molecular basis of the human DJS. In addition, several genetic variants in the *ABCC2* and *ABCC3* genes have been associated with altered protein expression levels, altered pharmacokinetics of drugs, or susceptibility to disease. It will be an important issue for future research to further explore and identify novel genotype–phenotype correlations involving ABCC2 and ABCC3.

ACKNOWLEDGMENTS

The author's study was supported by the Robert Bosch Foundation, Stuttgart, Germany; the IZEPHA Grant (#8-0-0/674), University of Tübingen; the Wilhelm Sander-Stiftung, München, Germany (no. 2010.059.1); and the 7th FP EU Initial Training Network program "FightingDrugFailure" (PITN-GA-2009-238132).

REFERENCES

Akamine Y, Miura M, Sunagawa S, Kagaya H, Yasui-Furukori N, Uno T (2010) Influence of drug-transporter polymorphisms on the pharmacokinetics of fexofenadine enantiomers. *Xenobiotica*, 40, 782–789.

Aronica E, Gorter JA, Ramkema M, Redeker S, Ozbas-Gerceker F, van Vliet EA, Scheffer GL, Scheper RJ, van der Valk P, Baayen JC, Troost D (2004) Expression and cellular distribution of multidrug resistance-related proteins in the hippocampus of patients with mesial temporal lobe epilepsy. *Epilepsia*, 45, 441–451.

Baldelli S, Merlini S, Perico N, Nicastri A, Cortinovis M, Gotti E, Remuzzi G, Cattaneo D (2007) C-440T/T-331C polymorphisms in the UGT1A9 gene affect the pharmacokinetics of mycophenolic acid in kidney transplantation. *Pharmacogenomics*, 8, 1127–1141.

Bart J, Hollema H, Groen HJ, de Vries EG, Hendrikse NH, Sleijfer DT, Wegman TD, Vaalburg W, van der Graaf WT (2004) The distribution of drug-efflux pumps, P-gp, BCRP, MRP1 and MRP2, in the normal blood–testis barrier and in primary testicular tumours. *Eur J Cancer*, 40, 2064–2070.

Belinsky MG, Dawson PA, Shchaveleva I, Bain LJ, Wang R, Ling V, Chen ZS, Grinberg A, Westphal H, Klein-Szanto A, Lerro A, Kruh GD (2005) Analysis of the in vivo functions of MRP3. *Mol Pharmacol*, 68, 160–168.

Borst P, de Wolf C, van de Wetering K (2007) Multidrug resistance-associated proteins 3, 4, and 5. *Pflugers Arch*, 453, 661–673.

Bosch TM, Doodeman VD, Smits PH, Meijerman I, Schellens JH, Beijnen JH (2006) Pharmacogenetic screening for polymorphisms in drug-metabolizing enzymes and drug transporters in a Dutch population. *Mol Diagn Ther*, 10, 175–185.

Büchler M, König J, Brom M, Kartenbeck J, Spring H, Horie T, Keppler D (1996) cDNA cloning of the hepatocyte canalicular isoform of the multidrug resistance protein, cMRP, reveals a novel conjugate export pump deficient in hyperbilirubinemic mutant rats. *J Biol Chem*, 271, 15091–15098.

Campa D, Vodicka P, Pardini B, Novotny J, Forsti A, Hemminki K, Barale R, Canzian F (2008) Could polymorphisms in ATP-binding cassette C3/multidrug resistance associated protein 3 (ABCC3/MRP3) modify colorectal cancer risk? *Eur J Cancer*, 44, 854–857.

Cascorbi I, Haenisch S (2010) Pharmacogenetics of ATP-binding cassette transporters and clinical implications. *Methods Mol Biol*, 596, 95–121.

Cebecauerova D, Jirasek T, Budisova L, Mandys V, Volf V, Novotna Z, Subhanova I, Hrebicek M, Elleder M, Jirsa M (2005) Dual hereditary jaundice: simultaneous occurrence of mutations causing Gilbert's and Dubin–Johnson syndrome. *Gastroenterology*, 129, 315–320.

Chen ZS, Kawabe T, Ono M, Aoki S, Sumizawa T, Furukawa T, Uchiumi T, Wada M, Kuwano M, Akiyama SI (1999) Effect of multidrug resistance-reversing agents on transporting activity of human canalicular multispecific organic anion transporter. *Mol Pharmacol*, 56, 1219–1228.

Choi JH, Ahn BM, Yi J, Lee JH, Lee JH, Nam SW, Chon CY, Han KH, Ahn SH, Jang IJ, et al. (2007) MRP2 haplotypes confer differential susceptibility to toxic liver injury. *Pharmacogenet Genomics*, 17, 403–415.

Chu X, Strauss JR, Mariano MA, Li J, Newton DJ, Cai X, Wang RW, Yabut J, Hartley DP, Evans DC, Evers R (2006) Characterization of mice lacking the multidrug resistance protein MRP2 (ABCC2). *J Pharmacol Exp Ther*, 317, 579–589.

Corpechot C, Ping C, Wendum D, Matsuda F, Barbu V, Poupon R (2006) Identification of a novel 974C→G nonsense mutation of the MRP2/ABCC2 gene in a patient with Dubin–Johnson syndrome and analysis of the effects of rifampicin and ursodeoxycholic acid on serum bilirubin and bile acids. *Am J Gastroenterol*, 101, 2427–2432.

Cui Y, König J, Buchholz U, Spring H, Leier I, Keppler D (1999) Drug resistance and ATP-dependent conjugate transport mediated by the apical multidrug resistance protein, MRP2, permanently expressed in human and canine cells. *Mol Pharmacol*, 55, 929–937.

Cui Y, König J, Keppler D (2001) Vectorial transport by double-transfected cells expressing the human uptake transporter SLC21A8 and the apical export pump ABCC2. *Mol Pharmacol*, 60, 934–943.

Daly AK, Aithal GP, Leathart JB, Swainsbury RA, Dang TS, Day CP (2007) Genetic susceptibility to diclofenac-induced hepatotoxicity: contribution of UGT2B7, CYP2C8, and ABCC2 genotypes. *Gastroenterology*, 132, 272–281.

Daub H, Olsen JV, Bairlein M, Gnad F, Oppermann FS, Körner R, Greff Z, Kéri G, Stemmann O, Mann M (2008) Kinase-selective enrichment enables quantitative phosphoproteomics of the kinome across the cell cycle. *Mol Cell*, 31, 438–448.

Deeley RG, Westlake C, Cole SP (2006) Transmembrane transport of endo- and xenobiotics by mammalian ATP-binding cassette multidrug resistance proteins. *Physiol Rev*, 86, 849–899.

den Dunnen JT, Antonarakis SE (2000) Mutation nomenclature extensions and suggestions to describe complex mutations: a discussion. *Hum Mutat*, 15, 7–12.

Dephoure N, Zhou C, Villen J, Beausoleil SA, Bakalarski CE, Elledge SJ, Gygi SP (2008) A quantitative atlas of mitotic phosphorylation. *Proc Natl Acad Sci U S A*, 105, 10762–10767.

Doerfel C, Rump A, Sauerbrey A, Gruhn B, Zintl F, Steinbach D (2006) In acute leukemia, the polymorphism -211C>T in the promoter region of the multidrug resistance-associated protein 3 (MRP3) does not determine the expression level of the gene. *Pharmacogenet Genomics*, 16, 149–150.

Donner MG, Keppler D (2001) Up-regulation of basolateral multidrug resistance protein 3 (MRP3) in cholestatic rat liver. *Hepatology*, 34, 351–359.

Dubin IN, Johnson FB (1954) Chronic idiopathic jaundice with unidentified pigment in liver cells; a new clinicopathologic entity with report of 12 cases. *Medicine*, 33, 155–179.

Elens L, Yombi JC, Lison D, Wallemacq P, Vandercam B, Haufroid V (2009) Association between ABCC2 polymorphism and lopinavir accumulation in peripheral blood mononuclear cells of HIV-infected patients. *Pharmacogenomics*, 10, 1589–1597.

Eloranta JJ, Meier PJ, Kullak-Ublick GA (2005) Coordinate transcriptional regulation of transport and metabolism. *Methods Enzymol*, 400, 511–530.

Evers R, Kool M, van Deemter L, Janssen H, Calafat J, Oomen LC, Paulusma CC, Oude Elferink RP, Baas F, Schinkel AH, Borst P (1998) Drug export activity of the human canalicular multispecific organic anion transporter in polarized kidney MDCK cells expressing *cMOAT* (*MRP2*) cDNA. *J Clin Invest*, 101, 1310–1319.

Fernandez SB, Hollo Z, Kern A, Bakos E, Fischer PA, Borst P, Evers R (2002) Role of the N-terminal transmembrane region of the multidrug resistance protein MRP2 in routing to the apical membrane in MDCKII cells. *J Biol Chem*, 277, 31048–31055.

Fromm MF, Kauffmann HM, Fritz P, Burk O, Kroemer HK, Warzok RW, Eichelbaum M, Siegmund W, Schrenk D (2000) The effect of rifampin treatment on intestinal expression of human MRP transporters. *Am J Pathol*, 157, 1575–1580.

Fukushima-Uesaka H, Saito Y, Maekawa K, Hasegawa R, Suzuki K, Yanagawa T, Kajio H, Kuzuya N, Noda M, Yasuda K, Tohkin M, Sawada J (2007) Genetic variations of the ABC transporter gene ABCC3 in a Japanese population. *Drug Metab Pharmacokinet*, 22, 129–135.

Geier A, Wagner M, Dietrich CG, Trauner M (2007) Principles of hepatic organic anion transporter regulation during cholestasis, inflammation and liver regeneration. *Biochim Biophys Acta*, 1773, 283–308.

Gradhand U, Tegude H, Burk O, Eichelbaum M, Fromm MF, Konig J (2007) Functional analysis of the polymorphism -211C>T in the regulatory region of the human ABCC3 gene. *Life Sci*, 80, 1490–1494.

Haenisch S, May K, Wegner D, Caliebe A, Cascorbi I, Siegmund W (2008) Influence of genetic polymorphisms on intestinal expression and rifampicin-type induction of ABCC2 and on bioavailability of talinolol. *Pharmacogenet Genomics*, 18, 357–365.

Haenisch S, Zimmermann U, Dazert E, Wruck CJ, Dazert P, Siegmund S, Kroemer HK, Warzok RW, Cascorbi I (2007) Influence of polymorphisms of ABCB1 and ABCC2 on mRNA and protein expression in normal and cancerous kidney cortex. *Pharmacogenomics J*, 7(1):56–65.; Erratum, *Pharmacogenomics J*, 7, 74.

Haimeur A, Conseil G, Deeley RG, Cole SP (2004) The MRP-related and BCRP/ABCG2 multidrug resistance proteins: biology, substrate specificity and regulation. *Curr Drug Metab*, 5, 21–53.

Hashimoto K, Uchiumi T, Konno T, Ebihara T, Nakamura T, Wada M, Sakisaka S, Maniwa F, Amachi T, Ueda K, Kuwano M (2002) Trafficking and functional defects by mutations of the ATP-binding domains in MRP2 in patients with Dubin–Johnson syndrome. *Hepatology*, 36, 1236–1245.

Haufroid V (2010) Genetic polymorphisms of ATP-binding cassette transporters ABCB1 and ABCC2 and their impact on drug disposition. *Curr Drug Targets*, 12, 631–646.

Hegedus T, Sessler T, Scott R, Thelin W, Bakos E, Váradi A, Szabó K, Homolya L, Milgram SL, Sarkadi B (2003) C-terminal phosphorylation of MRP2 modulates its interaction with PDZ proteins. *Biochem Biophys Res Commun*, 302, 454–461.

Hirouchi M, Suzuki H, Itoda M, Ozawa S, Sawada J, Ieiri I, Ohtsubo K, Sugiyama Y (2004) Characterization of the cellular localization, expression level, and function of SNP variants of MRP2/ABCC2. *Pharm Res*, 21, 742–748.

Ho WF, Koo SH, Yee JY, Lee JD (2008) Genetic variations of the ABCC2 gene in the Chinese, Malay, and Indian populations of Singapore. *Drug Metab Pharmacokinet*, 23, 385–391.

Hoffmann K, Gastens AM, Volk HA, Loscher W (2006) Expression of the multidrug transporter MRP2 in the blood–brain barrier after pilocarpine-induced seizures in rats. *Epilepsy Res*, 69, 1–14.

Hosokawa S, Tagaya O, Mikami T, Nozaki Y, Kawaguchi A, Yamatsu K, Shamoto M (1992) A new rat mutant with chronic conjugated hyperbilirubinemia and renal glomerular lesions. *Lab Anim Sci*, 42, 27–34.

Hrebicek M, Jirasek T, Hartmannova H, Noskova L, Stranecky V, Ivanek R, Kmoch S, Cebecauerová D, Vítek L, Mikulecký M, et al. (2007) Rotor-type hyperbilirubinaemia has no defect in the canalicular bilirubin export pump. *Liver Int*, 27, 485–491.

Innocenti F, Kroetz DL, Schuetz E, Dolan ME, Ramirez J, Relling M, Chen P, Das S, Rosner GL, Ratain MJ (2009) Comprehensive pharmacogenetic analysis of irinotecan neutropenia and pharmacokinetics. *J Clin Oncol*, 27, 2604–2614.

Ito K, Suzuki H, Hirohashi T, Kume K, Shimizu T, Sugiyama Y (1997) Molecular cloning of canalicular multispecific organic anion transporter defective in EHBR. *Am J Physiol*, 272, G16–G22.

Ito S, Ieiri I, Tanabe M, Suzuki A, Higuchi S, Otsubo K (2001) Polymorphisms of the ABC transporter genes, MDR1, MRP1 and MRP2/cMOAT, in healthy Japanese subjects. *Pharmacogenetics*, 11, 175–184.

Itoda M, Saito Y, Soyama A, Saeki M, Murayama N, Ishida S, Sai K, Nagano M, Suzuki H, Sugiyama Y, Ozawa S, Sawada J (2002) Polymorphisms in the ABCC2 (cMOAT/MRP2) gene found in 72 established cell lines derived from Japanese individuals: an association between single nucleotide polymorphisms in the 5′-untranslated region and exon 28. *Drug Metab Dispos*, 30, 363–364.

Jansen PLM, Peters WHM, Lamers WH (1985) Hereditary chronic conjugated hyperbilirubinemia in mutants rats caused by defective hepatic anion transport. *Hepatology*, 5, 573–579.

Jedlitschky G, Hoffmann U, Kroemer HK (2006) Structure and function of the MRP2 (ABCC2) protein and its role in drug disposition. *Expert Opin Drug Metab Toxicol*, 2, 351–366.

Kajihara S, Hisatomi A, Mizuta T, Hara T, Ozaki I, Wada I, Yamamoto K (1998) A splice mutation in the human canalicular multispecific organic anion transporter gene causes Dubin–Johnson syndrome. *Biochem Biophys Res Comm*, 253, 454–457.

Kanai N, Lu R, Satriano JA, Bao Y, Wolkoff AW, Schuster VL (1995) Identification and characterization of a prostaglandin transporter. *Science*, 268, 866–869.

Kanda D, Takagi H, Kawahara Y, Yata Y, Takakusagi T, Hatanaka T, Yoshinaga T, Iesaki K, Kashiwabara K, Higuchi T, et al. (2009) Novel large-scale deletion (whole exon 7) in the ABCC2 gene in a patient with the Dubin–Johnson syndrome. *Drug Metab Pharmacokinet*, 24, 464–468.

Kartenbeck J, Leuschner U, Mayer R, Keppler D (1996) Absence of the canalicular isoform of the *MRP* gene-encoded conjugate export pump from the hepatocytes in Dubin–Johnson syndrome. *Hepatology*, 23, 1061–1066.

Keitel V, Kartenbeck J, Nies AT, Spring H, Brom M, Keppler D (2000) Impaired protein maturation of the conjugate export pump multidrug resistance protein 2 as a consequence of a deletion mutation in Dubin–Johnson syndrome. *Hepatology*, 32, 1317–1328.

Keitel V, Nies AT, Brom M, Hummel-Eisenbeiss J, Spring H, Keppler D (2003) A common Dubin–Johnson syndrome mutation impairs protein maturation and transport activity of MRP2 (ABCC2). *Am J Physiol Gastrointest Liver Physiol*, 284, G165–G174.

Keppler D (2011) Multidrug resistance proteins (MRPs, ABCCs): importance for pathophysiology and drug therapy. In: Fromm MF, Kim RB, editors. *Drug Transporters*. Berlin Heidelberg: Springer-Verlag. pp 299–323.

Keppler D, Kartenbeck J (1996) The canalicular conjugate export pump encoded by the *CMRP/CMOAT* gene. In: Boyer JL, Ockner RK, editors. *Progress in Liver Diseases*. Philadelphia: Saunders. pp 55–67.

Klaassen CD, Aleksunes LM (2010) Xenobiotic, bile acid, and cholesterol transporters: function and regulation. *Pharmacol Rev*, 62, 1–96.

Kobayashi K, Ito K, Takada T, Sugiyama Y, Suzuki H (2008) Functional analysis of nonsynonymous single nucleotide polymorphism type ATP-binding cassette transmembrane transporter subfamily C member 3. *Pharmacogenet Genomics*, 18, 823–833.

König J, Hartel M, Nies AT, Martignoni ME, Guo J, Büchler MW, Friess H, Keppler D (2005) Expression and localization of human multidrug resistance protein (ABCC) family members in pancreatic carcinoma. *Int J Cancer*, 115, 359–367.

König J, Nies AT, Cui Y, Keppler D (2003) MRP2, the apical export pump for anionic conjugates. In: Holland IB, Kuchler K, Higgins C, Cole SPC, editors. *ABC Proteins: From Bacteria to Man*. London: Academic Press. pp 423–443.

König J, Nies AT, Cui Y, Leier I, Keppler D. (1999a). Conjugate export pumps of the multidrug resistance protein (MRP) family: localization, substrate specificity, and MRP2-mediated drug resistance. *Biochim Biophys Acta*, 1461, 377–394.

König J, Rost D, Cui Y, Keppler D (1999b) Characterization of the human multidrug resistance protein isoform MRP3 localized to the basolateral hepatocyte membrane. *Hepatology*, 29, 1156–1163.

Kool M, van der Linden M, de Haas M, Scheffer GL, de Vree JM, Smith AJ, Jansen G, Peters GJ, Ponne N, Scheper RJ, et al. (1999) MRP3, an organic anion transporter able to transport anti-cancer drugs. *Proc Natl Acad Sci U S A*, 96, 6914–6919.

Kopplow K, Letschert K, König J, Walter B, Keppler D (2005) Human hepatobiliary transport of organic anions analyzed by quadruple-transfected cells. *Mol Pharmacol*, 68, 1031–1038.

Kubitz R, Häussinger D (2007) Osmoregulation of bile formation. *Methods Enzymol*, 428, 313–324.

Lagas JS, Vlaming ML, Schinkel AH (2009) Pharmacokinetic assessment of multiple ATP-binding cassette transporters: the power of combination knockout mice. *Mol Interv*, 9, 136–145.

Lang T, Hitzl M, Burk O, Mornhinweg E, Keil A, Kerb R, Klein K, Zanger UM, Eichelbaum M, Fromm MF (2004) Genetic polymorphisms in the multidrug resistance-associated protein 3 (ABCC3, MRP3) gene and relationship to its mRNA and protein expression in human liver. *Pharmacogenetics*, 14, 155–164.

Lee JH, Chen HL, Chen HL, Ni YH, Hsu HY, Chang MH (2006) Neonatal Dubin–Johnson syndrome: long-term follow-up and MRP2 mutations study. *Pediatr Res*, 59, 584–589.

Lee YMA, Cui Y, König J, Risch A, Jäger B, Drings P, Bartsch H, Keppler D, Nies AT (2004) Identification and functional characterization of the natural variant MRP3-Arg^{1297}His of human multidrug resistance protein 3 (MRP3/ABCC3). *Pharmacogenetics*, 14, 213–223.

Leschziner G, Zabaneh D, Pirmohamed M, Owen A, Rogers J, Coffey AJ, Balding DJ, Bentley DB, Johnson MR (2006) Exon sequencing and high resolution haplotype analysis of ABC transporter genes implicated in drug resistance. *Pharmacogenet Genomics*, 16, 439–450.

Levesque E, oit-Biancamano MO, Delage R, Couture F, Guillemette C (2008) Pharmacokinetics of mycophenolate mofetil and its glucuronide metabolites in healthy volunteers. *Pharmacogenomics*, 9, 869–879.

Machida I, Wakusawa S, Sanae F, Hayashi H, Kusakabe A, Ninomiya H, Yano M, Yoshioka K (2005) Mutational analysis of the MRP2 gene and long-term follow-up of Dubin–Johnson syndrome in Japan. *J Gastroenterol*, 40, 366–370.

Meier Y, Pauli-Magnus C, Zanger UM, Klein K, Schaeffeler E, Nussler AK, Eichelbaum M, Meier PJ, Stieger B (2006) Interindividual variability of canalicular ATP-binding-cassette (ABC)-transporter expression in human liver. *Hepatology*, 44, 62–74.

Meyer zu Schwabedissen HE, Jedlitschky G, Gratz M, Haenisch S, Linnemann K, Fusch C, Cascorbi I, Kroemer HK (2005) Variable expression of MRP2 (ABCC2) in human placenta: influence of gestational age and cellular differentiation. *Drug Metab Dispos*, 33, 896–904.

Minami S, Ito K, Honma M, Ikebuchi Y, Anzai N, Kanai Y, Nishida T, Tsukita S, Sekine S, Horie T, Suzuki H (2009) Posttranslational regulation of ABCC2 expression by SUMOylation system. *Am J Physiol Gastrointest Liver Physiol*, 296, G406–G413.

Miura M, Satoh S, Inoue K, Kagaya H, Saito M, Inoue T, Suzuki T, Habuchi T (2007) Influence of SLCO1B1, 1B3, 2B1 and ABCC2 genetic polymorphisms on mycophenolic acid pharmacokinetics in Japanese renal transplant recipients. *Eur J Clin Pharmacol*, 63, 1161–1169.

Miura M, Satoh S, Inoue K, Saito M, Habuchi T, Suzuki T (2009) Telmisartan pharmacokinetics in Japanese renal transplant recipients. *Clin Chim Acta*, 399, 83–87.

Moitra K, Dean M (2011) Evolution of ABC transporters by gene duplication and their role in human disease. *J Biol Chem*, 392, 29–37.

Mor-Cohen R, Zivelin A, Rosenberg N, Goldberg I, Seligsohn U (2005) A novel ancestral splicing mutation in the multidrug resistance protein 2 gene causes Dubin–Johnson syndrome in Ashkenazi Jewish patients. *Hepatol Res*, 31, 104–111.

Mor-Cohen R, Zivelin A, Rosenberg N, Shani M, Muallem S, Seligsohn U (2001) Identification and functional analysis of two novel mutations in the multidrug resistance protein 2 gene in Israeli patients with Dubin–Johnson syndrome. *J Biol Chem*, 276, 36923–36930.

Moriya Y, Nakamura T, Horinouchi M, Sakaeda T, Tamura T, Aoyama N, Shirakawa T, Gotoh A, Fujimoto S, Matsuo M, Kasuga M, Okumura K (2002) Effects of polymorphisms of MDR1, MRP1, and MRP2 genes on their mRNA expression levels in duodenal enterocytes of healthy Japanese subjects. *Biol Pharm Bull*, 25, 1356–1359.

Müller P, Asher N, Heled M, Cohen SB, Risch A, Rund D (2008) Polymorphisms in transporter and phase II metabolism genes as potential modifiers of the predisposition to and treatment outcome of de novo acute myeloid leukemia in Israeli ethnic groups. *Leuk Res*, 32, 919–929.

Müller PJ, Dally H, Klappenecker CN, Edler L, Jager B, Gerst M, Spiegelhalder B, Tuengerthal S, Fischer JR, Drings P, Bartsch H, Risch A (2009) Polymorphisms in ABCG2, ABCC3 and CNT1 genes and their possible impact on chemotherapy outcome of lung cancer patients. *Int J Cancer*, 124, 1669–1674.

Naesens M, Kuypers DR, Verbeke K, Vanrenterghem Y (2006) Multidrug resistance protein 2 genetic polymorphisms influence mycophenolic acid exposure in renal allograft recipients. *Transplantation*, 82, 1074–1084.

Nezasa K, Tian X, Zamek-Gliszczynski MJ, Patel NJ, Raub TJ, Brouwer KL (2006) Altered hepatobiliary disposition of 5 (and 6)-carboxy-2',7'-dichlorofluorescein in ABCG2 (BCRP1) and ABCC2 (MRP2) knockout mice. *Drug Metab Dispos*, 34, 718–723.

Niemi M, Arnold KA, Backman JT, Pasanen MK, Godtel-Armbrust U, Wojnowski L, Zanger UM, Neuvonen PJ, Eichelbaum M, Kivistö KT, Lang T (2006) Association of genetic polymorphism in ABCC2 with hepatic multidrug resistance-associated protein 2 expression and pravastatin pharmacokinetics. *Pharmacogenet Genomics*, 16, 801–808.

Nies AT, Keppler D (2007) The apical conjugate efflux pump ABCC2 (MRP2). *Pflugers Arch*, 453, 643–659.

Nies AT, Rius M, König J, Herold-Mende C, Keppler D (2006) Localization and function of ATP-binding cassette C (ABCC/MRP) efflux pumps in human brain. In: Häussinger D, Kircheis G, Schliess F, editors. *Hepatic Encephalopathy and Nitrogen Metabolism*. Dordrecht: Springer. pp 112–123.

Nies AT, Schwab M, Keppler D (2008) Interplay of conjugating enzymes with OATP uptake transporters and ABCC/MRP efflux pumps in the elimination of drugs. *Expert Opin Drug Metab Toxicol*, 4, 545–568.

O'Brien C, Cavet G, Pandita A, Hu X, Haydu L, Mohan S, Toy K, Rivers CS, Modrusan Z, Amler LC, Lackner MR (2008) Functional genomics identifies ABCC3 as a mediator of taxane resistance in HER2-amplified breast cancer. *Cancer Res*, 68, 5380–5389.

Paulusma CC, Bosma PJ, Zaman GJR, Bakker CTM, Otter M, Scheffer GL, Scheper RJ, Borst P, Oude Elferink RP (1996) Congenital jaundice in rats with a mutation in a multidrug resistance associated-protein gene. *Science*, 271, 1126–1128.

Paulusma CC, Kool M, Bosma PJ, Scheffer GL, ter Borg F, Scheper RJ, Tytgat GN, Borst P, Baas F, Oude Elferink RP (1997) A mutation in the human canalicular multispecific organic anion transporter gene causes the Dubin–Johnson syndrome. *Hepatology*, 25, 1539–1542.

Rau T, Erney B, Gores R, Eschenhagen T, Beck J, Langer T (2006) High-dose methotrexate in pediatric acute lymphoblastic leukemia: impact of ABCC2 polymorphisms on plasma concentrations. *Clin Pharmacol Ther*, 80, 468–476.

Rosenberg MF, Oleschuk CJ, Wu P, Mao Q, Deeley RG, Cole SP, Ford RC (2010) Structure of a human multidrug transporter in an inward-facing conformation. *J Struct Biol*, 170, 540–547.

Rost D, König J, Weiss G, Klar E, Stremmel W, Keppler D (2001) Expression and localization of the multidrug resistance proteins MRP2 and MRP3 in human gallbladder epithelia. *Gastroenterology*, 121, 1203–1208.

Sai K, Saito Y, Itoda M, Fukushima-Uesaka H, Nishimaki-Mogami T, Ozawa S, Maekawa K, Kurose K, Kaniwa N, Kawamoto M, et al. (2008) Genetic variations and haplotypes of ABCC2 encoding MRP2 in a Japanese population. *Drug Metab Pharmacokinet*, 23, 139–147.

Saito S, Iida A, Sekine A, Miura Y, Ogawa C, Kawauchi S, Higuchi S, Nakamura Y (2002) Identification of 779 genetic variations in eight genes encoding members of the ATP-binding cassette, subfamily C (ABCC/MRP/CFTR). *J Hum Genet*, 47, 147–171.

Sandusky GE, Mintze KS, Pratt SE, Dantzig AH (2002) Expression of multidrug resistance-associated protein 2 (MRP2) in normal human tissues and carcinomas using tissue microarrays. *Histopathology*, 41, 65–74.

Sasaki M, Suzuki H, Ito K, Abe T, Sugiyama Y (2002) Transcellular transport of organic anions across a double-transfected Madin-Darby canine kidney II cell monolayer expressing both human organic anion-transporting polypeptide (OATP2/SLC21A6) and multidrug resistance-associated protein 2 (MRP2/ABCC2). *J Biol Chem*, 277, 6497–6503.

Schaub TP, Kartenbeck J, König J, Spring H, Dörsam H, Stähler G, Störkel S, Thon WF, Keppler D (1999) Expression of the MRP2 gene-encoded conjugate export pump in human kidney proximal tubules and in renal-cell carcinoma. *J Am Soc Nephrol*, 10, 1159–1169.

Scheffer GL, Kool M, de Haas M, de Vree JM, Pijnenborg AC, Bosman DK, Elferink RP, van der Valk P, Borst P, Scheper RJ (2002a) Tissue distribution and induction of human multidrug resistance protein 3. *Lab Invest*, 82, 193–201.

Scheffer GL, Pijnenborg AC, Smit EF, Muller M, Postma DS, Timens W, van der Valk P, de Vries EG, Scheper RJ (2002b) Multidrug resistance related molecules in human and murine lung. *J Clin Pathol*, 55, 332–339.

Shani M, Seligsohn U, Gilon E, Sheba C, Adam A (1970) Dubin–Johnson syndrome in Israel. I. Clinical, laboratory, and genetic aspects of 101 cases. *Quarterly J Med*, 39, 549–567.

Shoda J, Suzuki H, Suzuki H, Sugiyama Y, Hirouchi M, Utsunomiya H, Oda K, Kawamoto T, Matsuzaki Y, Tanaka N (2003) Novel mutations identified in the human multidrug resistance-associated protein 2 (MRP2/ABCC2) gene in a Japanese patient with Dubin–Johnson syndrome. *Hepatol Res*, 27, 323–326.

Sprinz H, Nelson RS (1954) Persistent nonhemolytic hyperbilirubinemia associated with lipochrome-like pigment in liver cells; report of four cases. *Ann Intern Med*, 41, 952–962.

St-Pierre MV, Serrano MA, Macias RI, Dubs U, Hoechli M, Lauper U, Meier PJ, Marin JJ (2000) Expression of members of the multidrug resistance protein family in human term placenta. *Am J Physiol Regul Integr Comp Physiol*, 279, R1495–R1503.

Szakacs G, Paterson JK, Ludwig JA, Booth-Genthe C, Gottesman MM (2006) Targeting multidrug resistance in cancer. *Nat Rev Drug Discov*, 5, 219–234.

Takikawa H, Sano N, Narita T, Uchida Y, Yamanaka M, Horie T, Mikami T, Tagaya O (1991) Biliary excretion of bile acid conjugates in a hyperbilirubinemic mutant Sprague-Dawley rat. *Hepatology*, 14, 352–360.

Tate G, Li M, Suzuki T, Mitsuya T (2002) A new mutation of the ATP-binding cassette, sub-family C, member 2 (ABCC2) gene in a Japanese patient with Dubin–Johnson syndrome. *Genes Genet Syst*, 77, 117–121.

Toh S, Wada M, Uchiumi T, Inokuchi A, Makino Y, Horie Y, Adachi Y, Sakisaka S, Kuwano M (1999) Genomic structure of the canalicular multispecific organic anion-transporter gene (MRP2/cMOAT) and mutations in the ATP-binding-cassette region in Dubin–Johnson syndrome. *Am J Hum Genet*, 64, 739–746.

Tsujii H, König J, Rost D, Stöckel B, Leuschner U, Keppler D (1999) Exon–intron organization of the human multidrug-resistance protein 2 (MRP2) gene mutated in Dubin–Johnson syndrome. *Gastroenterology*, 117, 653–660.

van de Steeg E, Stránecký V, Hartmannová H, Nosková L, Hřebíček M, Wagenaar E, van Esch A, de Waart DR, Oude Elferink RP, Kenworthy KE, Sticová E, al-Edreesi M, Knisely AS, Kmoch S, Jirsa M, Schinkel AH (2012). Complete OATP1B1 and OATP1B3 deficiency causes human Rotor syndrome by interrupting conjugated bilirubin reuptake into the liver. *J Clin Invest*, 122, 519–528.

Vlaming ML, Mohrmann K, Wagenaar E, de Waart DR, Oude Elferink RP, Lagas JS, van Tellingen O, Vainchtein LD, Rosing H, Beijnen JH, Schellens JH, Schinkel AH (2006). Carcinogen and anti-cancer drug transport by MRP2 in vivo: studies using MRP2 (ABCC2) knockout mice. *J Pharmacol Exp Ther*, 318, 319–327.

Wada M, Toh S, Taniguchi K, Nakamura T, Uchiumi T, Kohno K, Yoshida I, Kimura A, Sakisaka S, Adachi Y, Kuwano M (1998) Mutations in the canalicular multispecific organic anion transporter (cMOAT) gene, a novel ABC transporter, in patients with hyperbilirubinemia II/Dubin–Johnson syndrome. *Hum Mol Genet*, 7, 203–207.

Wakusawa S, Machida I, Suzuki S, Hayashi H, Yano M, Yoshioka K (2003) Identification of a novel 2026G→C mutation of the MRP2 gene in a Japanese patient with Dubin–Johnson syndrome. *J Hum Genet*, 48, 425–429.

Wolkoff AW, Wolpert E, Pascasio FN, Arias IM (1976) Rotor's syndrome. A distinct inheritable pathophysiologic entity. *Am J Med*, 60, 173–179.

Zamek-Gliszczynski MJ, Nezasa K, Tian X, Bridges AS, Lee K, Belinsky MG, Kruh GD, Brouwer KL (2006) Evaluation of the role of multidrug resistance-associated protein (MRP) 3 and MRP4 in hepatic basolateral excretion of sulfate and glucuronide metabolites of acetaminophen, 4-methylumbelliferone, and harmol in ABCC3-/- and ABCC4-/- mice. *J Pharmacol Exp Ther*, 319, 1485–1491.

Zelcer N, Saeki T, Reid G, Beijnen JH, Borst P (2001) Characterization of drug transport by the human multidrug resistance protein 3 (ABCC3). *J Biol Chem*, 276, 46400–46407.

Zeng H, Bain LJ, Belinsky MG, Kruh GD (1999) Expression of multidrug resistance protein-3 (multispecific organic anion transporter-D) in human embryonic kidney 293 cells confers resistance to anticancer agents. *Cancer Res*, 59, 5964–5967.

Zhang WX, Chen B, Jin Z, Yu Z, Wang X, Chen H, Mao A, Cai W (2008) Influence of uridine diphosphate (UDP)-glucuronosyltransferases and ABCC2 genetic polymorphisms on the pharmacokinetics of mycophenolic acid and its metabolites in Chinese renal transplant recipients. *Xenobiotica*, 38, 1422–1436.

Zhang Y, Schuetz JD, Elmquist WF, Miller DW (2004) Plasma membrane localization of multidrug resistance-associated protein homologs in brain capillary endothelial cells. *J Pharmacol Exp Ther*, 311, 449–55.

Zhang Y, Zhao T, Li W, Vore M (2010) The 5′-untranslated region of multidrug resistance associated protein 2 (MRP2; ABCC2) regulates downstream open reading frame expression through translational regulation. *Mol Pharmacol*, 77, 237–246.

CHAPTER 16

MRP4 (ABCC4)

Satish B. Cheepala
Mugdha Sukthankar
John D. Schuetz

16.1 INTRODUCTION

During a functional screen for other ABCC/MRP drug resistant proteins, ABCC4/MRP4 was identified as a member of the MRP family that was not associated with acquired resistance to doxorubicin or cisplatin (Kool et al., 1997). ABCC4 remained an orphan drug transporter until 1999, when it was identified as the first transporter that effluxed nucleoside monophosphate derivatives (Schuetz et al., 1999). This finding provided the opportunity to test the idea that a specific transporter was required for the cellular efflux of natural and drug-derived nucleotide monophosphates, a phenomenon that had been initially described for cAMP (Ho and Sutherland, 1975).

The *MRP4* gene, located on chromosome 13q32.1, encodes the smallest (i.e., 1325 amino acids) member of the ABCC subfamily and is composed of 31 exons (Fig. 16.1). Two additional exons (exons 1a and 1b) are within the large intron 1 (Lamba et al., 2003). It has been speculated that to reduce MRP4 expression, some tissues may use an unknown mechanism to select these exons more frequently (Lamba et al., 2003) as transcripts containing either one or both of these exons producing premature termination codons, with the resulting transcripts eliminated by a process of "nonsense-mediated decay." It is unknown if the nonsense-mediated decay of MRP4 is regulated or occurs in a tissue-specific fashion. The domain organization of MRP4 is typical of ABC transporters, with a core structure composed of two membrane-spanning domains, each consisting of six transmembrane helices with two cytosolic ATP-binding domains, which bind and hydrolyze ATP to power substrate transport (Sauna et al., 2004; Deeley et al., 2006; Borst et al., 2007; Lamba et al., 2003; Fig. 16.1).

16.2 LOCALIZATION OF MRP4

MRP4 appears ubiquitously expressed in many tissues, including the prostate (Lee et al., 1998, 2000), liver (Lee et al., 1998), testis (Lee et al., 1998), ovary (Lee et al.,

Pharmacogenomics of Human Drug Transporters: Clinical Impacts, First Edition.
Edited by Toshihisa Ishikawa, Richard B. Kim and Jörg König

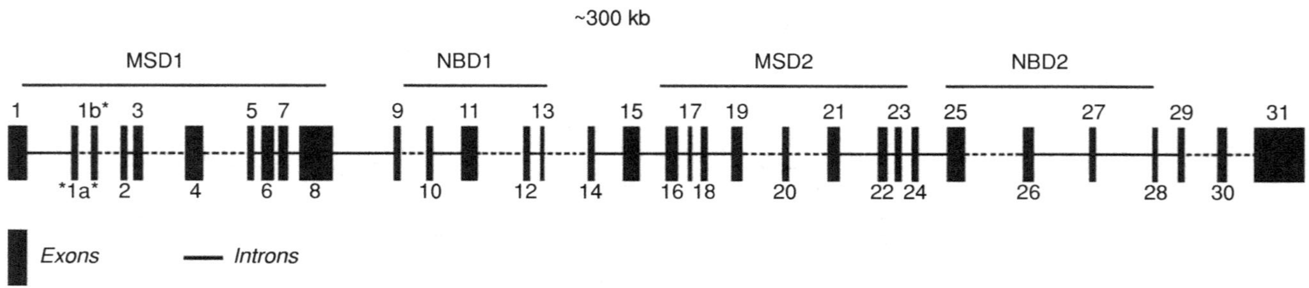

Figure 16.1 ABCC4 has 33 exons, but exons 1a and/or 1b would not encode functional ABCC4 (relative distance between exons is indicated by lines).

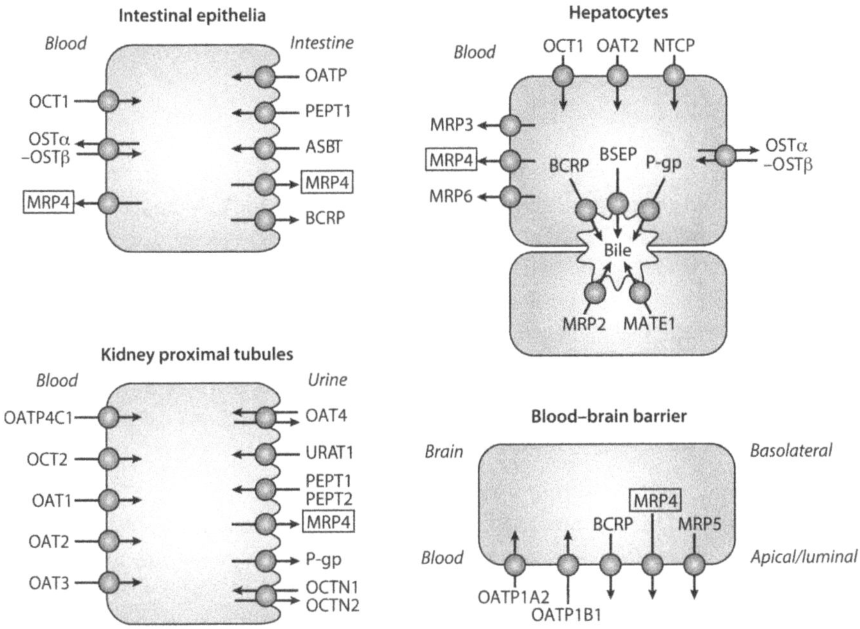

Figure 16.2 MRP4 localization relative to other transporters. (a) Intestinal epithelia, (b) hepatocytes, (c) kidney proximal tubules, and (d) brain capillary endothelial cells are represented.

1998), brain (Kool et al., 1997), kidney (Lee et al., 1998), and adrenal gland (Chen et al., 2001). Among the ABCC subfamily, MRP4 is unique, because it is capable of localizing to either apical or basolateral membranes in cells that polarize. Recent studies have identified protein interactions in MRP4 c terminus that contributes to its localization (Hoque and Cole, 2008; Hoque et al., 2009). The apical and basolateral locations of transporters in the liver, kidney, intestine, and blood–brain barrier are depicted in Figure 16.2.

16.2.1 Prostate

MRP4 has been detected in prostate in the basolateral membrane of tubuloacinar cells (Lee et al., 2000). MRP4 is prominently expressed in the basal cells of the prostatic glandular epithelium where it localizes to the basolateral membrane. On the basis of this location, it was speculated to efflux xenobiotics out of prostatic epithelial cells.

There have been conflicting reports about the expression of ABCC4/MRP4 in malignant and benign prostate cancer. A gene expression profile study shows reduced MRP4 expression as prostate cancer progresses from benign to malignant (Tomlins et al., 2007). Because MRP4 is expressed in normal prostate epithelia (Lee et al., 2000; Cai et al., 2007), this finding suggests that loss of MRP4 expression might contribute to disease progression. However, this finding contrasts with a recent study showing that MRP4 expression is greater in malignant prostate cancer than in benign

(Ho et al., 2008). The reason for the discrepancy between these studies is not known at this time.

16.2.2 Liver

MRP4 is localized at the sinusoidal (i.e., basolateral) membrane of hepatocytes, a location that would permit the transport of both endogenous and exogenous molecules into the blood from the liver. The protective role of MRP4 against bile acid overload was revealed in a model of cholestasis in the MRP4-knockout mouse (Mennone et al., 2006). After bile duct ligation, these MRP4-knockout mice had elevated transaminases and more extensive morphological evidence of liver damage compared to wild-type mice.

16.2.3 Kidney

In mouse kidney, MRP4 is more abundant in females than in males, a finding that appears due to the suppressive influence of androgens (Maher et al., 2006). However, it is unknown if this gender-predominant expression occurs in humans. Immunofluroscence microscopic analyses of human kidney indicates that the distal tubules are devoid of MRP4 staining whereas the proximal tubules exhibit prominent MRP4 staining on the brush border (van Aubel et al., 2002b). This is an ideal location for MRP4 to export anionic drugs into the urine. The secretory pathway of potentially toxic anionic drugs, appears to be rate limiting, with the nephrotoxic side effects of antiviral drug treatment appears to be related to renal secretion by MRP4 (Adachi et al., 2002; Russel et al., 2002).

The release of cAMP into the urine has been proposed as an important step in generating adenosine which modulates renal fluid homeostasis. Currently, it is unknown if efflux of cAMP by MRP4 influences renal water and salt homeostasis (van Aubel et al., 2002b; Hofer and Lefkimmiatis, 2007).

16.2.4 Brain

Studies of the MRP4-knockout mouse revealed MRP4's role in the blood–brain barrier and how its ability to localize apically and basolaterally has been exploited to protect the brain. In the brain capillaries, MRP4 is expressed luminally whereas in the choroid plexus epithelium MRP4 localizes to the basolateral membrane (Leggas et al., 2004). As expected, in MRP4-knockout mice, brain and cerebrospinal fluid concentrations of the anticancer drug, topotecan, were higher than in the wild type (Leggas et al., 2004). This study also established that MRP4's expression confers resistance to the camptothecin and topotecan (Leggas et al., 2004), and that penetration of the antiviral drug, adefovir, into the brain is increased in the MRP4-knockout mouse (Belinsky et al., 2007). These findings are likely to be relevant to humans, as MRP4 is expressed in the CNS at the capillary endothelium (Nies et al., 2004).

Oseltamivir, an ester-type prodrug of Ro 64-0802, is a potent and selective inhibitor of viral neuraminidase, a key enzyme in the release of influenza virus from host cells. Recent studies show that oseltamivir's brain penetration is limited by

MRP4-mediated active efflux (Ose et al., 2009). This finding is notable, because one of the side effects of oseltamivir is neurotoxicity. Based on this finding, one might speculate that defects in human MRP4 function (see Genetic Polymorphisms of MRP4 below) might contribute to the neurotoxicity of oseltamivir. As reduced brain penetration of antiviral nucleoside and anticancer drugs severely impairs therapy in many diseases of the CNS, the development of specific MRP4 inhibitors might increase drug penetration into the brain and improves therapeutic efficacy of many drugs that are substrates of MRP4.

16.2.5 Hematopoietic Cells

MRP4 is expressed on the plasma membrane of $CD34^+$ hematopoietic progenitor cells and its expression significantly increases as $CD34^+$ hematopoietic progenitors differentiate into megakaryocytes (Oevermann et al., 2009). Additionally, MRP4 expression is significantly decreased during differentiation into mature leukocytes. MRP4 is also expressed in acute myeloid leukemia blast cells, with the highest expression being in the least differentiated subtype of the disease (Guo et al., 2009). In total, these findings are intriguing and suggest that MRP4 function may play a role in hematopoietic differentiation.

Unexpectedly, in platelets, MRP4 primarily localizes to an intracellular organelle, the dense granules (Jedlitschky et al., 2004). On the basis of these observations, it was proposed that MRP4 contributes to the dense granule function playing a role in the intracellular storage and the release of mediators of platelet aggregation (ADP, Serotonin, etc.). One potential explanation for this atypical intracellular localization of MRP4 is that this study immobilized platelets on slides coated with collagen. This effectively initiates platelet activation due to collagen interaction with its receptor, GPV1 (Jedlitschky et al., 2004). Platelet activation mobilizes the intracellular dense (containing P-selectin) granules to the plasma membrane which may be the reason why MRP4 appears to be in the dense granules.

16.2.6 Intestine

In the intestine, MRP4 appears to localize both the apical and basolateral membranes; however, it appears more abundant apically (Zimmermann et al., 2005; Li et al., 2007). As one moves down the gut, the abundance increases showing threefold greater amount in the colon versus the ileum (Zimmermann et al., 2005).

16.3 ROLE OF MRP4 IN DRUG EFFLUX

16.3.1 Nucleotide Drug Efflux

Nucleoside analogs have an essential role in front-line therapy for many human diseases, including solid and hematologic malignancies, HIV, and other viral infections, as well as autoimmune and inflammatory diseases, such as ulcerative colitis and

Crohn's disease. Conversion of these nucleoside analogs into nucleotides is an essential determinant of drug efficacy. For example, the antiretroviral drug, azidothymidine (AZT), is taken up into a cell and converted by nucleoside kinases to mono-, di-, and triphosphate forms. The formation of AZT triphosphate is critical to AZT effectiveness; however, the export of any one of these nucleotides would undoubtedly reduce AZT efficacy.

Prior to the identification of MRP4/ABCC4 as a mediator of nucleotide efflux, the process was studied and noted that nucleoside drug efficacy might be impaired by active efflux of their nucleotide monophosphate forms. This concept was confirmed when Plagemann first demonstrated in 1976 that the nucleoside monophosphate form of tricyclic 7-deazapurine (TCN) was actively effluxed (Plagemann, 1976). The export TCN monophosphate was related to its cytotoxicity (Plagemann, 1976). Notably, TCN monophosphate was effluxed from cells and its export revealed properties such as saturable kinetics, inhibition by depletion of intracellular ATP, and inhibition by the broad range organic anion transport inhibitor, probenecid. This finding supported the concept of a specific exporter of nucleoside monophosphates (Plagemann and Erbe, 1977).

16.3.2 Antiviral Drug Efflux

16.3.2.1 3′-azido-3′dioxythymidine (AZT) In the 1980s, a number of cell lines were found to secrete AZT monophosphate, but with varying efficiency (Frick et al., 1988). In fact, a reciprocal relationship between intracellular AZT monophosphate concentration and extracellular AZT monophosphate was found such that, as intracellular pools of AZT monophosphate declined, extracellular AZT monophosphate accumulated. As with TCN, di- and triphosphate nucleotide derivatives of AZT were not found in the extracellular medium (Frick et al., 1988). Furthermore, AZT-monophosphate secretion was not associated with cell death or cell leakage of metabolites (Fridland et al., 1990).

16.3.2.2 9-(2-phosphonylmethoxyethyl) adenine (PMEA) Further confirmatory evidence of nucleoside monophosphate drug efflux was revealed by the development of antiviral drugs that were chemically still stable monophosphorylated nucleotide analogs such as PMEA. Cells with acquired resistance to PMEA's cytotoxicity accumulated less drug (Robbins et al., 1995; Hatse et al., 1998). Moreover, additional cell lines with acquired resistance to PMEA had enhanced PMEA export that was attenuated by inhibitors of drug efflux (e.g., indomethacin) (Hatse et al., 1998; Sampath et al., 2002). In total, these facts indicated the existence of a biological export process for monophosphorylated nucleoside drugs that required cellular energy and was inhibited by known efflux inhibitors.

MRP4's role in the active efflux of several antiviral drugs such as zidovudine, ganciclovir, adefovir, and tenofovir has been summarized in a recent review (Errasti-Murugarren and Pastor-Anglada, 2010; Table 16.1). Because these drugs can cause nephrotoxicity, it is important to know how these drugs are eliminated by the kidneys. The antiviral drug, adefovir, was first shown to be excreted into the urine by a probenecid-sensitive transport mechanism (Cundy, 1999). The confirmation that

TABLE 16.1 Physiological and Xenobiotic Substrates of MRP4 (ABCC4)

Xenobiotics	Km (μM)	Reference	Physiological substrates	Km (μM)	Reference
Methotrexate	220, 1300	(van Aubel et al., 2002; van Aubel et al., 2005)	cGMP	10	(van Aubel et al., 2005)
Leucovorin	640	(Chen et al., 2002)	cAMP	45	(Chen et al., 2001)
Topotecan	ND	(Leggas et al., 2004)	ADP	ND	(Jedlitschky et al., 2004)
Adefovir	>1000	(Imaoka et al., 2007)	Urate	1500	(van Aubel et al., 2005)
Ceftizoxime	18	(Ci et al., 2007)	PGE1	2.1	(Reid et al., 2003)
Tenofovir	>1000	(Ray et al., 2006)	PGE2	3.4, 3.5	(Reid et al., 2003)
Cefazolin	80	(Takenaka et al., 2007)	PGF2a	12.6	(Rius et al., 2005)
Cefmetazole	ND	(Uchida et al., 2007)	TXB2	9.9	(Rius et al., 2005)
Hydrochlorothiazide	ND	(Hasegawa et al., 2007)	LTB4	5.2–5.6	(Rius et al., 2008)
Furosemide	ND	(Hasegawa et al., 2007)	LTC4	0.1–0.3	(Rius et al., 2008)
Olmesartan	26	(Yamada et al., 2007)	Folic acid	170	(Chen et al., 2002)
Edaravone Glucuronide	9.8	(Mizuno et al., 2007)	Cholate	14.8	(Rius et al., 2006)
PAH	160	(Smeets et al., 2004)	Cholylglycine	25.8	(Rius et al., 2006)
Nucleoside Derived Drugs			Deoxycholylglycine	6.7	(Rius et al., 2006)
Cladribine	ND	(de Wolf et al., 2007, Reid et al., 2003)	Chenodeoxycholylglycine	3.6	(Rius et al., 2006)
6-mercaptopurine	ND	(Zhang et al., 2004, Ritter et al., 2005, de Wolf et al., 2007)	$E_2$17bG	30	(Zelcer et al., 2003)
6-thioguanine	ND				

PAH, p-aminohippuric acid; LT, leukotrienes; PG, prostaglandin; TXB2, thromboxane B2; $E_2 17\beta G$, estradiol 17β-D-glucuronide; cAMP, cyclic adenosine monophosphate; cGMP, cyclic guanosine monophosphate.

MRP4 effluxed the PMEA-derived drugs, adefovir and tenofovir, from the kidneys was provided by studies of the MRP4-knockout mice, which showed reduced renal clearance of adefovir and tenofovir along with the elevated concentrations of these drugs in the kidneys of the MRP4-knockout mice (Imaoka et al., 2007). Notably, accumulation of the monophosphorylated form of adefovir (i.e., PMEA), which is associated with nephrotoxicity is significantly greater in the kidneys of the MRP4-knockout mouse than in the wild-type mouse. This difference is not attributable to altered nucleotide metabolism because the ratio of adefovir to its metabolites is almost identical in MRP4-knockout and wild-type mice. Given these facts and the evidence that nephrotoxicity is frequently dose limiting for PMEA-derived drugs (e.g., adefovir, tenofovir), renal toxicity may be due to concomitant use of MRP4 inhibitors (e.g., NSAIDS) or due to functionally impaired alleles of MRP4 (Abla et al., 2008; Krishnamurthy et al., 2008).

16.3.3 Anticancer Drug Efflux

16.3.3.1 6-mercaptopurine (6-MP) and 6-thioguanine (6-TG) Some cancer chemotherapeutic drugs, such as 6-MP and 6-TG, form nucleoside monophosphates that are MRP4 substrates (Elion, 1989; Wielinga et al., 2002, 2003; Peng et al., 2008). The possibility that MRP4 might impair the antileukemic efficacy of thiopurines is provided by studies in a T-cell lymphoblastic leukemia cell line that acquired MRP4 overexpression concurrent with its resistance to 6-MP and 6-TG (Peng et al., 2008). This suggests that MRP4-mediated export of 6-MP-derived nucleotide monophosphate derivatives might account for impaired therapeutic response. As 6-MP and 6-TG are widely used during maintenance therapy for lymphoblastic leukemia, it is possible that relapse might occur secondary to acquired MRP4 overexpression.

16.3.3.2 Ganciclovir The guanine analog, ganciclovir, is used in gene therapy to kill cancer cells that have been modified by herpes simplex virus thymidine kinase (HSV-TK), which converts ganciclovir to its active nucleotide form. However, the cytotoxic effects of ganciclovir on tumors modified with HSV-TK are variable, even when equal levels of the virally derived thymidine kinase are found within the cells (Beck et al., 1995; Yang et al., 1998). One explanation was variable efflux of ganciclovir nucleotide. In support of this was the finding that enhanced levels of MRP4 dramatically decreases ganciclovir ability to kill these HSV-TK modified cells, a finding that was directly related to the enhanced efflux of ganciclovir nucleotides (Adachi et al., 2002; Sampath et al., 2002).

16.4 MRP4 SUBSTRATES OVERLAP WITH OTHER ABC TRANSPORTERS

16.4.1 Substrates of ABCC2 and ABCG2

Some MRP2 substrates such as methotrexate and estradiol-17β-D-glucuronide are also transported by MRP4 (Chen et al., 2001; van Aubel et al., 2002a). In some cases,

MRP4 is more effective than MRP2, as shown in the example of *p*-aminohippurate (Smeets et al., 2004). MRP4 is also capable of transporting some loop diuretics and thiazides, such as furosemide and hydrochlorothiazide. Studies in MRP4-knockout mice demonstrate that although the total clearance of hydrochlorothiazide is similar to that of wild-type mice, the renal clearance of hydrochlorothiazide and furosemide was reduced by approximately 40% in the knockout mice (Hasegawa et al., 2007). In contrast, absence of ABCG2 had no impact on excretion of these diuretics. Thus far, these findings indicate that MRP4, along with unknown transporters, accounts for the renal excretion of the diuretics, hydrochlorothiazide and furosemide (Table 16.1).

16.4.2 Cephalosporins

Most cephalosporins undergo active secretion into the urine in an unchanged form. As MRP4 is highly expressed in the renal proximal tubules (van Aubel et al., 2002a; Leggas et al., 2004), it is likely to participate in tubular secretion. Furthermore, it seemed unlikely that cephalosporins could be luminally effluxed by facilitated diffusion, because probenecid, an inhibitor of organic anion transporters, reduced renal elimination (Lepsy et al., 2003). While membrane vesicle transport experiments showed that cephalosporins such as ceftizoxime, cefazolin, cefmetazole, and cefotaxime were MRP4 substrates, the MRP4-knockout mouse provided a definitive evidence that ceftizoxime and cefazolin are actively effluxed by MRP4 (Ci et al., 2007).

16.4.3 Edaravone

Edaravone, a potent radical scavenger, has been suggested to protect against cerebral and myocardial ischemia–reperfusion injury in various experimental models (Watanabe et al., 1994; Yamamoto et al., 1996). However, acute renal failure or dysfunction was reported in patients after edaravone treatment (Hishida, 2007). Because the renal clearance of edaravone sulfate and edaravone glucuronide was greater than their glomerular filtration, it was suggested that edaravone undergoes tubular secretion. While both ABCG2 and MRP4 export edaravone metabolites, vesicle transport studies demonstrated that MRP4 transported edaravone glucuronide whereas ABCG2 transported edaravone sulfate. These studies were extended to show that maximal excretion of edaravone glucuronide required MRP4 (Mizuno et al., 2007).

16.5 ROLES OF MRP4 IN PHYSIOLOGICAL SUBSTRATE EFFLUX

16.5.1 cAMP

MRP4 has recently been identified as a strong regulator of intracellular concentrations of the cyclic nucleotide cAMP (Li et al., 2007; Sassi et al., 2008). This intracellular secondary messenger is synthesized from ATP by adenylyl cyclase in response to receptor activation, and has a critical role in regulating signaling cascades. The initiation of these cAMP-signaling cascades can be mediated by either local membrane changes or overall intracellular changes in cAMP concentrations. Since the discovery of cAMP, research has focused on identifying pathways in its synthesis, metabolism, and signaling pathways. In the 1970s, cAMP was implicated as a regulator of cell growth. Increasing intracellular cAMP concentrations induced cell death and cell cycle arrest in both normal and malignant lymphoid cells (Abell and Monahan, 1973; Pastan, 1975; Pastan et al., 1975; Friedman et al., 1976). Tight regulation of intracellular cyclic nucleotide concentration is vital, because uncontrolled intracellular cyclic nucleotide production can cause hyperactivation of pathways leading to cytotoxicity (Beavo, 1995). Increased intracellular cyclic nucleotide concentrations are believed to be restored to normal via hydrolysis by phosphodiesterases (Jackson and Raghvendra, 2004). However, the capability to reduce intracellular cAMP concentrations by phosphodiesterases varies substantially among various cell types, and efflux has been proposed as a regulatory mechanism (Orlov and Maksimova, 1999). This is supported by membrane vesicle transport studies that show MRP4 has a relatively high affinity for both cAMP and cGMP, with affinity constants of 45 μM and 10 μM, respectively (Chen et al., 2001). However, in other cases, MRP4 has been reported to have a low affinity for cyclic nucleotides (>500 μM) and, therefore, cyclic nucleotide efflux may only be a backup or overflow system (Wielinga et al., 2003). This proposition seems supported by studies with MRP4-expressing cells that show, without external activation of adenylate cyclase, that extracellular cAMP concentrations are not significantly different than those in non–MRP4-expressing cells in the absence of adenylate cyclase activation (Lai and Tan, 2002). However, upon adenylate cyclase activation, an almost 15-fold increase in the ratio of extracellular to intracellular cAMP occurs in the MRP4-expressing cells compared to cells lacking MRP4.

It is possible that MRP4 alters local concentrations of cAMP within membrane microdomains, without dramatically affecting the overall intracellular concentration of cAMP (Li et al., 2007). This seems likely, because cAMP is poorly diffusible (Rich et al., 2000). This is supported by recent evidence showing MRP4 localizing with caveolins (Sassi et al., 2008), integral membrane proteins that can act as a scaffold for membrane proteins in microdomains. By concentrating MRP4 in microdomains, caveolins might facilitate MRP4's potential to regulate local membrane cAMP concentrations. This view is consistent with an evidence that MRP4 inhibition produced a preferential rise in cAMP accumulation in local regions of the plasma membrane (Li et al., 2007). Collectively, MRP4 is likely a regulator of both local concentrations of cAMP and global cellular levels when intracellular concentrations rise above a certain threshold.

The role of MRP4 in cyclic nucleotide-mediated cell signaling processes has recently been evaluated in rat and human smooth muscle cell proliferation models (Sassi et al., 2008). These authors showed that MRP4 was highly expressed in human and rodent coronary smooth muscle endothelial cells. Importantly, siRNA knockdown of MRP4 significantly increased intracellular cAMP and cGMP concentrations, and this increase was sufficient to block endothelial cell proliferation and prevent neointimal growth in injured rat carotid arteries. This antiproliferative effect of MRP4 inhibition was related to elevated intracellular cAMP producing activation of the protein kinase A/CREB-signaling pathway (Sassi et al., 2008).

16.5.2 Arachidonate Metabolites

Prostanoids are arachidonate metabolites that are potent mediators of inflammatory responses and studies show that MRP4 effluxes some prostanoids (Table 16.1). Specifically, MRP4, unlike MRP1–3 and 5, is a high-affinity transporter of PGE_1 and PGE_2 (Reid et al., 2003). Because nonsteroidal anti-inflammatory drugs (NSAID) are potent inhibitors of MRP4, release of prostaglandins from cells might be regulated by MRP4 to affect inflammatory response. Recent studies with the MRP4-knockout mice strongly support this idea because these mice have fewer plasma PGE metabolites (consistent with reduced production extracellular PGE_2) and a greater inflammatory pain threshold than wild-type mice (Lin et al., 2008) suggesting MRP4 appears to play an indispensable role in regulating PGE_2 (Lin et al., 2008). The proinflammatory leukotrienes, LTB4 and LTC4, are also synthesized from arachidonate by the lipoxygenase pathway. While LTC4 is effluxed by MRP1 and has a role in inflammatory response, (Cole and Deeley, 2006) recent studies show that MRP4 transports LTB4 (Rius et al., 2005).

16.5.3 Bile Salts

Cholestatic liver injury causes adaptive changes in expression of some bile acid and organic solute transporters that may provide protection against bile acid overload (Trauner and Boyer, 2003; Boyer, 2004). Hepatic upregulation of MRP4 might act as a compensatory mechanism to reduce intracellular hepatic bile acids that accumulate in a murine model of cholestasis (mice lacking the bile acid-activated nuclear receptor, FXR (Sinal et al., 2000)), as MRP4 was the only transporter with dramatically increased levels (Schuetz et al., 2001; Wagner et al., 2003). Based on these findings, bile acid conjugates were shown to be the substrates of human MRP4 (Rius et al., 2003; Zelcer et al., 2003). MRP4 might also protect humans from extensive cholestatic injury, because children diagnosed with progressive familial intrahepatic cholestasis subtypes 2 and 3 (PFIC2 and PFIC3) showed elevated MRP4 mRNA and protein (Keitel et al., 2005). Further studies have confirmed that MRP4 mRNA and protein upregulation is an important adaptive response to increased hepatic bile salt concentrations in cholestasis (Marschall et al., 2005).

16.6 ROLE OF GENETIC POLYMORPHISMS OF MRP4

As described above, *MRP4* gene is among the most polymorphic gene among the ABC transporters (Saito et al., 2002; Table 16.2; Fig. 16.3). This may be related to its size as it spans more than 300 kb and has 31 exons and 30 introns, with two additional highly conserved exons, 1a and 1b, in the large intron 1. Notably, during splicing of the primary MRP4 transcript, inclusion of exon 1a, exon 1b, or both will result in the formation of a MRP4 cDNA that includes premature termination codons, which

TABLE 16.2 MRP4 (ABCC4) Single-nucleotide Polymorphisms. Location, Allele Frequency, and Functional Effects

Gene	Nucleotide position and change	Amino acid change	Effect	Frequency				Reference
				CEU	YRI	JPT	CHB	
ABCC4 (MRP4)	52 A>C	I18L	NA	0.032	0.021	0.32	0.44	
	232 C>G	P78A	↓ AZT uptake	0	0	0	0.011	
	511T>G	C171G	NA	0	0	0	NA	
	559 G>T	G187W	↓ AZT uptake and PMEA uptake	0	0	0	NA	(Abla et al., 2008)
	877A>G	K293E	NA	0	0	0	NA	
	912G>T	K304N	NA	0.023	0.087	0.295	0.156	
	1208 C>T	P403L	↓ AZT uptake	0.009	0.003	0.005	0.007	(Ekaitz Errasti-Murrugarrent, 2010)
	1460 G>A	G487E	NA	NA	NA	NA	NA	
	1492A>G	L498E	↓ AZT uptake and PMEA uptake	NA	NA	NA	0.007	
	1592G>A	N531G	NA	NA	NA	NA	NA	
	1667 A>G	Y556C	↑ PMEA uptake	NA	NA	NA	NA	(Janke et al.)
	2230A>G	M744V	NA	0	0.078	0.004	0.011	

2269 G>A	E857K	↓ Activity	0.025	0.049	0.32	0.084	(Krishnamurthy et al., 2008)
2326 G>A	V776I	↑ 6-mercaptopurine uptake	NA	NA	NA	NA	
2560G>T	V854F	NA	NA	NA	NA	NA	(Abla et al., 2008)
2569A>G	I866V	NA	NA	NA	NA	NA	
2867 G>C	C956S	↑ PMEA uptake	NA	NA	NA	NA	(Abla et al., 2008)
3425C>T	T1142M	NA	NA	NA	NA	NA	
3463 A>G	Unchanged	May affect protein expression	NA	NA	NA	NA	(Kiser et al., 2008)
3724 G>A	Altered mRNA splicing	NA	NA	NA	NA	NA	(Anderson et al., 2006)
4131 T>G	Altered mRNA splicing	AZT uptake	NA	NA	NA	NA	(Anderson et al., 2006)

NA, allele frequency data not available.

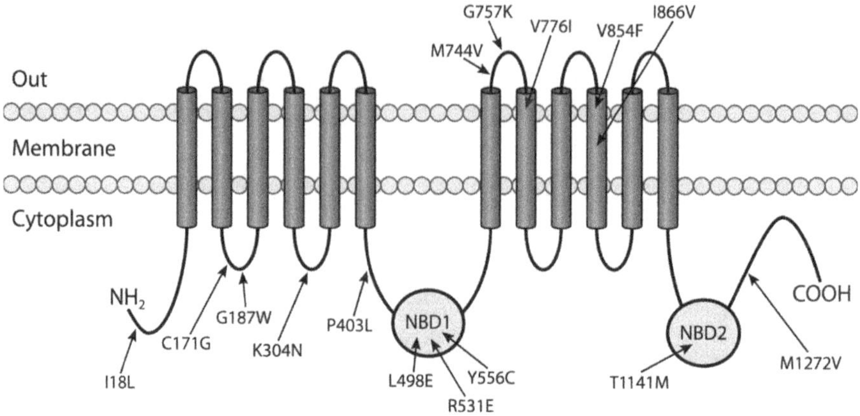

Figure 16.3 Localization of nonsynonymous polymorphisms in MRP4 based on a predicted membrane topology. Arrows indicate approximate location of the polymorphisms.

undergo nonsense-mediated decay (Lamba et al., 2003). Attempts to determine if there was a genetic basis for selection of exon 1a or exon 1b were not successful suggesting genetic variation in the *MRP4* gene did not account for the inclusion of exons 1a and 1b or the variation in MRP4 expression (Lamba et al., 2003).

16.6.1 Nonsynonymous SNPs

Recently, several studies evaluated the role of *MRP4* SNPs on drug disposition. In a cohort of healthy individuals from different ethnic groups, 98 variants were found in the coding and flanking intronic regions of the *MRP4* gene. Among these, 43 variants were in the coding region of the gene, with 22 being nonsynonymous changes producing an amino acid (Abla et al., 2008). The nonsynonymous SNPs are distributed throughout the MRP4, but intriguingly a majority of these are in domains that are, in fact, important for ABC transporter's function: the intracellular loops and nucleotide-binding domains. The nonsynonymous variants G187W and G487E are in an intracellular loop and nucleotide-binding domain, respectively. These variants are significantly less able to transport AZT and PMEA than the reference MRP4 allele. While reduced expression of the G187W mutant appears to account for reduced function, the mechanism accounting for loss of function in G487E is unknown (Abla et al., 2008).

A comprehensive study of MRP4 variability was conducted on 95 liver samples from cholestatic patients or their healthy peers. Significant increases in *MRP4*'s mRNA and protein expression were seen in the cholestatic livers (Gradhand et al., 2007). To determine if the variability in MRP4 expression was due to genetic variation, 2 kb of the 5′ flanking region, all exons, and the exon/intron boundaries of the *MRP4* genes of all 95 patients were sequenced. These investigators tallied 74 variations distributed in the exons, exon/intron boundaries, promoter, and the 3′ untranslated region of *MRP4*. Nonsynonymous changes were identified in exons 1, 5, 12, 13, 19, 21, and 28 to produce the following amino acid changes: I18L, N531E, Y556C,

V1776I, V854F, I866V, and T1142M. However, among these nonsynonymous SNPs, none of the polymorphisms were associated with altered MRP4 protein expression or its basolateral localization in human liver.

Thiopurines are effective immunosuppressants and anticancer agents, but intracellular accumulation of their active metabolites (i.e., 6-TG nucleotides [6-TGN]) causes dose-limiting hematopoietic toxicity. Typically, the susceptibility to thiopurine cytotoxic effects (e.g., bone marrow suppression) is dependent on interindividual variation in the thiopurine S-methyltransferase gene (Elion, 1989). However, some patients receiving mercaptopurines are highly sensitive for unknown reasons (Gearry et al., 2004; Gisbert and Gomollon, 2008). MRP4 transports mercaptopurine monophosphates (Wielinga et al., 2002) and cells engineered to overexpress MRP4 are resistant to thiopurine cytotoxicity (Chen et al., 2001; Lai and Tan, 2002). Hematopoietic toxicity is one possibility from mercaptopurines, and MRP4-knockout mice are highly susceptible to hematopoietic toxicity due to accumulation of 6-TGN in their myelopoieitic cells. Based upon these findings and the anecdotal reports of increased thiopurine sensitivity among Japanese patients, a genetic analysis of MRP4 alleles was conducted. A nonsynonymous SNP in human *MRP4* (G2269A (E757K)) was identified that had reduced the function due to impaired membrane localization. In the Japanese population, this variant allele is very common (>18%) (Krishnamurthy et al., 2008).

More recently, the clinical relevance of this MRP4 missense allele was corroborated in a study of Japanese patients with irritable bowel disease (IBD) (Ban et al., 2010). In samples from 279 patients (44 healthy volunteers, 235 IBD), the *MRP4* G2269A variant allele had an allele frequency comparable to the Japanese population. However, among the IBD patients treated with azathioprine/6-MP, those patients with the *MRP4* G2269A variant allele had greater hematopoietic toxicity as indicated by the significantly lower white blood cell counts and higher 6-TGN levels. These authors proposed that the increased thiopurine sensitivity among the Japanese IBD patients is caused by the nonfunctional *MRP4* G2269A allele (Ban et al., 2010).

16.6.2 Intronic SNPs

While cyclophosphamide-based combination chemotherapy is effective for patients with breast cancer, it often results in adverse drug reactions. To identify those at risk for such reactions, a case control study of 403 patients with breast cancer who received cyclophosphamide combination therapy was conducted. Among the 184 patients who experienced adverse drug reactions (i.e., gastrointestinal toxicity and leucopenia/neutropenia), a genetic polymorphism in the intronic region of the *MRP4* gene (rs 9561778) was significantly associated (Low et al., 2009). It has been speculated that this intronic MRP4 SNP might be useful in predicting the risk of adverse drug reactions in patients receiving cyclophosphamide. However, because there is no direct evidence for MRP4 transport of cyclophosphamide, it seems unlikely that this SNP produces an alteration in MRP4 function. A more likely scenario is that this nonsynonymous SNP is linked to a causative genetic defect that is independent of MRP4.

16.6.3 Synonymous SNPs

Synonymous SNPs in the *MRP4* gene are also closely associated with variation in the intracellular drug concentration and disposition. In HIV patients receiving the antiviral drugs tenofovir, zidovudine, and lamivudine, there was a 20% increase in the lamivudine–triphosphate concentration of the patient carrying the 4131 T>G variant, and those with the 3724G>A variant trended toward elevated AZT-triphosphate concentrations, suggesting an impaired MRP4 efflux function (Anderson et al., 2006; Kiser et al., 2008). Notably, a synonymous SNP in exon 26 (C3435T) of the *MDR1* gene alters the substrate specificity and multidrug resistance of the protein in human cancer cells. Because this substitution enforced use of a rare codon during MDR translation, it was proposed that the infrequently used codons affected the cotranslational folding of *p*-glycoprotein (the gene product of MDR1) producing a protein with altered substrate specificity (Kimchi-Sarfaty et al., 2007). Therefore, one could speculate that synonymous SNPs in the *MRP4* gene might also alter protein folding secondary to changes in codons used for translation.

Genetic polymorphisms in the regulatory regions of the *MRP4* gene might also affect therapeutic response by affecting the amount of MRP4 expressed. In patients with childhood acute lymphoblastic leukemia treated with methotrexate, the possession of an allele that tags the MRP4 promoter was closely associated with improved event-free survival (Ansari et al., 2009). Consistent with this, *in vitro* studies of the MRP4 promoter function revealed that variant *MRP4* promoter had twofold greater activity than the other haplotypes. This study supports the idea that genetic variants affecting the regulation of MRP4 may have a role in therapeutic response.

16.7 CONCLUSION

This chapter reveals how MRP4's expression, tissue localization, and genetic variants affect therapeutic response to numerous drugs. Moreover, its broad substrate specificity has made it an important transporter that should be considered when designing treatment regimens against diseases. The MRP4-knockout mouse model is an excellent tool to investigate MRP's role in transporting physiological substrates and drugs in various tissues. In conclusion, studies of nucleoside drugs' efflux, tissue locations, and genetic variations have provided novel insights into MRP4 since its discovery.

ACKNOWLEDGMENT

This work was supported by NIH Research Grants CA77545, GM60346, ES058571, and HL67366; by Cancer Center Support Grant P30 CA21745; and by the American Lebanese Syrian Associated Charities (ALSAC).

REFERENCES

Abell CW, Monahan TM (1973) The role of adenosine 3′,5′-cyclic monophosphate in the regulation of mammalian cell division. *J Cell Biol*, 59, 549–558.

Abla N, Chinn LW, Nakamura T, Liu L, Huang CC, Johns SJ, Kawamoto M, Stryke D, Taylor TR, Ferrin TE, Giacomini KM, Kroetz DL (2008) The human multidrug resistance protein 4 (MRP4, ABCC4): functional analysis of a highly polymorphic gene. *J Pharmacol Exp Ther*, 325, 859–868.

Adachi M, Reid G, Schuetz JD (2002) Therapeutic and biological importance of getting nucleotides out of cells: a case for the ABC transporters, MRP4 and 5. *Adv Drug Deliv Rev*, 54, 1333–1342.

Anderson PL, Lamba J, Aquilante CL, Schuetz E, Fletcher CV (2006) Pharmacogenetic characteristics of indinavir, zidovudine, and lamivudine therapy in HIV-infected adults: a pilot study. *J Acquir Immune Defic Syndr*, 42, 441–449.

Ansari M, Sauty G, Labuda M, Gagne V, Laverdiere C, Moghrabi A, Sinnett D, Krajinovic M (2009) Polymorphisms in multidrug resistance-associated protein gene 4 is associated with outcome in childhood acute lymphoblastic leukemia. *Blood*, 114, 1383–1386.

Ban H, Andoh A, Imaeda H, Kobori A, Bamba S, Tsujikawa T, Sasaki M, Saito Y, Fujiyama Y (2010) The multidrug-resistance protein 4 polymorphism is a new factor accounting for thiopurine sensitivity in Japanese patients with inflammatory bowel disease. *J Gastroenterol*, 45, 1014–1021.

Beavo JA (1995) Cyclic nucleotide phosphodiesterases: functional implications of multiple isoforms. *Physiol Rev*, 75, 725–748.

Beck C, Cayeux S, Lupton SD, Dorken B, Blankenstein T (1995) The thymidine kinase/ganciclovir-mediated "suicide" effect is variable in different tumor cells. *Hum Gene Ther*, 6, 1525–1530.

Belinsky MG, Guo P, Lee K, Zhou F, Kotova E, Grinberg A, Westphal H, Shchaveleva I, Klein-Szanto A, Gallo JM, Kruh GD (2007) Multidrug resistance protein 4 protects bone marrow, thymus, spleen, and intestine from nucleotide analogue-induced damage. *Cancer Res*, 67, 262–268.

Borst P, De Wolf C, Van De Wetering K (2007) Multidrug resistance-associated proteins 3, 4, and 5. *Pflugers Arch*, 453, 661–673.

Boyer JL (2004) *Mechanism of Bile Formation: An Introduction* New York: Kluwere Academic/Plenum.

Cai C, Omwancha J, Hsieh CL, Shemshedini L (2007) Androgen induces expression of the multidrug resistance protein gene MRP4 in prostate cancer cells. *Prostate Cancer Prostatic Dis*, 10, 39–45.

Chen ZS, Lee K, Kruh GD (2001) Transport of cyclic nucleotides and estradiol 17-beta-D-glucuronide by multidrug resistance protein 4. Resistance to 6-mercaptopurine and 6-thioguanine. *J Biol Chem*, 276, 33747–33754.

Ci L, Kusuhara H, Adachi M, Schuetz JD, Takeuchi K, Sugiyama Y (2007) Involvement of MRP4 (ABCC4) in the luminal efflux of ceftizoxime and cefazolin in the kidney. *Mol Pharmacol*, 71, 1591–1597.

Cole SP, Deeley RG (2006) Transport of glutathione and glutathione conjugates by MRP1. *Trends Pharmacol Sci*, 27, 438–446.

Cundy KC (1999) Clinical pharmacokinetics of the antiviral nucleotide analogues cidofovir and adefovir. *Clin Pharmacokinet*, 36, 127–143.

Deeley RG, Westlake C, Cole SP (2006) Transmembrane transport of endo- and xenobiotics by mammalian ATP-binding cassette multidrug resistance proteins. *Physiol Rev*, 86, 849–899.

Elion GB (1989) The purine path to chemotherapy. *Science*, 244, 41–47.

Errasti-Murugarren E, Pastor-Anglada MA (2010) Drug transporter pharmacogenetics in nucleoside-based therapies. *Pharmacogenomics*, 11, 809–841.

Frick LW, Nelson DJ, St Clair MH, Furman PA, Krenitsky TA (1988) Effects of 3′-azido-3′-deoxythymidine on the deoxynucleotide triphosphate pools of cultured human cells. *Biochem Biophys Res Commun*, 154, 124–129.

Fridland A, Connelly MC, Ashmun R (1990) Relationship of deoxynucleotide changes to inhibition of DNA synthesis induced by the antiretroviral agent 3′-azido-3′-deoxythymidine and release of its monophosphate by human lymphoid cells (CCRF-CEM). *Mol Pharmacol*, 37, 665–670.

Friedman DL, Johnson RA, Zeilig CE (1976) The role of cyclic nucleotides in the cell cycle. *Adv Cyclic Nucleotide Res*, 7, 69–114.

Gearry RB, Barclay ML, Burt MJ, Collett JA, Chapman BA (2004) Thiopurine drug adverse effects in a population of New Zealand patients with inflammatory bowel disease. *Pharmacoepidemiol Drug Saf*, 13, 563–567.

Gisbert JP, Gomollon F (2008) Thiopurine-induced myelotoxicity in patients with inflammatory bowel disease: a review. *Am J Gastroenterol*, 103, 1783–1800.

Gradhand U, Lang T, Schaeffeler E, Glaeser H, Tegude H, Klein K, Fritz P, Jedlitschky G, Kroemer HK, Bachmakov I, Anwald B, Kerb R, Zanger UM, Eichelbaum M, Schwab M, Fromm MF (2007) Variability in human hepatic MRP4 expression: influence of cholestasis and genotype. *Pharmacogenomics J*, 8, 42–52.

Guo Y, Kock K, Ritter CA, Chen ZS, Grube M, Jedlitschky G, Illmer T, Ayres M, Beck JF, Siegmund W, Ehninger G, Gandhi V, Kroemer HK, Kruh GD, Schaich M (2009) Expression of ABCC-type nucleotide exporters in blasts of adult acute myeloid leukemia: relation to long-term survival. *Clin Cancer Res*, 15, 1762–1769.

Hasegawa M, Kusuhara H, Adachi M, Schuetz JD, Takeuchi K, Sugiyama Y (2007) Multidrug resistance-associated protein 4 is involved in the urinary excretion of hydrochlorothiazide and furosemide. *J Am Soc Nephrol*, 18, 37–45.

Hatse S, De Clercq E, Balzarini J (1998) Enhanced 9-(2-phosphonylmethoxyethyl)adenine secretion by a specific, indomethacin-sensitive efflux pump in a mutant 9-(2-phosphonylmethoxyethyl)adenine-resistant human erythroleukemia K562 cell line. *Mol Pharmacol*, 54, 907–917.

Hishida A (2007) Clinical analysis of 207 patients who developed renal disorders during or after treatment with edaravone reported during post-marketing surveillance. *Clin Exp Nephrol*, 11, 292–296.

Ho LL, Kench JG, Handelsman DJ, Scheffer GL, Stricker PD, Grygiel JG, Sutherland RL, Henshall SM, Allen JD, Horvath LG (2008) Androgen regulation of multidrug resistance-associated protein 4 (MRP4/ABCC4) in prostate cancer. *Prostate*, 68, 1421–1429.

Ho R, Sutherland EW (1975) cAMP-mediated feedback regulation in target cells. *Adv Cyclic Nucleotide Res*, 5, 533–548.

Hofer AM, Lefkimmiatis K (2007) Extracellular calcium and cAMP: second messengers as “third messengers”? *Physiology (Bethesda)*, 22, 320–327.

Hoque MT, Cole SP (2008) Downregulation of Na+/H +exchanger regulatory factor 1 increases expression and function of multidrug resistance protein 4. *Cancer Res*, 68, 4802–4809.

Hoque MT, Conseil G, Cole SP (2009) Involvement of NHERF1 in apical membrane localization of MRP4 in polarized kidney cells. *Biochem Biophys Res Commun*, 379, 60–64.

Imaoka T, Kusuhara H, Adachi M, Schuetz JD, Takeuchi K, Sugiyama Y (2007) Functional involvement of multidrug resistance-associated protein 4 (MRP4/ABCC4) in the renal elimination of the antiviral drugs adefovir and tenofovir. *Mol Pharmacol*, 71, 619–627.

Jackson EK, Raghvendra DK (2004) The extracellular cyclic AMP-adenosine pathway in renal physiology. *Annu Rev Physiol*, 66, 571–599.

Janke D, Mehralivand S, Strand D, Godtel-Ambrust U, Habermeier A, Gradhand U, Fischer C, Tollat MR, Fritz, Zanger UM, Schwab M, Fromm MF, Nurnberg P, Wojnowski L, Closs EI, Lang T (2008) 6-mercaptopurine and 9-(2-phosphonyl-methoxyethyl)adenine (PMEA) transport altered by two missense mutations in the drug transporter gene ABCC4. *Hum Mut*, 29, 659–669.

Jedlitschky G, Tirschmann K, Lubenow LE, Nieuwenhuis HK, Akkerman JW, Greinacher A, Kroemer HK (2004) The nucleotide transporter MRP4 (ABCC4) is highly expressed in human platelets and present in dense granules, indicating a role in mediator storage. *Blood*, 104, 3603–3610.

Keitel V, Burdelski M, Warskulat U, Kuhlkamp T, Keppler D, Haussinger D, Kubitz R (2005) Expression and localization of hepatobiliary transport proteins in progressive familial intrahepatic cholestasis. *Hepatology*, 41, 1160–1172.

Kimchi-Sarfaty C, Oh JM, Kim IW, Sauna ZE, Calcagno AM, Ambudkar SV, Gottesman MM (2007) A “silent” polymorphism in the MDR1 gene changes substrate specificity. *Science*, 315, 525–528.

Kiser JJ, Aquilante CL, Anderson PL, King TM, Carten ML, Fletcher CV (2008) Clinical and genetic determinants of intracellular tenofovir diphosphate concentrations in HIV-infected patients. *J Acquir Immune Defic Syndr*, 47, 298–303.

Kool M, De Haas M, Scheffer GL, Scheper RJ, Van Eijk MJ, Juijn JA, Baas F, Borst P (1997) Analysis of expression of cMOAT (MRP2), MRP3, MRP4, and MRP5, homologues of the multidrug resistance-associated protein gene (MRP1), in human cancer cell lines. *Cancer Res*, 57, 3537–3547.

Krishnamurthy P, Schwab M, Takenaka K, Nachagari D, Morgan J, Leslie M, Du W, Boyd K, Cheok M, Nakauchi H, Marzolini C, Kim RB, Poonkuzhali B, Schuetz E, Evans W, Relling M, Schuetz JD (2008) Transporter-mediated protection against thiopurine-induced hematopoietic toxicity. *Cancer Res*, 68, 4983–4989.

Lai L, Tan TM (2002) Role of glutathione in the multidrug resistance protein 4 (MRP4/ABCC4)-mediated efflux of cAMP and resistance to purine analogues. *Biochem J*, 361, 497–503.

Lamba JK, Adachi M, Sun D, Tammur J, Schuetz EG, Allikmets R, Schuetz JD (2003) Nonsense-mediated decay downregulates conserved alternatively spliced ABCC4 transcripts bearing nonsense codons. *Hum Mol Genet*, 12, 99–109.

Lee K, Belinsky MG, Bell DW, Testa JR, Kruh GD (1998) Isolation of MOAT-B, a widely expressed multidrug resistance-associated protein/canalicular multispecific organic anion transporter-related transporter. *Cancer Res*, 58, 2741–2747.

Lee K, Klein-Szanto AJ, Kruh GD (2000) Analysis of the MRP4 drug resistance profile in transfected NIH3T3 cells. *J Natl Cancer Inst*, 92, 1934–1940.

Leggas M, Adachi M, Scheffer GL, Sun D, Wielinga P, Du G, Mercer KE, Zhuang Y, Panetta JC, Johnston B, Scheper RJ, Stewart CF, Schuetz JD (2004) MRP4 confers resistance to topotecan and protects the brain from chemotherapy. *Mol Cell Biol*, 24, 7612–7621.

Lepsy CS, Guttendorf RJ, Kugler AR, Smith DE (2003) Effects of organic anion, organic cation, and dipeptide transport inhibitors on cefdinir in the isolated perfused rat kidney. *Antimicrob Agents Chemother*, 47, 689–696.

Li C, Krishnamurthy PC, Penmatsa H, Marrs KL, Wang XQ, Zaccolo M, Jalink K, Li M, Nelson DJ, Schuetz JD, Naren AP (2007) Spatiotemporal coupling of cAMP transporter to CFTR chloride channel function in the gut epithelia. *Cell*, 131, 940–951.

Lin ZP, Zhu YL, Johnson DR, Rice KP, Nottoli T, Hains BC, Mcgrath J, Waxman SG, Sartorelli AC (2008) Disruption of cAMP and prostaglandin E2 transport by multidrug resistance protein 4 deficiency alters cAMP-mediated signaling and nociceptive response. *Mol Pharmacol*, 73, 243–251.

Low SK, Kiyotani K, Mushiroda T, Daigo Y, Nakamura Y, Zembutsu H (2009) Association study of genetic polymorphism in ABCC4 with cyclophosphamide-induced adverse drug reactions in breast cancer patients. *J Hum Genet*, 54, 564–571.

Maher JM, Cheng X, Tanaka Y, Scheffer GL, Klaassen CD (2006) Hormonal regulation of renal multidrug resistance-associated proteins 3 and 4 (MRP3 and MRP4) in mice. *Biochem Pharmacol*, 71, 1470–1478.

Marschall HU, Wagner M, Zollner G, Fickert P, Diczfalusy U, Gumhold J, Silbert D, Fuchsbichler A, Benthin L, Grundstrom R, Gustafsson U, Sahlin S, Einarsson C, Trauner M (2005) Complementary stimulation of hepatobiliary transport and detoxification systems by rifampicin and ursodeoxycholic acid in humans. *Gastroenterology*, 129, 476–485.

Mennone A, Soroka CJ, Cai SY, Harry K, Adachi M, Hagey L, Schuetz JD, Boyer JL (2006) Mrp4-/- mice have an impaired cytoprotective response in obstructive cholestasis. *Hepatology*, 43, 1013–1021.

Mizuno N, Takahashi T, Kusuhara H, Schuetz JD, Niwa T, Sugiyama Y (2007) Evaluation of the role of breast cancer resistance protein (BCRP/ABCG2) and multidrug resistance associated protein 4 (MRP4/ABCC4) in the urinary excretion of sulfate and glucuronide metabolites of edaravone (MCI-186; 3-methyl-1-phenyl-2-pyrazolin-5-one). *Drug Metab Dispos*, 35, 2045–2052.

Nies AT, Jedlitschky G, Konig J, Herold-Mende C, Steiner HH, Schmitt HP, Keppler D (2004) Expression and immunolocalization of the multidrug resistance proteins, MRP1–MRP6 (ABCC1–ABCC6), in human brain. *Neuroscience*, 129, 349–360.

Oevermann L, Scheitz J, Starke K, Köck K, Kiefer T, Dölken G, Nießen J, Greinacher A, Siegmund W, Zygmunt M, Kroemer HK, Jedlitschky G, Ritter CA (2009) Hematopoietic stem cell differentiation affects expression and function of MRP4 (ABCC4), a transport protein for signaling molecules and drugs. *Int J Cancer*, 124, 2303–2311.

Orlov SN, Maksimova NV (1999) Efflux of cyclic adenosine monophosphate from cells: mechanisms and physiological implications. *Biochemistry (Mosc)*, 64, 127–135.

Ose A, Ito M, Kusuhara H, Yamatsugu K, Kanai M, Shibasaki M, Hosokawa M, Schuetz JD, Sugiyama Y (2009) Limited brain distribution of [3R,4R,5S]-4-acetamido-5-amino-3-(1-ethylpropoxy)-1-cyclohexene-1-carboxylate phosphate (Ro 64-0802), a pharmacologically active form of oseltamivir, by active efflux across the blood–brain barrier mediated by organic anion transporter 3 (Oat3/Slc22a8) and multidrug resistance-associated protein 4 (MRP4/ABCC4). *Drug Metab Dispos*, 37, 315–321.

Pastan I (1975) Cyclic AMP and the malignant transformation of cells. *Adv Metab Disord*, 8, 377–383.

Pastan IH, Johnson GS, Anderson WB (1975) Role of cyclic nucleotides in growth control. *Annu Rev Biochem*, 44, 491–522.

Peng X-X, Shi Z, Damaraju VL, Huang X-C, Kruh GD, Wu H-C, Zhou Y, Tiwari A, Fu L, Cass CE, Chen Z-S (2008) Upregulation of MRP4 and downregulation of influx transporters in human leukemic cells with acquired resistance to 6-mercaptopurine. *Leuk Res*, 32, 799–809.

Plagemann PG (1976) Transport, phosphorylation, and toxicity of a tricyclic nucleoside in cultured Novikoff rat hepatoma cells and other cell lines and relase of its monophosphate by the cells. *J Natl Cancer Inst*, 57, 1283–1295.

Plagemann PG, Erbe J (1977) Exit transport of a cyclic nucleotide from mouse L-cells. *J Biol Chem*, 252, 2010–2016.

Reid G, Wielinga P, Zelcer N, Van Der Heijden I, Kuil A, De Haas M, Wijnholds J, Borst P (2003) The human multidrug resistance protein MRP4 functions as a prostaglandin efflux transporter and is inhibited by nonsteroidal anti-inflammatory drugs. *Proc Natl Acad Sci USA*, 100, 9244–9249.

Rich TC, Fagan KA, Nakata H, Schaack J, Cooper DM, Karpen JW (2000) Cyclic nucleotide-gated channels colocalize with adenylyl cyclase in regions of restricted cAMP diffusion. *J Gen Physiol*, 116, 147–161.

Rius M, Nies AT, Hummel-Eisenbeiss J, Jedlitschky G, Keppler D (2003) Cotransport of reduced glutathione with bile salts by MRP4 (ABCC4) localized to the basolateral hepatocyte membrane. *Hepatology*, 38, 374–384.

Rius M, Thon WF, Keppler D, Nies AT (2005) Prostanoid transport by multidrug resistance protein 4 (MRP4/ABCC4) localized in tissues of the human urogenital tract. *J Urol*, 174, 2409–2414.

Robbins BL, Connelly MC, Marshall DR, Srinivas RV, Fridland A (1995) A human T lymphoid cell variant resistant to the acyclic nucleoside phosphonate 9-(2-phosphonylmethoxyethyl)adenine shows a unique combination of a phosphorylation defect and increased efflux of the agent. *Mol Pharmacol*, 47, 391–397.

Russel FG, Masereeuw R, Van Aubel RA (2002) Molecular aspects of renal anionic drug transport. *Annu Rev Physiol*, 64, 563–594.

Saito S, Iida A, Sekine A, Miura Y, Ogawa C, Kawauchi S, Higuchi S, Nakamura Y (2002) Identification of 779 genetic variations in eight genes encoding members of the ATP-binding cassette, subfamily C (ABCC/MRP/CFTR. *J Hum Genet*, 47, 147–171.

Sampath J, Adachi M, Hatse S, Naesens L, Balzarini J, Flatley RM, Matherly LH, Schuetz JD (2002) Role of MRP4 and MRP5 in biology and chemotherapy. *AAPS PharmSci*, 4, E14.

Sassi Y, Lipskaia L, Vandecasteele G, Nikolaev VO, Hatem SN, Cohen Aubart F, Russel FG, Mougenot N, Vrignaud C, Lechat P, Lompre AM, Hulot JS (2008) Multidrug resistance-associated protein 4 regulates cAMP-dependent signaling pathways and controls human and rat SMC proliferation. *J Clin Invest*, 118, 2747–2757.

Sauna ZE, Nandigama K, Ambudkar SV (2004) Multidrug resistance protein 4 (ABCC4)-mediated ATP hydrolysis: effect of transport substrates and characterization of the post-hydrolysis transition state. *J Biol Chem*, 279, 48855–48864.

Schuetz EG, Strom S, Yasuda K, Lecureur V, Assem M, Brimer C, Lamba J, Kim RB, Ramachandran V, Komoroski BJ, Venkataramanan R, Cai H, Sinal CJ, Gonzalez FJ, Schuetz JD (2001) Disrupted bile acid homeostasis reveals an unexpected interaction among nuclear hormone receptors, transporters, and cytochrome P450. *J Biol Chem*, 276, 39411–39418.

Schuetz JD, Connelly MC, Sun D, Paibir SG, Flynn PM, Srinivas RV, Kumar A, Fridland A (1999) MRP4: a previously unidentified factor in resistance to nucleoside-based antiviral drugs. *Nat Med*, 5, 1048–1051.

Sinal CJ, Tohkin M, Miyata M, Ward JM, Lambert G, Gonzalez FJ (2000) Targeted disruption of the nuclear receptor FXR/BAR impairs bile acid and lipid homeostasis. *Cell*, 102, 731–744.

Smeets PH, van Aubel RA, Wouterse AC, van den Heuvel JJ, Russel FG (2004) Contribution of multidrug resistance protein 2 (MRP2/ABCC2) to the renal excretion of *p*-aminohippurate (PAH) and identification of MRP4 (ABCC4) as a novel PAH transporter. *J Am Soc Nephrol*, 15, 2828–2835.

Tomlins SA, Mehra R, Rhodes DR, Cao X, Wang L, Dhanasekaran SM, Kalyana-Sundaram S, Wei JT, Rubin MA, Pienta KJ, Shah RB, Chinnaiyan AM (2007) Integrative molecular concept modeling of prostate cancer progression. *Nat Genet*, 39, 41–51.

Trauner M, Boyer JL (2003) Bile salt transporters: molecular characterization, function, and regulation. *Physiol Rev*, 83, 633–671.

van Aubel RA, Smeets PH, Peters JG, Bindels RJ, Russel FG (2002a) The *MRP4/ABCC4* gene encodes a novel apical organic anion transporter in human kidney proximal tubules: putative efflux pump for urinary cAMP and cGMP. *J Am Soc Nephrol*, 13, 595–603.

van Aubel RA, Smeets PH, Peters JG, Bindels RJ, Russel FG (2002b) The *MRP4/ABCC4* gene encodes a novel apical organic anion transporter in human kidney proximal tubules: putative efflux pump for urinary cAMP and cGMP. *J Am Soc Nephrol*, 13, 595–603.

Wagner M, Fickert P, Zollner G, Fuchsbichler A, Silbert D, Tsybrovskyy O, Zatloukal K, Guo GL, Schuetz JD, Gonzalez FJ, Marschall HU, Denk H, Trauner M (2003) Role of farnesoid X receptor in determining hepatic ABC transporter expression and liver injury in bile duct-ligated mice. *Gastroenterology*, 125, 825–838.

Watanabe T, Yuki S, Egawa M, Nishi H (1994) Protective effects of MCI-186 on cerebral ischemia: possible involvement of free radical scavenging and antioxidant actions. *J Pharmacol Exp Ther*, 268, 1597–1604.

Wielinga PR, Reid G, Challa EE, Van Der Heijden I, Van Deemter L, De Haas M, Mol C, Kuil AJ, Groeneveld E, Schuetz JD, Brouwer C, De Abreu RA, Wijnholds J, Beijnen JH, Borst P (2002) Thiopurine metabolism and identification of the thiopurine metabolites transported by MRP4 and MRP5 overexpressed in human embryonic kidney cells. *Mol Pharmacol*, 62, 1321–1331.

Wielinga PR, Van Der Heijden I, Reid G, Beijnen JH, Wijnholds J, Borst P (2003) Characterization of the MRP4- and MRP5-mediated transport of cyclic nucleotides from intact cells. *J Biol Chem*, 278, 17664–17671.

Yamamoto Y, Kuwahara T, Watanabe K (1996) Antioxidant activity of 3-methyl-1-phenyl-2-pyrazolin-5-one. *Redox Report*, 2, 333–338.

Yang L, Hwang R, Chiang Y, Gordon EM, Anderson WF, Parekh D (1998) Mechanisms for ganciclovir resistance in gastrointestinal tumor cells transduced with a retroviral vector containing the herpes simplex virus thymidine kinase gene. *Clin Cancer Res*, 4, 731–741.

Zelcer N, Reid G, Wielinga P, Kuil A, Van Der Heijden I, Schuetz JD, Borst P (2003) Steroid and bile acid conjugates are substrates of human multidrug-resistance protein (MRP) 4 (ATP-binding cassette C4). *Biochem J*, 371, 361–367.

Zimmermann C, Gutmann H, Hruz P, Gutzwiller J-P, Beglinger C, Drewe J (2005) Mapping of multidrug resistance-associated protein isoform 1 to 5 mRNA expression along the human intestinal tract. *Drug Metab Dispos*, 33, 219–224.

CHAPTER 17

MRP8 (ABCC11)

Yu Toyoda
Toshihisa Ishikawa

17.1 INTRODUCTION

Cell secretion is an important physiological process that ensures smooth metabolic activities and tissue repair as well as growth and immunological functions in the body. Apocrine secretion occurs when the secretory process is accomplished with a partial loss of cell cytoplasm. The secretory materials may be contained within the secretory vesicles or dissolved in the cytoplasm and are released during secretion as cytoplasmic fragments into the glandular lumen or interstitial space. Hitherto apocrine secretory mechanisms have not been well characterized. Although the biochemical and physiological pathways that regulate the apocrine secretory process are not clearly known, our recent finding that the nonsynonymous SNP 538G>A (rs17822931; Gly180Arg) in the *ABCC11* gene determines the type of earwax in humans has shed light on the novel function of this ABC transporter in apocrine glands. Furthermore, the wild type (Gly180) of ABCC11 is associated with wet-type earwax, axillary osmidrosis, and colostrum secretion from the mammary gland as well as the potential susceptibility to breast cancer. The SNP (538G>A) in the *ABCC11* gene is thus suggested to be a clinical biomarker for the prediction of chemotherapeutic efficacy. In this chapter, we provide an overview on the discovery and characterization of genetic polymorphisms in the human *ABCC11*, focusing on the impact of *ABCC11* 538G>A on the apocrine phenotype, patients' response to nucleoside-based chemotherapy, and the potential risk of breast cancer.

17.2 DISCOVERY OF THE HUMAN *ABCC11* GENE

In 2001, three research groups, including ours, independently cloned two novel ABC transporters, named ABCC11 and ABCC12, from the cDNA library of human adult liver (Bera et al., 2001; Tammur et al., 2001; Yabuuchi et al., 2001). These two genes have been found to be located on human chromosome 16q12.1 in a tail-to-head orientation with a separation distance of about 20 kb (Fig. 17.1). The predicted amino acid sequences of both the gene products show a high similarity to those of ABCC4

Pharmacogenomics of Human Drug Transporters: Clinical Impacts, First Edition.
Edited by Toshihisa Ishikawa, Richard B. Kim and Jörg König

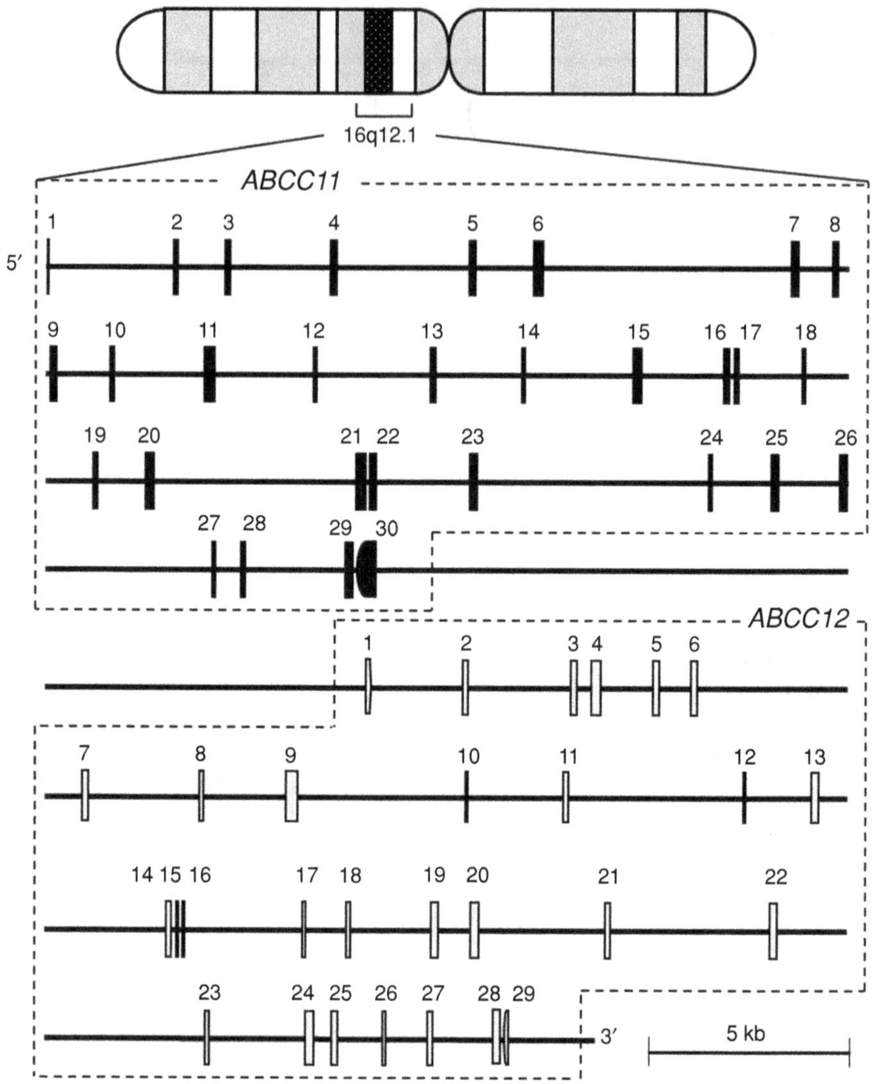

Figure 17.1 Schematic illustration of the genomic structures of *ABCC11* and *ABCC12* genes on human chromosome 16q12.1. The cytogenetic location of the *ABCC11* gene as well as the structures of exons and introns was analyzed by BLAST searches on the human genome. A nonsynonymous SNP: 538G>A (Gly180Arg), an earwax determinant, is located in exon 4.

and ABCC5, suggesting that they have the typical structure of "full" ABC transporters (Fig. 17.2). Interestingly, however, there is no putative mouse or rat orthologous gene corresponding to human *ABCC11* (Shimizu et al., 2003). This fact indicates that *ABCC11* is not an orthologous gene but rather a paralogous gene generated by gene duplication in the human genome. In contrast, *ABCC12* and its orthologous genes are found in several different species including humans, primates, and rodents (Shimizu et al., 2003).

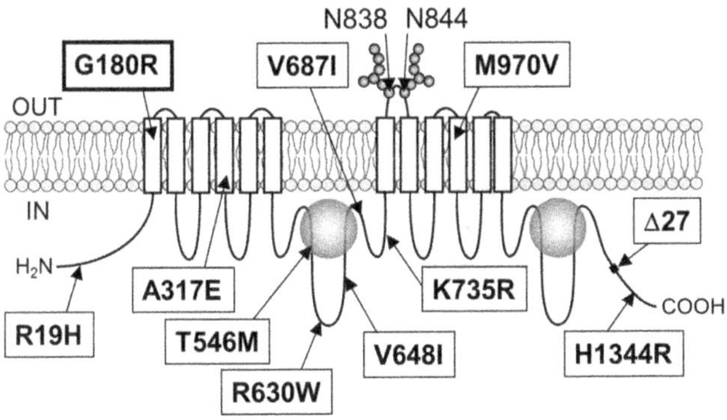

Figure 17.2 Schematic illustration of ABCC11 structure and hitherto known nonsynonymous SNPs. ABCC11 has a total of 12 transmembrane (TM) regions and two intracellular ATP-binding cassettes. Asn838 and Asn844 residing in an extracellular loop between transmembrane helices TM7 and TM8 are *N*-linked glycosylation sites in the ABCC11 WT protein. Locations of hitherto reported nonsynonymous SNPs and Δ27 (a rare deletion mutation) are indicated in the putative structure of ABCC11. G180R and Δ27 are related to the formation of dry-type earwax.

Transcript analyses suggest that human ABCC11 mRNA is ubiquitously expressed in human adult and fetal tissues (Tammur et al., 2001; Yabuuchi et al., 2001). In addition, we and Bera et al. demonstrated high levels of ABCC11 mRNA in breast cancer tissues (Bera et al., 2001; Yabuuchi et al., 2001). The increased expression of ABCC11 wild-type (WT) in breast cancer might be associated with low efficacy levels of chemotherapy, as is discussed later in this chapter.

When transfected exogenously, the ABCC11 WT protein was localized in the apical membrane of Madin-Darby canine kidney cells strain II (MDCK II cells) (Bortfeld et al., 2006). The substrate specificity of ABCC11 WT was characterized in more detail by an *in vitro* transport assay with plasma membrane vesicles prepared from pig LLC-PK1 cells transfected with an ABCC11 WT expression vector (Chen et al., 2005). The results of this assay demonstrated that the ABCC11 WT is able to transport a variety of lipophilic anions including cyclic nucleotides, glutathione conjugates such as leukotriene C_4 (LTC_4) and S-(2,4-dinitrophenyl)-glutathione (DNP-SG), steroid sulfates such as estrone 3-sulfate ($E_1$3S) and dehydroepiandrosterone 3-sulfate (DHEAS), glucuronides such as estradiol 17-β-D-glucuronide ($E_2$17βG), monoanionic bile acids glycocholate and taurocholate, as well as folic acid and its analog methotrexate (MTX) (Chen et al., 2005; Bortfeld et al., 2006).

17.3 GENETIC POLYMORPHISMS AND PHYSIOLOGICAL FUNCTION OF ABCC11

Hitherto more than 10 nonsynonymous single nucleotide polymorphisms (SNPs) have been reported in the human *ABCC11* gene (Fig. 17.2). Among those SNPs, we have recently found that one SNP (rs17822931; 538G>A, Gly180Arg) determines

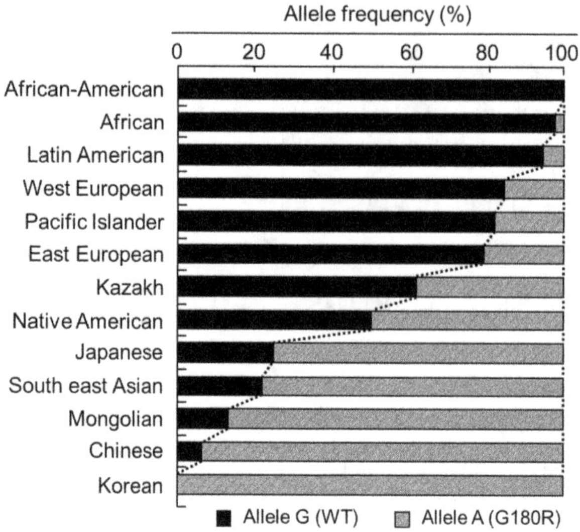

Figure 17.3 The allele frequencies of the wild type (WT; Gly180) and 538G>A (Arg180) variant of human *ABCC11* among different ethnic populations. Data are from Yoshiura et al. (2006).

the human earwax type (Yoshiura et al., 2006). Interestingly, this SNP exhibits wide ethnic differences in allelic frequency (Toyoda et al., 2008). In Mongoloid populations in Asia, the frequency of the 538A allele is predominantly high whereas the frequency of this allele is low among Caucasians and Africans (Yoshiura et al., 2006; Toyoda et al., 2008) (Fig. 17.3). The frequency of the 538A allele exhibits a north–south and east–west downward geographical gradient with the highest peak in northeastern Asia. It is suggested that the 538A allele arose in northeast Asia and thereafter spread throughout the world (Yoshiura et al., 2006), apparently reflecting the intercontinental migration of *Homo sapiens* (Toyoda et al., 2008). A similar west–east downward geographical gradient was observed in the frequency of the 2677G (Ala893) allele of the *ABCB1* (*P-glycoprotein/MDR1*) gene, as well (Sakurai et al., 2007).

Earwax (cerumen) is a secretory product of the ceruminous apocrine glands, which can be classified into two phenotypes in humans, wet (sticky) and dry. The dry type is most commonly found in the Asian population, especially among Koreans, Japanese, and Chinese whereas the wet type is the dominant phenotype for many Africans and Caucasians. The 538A/A genotype gives the dry phenotype whereas both 538G/A and G/G genotypes give the wet phenotype. This relationship is consistent with observations that earwax type is a Mendelian trait and that the wet phenotype is dominant to the dry one.

Immunohistochemical studies with cerumen gland-containing tissue specimens revealed that the ABCC11 WT protein with Gly180 was expressed in the cerumen gland (Toyoda et al., 2009). The cerumen gland is one of the apocrine glands. Not only in the external auditory canal, apocrine glands can also be found in axillary region and breast; those physical characteristics also are concerned with apocrine glands. In fact, there is a positive association among the wet earwax type, axillary osmidrosis (Yoo et al., 2006), and colostrum secretion from the breast (Miura et al., 2007).

Apocrine secretion occurs when the secretory process is accomplished with a partial loss of cell cytoplasm. The secretory materials may be contained within the secretory vesicles or dissolved in the cytoplasm and then released during excretion as cytoplasmic fragments into the glandular lumen or interstitial space (Gesase and Satoh, 2003). Hitherto apocrine secretory mechanisms have not been well characterized (Gesase and Satoh, 2003). Although the biochemical and physiological pathways that regulate the apocrine secretory process are not clearly known, our recent findings (Yoshiura et al., 2006; Toyoda et al., 2009; Inoue et al., 2010) that the SNP (538G>A, Gly180Arg) in the *ABCC11* gene determines both earwax types and axillary osmidrosis have shed light on the novel function of this ABC transporter in the apocrine glands.

17.4 STRONG ASSOCIATION BETWEEN AXILLARY OSMIDROSIS AND THE GENOTYPE OF *ABCC11* 538G>A

In Japan, axillary osmidrosis is recognized as a disease that is covered by the national health insurance system. Axillary osmidrosis is often perceived, especially by young women, as a distressing and troublesome problem. Certain people display an excessive fear, aversion, or psychological hypersensitivity to smells or odors. They tend to opt for aggressive surgical treatments and are sometimes categorized as having osmophobia. Interestingly, an association between wet-type earwax and axillary osmidrosis had already been recognized more than half a century ago (Matsunaga, 1962). Hence, the wet type of earwax has frequently been used as one of the diagnostic criteria and characteristics in clinical sites. This relationship, however, had only been based on the observations of those two respective phenotypes. Therefore, there has been a need for objective evidence for diagnosis of axillary osmidrosis to prevent unnecessary treatments for such patients.

Recently, we and collaborators have revealed that the ABCC11 WT allele is intimately associated with axillary osmidrosis as well as the wet type of earwax (Table 17.1). Several studies have already concluded that the genotypes at *ABCC11* 538G>A would be a useful biomarker for the diagnosis of axillary osmidrosis (Nakano et al., 2009; Toyoda et al., 2009; Inoue et al., 2010; Martin et al., 2010). This important clinical implication would provide an accurate and practical criterion for guidance of appropriate treatment and psychological management of patients.

TABLE 17.1 Association of Earwax Type with *ABCC11* Genotype and Axillary Osmidrosis in Japanese Subjects

Earwax type		Genotype at *ABCC11* 538G>A G/G	G/A	A/A	Axillary osmidrosis patients
Dry	262	0	0	262	2
Wet	300	23	277	0	182

Data are calculated from Inoue et al. (2010) and Nakano et al. (2009).

Sweat produced by the axillary apocrine glands is odorless. Secretions from the apocrine glands, however, can be converted to odoriferous compounds by bacteria (*Corynebacteria*), which results in the formation of the unique "human axillary odor" (Shehadeh and Kligman, 1963). In axillary osmidrosis patients (538G/G homozygote or G/A heterozygote), significantly more numerous and larger-sized axillary apocrine glands were observed as compared with the subjects carrying the A/A homozygote. Indeed, the 538G allele in the *ABCC11* gene is associated with axillary osmidrosis (Nakano et al., 2009; Toyoda et al., 2009; Inoue et al., 2010; Martin et al., 2010) and ABCC11 WT (Gly180) would be responsible for the secretion of preodoriferous compounds from the axillary apocrine gland. In primates, the axillary odors may play a role in olfactory communication, although no documented behavioral or endocrine changes resulting from volatiles produced in the axillae have been reported to occur in humans. Previous studies have described the presence of androgen steroids in the axillary area. Androsterone sulfate (AS) and DHEAS were detected in an extract of axillary hairs in addition to high levels of cholesterol (Julesz, 1968). It was also demonstrated that, following injection of radioactive pregnenolone or progesterone, the steroid secretion was concentrated in the axillary area (Brooksbank, 1970). In these studies, the axillary sweat collected from the skin surface, however, represents a mixture of materials from apocrine, eccrine, and sebaceous glands, in addition to desquamating epidermal cells. In this respect, Labows et al. demonstrated that pure apocrine secretions contained at least two androgen steroids, AS and DHEAS, in addition to cholesterol (Labows et al., 1979). It is strongly suggested that one of the physiological functions of ABCC11 WT is the active transport of steroid metabolites, such as AS and DHEAS, into the lumen of apocrine glands.

17.5 ENDOPLASMIC RETICULUM-ASSOCIATED DEGRADATION OF THE SNP VARIANT OF ABCC11

Why does one SNP (538G>A) in the human *ABCC11* gene affect the function of apocrine glands? To address this question, we have recently provided evidence that proteasomal degradation of the SNP variant (Arg180) of ABCC11 is the underlying molecular mechanism (Toyoda et al., 2009). ABCC11 WT with Gly180 is an *N*-linked glycosylated protein, which is localized in intracellular granules and large vacuoles as well as at the luminal membrane of secretory cells in the cerumen apocrine gland (Toyoda et al., 2009). *N*-linked glycosylation occurs at both Asn838 and Asn844 in the extracellular loop between transmembrane domains 7 (TM7) and 8 (TM8) of the ABCC11 WT protein. In contrast, the SNP variant (Arg180) lacks *N*-linked glycosylation and readily undergoes proteasomal degradation, most probably via ubiquitination. As a consequence, no granular or vacuolar localization was detected in the cerumen apocrine glands of people having the SNP variant as homozygote.

Morphological differences were previously reported between the secretory cells of wet and dry types of human ceruminous glands (Shugyo et al., 1988). In the wet-type glands, the *Golgi* apparatus was reportedly well developed whereas it was generally small in the corresponding cells of the dry type. Furthermore, intracellular granules were abundantly observed in the wet-type gland in close relationship to their well-developed *Golgi* apparatus whereas intracellular granules were rare in the dry-type gland.

The endoplasmic reticulum (ER) and *Golgi* apparatus are the sites of synthesis and maturation of proteins destined for the plasma membrane, the secretory and endocytic organelles, and secretion (Ellgaard et al., 1999; Helenius and Aebi, 2004). Efficient quality control systems have evolved to prevent incompletely folded proteins from moving along the secretory pathway. Accumulation of misfolded proteins in the ER would detrimentally affect cellular functions. Therefore, misfolded proteins may be removed from the ER by retrotranslocation to the cytosol compartment where they are degraded by the ubiquitin–proteasome system. This process is known as endoplasmic reticulum-associated degradation (ERAD) (Mori, 2000; Ellgaard and Helenius, 2001; Hampton, 2002; Kleizen and Braakman, 2004). It is likely that the product of the SNP variant (Arg180) is recognized as a misfolded protein in the ER and readily undergoes proteasomal degradation. We consider that an electrostatic charge (either positive or negative) at amino acid 180 in theTM1 interferes with correct folding of the *de novo* synthesized ABCC11 protein in the ER (Toyoda et al., 2009). This ERAD processing of the SNP variant (Arg180) of ABCC11 may greatly influence the activity of ceruminous apocrine glands and determine the type of human earwax. Similar ERAD processing is considered to take place for the SNP variant (Arg180) of ABCC11 in axillary and mammary apocrine glands. In Figure 17.4, we schematically illustrate the impact of this SNP on the cellular localization and function of ABCC11 in secretory cells of the apocrine gland. Asn838 and Asn844 are glycosylation target sites in human ABCC11. The *N*-linked glycans are thought to be subjected to extensive modification as glycoproteins mature and move through the ER via the *Golgi* apparatus to their final destinations as, for example, intracellular granules and large vacuoles of secretory cells in the apocrine gland.

17.6 ABCC11 WILD-TYPE ALLELE AND BREAST CANCER RISK

In 1971, Nicholas L. Petrakis first reported that international mortality and frequency rates for breast cancer seemed to be associated with the frequency of the allele for wet-type earwax (Petrakis, 1971). Caucasians and African-Americans in the US as well as Germans exhibited approximately fourfold higher rates of breast cancer mortality as compared with Japanese and Taiwanese women (Petrakis, 1971). Nevertheless, the phenotypic association of the wet type of earwax with breast cancer has remained controversial (Petrakis, 1971; Ing et al., 1973).

At the present time, it is not well understood whether the ABCC11 WT really contributes to breast cancer risk. Therefore, we have most recently carried out genotyping of the SNP 538G>A (Gly180Arg) for a total of 543 Japanese women to examine the association between the frequency rate of breast cancer and the allelic frequency of the G allele (WT). Using blood samples from patients with invasive breast cancer ($n = 270$) and control volunteers ($n = 273$), we have genotyped the SNP 538G>A in the *ABCC11* gene. The frequency of the 538G allele in the breast cancer patients was higher than that in the control volunteers. The odds ratio for

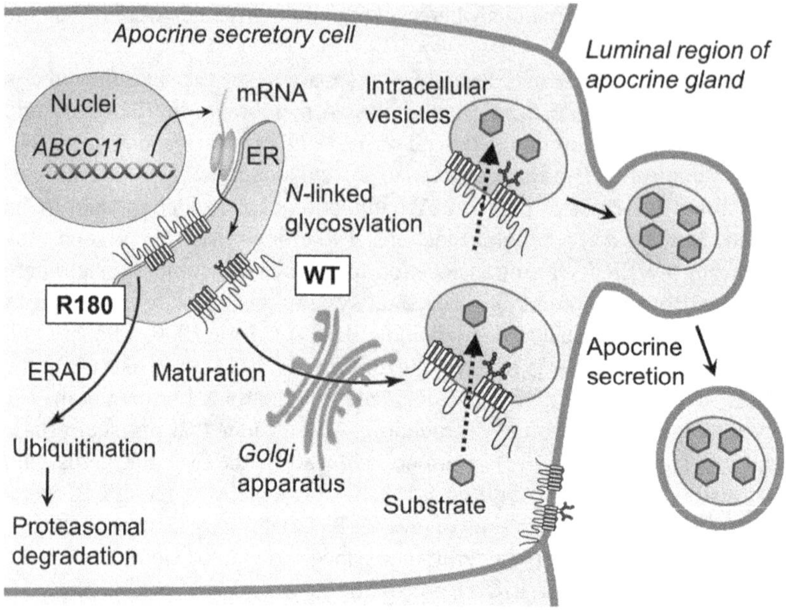

Figure 17.4 Schematic illustration of intracellular sorting of ABCC11 WT and proteasomal degradation of the R180 (Arg180) variant in secretory cells of the ceruminous apocrine gland. *De novo* synthesized ABCC11 WT is *N*-linked glycosylated at Asn838 and Asn844 in the ER, further processed in the *Golgi* apparatus, and destined for the membranes of intracellular granules and vacuoles. Ceruminous components are thought to be transported by ABCC11 WT and sequestered in intracellular granules and vacuoles. SNP variant R180 lacking *N*-linked glycosylation is recognized as a misfolded protein in the ER and readily undergoes ubiquitination and proteasomal degradation (ERAD pathway). ER, endoplasmic reticulum; ERAD, ER-associated degradation.

the women with genotypes (G/G + G/A) to develop breast cancer was estimated as 1.63 ($p = .026$), suggesting that the 538G allele in the *ABCC11* gene is moderately associated with the risk of breast cancer (Ota et al., 2010). Figure 17.5 shows that the relative ratio of breast cancer patients carrying the homozygous 538G/G allele was 1.77-fold greater than that of the corresponding healthy volunteers. This relative ratio was even greater than that (1.41-fold) for breast cancer patients carrying the heterozygous 538G/A allele. The G allele appears to be positively related to breast cancer frequency in the groups of Japanese women studied. In contrast, no significant association with breast cancer risk was observed in Europeans (Lang et al., 2011).

We initially thought that some genetically determined variation(s) in the apocrine system might influence susceptibility to breast cancer, although the genetic determinant (538G>A SNP in *ABCC11*) was not known at that time. It is hypothesized that the function of ABCC11 *per se*, or metabolites transported by ABCC11, may stimulate the proliferation of apocrine gland cells to enhance the risk of mastopathy (Fig. 17.6). This hypothesis is supported by the evidence that the apocrine glands are large in individuals carrying the WT allele of the *ABCC11* gene. So far as the cell

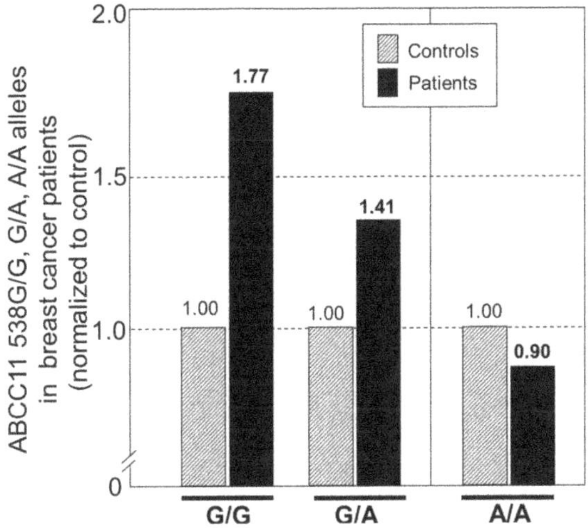

Figure 17.5 Relative ratios of breast cancer patients carrying 538G/G, 538G/A, and 538 A/A alleles in *ABCC11* gene. The genotypes of ABCC11 538G>A were compared for Japanese breast cancer patients and control subjects (female volunteers). Data are from Ota et al. (2010).

cycle machinery is operating normally, proliferation of apocrine gland cells should be controlled to a certain extent. When a somatic mutation has occurred in *BRCA1*, *BRCA2*, *p53*, or *p21*, however, it can lead to deleterious and unregulated proliferation of those cells (Fig. 17.6).

17.7 RELEVANCE TO DRUG RESISTANCE IN CANCER CHEMOTHERAPY

The potential involvement of ABCC11 in drug resistance of breast cancer has recently been reported. For example, ABCC11 mRNA is found to be highly expressed in breast tumors (Bera et al., 2001; Yabuuchi et al., 2001; Bieche et al., 2004), in

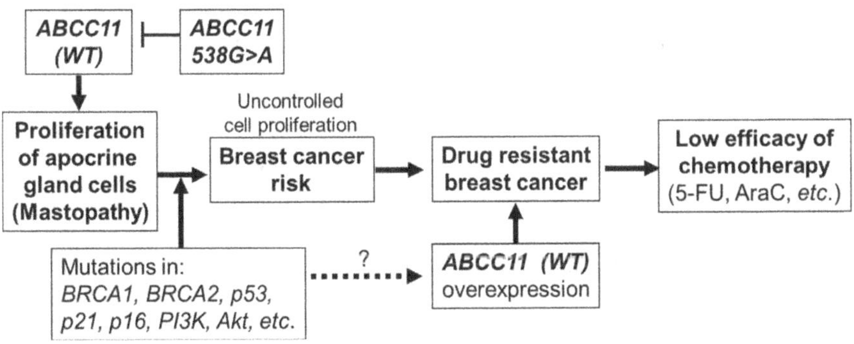

Figure 17.6 The potential impact of *ABCC11 WT* (538G) on the apocrine phenotype, patients' response to nucleoside-based chemotherapy, and the potential risk of mastopathy and breast cancer. BRCA-1, breast cancer-1; BRCA-2, breast cancer-2; PI3K, phosphatidylinositol 3-kinase; ERα (+), estrogen receptor α-positive; 5-FU, 5-fluorouracil; AraC, cytarabine.

particular, in invasive ductal adenocarcinomas (Available at: https://www.oncomine.org/resource/login.html. Accessed on 10 Jan 2013). Its expression is reportedly regulated by estrogen receptor-β (Honorat et al., 2008) and induced by 5-fluorouracil (5-FU) (Oguri et al., 2007). Furthermore, it has been reported that ABCC11 is directly involved in 5-FU resistance by means of the efflux transport of the active metabolite 5-fluoro-2′-deoxyuridine 5′-monophosphate (FdUMP) (Guo et al., 2003; Kruh et al., 2007; Oguri et al., 2007). It is of great interest to investigate whether the expression of ABCC11 WT (538G) is related to drug resistance of breast cancer and high rates of mortality. Further clinical studies, including protein expression studies in tumors, will be needed to clarify the potential contribution of ABCC11 to breast cancer risk and prognosis, including drug resistance and chemosensitivity.

Because of their structural similarities, it could be anticipated that substrate specificity of ABCC11 would be related to those of ABCC4 and ABCC5. This indeed has been the case. Ectopic expression of ABCC11 in mammalian cells enhances the cellular efflux of cyclic nucleotides and confers resistance to certain anticancer and antiviral nucleotide analogs (Guo et al., 2003). In fact, it has been reported that the ABCC11 WT has an ability to efflux cyclic nucleotides (e.g., cGMP and cAMP) and confers resistance to several antiviral and anticancer nucleotide analogs, such as 5′-FdUMP and 9′-(2′-phosohonylmethoxynyl)adenine (PMEA) (Guo et al., 2003; Kruh et al., 2007; Oguri et al., 2007).

Therapy with nucleoside-derived drugs is characterized by interindividual variability (Heinemann et al., 1988; Abbruzzese et al., 1991). Genetic variants that affect protein products involved in all steps leading to a drug's action may be the major contributors to this heterogeneity of responses to nucleoside-based treatments. In particular, variants of drug-metabolizing enzymes and transporters might affect the amount of drug needed for an efficient therapeutic response (Errasti-Murugarren and Pastor-Anglada, 2010).

Successful treatment of cancer remains a therapeutic challenge, with a high percentage of patients suffering from drug resistance or relapsed disease. One of such examples involves antileukemia treatment with nucleoside analos, such as cytarabine (Ara-C). Guo et al. have recently presented evidence that the expression of ABCC11 WT is an important factor affecting acute myeloid leukemia patient survival. It is very likely that the cause of treatment failure in those patients with high expression of ABCC11 WT is an increased extrusion of Ara-C from blast cells mediated by the transporter (Guo et al., 2009).

Uemura et al. have recently found that both gene and protein expression of ABCC11 were higher in pemetrexed (MTA)-resistant cells than in the parental cells (Uemura et al., 2010). The MTA-resistant cells showed cross-resistance to methotrexate (MTX), which is a substrate for ABCC11, and intracellular MTX accumulation in MTA-resistant cells was lower than that in the parental cells. They then tested the effect of decreasing the expression of ABCC11 by siRNA and found that decreased expression of ABCC11 enhanced MTA cytotoxicity and increased intracellular MTX accumulation in MTA-resistant cells. These findings suggest that ABCC11 confers resistance to MTA by enhancing the efflux of the intracellular anticancer drug.

They further analyzed the relationship between the *ABCC11* gene expression and MTA sensitivity of 13 adenocarcinoma cell lines. In contrast to their expectation, there was no correlation. Instead, the 13 lung adenocarcinoma cell lines could be classified into 3 groups based on the genotypes of the ABCC11 SNP (538G>A); G/G, G/A, and A/A. The A/A group showed a significant reduction in the IC_{50} value of MTA compared with the combined G/G and G/A groups, indicating that *ABCC11* 538G>A is an important determinant of MTA sensitivity. These results suggest that *ABCC11* 538G>A may be one of the biomarkers for selection of MTA treatment in adenocarcinomas. This finding, however, should be carefully evaluated by clinical studies to determine whether *ABCC11* 538G>A is truly a clinically important biomarker for the prediction of chemotherapeutic efficacy.

17.8 REGULATION OF ABCC11 GENE EXPRESSION

In 2004, Bieche et al. reported that ABCC11 was upregulated in estrogen receptor α-positive breast tumors, as compared with normal breast tissue. (Bieche et al., 2004). Park et al. investigated the mRNA levels of ABC transporter genes in breast cancer patients who underwent sequential weekly paclitaxel/FEC (5-fluorouracil, epirubicin, and cyclophosphamide) neoadjuvant chemotherapy (Park et al., 2006). Their analysis showed that the expression of ABCC11 was increased (fold ratio = 2.71) in those patients with residual disease as compared with the patients having no pathologic evidence of any residual invasive cancer cells in the breast.

More recently, Honorat et al. have demonstrated that endogenous ABCC11 mRNA levels in breast cell lines are correlated with their estrogen receptor α-status (Honorat et al., 2008). Interestingly, they found that ABCC11 expression was reduced *in vitro* by estradiol treatments. Furthermore, this estradiol-dependent downregulation of ABCC11 expression was blocked by cotreatment with tamoxifen, an estradiol antagonist. These findings suggest that the ABCC11 expression is regulated directly or indirectly by estrogen receptor α and that the prolonged exposure of breast cancer cells to tamoxifen can lead to upregulation of ABCC11.

Hauswald et al., on the other hand, have shown that some of the histone deacetylase inhibitors induced the expression of several ABC transporters, including the *ABCC11* gene, to render acute myeloid leukemia cells resistant to a broad spectrum of drugs (Hauswald et al., 2009). Molecular mechanisms underlying the induction remain to be elucidated. Since histone deacetylase inhibitors can be utilized in combination with conventional anticancer drugs in clinical trials, such induction of the ABCC11 WT may affect the efficacy of nucleoside-based chemotherapy.

17.9 CONCLUSION

In this chapter, we have addressed the potential impact of *ABCC11* 538G>A on the apocrine phenotype, patients' response to nucleoside-based chemotherapy, and the potential risk of breast cancer. Whether the SNP (538G>A) of the *ABCC11* gene can be considered a clinically important biomarker for the prediction of chemotherapeutic efficacy, however, should be carefully evaluated by clinical studies. In addition, we need to further explore the clinical significance of ABCC11 so as to elucidate whether there are any other diseases that involve apocrine secretion.

ACKNOWLEDGMENTS

The authors' study was supported by the Japan Science and Technology Agency (JST) research project named "Development of the world's fastest SNP detection system." Yu Toyoda is a research fellow of the Japanese Society for the Promotion of Science (JSPS).

REFERENCES

Abbruzzese JL, Grunewald R, Weeks EA, Gravel D, Adams T, Nowak B, Mineishi S, Tarassoff P, Satterlee W, Raber MN (1991) A phase I clinical, plasma, and cellular pharmacology study of gemcitabine. *J Clin Oncol*, 9, 491–498.

Bera TK, Lee S, Salvatore G, Lee B, Pastan I (2001) MRP8, a new member of ABC transporter superfamily, identified by EST database mining and gene prediction program, is highly expressed in breast cancer. *Mol Med*, 7, 509–516.

Bieche I, Girault I, Urbain E, Tozlu S, Lidereau R (2004) Relationship between intratumoral expression of genes coding for xenobiotic-metabolizing enzymes and benefit from adjuvant tamoxifen in estrogen receptor alpha-positive postmenopausal breast carcinoma. *Breast Cancer Res*, 6, R252–R263.

Bortfeld M, Rius M, Konig J, Herold-Mende C, Nies AT, Keppler D (2006) Human multidrug resistance protein 8 (MRP8/ABCC11), an apical efflux pump for steroid sulfates, is an axonal protein of the CNS and peripheral nervous system. *Neuroscience*, 137, 1247–1257.

Brooksbank BW (1970) Labelling of steroids in axillary sweat after administration of 3H-delta-5-pregnenolone and 14C-progesterone to a healthy man. *Experientia*, 26, 1012–1014.

Chen ZS, Guo Y, Belinsky MG, Kotova E, Kruh GD (2005) Transport of bile acids, sulfated steroids, estradiol 17-beta-D-glucuronide, and leukotriene C4 by human multidrug resistance protein 8 (ABCC11). *Mol Pharmacol*, 67, 545–557.

Ellgaard L, Helenius A (2001) ER quality control: towards an understanding at the molecular level. *Curr Opin Cell Biol*, 13, 431–437.

Ellgaard L, Molinari M, Helenius A (1999) Setting the standards: quality control in the secretory pathway. *Science*, 286, 1882–1888.

Errasti-Murugarren E, Pastor-Anglada M (2010) Drug transporter pharmacogenetics in nucleoside-based therapies. *Pharmacogenomics*, 11, 809–841.

Gesase AP, Satoh Y (2003) Apocrine secretory mechanism: recent findings and unresolved problems. *Histol Histopathol*, 18, 597–608.

Guo Y, Kock K, Ritter CA, Chen ZS, Grube M, Jedlitschky G, Illmer T, Ayres M, Beck JF, Siegmund W, et al. (2009) Expression of ABCC-type nucleotide exporters in blasts of adult acute myeloid leukemia: relation to long-term survival. *Clin Cancer Res*, 15, 1762–1769.

Guo Y, Kotova E, Chen ZS, Lee K, Hopper-Borge E, Belinsky MG, Kruh GD (2003) MRP8, ATP-binding cassette C11 (ABCC11), is a cyclic nucleotide efflux pump and a resistance factor for fluoropyrimidines 2′,3′-dideoxycytidine and 9′-(2′-phosphonylmethoxyethyl)adenine. *J Biol Chem*, 278, 29509–29514.

Hampton RY (2002) ER-associated degradation in protein quality control and cellular regulation. *Curr Opin Cell Biol*, 14, 476–482.

Hauswald S, Duque-Afonso J, Wagner MM, Schertl FM, Lubbert M, Peschel C, Keller U, Licht T (2009) Histone deacetylase inhibitors induce a very broad, pleiotropic anticancer drug resistance phenotype in acute myeloid leukemia cells by modulation of multiple ABC transporter genes. *Clin Cancer Res*, 15, 3705–3715.

Heinemann V, Hertel LW, Grindey GB, Plunkett W (1988) Comparison of the cellular pharmacokinetics and toxicity of 2',2'-difluorodeoxycytidine and 1-beta-D-arabinofuranosylcytosine. *Cancer Res*, 48, 4024–4031.

Helenius A, Aebi M (2004) Roles of N-linked glycans in the endoplasmic reticulum. *Annu Rev Biochem*, 73, 1019–1049.

Honorat M, Mesnier A, Vendrell J, Guitton J, Bieche I, Lidereau R, Kruh GD, Dumontet C, Cohen P, Payen L (2008) ABCC11 expression is regulated by estrogen in MCF7 cells, correlated with estrogen receptor alpha expression in postmenopausal breast tumors and overexpressed in tamoxifen-resistant breast cancer cells. *Endocr Relat Cancer*, 15, 125–138.

Ing R, Petrakis L, Ho HC (1973) Evidence against association between wet cerumen and breast cancer. *Lancet*, 1, 41.

Inoue Y, Mori T, Toyoda Y, Sakurai A, Ishikawa T, Mitani Y, Hayashizaki Y, Yoshimura Y, Kurahashi H, Sakai Y (2010) Correlation of axillary osmidrosis to a SNP in the ABCC11 gene determined by the Smart Amplification Process (SmartAmp) method. *J Plast Reconstr Aesthet Surg*, 63, 1369–1374.

Julesz M (1968) New advances in the field of androgenic steroidogenesis of the human skin. *Acta Med Acad Sci Hung*, 25, 273–285.

Kleizen B, Braakman I (2004) Protein folding and quality control in the endoplasmic reticulum. *Curr Opin Cell Biol*, 16, 343–349.

Kruh GD, Guo Y, Hopper-Borge E, Belinsky MG, Chen ZS (2007) ABCC10, ABCC11, and ABCC12. *Pflugers Arch*, 453, 675–684.

Labows JN, Preti G, Hoelzle E, Leyden J, Kligman A (1979) Steroid analysis of human apocrine secretion. *Steroids*, 34, 249–258.

Lang T, Justenhoven C, Winter S, Baisch C, Hamann U, Harth V, Ko YD, Rabstein S, Spickenheuer A, Pesch B, et al. (2011) The earwax-associated SNP c.538G>A (G180R) in ABCC11 is not associated with breast cancer risk in Europeans. *Breast Cancer Res Treat*, 129, 993–999.

Martin A, Saathoff M, Kuhn F, Max H, Terstegen L, Natsch A (2010) A functional ABCC11 allele is essential in the biochemical formation of human axillary odor. *J Invest Dermatol*, 130, 529–540.

Matsunaga E (1962) The dimorphism in human normal cerumen. *Ann Hum Genet*, 25, 273–286.

Miura K, Yoshiura K, Miura S, Shimada T, Yamasaki K, Yoshida A, Nakayama D, Shibata Y, Niikawa N, Masuzaki H (2007) A strong association between human earwax-type and apocrine colostrum secretion from the mammary gland. *Hum Genet*, 121, 631–633.

Mori K (2000) Tripartite management of unfolded proteins in the endoplasmic reticulum. *Cell*, 101, 451–454.

Nakano M, Miwa N, Hirano A, Yoshiura K, Niikawa N (2009) A strong association of axillary osmidrosis with the wet earwax type determined by genotyping of the ABCC11 gene. *BMC Genet*, 10, 42.

Oguri T, Bessho Y, Achiwa H, Ozasa H, Maeno K, Maeda H, Sato S, Ueda R (2007) MRP8/ABCC11 directly confers resistance to 5-fluorouracil. *Mol Cancer Ther*, 6, 122–127.

Ota I, Sakurai A, Toyoda Y, Morita A, Sasaki T, Chishima T, Yamakado M, Kawai Y, Ishidao T, Lezhava A, et al. (2010) Association between breast cancer risk and the wild-type allele of human ABC transporter ABCC11. *Anticancer Res*, 30, 5189–5194.

Park S, Shimizu C, Shimoyama T, Takeda M, Ando M, Kohno T, Katsumata N, Kang YK, Nishio K, Fujiwara Y (2006) Gene expression profiling of ATP-binding cassette (ABC) transporters as a predictor of the pathologic response to neoadjuvant chemotherapy in breast cancer patients. *Breast Cancer Res Treat*, 99, 9–17.

Petrakis NL (1971) Cerumen genetics and human breast cancer. *Science*, 173, 347–349.

Sakurai A, Onishi Y, Hirano H, Seigneuret M, Obanayama K, Kim G, Liew EL, Sakaeda T, Yoshiura K, Niikawa N, Sakurai M, Ishikawa T (2007) Quantitative structure-activity relationship analysis and molecular dynamics simulation to functionally validate nonsynonymous polymorphisms of human ABC transporter ABCB1 (P-glycoprotein/MDR1). *Biochemistry*, 46, 7678–7693.

Shehadeh NH, Kligman AM (1963) The effect of topical antibacterial agents on the bacterial flora of the axilla. *J Invest Dermatol*, 40, 61–71.

Shimizu H, Taniguchi H, Hippo Y, Hayashizaki Y, Aburatani H, Ishikawa T (2003) Characterization of the mouse Abcc12 gene and its transcript encoding an ATP-binding cassette transporter, an orthologue of human ABCC12. *Gene*, 310, 17–28.

Shugyo Y, Sudo N, Kanai K, Yamashita T, Kumazawa T, Kanamura S (1988) Morphological differences between secretory cells of wet and dry types of human ceruminous glands. *Am J Anat*, 181, 377–384.

Tammur J, Prades C, Arnould I, Rzhetsky A, Hutchinson A, Adachi M, Schuetz JD, Swoboda KJ, Ptacek LJ, Rosier M, Dean M, Allikmets R (2001) Two new genes from the human ATP-binding cassette transporter superfamily, ABCC11 and ABCC12, tandemly duplicated on chromosome 16q12. *Gene*, 273, 89–96.

Toyoda Y, Hagiya Y, Adachi T, Hoshijima K, Kuo MT, Ishikawa T (2008) MRP class of human ATP binding cassette (ABC) transporters: historical background and new research directions. *Xenobiotica*, 38, 833–862.

Toyoda Y, Sakurai A, Mitani Y, Nakashima M, Yoshiura K, Nakagawa H, Sakai Y, Ota I, Lezhava A, Hayashizaki Y, Niikawa N, Ishikawa T (2009) Earwax, osmidrosis, and breast cancer: why does one SNP (538G>A) in the human ABC transporter ABCC11 gene determine earwax type? *FASEB J*, 23, 2001–2013.

Uemura T, Oguri T, Ozasa H, Takakuwa O, Miyazaki M, Maeno K, Sato S, Ueda R (2010) ABCC11/MRP8 confers pemetrexed resistance in lung cancer. *Cancer Sci*, 101, 2404–2410.

Yabuuchi H, Shimizu H, Takayanagi S, Ishikawa T (2001) Multiple splicing variants of two new human ATP-binding cassette transporters, ABCC11 and ABCC12. *Biochem Biophys Res Commun*, 288, 933–939.

Yoo WM, Pae NS, Lee SJ, Roh TS, Chung S, Tark KC (2006) Endoscopy-assisted ultrasonic surgical aspiration of axillary osmidrosis: a retrospective review of 896 consecutive patients from 1998 to 2004. *J Plast Reconstr Aesthet Surg*, 59, 978–982.

Yoshiura K, Kinoshita A, Ishida T, Ninokata A, Ishikawa T, Kaname T, Bannai M, Tokunaga K, Sonoda S, Komaki R, et al. (2006) A SNP in the ABCC11 gene is the determinant of human earwax type. *Nat Genet*, 38, 324–330.

CHAPTER **18**

FUTURE PERSPECTIVES

Toshihisa Ishikawa
Joseph Ware

18.1 INTRODUCTION

Since initial sequencing and analysis of the human genome was accomplished (Lander et al., 2001; Venter et al., 2001), a huge effort has been put into medical research focused on associating genomic variations with individual phenotypes. Personalized medicine is often defined as "the right treatment for the right person at the right time." While the market for diagnostic tests and therapies that leverage this new science is growing, the biggest opportunities exist outside of the traditional healthcare sector. The personalized medicine market is projected to grow by 11.56% annually and is expected to reach US $148.4 billion by 2015 (Jain, 2011). While personalized medicine is already being considered in drug development strategies, it is still at an early stage with respect to clinical applications that support patient-specific therapy. It is expected that the rapidly growing knowledge on individual genome variations obtained by genome sequencing (1000 Genome Project Consortium, 2010) as well as genome-wide association studies (GWAS) (Rosenberg et al., 2010) will lead to much better understanding of personal health risks for disease (Ashley et al., 2010) and drug-induced adverse reactions (Daly, 2010).

The number of polymorphisms identified in genes encoding drug metabolizing enzymes, drug transporters, and receptors are rapidly increasing. In many cases, these genetic factors majorly impact the pharmacokinetics and pharmacodynamics of a particular drug and thereby influence the sensitivity to such drug in an individual patient with a certain genotype. The highest impact is seen for drugs with a narrow therapeutic index, with important examples emerging from treatment with antidepressants, oral anticoagulants, and chemotherapeutics, which are metabolized by CYP2D6/CYP2C9, VKORC1, and TPMT, respectively. To apply the ever increasing amounts of pharmacogenomics knowledge to clinical practice, specific dosage recommendations based on genotypes will have to be developed to guide the clinician; and these recommendations will have to be evaluated by the use of prospective clinical studies. Such efforts will lead to the development of personalized medicines,

Pharmacogenomics of Human Drug Transporters: Clinical Impacts, First Edition.
Edited by Toshihisa Ishikawa, Richard B. Kim and Jörg König

which would be expected to exhibit higher efficacy with fewer adverse drug reactions, thereby improving the therapeutic index for drugs whose pharmacokinetics, pharmacodynamics, and safety are influenced by pharmacogenetics.

The rapid growth of personalized medicine is being supported by emerging new technologies together with accumulating knowledge of pharmacogenomics. Basic technologies of molecular diagnostics play a role in expanding pharmacogenomic information, particularly with respect to SNP genotyping. Diagnosis is thus integrated with therapy for selecting treatments as well for monitoring results. Cost-effective methods should be developed for genotyping, however, and it would be desirable to include this information in the patient's record as guidance for physicians to individualize the treatment. The accurate measurement of allele frequency variations among population groups with different sensitivities to diseases and/or different responses to drugs is fundamental to genetic epidemiology. Genotyping errors can markedly influence the biological conclusions of a study. Thus, it is critically important to choose the appropriate method for SNP detection.

18.2 NEXT GENERATION SEQUENCING TECHNOLOGY

The proliferation of large-scale DNA-sequencing projects in recent years has driven a search for alternative methods that would reduce the time and costs. Currently, the rate of technological advancements in next-generation sequencing (NGS) platforms is projected to achieve rapid drops in sequencing costs. NGS technologies are expected to reduce the cost for whole genome analysis to $1,000 (1000 Genome Project Consortium, 2010; Rubenstein, 2010).

The role of sequencing in life sciences and clinical applications will change dramatically over the next few years. In fact, these NGS technologies are opening up a new era in healthcare and drug discovery/development. As a result, the sequencing of complete diploid human genomes is becoming increasingly affordable. Datasets thus obtained will provide support for researchers and clinicians to develop improved prevention and intervention strategies. Nevertheless, several issues (cost, ethics, standardization of sample preparation, etc.) still remain to be resolved to enable complete genome sequencing as a standard component of a patient's medical care.

Personalized medicine may be supported, in part, by the latest efforts in personal genome sequencing (Fig. 18.1). Analysis of a patient's full genome sequence may facilitate various clinical assessments, risk prediction for disease, and genetic counseling. Disease and risk analyses focus on the prediction of genetic risk of variants associated with Mendelian diseases, recognized drug responses, and pathogenicity for novel variants. Although challenges remain, it is strongly suggested that whole-genome sequencing can yield useful and clinically relevant information for individual patients.

While the scientific community has largely accepted the utility of sequencing for research purposes, the use of the NGS technology in a clinical setting has yet to be fully addressed or accepted by the medical community. The formation of collaborations to bridge the gap between research and medicine will provide a means

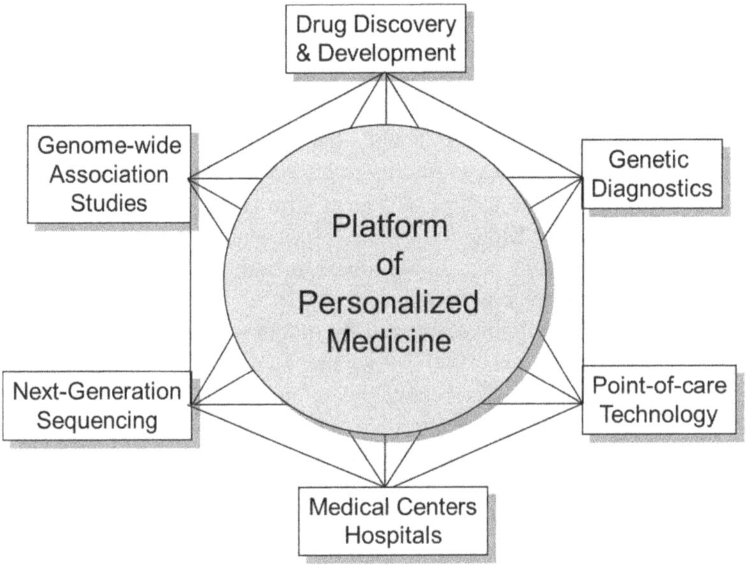

Figure 18.1 Schematic illustration representing the relationships among next-generation sequencing, GWAS, drug discovery and development, genetic diagnostics, personal medicine, and point-of-care.

not only to demonstrate the clinical utility of sequencing but also to pave the way for potential use of this technology by a new class of physicians and their patients.

18.3 GENOME-WIDE ASSOCIATION STUDIES

SNP maps and candidate-gene strategies are based on existing knowledge of a medication's mechanisms of action and pathways of metabolism and disposition. The candidate-gene strategy has the advantage of focusing resources on a manageable number of genes and polymorphisms that are likely to be important; however, the limitation is that the knowledge of a medication's pharmacokinetics and mechanisms of action is incomplete.

GWAS have broken a bottleneck in the understanding of common complex diseases and drug-induced adverse reactions by successfully identifying SNPs associated with many human traits (McCathy et al., 2008; Rosenberg et al., 2010). GWAS use SNP arrays in automated systems that analyze about 500,000 to one million sites where SNPs tend to occur without a priori knowledge of underlying molecular mechanisms. Scientists have used these SNP arrays over the past decade to compare DNA samples between healthy subjects and patients and have consequently identified thousands of SNPs that are associated with common complex diseases.

Recently, GWAS have revealed a strong association for an increased risk of statin-induced myopathy with SNPs (rs4263657 and rs4149056) in the *SLCO1B1* gene ($p = 4 \times 10^{-9}$) (The SEARCH Collaborative Group, 2008). The *SLCO1B1* gene encodes the organic anion-transporting polypeptide OATP1B1 that mediates the hepatic uptake of statins (see Chapter 6 for more details). The SNP (rs4263657) in the noncoding region of the gene is in nearly complete linkage disequilibrium with a nonsynonymous SNP (rs4149056). While the incidence of myopathy has been reported to range from approximately one to seven cases per 10,000 patients per year with standard doses of statins (e.g., 20 to 40 mg of simvastatin daily), it increases with higher doses (e.g., 80 mg of simvastatin daily) (Armitage, 2007; Law and Rudnicka, 2006; Molokhia et al., 2008). Statin-induced myopathy is characterized by severe muscle pain or weakness in association with elevated creatine kinase levels, which may persist for one year or longer. Moreover, in the worst-case situation, muscle breakdown and myoglobin release produce renal failure and death (Thompson et al., 2003; Joy and Hegele, 2009). More than 60% of statin-induced myopathy cases have been attributed to the c521T>C variant (rs4149056), which causes an amino acid alteration (V174A) (Niemi et al., 2011). In 2004, Niemi et al. reported that high plasma concentrations of pravastatin were associated with haplotypes in the *SCL1B1* gene (Niemi et al., 2004). Recently, the prescribing information for rosuvastatin has been modified to reflect the significance of dose selection with respect to the observed pharmacokinetic differences between patients of Asian descent versus those of other ethnicities (Rosuvastatin U.S. Prescribing Information, http://www1.astrazeneca-us.com/pi/crestor.pdf).

In addition, more recently, large meta-analyses of GWAS have revealed that SNPs in the *SLC2A9* (GLUT9) and *ABCG2* genes are strongly associated with the phenotype of gout (Dehghan et al., 2008; Kolz et al., 2009; Stark et al., 2009). Since serum levels of uric acid are highly heritable, the involvement of genetic factors in gout has been previously speculated. Several laboratories have independently found that the SNP 421C>A in the *ABCG2* gene is one of the major genetic factors for elevated serum uric acid levels and an increased risk of gout (Matsuo et al., 2009; Woodward et al., 2009). ABCG2 expressed on the apical side of the proximal tubular cells in human kidney plays a pivotal role in renal excretion of serum uric acid. Since the expression levels of the Q141K variant are reduced by ubiquitin-mediated proteasomal degradation (Tamura et al., 2007; Nakagawa et al., 2008; Furukawa et al., 2009), renal excretion of serum uric acid via ABCG2 is impaired in persons who are carrying the 421A allele (Q141K variant). As a consequence, serum uric acid levels are elevated, which enhances the risk of gout.

GWAS are successful in identifying risk-associated loci, but the pinpointing of causal variants in subsequent fine-mapping studies remains a challenge. A conventional fine-mapping effort starts by sequencing dozens of randomly selected samples at susceptibility loci to discover candidate variants. These are then placed on custom arrays and examined by the use of algorithms to find the causal variants. The high cost of genotyping SNP arrays in large datasets remains a major barrier to widespread utilization of such studies. To solve such a cost-related problem, Craig et al. (2009) have recently proposed a rapid and inexpensive GWAS approach using pooled whole blood, whereby whole blood from both cases and controls is pooled prior to DNA extraction for genotyping. The pooling-of-blood (POB) method may be easily applied to array-based GWAS, which aim to identify common variants of small effect (Craig et al., 2009).

18.4 POINT-OF-CARE TECHNOLOGY

The development of PCR in 1985 marked a major step forward in molecular biology, and PCR has since become a fundamental tool supporting numerous research and diagnostic applications (Saiki et al., 1985). Today, PCR largely dominates molecular testing. In pharmacogenomics, for example, PCR is used in combination with restriction fragment length polymorphism analysis (RFLP), one of the first PCR-based methods applied to SNP analysis (Saperstein and Nickerson, 1991). Current systems for clinical diagnostic applications are mainly PCR-based, can only be used in hospitals, and are still relatively complex and expensive.

More recently, new methods have emerged that enable nucleic acid amplification under isothermal conditions. Table 18.1 provides a list of current isothermal nucleic acid amplification methods, including nucleic acid sequence-based amplification (NASBA) (Gracias and McKillip, 2007), transcription-mediated amplification (TMA) (Hofmann et al., 2005), signal-mediated amplification of RNA technology (SMART) (Wharam et al., 2007), helicase-dependent amplification (HDA) (Vincent et al., 2004; Jeong et al., 2009), recombinase polymerase amplification (RPA) (Piepenburg et al., 2006; Lutz et al., 2010), loop-mediated isothermal amplification (LAMP) (Curtis et al., 2009; Mori and Nakatomi, 2009), cross-priming amplification (CPA) (Fang et al., 2009), smart amplification (SmartAmp) (Mitani et al., 2007, 2009), rolling circle amplification (RCA) (Johne et al., 2009), and ramification amplification (RAM) (Yao et al., 2009). These methods are being used as a tool for the nucleic acid testing of infectious diseases at the point of care (Morlighem et al., 2011; Niemz et al., 2011). To effectively advance personalized medicine, it is

TABLE 18.1 Examples of Isothermal Nucleic Acid Amplification Methods

Methods	Target	Temperature (ºC)	Reaction time (min)	References
NASBA	RNA (DNA)	41	105	Gracias and McKillip (2007)
TMA	RNA (DNA)	60	140	Hofman et al. (2005)
SMART	RNA (DNA)	41	180	Wharam et al. (2007)
HAD	DNA	65	75–90	Jeong et al. (2009); Vincent et al. (2004)
RPA	DNA	30–42	20	Lutz et al. (2010); Piepenburg et al. (2006)
LAMP	DNA	60–65	60–90	Curtis et al. (2009); Mori and Nakatomi (2009)
CPA	DNA	65	60	Fang et al. (2009)
SmartAmp	DNA	60	30–60	Mitani et al. (2007), (2009)
RCA	DNA	65	60	Johne et al. (2009)
RAM	DNA	63	120–180	Yao et al. (2009)

NASBA, nucleic acid sequence-based amplification; TMA, transcription-mediated amplification; SMART, signal-mediated amplification of RNA technology; HAD, helicase-dependent amplification; RPA, recombinase polymerase amplification; LAMP, loop-mediated isothermal amplification; CPA, cross-priming amplification; SmartAmp, smart amplification; RCA, rolling circle amplification; RAM, ramification amplification necessary to be able to rapidly and conveniently test for patients' genetic polymorphisms and/or mutations. The SmartAmp method has recently been developed based on the principal concept that DNA amplification itself is the signal for detection of a genetic mutation or SNP (Mitani et al., 2009; Ishikawa and Hayashizaki, 2012). The molecular mechanism underlying isothermal DNA amplification in the SmartAmp reaction and its clinical applications for SNP typing are described in Chapter 5.

Isothermal amplification methods successfully compete with already established PCR methods to provide a preferable choice for new point-of-care testing. Moreover, isothermal amplification methods carry great advantages for the design of highly automated lab-on-chip solutions. Challenges remain, however, to integrate sample preparation with isothermal nucleic acid amplification and detection in a cost-effective, robust, and user-friendly format.

18.5 COMPANION DIAGNOSTICS

Innovative biotechnology and a small number of pharmaceutical companies have for the past 5–10 years integrated companion diagnostics into the drug development continuum. The US Food and Drug Administration (FDA) has recently approved two cancer drugs, that is, vemurafenib (Zelboraf) for BRAF-mutation-positive metastatic melanoma and crizotinib (Xalkori) for non-small-cell lung cancer driven by an anaplastic lymphoma kinase (ALK) fusion gene, together with their companion diagnostic tests (Schubert, 2011). This is in contrast to the typical dogma where pharmaceutical companies and diagnostics companies tended to work separately. The combination of diagnostics with therapy may have a great impact on the diagnostics industry, especially those companies that are linking their diagnostic know-how with therapy would be able to create a new market place. The companion diagnostic can facilitate the approval of a drug because it is effective in a large percentage of patients. For this reason, major pharmaceutical companies have adopted new programs to deal with companion diagnostic products by looking for collaborations with molecular diagnostic companies even in early phases, when they are considering FDA submissions of their products (Fig. 18.2).

In 2005, the US FDA published guidance for pharmaceutical companies that recommends the submission of pharmacogenomic data. The guidance was intended to facilitate scientific progress in the field of pharmacogenomics and to encourage the use of pharmacogenomic data in drug development. To promote and enable the use of pharmacogenomics in drug development, the FDA is encouraging that full reports of exploratory pharmacogenomic data be submitted during the investigational new drug (IND) phase (Fig. 18.2). The FDA is also encouraging full pharmacogenomic reports when only abbreviated or synopsis reports are required for the later phase of drug approval (Leighton et al., 2004; Ruano et al., 2004; Salerno and Lesko, 2004a, 2004b; Trepiccho et al., 2004). Furthermore, to improve drug safety, the FDA has started to update labels of previously approved drugs as new clinical and genetic evidence accrues (Frueh et al., 2008). The labels of medical products serve to inform prescribers and patients about the potential ways to improve the benefit versus risk

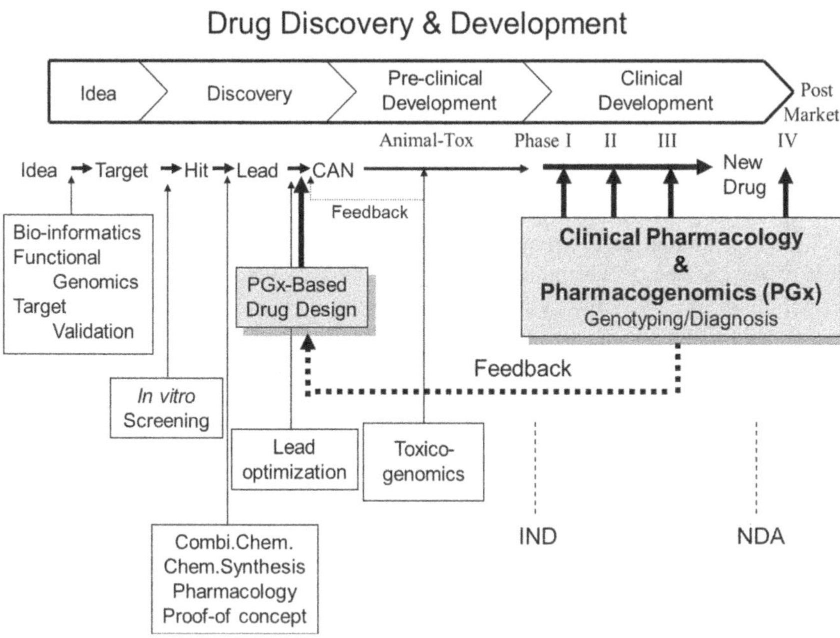

Figure 18.2 Schematic illustration for pharmacogenomics studies in the process of drug discovery and development.

ratio and/or optimize doses of the medical products. On August 16, 2007, the US FDA updated the label of warfarin to include the information on pharmacogenetic testing and to encourage the use of this information to determine the dosing for individual patients upon the initiation of warfarin therapy (Lesko, 2008; Kim et al., 2009).

Since personalized medicine is intrinsically linked to the development of diagnostic tests, the diagnostics side of the healthcare industry is expected to benefit a great deal from this shift, and this is especially so for those companies that have coordinated pharma and diagnostics divisions. Diagnostic tests enable the selection of one or several therapies. The AmpliChip CYP450 Test and other diagnostic kits can identify how individuals metabolize many of today's widely prescribed drugs. Although the AmpliChip CYP450 test is a valuable tool, its practical use in the field of medicine is still limited.

The fields of pharmacogenomics and pharmacogenetics have provided an exceptionally well-demonstrated platform for personalized medicine, which has ignited substantial growth in the availability of personalized medicine diagnostic products by emerging new technologies. Roche Diagnostics, Affymetrix Inc., HistoRx, Qiagen, Illumina Inc., Hologic Inc., and others are currently key players in the personalized medicine market. Their leadership may be challenged, however, by the emergence of new players working in this segment. By prolific segmentation of the market, new technologies would create a bigger platform for those players involved in the development of personalized medicine.

18.6 PHARMACOGENOMICS IN DRUG DISCOVERY AND DEVELOPMENT

Pharmacogenetics, which deals with heredity and response to drugs, is the scientific field that attempts to explain individual variability of drug responses and to search for the genetic basis for such variations or differences (Evans and Johnson, 2001). Interindividual variation in the rate of drug metabolism has been known for many years. Initially, the study of pharmacogenetics was only of academic interest, but today it is of major concern to the pharmaceutical industry as a means for documenting the metabolism of a new drug in development before registration. The knowledge of how a drug is metabolized and which enzymes are involved has provided a means for predicting drug-induced adverse effects and the rate at which individual patients might metabolize a specific drug.

Pharmacogenomics, on the other hand, broadly refers to the study of various DNA and RNA characteristics as related to drug response. Pharmacogenomics in drug discovery and development includes the use of genetic sequences and genomics information in patient management to enable therapy decisions. The genetic information includes the DNA sequence of the hosts (normal or diseased patients) or that of pathogens (viruses and microorganisms). Pharmacogenomic approaches will have an impact on all phases of drug development—from drug discovery to clinical trials.

Drug development is commonly described as progressing in phases I, II, and III. The first two phases (I and II) provide information about safety, pharmacokinetics, efficacy, and a broad range of doses, so that in phase III, the adequate and well-controlled trials needed to support marketing approval can be efficiently designed (e.g., with a reasonable range of doses, good patient selection criteria, appropriate endpoints). Typically, results of phase II studies that suggest genomic influences can lead to phase III trials that incorporate such findings into a pre-specified hypothesis, such as enriching the study with genomically defined individuals, determining a dose based on demonstrated variability in phase II studies, or defining a priori hypothesis testing of a primary endpoint in a genomic subset (Fig. 18.2). Genomic analyses are commonly carried out in phase III studies. In almost all cases, those post hoc analyses of efficacy will be considered exploratory; however, strong subset findings on safety (i.e., lack of efficacy) have on occasion been included in labeling. Pharmacogenomic information obtained from genomic investigations during the course of drug development and from post-marketing studies/trials can improve the effectiveness and safety of drugs. Corresponding examples of this are abacavir (antiretroviral drug for HIV treatment), clopidogrel (ADP-receptor antagonist to reduce atherothrombotic events), warfarin (anticoagulant to manage thromboembolic disorders), and irinotecan (topoisomerase I inhibitor for cancer therapy).

Evidence is accumulating to show that genetic polymorphisms in drug metabolizing enzymes, transporters, receptors, and other drug targets (e.g., toxicity targets) are linked to interindividual differences in the efficacy and toxicity of many medications (Andersson et al., 2005; Wilke et al., 2007; Sadee, 2008; Weiss et al., 2008). Drug discovery companies have long-accumulating expertise and potentials to modify lead compounds and drug candidates in order to screen and exclude from clinical trials patients of a certain genotype who are more prone to developing certain side effects. In particular, the contribution of polymorphic CYPs is widely evaluated as contributors to the metabolism of a lead compound. Assays of drug action typically evaluate biochemical and/or physiological activities. High-content screening by the use of cellular assays and emerging new technologies would help to bring together pharmacogenomics and medicinal chemistry (drug molecular design and chemical synthesis) by obtaining a broad range of information about the cellular physiology of drug action. Although it is not easy, our future goal is to accurately match therapeutic efficacy with biochemical and/or physiological activities.

To make the transition from a gene to an optimal therapeutic agent is a great challenge for drug discovery and development. Knowledge of genetic variations in a target enables early assessment of the clinical significance of polymorphisms through the appropriate design of both preclinical and clinical studies. A focused pharmacogenomic strategy at the preclinical phase of drug development can contribute to the decision-making process for full development of lead compounds. The availability of genomic samples in large phase IV trials provides a valuable resource for further understanding the molecular basis of disease heterogeneity by supplying data that feed back into the drug discovery process of target identification and validation for the next generation of improved medicines.

18.7 NEW CHALLENGES IN PHARMACOGENOMICS OF DRUG TRANSPORTERS

Validation of clinically important genetic polymorphisms and the development of new technologies to rapidly detect clinically important variants are critical issues for advancing pharmacogenomics-based drug discovery and development as well as personalized medicine. In the case of pharmacogenomics of drug transporters, it would be important to emphasize that the interdependencies of the substrate specificities of drug transporters and drug metabolizing enzymes of interest and their genotypes should be considered in the process of drug molecular design.

Since the first membrane transporter was cloned, our understanding of the role of transporters in drug disposition and response has grown enormously. The pharmacogenomics of membrane transporters (PMT) project is one of the leaders in the field of pharmacogenomics of drug transporters (Kroetz et al., 2010). This project is part of the Pharmacogenetics Research Network (PGRN) project sponsored by the National Institute of Health (NIH) (Giacomini et al., 2007). An overview of the scientific approaches of the PMT project is provided by Kroetz et al. in Chapter 4 of this book. The PMT has identified sequence variants in 129 membrane transporter genes including both the ATP-binding cassette (ABC) and the solute carrier (SLC) transporter families. SNPs discovered by the PMT are available on the Illumina 1M Duo Beadchip (Kroetz et al., 2010). These PMT SNPs will have an important role in informing current GWAS. Furthermore, the goal of the PMT's genomics and functional genomics studies is to generate testable hypotheses about the effects of genetic variants of drug transporters on clinical drug disposition and response. In this context, the PMT is expected to provide an important platform for proof-of-concept translational studies on human drug transporters.

18.8 FUNCTIONAL VALIDATION OF SNPs

In vitro studies of drug metabolism, transporters, or targets could help identify the need for human pharmacogenomics studies and contribute to the design and analysis of clinical studies. The following clinical pharmacology studies represent opportunities to integrate pharmacogenomic factors in the assessment of interindividual variability and its implications for subsequent clinical studies.

New technologies have been developed for *in vitro* evaluation of the impact of SNPs on the function and protein stability of drug transporters. Functional validation of "clinically important" SNPs and mutations in drug transporter genes will be a next important step. As the first steps in the development of new scientific tools and standardization of functional validation of drug transporters, the International Transporter Consortium (ITC) was created in 2007 by Kathleen Giacomini, Shiew-Mei Huang, Donald Tweedie, and Toshihisa Ishikawa (Huang et al., 2010). Thereafter, the ITC comprised scientists from academia, industry, and the FDA who are engaged in drug discovery, development, or regulatory approval. The FDA Critical Path Transporter Workshop was then held in October 2008, where participants discussed (1) an overview of key transporters, (2) current and future technical challenges, and (3) drug development and FDA examples. As a result of the Critical Path Workshop and enormous advances in the field of drug transporters, the ITC published a Whitepaper in 2010 (Giacomini et al., 2010).

It is one of our ultimate goals to translate the information of genetic polymorphisms into drug molecular design. In this regard, Ishikawa et al. have developed a new method for quantitative structure–activity relationship (QSAR) analysis, where high-speed screening and chemical fragmentation codes (CFCs) are used to quantitatively assess changes in the substrate specificity of a drug transporter (Sakurai et al., 2007; Ishikawa et al., 2010). The CFCs are a set of alphanumeric symbols, each representing a fragment of a chemical structure. The Markush TOPFRAG program is a tool for searching the chemical structures and structure information in Derwent's online databases. The extent of drug transport activity or transport inhibition is formulated as a linear combination of CFCs, each of which was weighed by the corresponding coefficient, C(i), as follows:

$$\text{Transport activity or inhibition} = \sum \text{C(i)} \times \text{Score(i)} + \text{Constant}$$

where the symbol (i) designates a specific CFC. The "score" means the presence [$\geq$1] or absence [0] of the corresponding CFC in the chemical structure of a test compound. Based on the CFCs and in comparison with the observed inhibition of transport activity for each test compound, chemical fragmentation coefficients, C(i), are calculated by multiple linear regression analysis (see Hirano et al., 2006; Saito et al., 2006; Sakurai et al., 2007; Ishikawa et al., 2010 as examples).

The QSAR analysis with CFCs would provide a practical tool to quantify the impact of genetic polymorphisms on the function of drug transporters. If the chemical structure of a drug is available, it will be possible to predict the effect of SNPs on the substrate specificity of drug transporter variants toward the drug by using the QSAR analysis method (Sakurai et al., 2007; Ishikawa et al., 2010).

18.9 COMPUTATIONAL SIMULATION ANALYSIS

To date, the best studied transporter polymorphisms have been those in the coding region of transporter genes. The amino acid sequence of proteins is generally believed to determine protein expression (stability and degradation), folding, and function. Some variants result in reduced trafficking of the transporter protein to the cell membrane, resulting from incorrect folding and/or endoplasmic reticulum-associated degradation (ERAD) (Nakagawa et al., 2011). Analysis of large numbers of SNPs in the coding regions of transporters demonstrated that genetic diversity is significantly higher in loop domains compared to trans-membrane domains, suggesting that there is selective pressure against amino acid changes in these regions (Leabman et al., 2003).

Amino acid changes may alter key domains necessary for substrate binding, ATP hydrolysis, or protein folding. Molecular dynamics (MD) simulation may be useful to gain insight into the impact of nonsynonymous polymorphisms on structural changes that may affect the activity of drug transporters. MD simulation has revealed that nonsynonymous polymorphisms (2677G>T, A, or C) at amino acid position 893 (Ala > Ser, Thr, or Pro) in the human ABCB1 (MDR1/P-glycoprotein) gene have an impact on both the activity and the substrate specificity (Sakurai et al., 2007; Ishikawa et al., 2010). Computational analysis of drug transporters would provide hypotheses for domain interactions in the transporter proteins, which could then be verified experimentally.

18.10 ETHNIC DIFFERENCES IN GENETIC POLYMORPHISMS OF DRUG TRANSPORTERS

In selected situations, pharmacogenomic studies on healthy volunteers as well as patients might support a decision to perform such prospective association studies. If the results of these studies are significant and potential health or economic benefits of therapy are considerable, a major clinical trial can be considered to assess the usefulness of a pharmacogenomics-based therapy.

Inherited components of the response to drugs are often polygenic. Furthermore, the drug response is probably affected by multiple genes, with each gene having multiple polymorphisms distributed in the general population. Racial differences add further confounding factors. Drug response might be predicted from a certain pattern of polymorphisms rather than only a single polymorphism, yet these patterns are likely to differ among ethnic groups. These factors could complicate predictions about drug responses across the general patient population, emphasizing the need to perform clinical pharmacogenomics studies on different ethnic populations.

We present herewith an example in this respect. Thiopurines are effective immunosuppressants and anticancer agents, but intracellular accumulation of their active metabolites, for example, 6-thioguanine nucleotides (6-TGN), causes dose-limiting hematopoietic toxicity (Weinshilboum and Sladek, 1980). Thiopurine S-methyltransferase (TPMT) deficiency has been reported to exacerbate thiopurine toxicity (Weinshilboum and Sladek, 1980). TPMT activity has been proven by numerous studies to have an inverse correlation to 6-TGN levels in erythrocytes and other hematopoietic tissues. Polymorphisms in the TPMT gene can lead to intermediate, low, or no TPMT activity; and patients with such reduced TPMT levels are at an increased risk of developing life-threatening hematologic toxicity. Population studies have shown that approximately 10% of Caucasians and African Americans inherit one nonfunctional TPMT allele called *3A. This nonfunctional allele is not common in Asians (Anstey et al., 1992; Yates et al., 1997; Black et al., 1998; Stolk et al., 1998; Clunie and Lennard, 2004). For largely unknown reasons, however, subsets of patients who suffer from dose-limiting hematopoietic toxicity are not TPMT deficient. Recent studies have revealed that ABCC4 protects against thiopurine-induced hematopoietic toxicity by actively exporting thiopurine nucleotides (Schuetz et al., 1999; Krishnamurthy et al., 2008). Interestingly, the *ABCC4* 2269G>A SNP is common in the Japanese population (>18%), which suggests that this nonsynonymous SNP could provide an explanation for the unsolved thiopurine toxicity that is not associated with genetic polymorphisms of TPMT (Ban et al., 2010).

Functional validation of genetic variants is a prerequisite to validate the clinical significance of genetic polymorphisms. Moreover, understanding of the dynamic complexity of the human genome, involvement of multiple genes in drug response, and ethnic differences in the prevalence of gene variants is important to assure the progress of personalized medicine toward practical clinical applications.

18.11 ETHICAL ISSUES TO BE RESOLVED

The value of DNA sample collection, and the information that the analysis of these samples can provide, will vary for different drugs and indications. Various technology platforms are available to rapidly characterize the contribution of established and evolving allelic variations of hundreds of metabolism and transporter genes as well as disease risk-related genes in clinical pharmacology studies. To design informative studies and interpret study results appropriately, careful attention should be given in clinical pharmacology studies to differences in the prevalence of gene variants among racially or ethnically distinct groups.

The collection and archiving of sensitive information on an individual's genetic makeup raises questions of privacy and security as well as ethical dilemmas with respect to disease prognosis and treatment choices. Polymorphisms relevant to drug response may overlap with disease susceptibility, and the divulging of such information could jeopardize an individual and his family. Thus, ethical issues also need to be resolved in personalized medicine. On the other hand, legal issues may force the inclusion of pharmacogenomics into clinical practice. Once the genetic component of a severe adverse drug effect is documented, doctors may be obliged to order genetic testing to avoid malpractice litigation.

18.12 CONCLUSION

Pharmacogenomics, the study of the influence of genetic factors on drug action and metabolism, is used for predicting drug efficacy and drug-induced adverse reactions.

This knowledge of the effects of polymorphisms in drug transporters is applicable in drug discovery and development as well as in the clinical use of drugs. Increases in efficacy and safety by the individualization of medical treatment may have benefits in financial terms, if information is presented to show that personalized medicine will be cost-effective in healthcare systems.

For the pharmaceutical companies, segmentation of the market may not be conducive for producing conventional blockbusters, but smaller and exclusive markets for personalized medicines would be profitable in terms of drug safety and efficacy. Development of personalized medicine, including point-of-care technology, requires the integration of various segments of biotechnology, clinical medicine, and pharmacology. Several players should be involved in the development of personalized therapy. Pharmaceutical and biotechnology companies would take leading roles in this venture in keeping with their future roles as healthcare enterprises.

REFERENCES

Andersson T, Flockert DA, Goldstein DB, Huang SM, Kroetz DL, Milos PM, Ratain MJ, Thummel K (2005) Drug-metabolizing enzymes: evidence for clinical utility of pharmacogenomic tests. *Clin Pharmacol Ther*, 78, 559–581.

Anstey A, Lennard L, Mayou SC, Kirby JD (1992) Pancytopenia related to azathioprine–an enzyme deficiency caused by a common genetic polymorphism: a review. *J Royal Soc Med*, 85, 752–756.

Armitage J (2007) The safety of statins in clinical practice. *Lancet*, 370, 1781–1790.

Ashley EA, Butte AJ, Wheeler MT, Chen R, Klein TE, Dewey FE, Dudley JT, Ormond KE, Pavlovic A, Morgan AA, et al. (2010) Clinical assessment incorporating a personal genome. *Lancet*, 375, 1525–1535.

Ban H, Andoh A, Imaeda H, Kobori A, Bamba S, Tsujikawa T, Sasaki M, Saito Y, Fujiyama Y (2010) The multidrug-resistance protein 4 polymorphism is a new factor accounting for thiopurine sensitivity in Japanese patients with inflammatory bowel disease. *J Gastroenterol*, 45, 1014–1021.

Black AJ, McLeod HL, Capell HA, Powrie RH, Matowe LK, Pritchard SC, Collie-Duguid ES, Reid DM (1998) Thiopurine methyltransferase genotype predicts therapy-limiting severe toxicity from azathioprine. *Ann Internal Med*, 129, 716–718.

Clunie GP, Lennard L (2004) Relevance of thiopurine methyltransferase status in rheumatology patients receiving azathioprine. *Rheumatology*, 43, 13–18.

Craig JE, Hewitt AW, McMellon AE, Henders AK, Ma L, Wallace L, Sharma S, Burdon KP, Visscher P, Montgomery GW, MacGrergor S (2009) Rapid inexpensive genome-wide association using pooled whole blood. *Genome Res*, 19, 2075–2080.

Curtis KA, Rudolph DL, Owen SM (2009) Sequence-specific detection method for reverse transcription, loop-mediated isothermal amplification of HIV-1. *J Med Virol*, 81, 966–972.

Daly AK (2010) Pharmacogenomics and human genetic polymorphisms. *Biochem J*, 429, 435–449.

Dehghan A, Kottgen A, Yang Q, Hwang SJ, Kao WL, Rivadeneira F, Boerwinkle E, Levy D, Hofman A, Astor BC, et al. (2008) Association of three genetic loci with uric acid concentration and risk of gout: a genome-wide association study. *Lancet*, 372, 1953–1961.

Evans WE, Johnson JA (2001) Pharmacogenomics: the inherited basis for interindividual differences in drug response. *Annu Rev Genomics Hum Genet*, 2, 9–39.

Fang R, Li X, Hu L, You Q, Li J, Wu J, Xu P, Zhong H, Luo Y, Mei J, Gao Q (2009) Cross-priming amplification for rapid detection of Mycobacterium tuberculosis in sputum specimens. *J Clin Microbiol*, 47, 845–857.

Frueh FW, Amur S, Mummaneni P, Epstein RS, Aubert RE, DeLuca TM, Verbrugge RR, Burckart GJ, Lesko LJ (2008) Pharmacogenomic biomarker information in drug labels approved by the United States food and drug administration: prevalence of related drug use. *Pharmacotherapy*, 28, 992–998.

Furukawa T, Wakabayashi K, Tamura A, Nakagawa H, Morishima Y, Osawa Y, Ishikawa T (2009) Major SNP (Q141K) variant of human ABC transporter ABCG2 undergoes lysosomal and proteasomal degradations. *Pharm Res*, 26, 469–479.

1000 Genome Project Consortium (2010) A map of human genome variation from population-scale sequencing. *Nature*, 467, 1061–1073.

Giacomini KM, Brett CM, Altman RB, Benowitz NL, Dolan ME, Flockhart DA, Johnson JA, Hayes DF, Klein T, Krauss RM, et al., for the Pharmacogenetics Research Network (2007) The pharmacogenetics research network: from SNP discovery to clinical drug response. *Clin Pharmacol Ther*, 81, 328–345.

Giacomini KM, Huang SM, Tweedie DJ, Benet LZ, Brouwer KL, Chu X, Dahlin A, Evers R, Fischer V, Hillgren KM, et al. (2010) Membrane Transporters in Drug Development: Report from the FDA Critical Path Initiative-Sponsored Workshop. *NatRev Drug Discov*, 9, 215–236.

Gracias KS, McKillip JL (2007) Nucleic acid sequence-based amplification (NASBA) in molecular bacteriology: a procedural guide. *J Rapid Methods Autom Microbiol*, 15, 295–309.

Hirano H, Kurata A, Onishi Y, Sakurai A, Saito H, Nakagawa H, Nagakura M, Tarui S, Kanamori Y, Kitajima M, Ishikawa T (2006) High-speed screening and QSAR analysis of human ATP-binding cassette transporter ABCB11 (bile salt export pump) To predict drug-induced intrahepatic cholestasis. *Mol Pharm*, 3, 252–265.

Hofmann WP, Dries V, Herrmann E, Gärtner B, Zeuzem S, Sarrazin C (2005) Comparison of transcription mediated amplification (TMA) and reverse transcription polymerase chain reaction (RT-PCR) for detection of hepatitis C virus RNA in liver tissue. *J Clin Virol*, 32, 289–293.

Huang S-M, Zhang L, Giacomini KM (2010) The international transporter consortium: a collaborative group of scientists from academia, industry and the FDA. *Clin Pharmacol Ther*, 87, 32–36.

Ishikawa T, Sakurai A, Hirano H, Lezhava A, Sakurai M, Hayashizaki Y (2010) Emerging new technologies in pharmacogenomics: rapid SNP detection, molecular dynamic simulation, and QSAR analysis methods to validate clinically important genetic variants of human ABC Transporter ABCB1 (P-gp/MDR1). *Pharmacol Ther*, 126, 69–81.

Ishikawa T, Hayashizaki Y (2012) Clinical SNP detection by the SmartAmp method. *Methods Mol Biol*, in press.

Jain KK (2011) *Personalized Medicine*, Basel, Switzerland: Jain PharmaBiotech, (http://pharmabiotech.ch/).

Jeong YJ, Park K, Kim DE (2009) Isothermal DNA amplification in vitro: the helicase-dependent amplification system. *Cell Mol Life Sci*, 66, 3325–3336.

Johne R, Müller H, Rector A, van Ranst M, Stevens H (2009) Rolling-circle amplification of viral DNA genomes using phi29 polymerase. *Trends Microbiol*, 17, 205–211.

Joy TR, Hegele RA (2009) Narrative review: statin-induced myopathy. *Ann Intern Med*, 150, 858–868.

Kim MJ, Huang SM, Meyer UA, Rahman A, Lesko LJ (2009) A regulatory science perspective on warfarin therapy: a pharmacogenetic opportunity. *J Clin Pharmacol*. 49, 138–146.

Kolz M, Johnson T, Sanna S, Teumer A, Vitart V, Perola M, Mangino M, Albrecht E, Wallace C, Farrall M, et al. (2009) Meta-analysis of 28,141 individuals identifies common variants within five new loci that influence uric acid concentrations. *PLoS Genet*, 5, e1000504.

Krishnamurthy P, Schwab M, Takenaka K, Nachagari D, Morgan J, Leslie M, Du W, Boyd K, Cheok M, Nakauchi H, et al. (2008) Transporter-mediated protection against thiopurine-induced hematopoietic toxicity. *Cancer Res*, 68, 4983–4989.

Kroetz DL, Yee SW, Giacomini KM (2010) The pharmacogenomics of membrane transporters project: Research at the interface of genomics and transporter pharmacology. *Clin Pharmacol Ther*, 87, 109–116.

Lander ES, Linton LM, Birren B, Nusbaum C, Zody MC, Baldwin J, Devon K, Dewar K, Doyle M, FitzHugh W, et al.; International Human Genome Sequencing Consortium (2001) Initial sequencing and analysis of the human genome. *Nature*, 409, 177–186.

Law M, Rudnicka AR (2006) Statin safety: a systematic review. *Am J Cardiol*, 97, 52C–60C.

Leabman MK, Huang CC, DeYoung J, Carlson EJ, Taylor TR, de la Cruz M, Johns SJ, Stryke D, Kawamoto M, Urban TJ, et al. (2003) Natural variation in human membrane transporter genes reveals evolutionary and functional constraints. *Proc Natl Acad Sci USA*, 100, 5896–5901.

Leighton JK, Degeorge J, Jacobson-Kram D, MacGregor J, Mendrick D, Worobec A (2004) Pharmacogenomic data submission to FDA: non-clinical case studies. *Pharmacogenomics*, 5, 507–511.

Lesko LJ (2008) The critical path of warfarin dosing: finding an optimal dosing strategy using pharmacogenetics. *Clin Pharmacol Ther*, 84, 301–303.

Lutz S, Weber P, Focke M, Faltin B, Hoffmann J, Müller C, Mark D, Roth G, Munday P, Armes N, et al. (2010) Microfluidic lab-on-a-foil for nucleic acid analysis based on isothermal recombinase polymerase amplification (RPA). *Lab Chip*, 10, 887–893.

Matsuo H, Takada T, Ichida K, Nakamura T, Nakayama A, Ikebuchi Y, Ito K, Kusanagi Y, Chiba T, Tadokoro S, et al. (2009) Common defects of ABCG2, a high-capacity urate exporter, cause gout: a function-based genetic analysis in a Japanese population. *Sci Transl Med*, 1, 5ra11.

McCathy MI, Abecasis GR, Cardon LR, Goldstein DB, Little J, Ioannidis JP, Hirschhorn JN (2008) Genome-wide association studies for complex traits: Consensus, Uncertainty and challenges. *Nat Rev Genet*, 9, 356–369.

Mitani Y, Lezhava A, Kawai Y, Kikuchi T, Oguchi-Katayama A, Kogo Y, Itoh M, Miyagi T, Takakura H, Hoshi K, et al. (2007) Rapid SNP diagnostics using asymmetric isothermal amplification and a new mismatch-suppression technology. *Nature Methods*, 4, 257–262.

Mitani Y, Lezhava A, Sakurai A, Horikawa A, Nagakura M, Hayashizaki Y, Ishikawa T (2009) Rapid and cost-effective SNP detection method: application of SmartAmp2 to pharmacogenomics research. *Pharmacogenomics*, 10, 1187–1197.

Molokhia M, McKeigue P, Curcin V, Majjeed A (2008) Statin induced myopathy and myalgia: time trend analysis and comparison of risk associated with statin class from 1991–2006. *PLoS ONE*, 3, e2522.

Mori Y, Nakatomi T (2009) Loop-mediated isothermal amplification (LAMP): a rapid, accurate, and cost-effective diagnostic method for infectious diseases. *J Infect Chemother*, 15, 62–69.

Morlighem JE, Harbers M, Traeger-Synodinos J, Lezhava A (2011) DNA amplification techniques in pharmacogenomics. *Pharmacogenomics*, 12, 845–860.

Nakagawa H, Tamura A, Wakabayashi K, Hoshijima K, Komada M, Yoshida T, Kometani S, Matsubara T, Mikuriya K, Ishikawa T (2008) Ubiquitin-mediated proteasomal degradation of non-synonymous SNP variants of human ABC transporter ABCG2. *Biochem J*, 411, 623–631.

Nakagawa H, Toyoda Y, Wakabayashi-Nakao K, Tamaki H, Osumi M, Ishikawa T (2011) Ubiquitin-mediated proteasomal degradation of ABC transporters: a new aspect of genetic polymorphisms and clinical impacts. *J Pharm Sci*, 100, 3602–3619.

Niemi M, Schaeffeler E, Lang T, Fromm MF, Neuvonen M, Kyklund C, Backman JT, Kerb R, Schwab M, Neuvonen PJ, Eichelbaum M, Kivistö KT. (2004) *Pharmacogenetics*, 14, 429–440.

Niemi M, Pasanen MK, Neuvonen PJ (2011) Organic anion transporting polypeptide 1B1: a genetically polymorphic transporter of major importance for hepatic drug uptake. *Pharmacol Rev*, 63, 157–181

Niemz A, Ferguson TM, Boyle DS (2011) Point-of-care nucleic acid testing for infectious diseases. *Trends Biotech*, 29, 240–250.

Piepenburg O, Williams CH, Stemple DL, Armes NA (2006) DNA detection using recombination proteins. *PLoS Biol*, 4, e204.

Rosenberg NA, Huasng L, Jewett EM, Szpiech ZA, Jankovic I, Boehnke M (2010) Genome-wide association studies in diverse populations. *Nat Rev Genet*, 11, 356–366.

Ruano G, Collins JM, Dorner AJ, Wang SJ, Guerciolini R, Huang SM (2004) Pharmacogenomic data submission to the FDA: clinical pharmacology case studies. *Pharmacogenomics*, 5, 513–517.

Rubenstein K (2010) Next-generation sequencing technologies: applications and markets. *Insight Pharma Reports*. Needham, MA, USA: Cambridge Healthcare Institute publication.

Sadee W (2008) Drug therapy and personalized health care: pharmacogenomics in perspective. *Pharm Res*, 25, 2713–2719.

Saiki R, Scharf S, Faloona F, Mullis K, Horn G, and Erlich H (1985) Enzymatic amplification of beta-globin genomic sequences and restriction site analysis for diagnosis of sickle cell anemia. *Science*, 230, 1350–1354

Saito H, Hirano H, Nakagawa H, Fukami T, Oosumi K, Murakami K, Kimura H, Kouchi T, Konomi M, Tao E, et al. (2006) A new strategy of high-speed screening and quantitative structure-activity relationship analysis to evaluate human ATP-binding cassette transporter ABCG2-drug interactions. *J Pharmacol Exp Ther*, 317, 1114–1124.

Sakurai A, Onishi Y, Hirano H, Seigneuret M, Obanayama K, Kim G, Liew EL, Sakaeda T, Yoshiura K, Niikawa N, et al. (2007) Quantitative SAR analysis and molecular dynamic simulation to functionally validate nonsynonymous polymorphisms of human ABC transporter ABCB1. *Biochemistry*, 46, 7678–7693.

Salerno RA, Lesko LJ (2004a) Pharmacogenomic data: FDA voluntary and required submission guidance. *Pharmacogenomics*, 5, 503–505.

Salerno RA, Lesko LJ (2004b) Pharmacogenomics in drug development and regulatory decision-making: the genomic data submission (GDS) proposal. *Pharmacogenomics*, 5, 25–30.

Saperstein DA, Nickerson JM (1991) Restriction fragment length polymorphism analysis using PCR coupled to restriction digests. *Biotechniques*, 10, 488–489.

Schubert C (2011) Cancer drugs find a companion with new diagnostic test. *Nat Med*, 17, 1157.

Schuetz JD, Connelly MC, Sun D, Paibir SG, Flynn PM, Srinivas RV, Kumar A, Fridland A (1999) MRP4: a previously unidentified factor in resistance to nucleoside-based antiviral drugs. *Nat Med*, 5, 1048–1051.

Stark K, Reinhard W, Grassl M, Erdmann J, Schunkert H, Illig T, Hengstenberg C (2009) Common polymorphisms influencing serum uric acid levels contribute to susceptibility to gout, but not to coronary artery disease. *PLoS One*, 4, e7729.

Stolk JN, Boerbooms AM, de Abreu RA, de Koning DG, van Beusekom HJ, Muller WH, van de Putte LB (1998) Reduced thiopurine methyltransferase activity and development of side effects of azathioprine treatment in patients with rheumatoid arthritis. *Arthritis Rheum*, 41, 1858–1866.

Tamura A, Wakabayashi K, Onishi Y, Takeda M, Ikegami Y, Sawada S, Tsuji M, Matsuda Y, Ishikawa T (2007) Re-evaluation and functional classification of non-synonymous single nucleotide polymorphisms of the human ATP-binding cassette transporter ABCG2. *Cancer Sci*, 98, 231–239.

The SEARCH Collaborative Group (2008). SLCO1B1 variants and statin-induced myopathy – a genomewide study. *New Eng J Med*, 359, 789–799.

Thompson PD, Clarkson P, Karas RH (2003) Statin-associated myopathy. *JAMA*, 289, 1681–1689.

Trepiccho WL, Williams GA, Essayan D, Hall ST, Harty LC, Shaw PM, Spear BB, Wang SJ, Watson ML (2004) Pharmacogenomic data submission to the FDA: clinical case studies. *Pharmacogenomics*, 5, 519–524.

Venter JC, Adams MD, Myers EW, Li PW, Mural RJ, Sutton GG, Smith HO, Yandell M, Evans CA, Holt RA, et al. (2001) The sequence of the human genome. *Science*, 291, 1304–1351.

Vincent M, Xu Y, Kong H (2004) Helicase-dependent isothermal DNA amplication. *EMBO Rep*, 5, 795–800.

Weinshilboum RM, Sladek SL (1980) Mercaptopurine pharmacogenetics: monogenic inheritance of erythrocyte thiopurine methyltransferase activity. *Am J Hum Genet*, 32, 651–662.

Weiss ST, McLeod HL, Flockhert DA, Dolan ME, Benowitz NL, Johnson JA, Ratain MJ, Giacomini KM (2008) Creating and evaluating genetic tests predictive of drug response. *Nat Rev Drug Discov*, 7, 568–574.

Wharam SD, Hall MJ, Wilson WH (2007) Detection of virus mRNA within infected host cells using an isothermal nucleic acid amplification assay: marine cyanophage gene expression within *Synechococcus sp. Viol J*, 4, 52.

Wilke RA, Lin DW, Roden DM, Watkins PB, Flockhart D, Zineh I, Giacomini KM, Krauss RM (2007) Identifying genetic risk factors for serious adverse drug reactions: current progress and challenges. *Nat Rev Drug Discov*, 6, 904–916.

Woodward OM, Kottgen A, Coresh J, Boerwinkle E, Guggino WB, Kottgen M (2009) Identification of a urate transporter, ABCG2, with a common functional polymorphism causing gout. *Proc Natl Acad Sci USA*, 106, 10338–10342.

Yao B, Huang H, Sun C, Wang Z, Fan Y, Chang Q, Li S, Xi J (2009) Quantitative analysis of zeptomole microRNAs based on isothermal ramification amplification. *RNA*, 15, 1787–1794.

Yates CR, Krynetski EY, Loennechen T, Fessing MY, Tai HL, Pui CH, Relling MV, Evans WE (1997) Molecular diagnosis of thiopurine S-methyltransferase deficiency: genetic basis for azathioprine and mercaptopurine intolerance. *Ann Internal Med*, 126, 608–614.

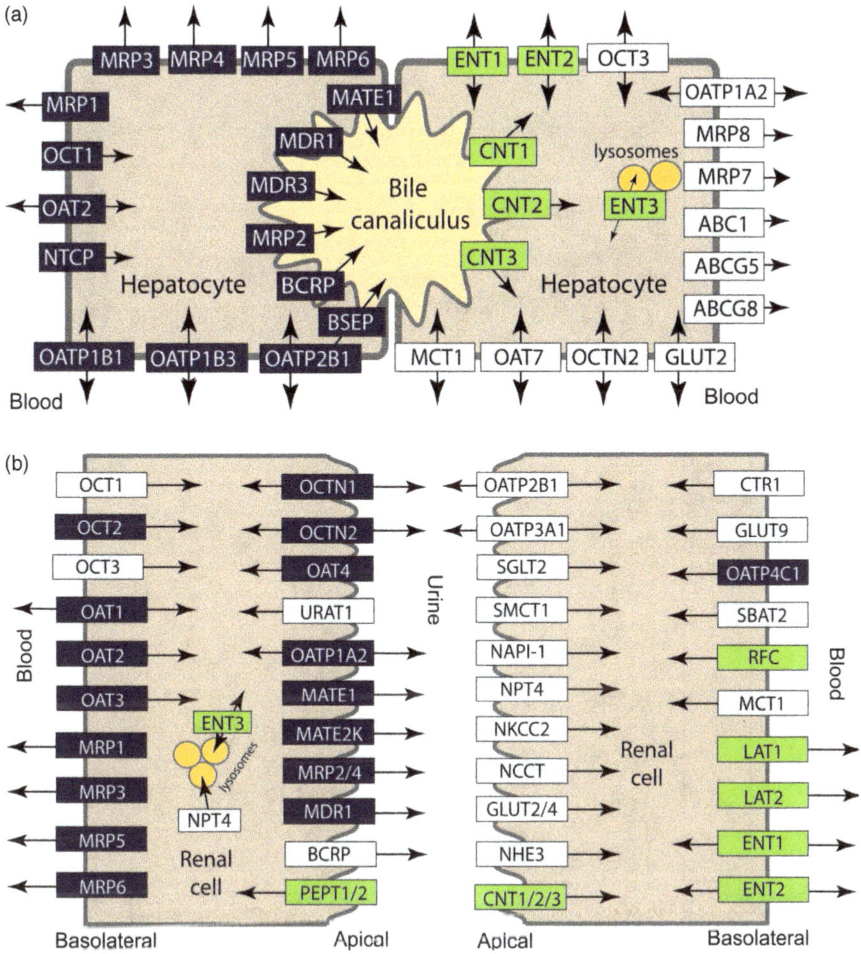

Figure 4.2 Membrane transporters that are the focus of the PMT project. Transporters on the basolateral and canalicular membranes of the hepatocyte (a) and on the apical and basolateral membranes of the renal epithelial cell (b) that play important roles in the uptake and efflux of drugs and other xenobiotics. The figure shows the key 60 membrane transporters in the kidney and/or liver which are the focus of the PMT project. Transporters that are well-studied, known to play many roles in the uptake or efflux of multiple clinical drugs and/or found at high expression levels are shaded dark. Transporters that recognize specific class of drugs are indicated by a medium shade. Many of the remaining transporters play specialized roles or may be drug targets (light shade).

Pharmacogenomics of Human Drug Transporters: Clinical Impacts, First Edition.
Edited by Toshihisa Ishikawa, Richard B. Kim and Jörg König

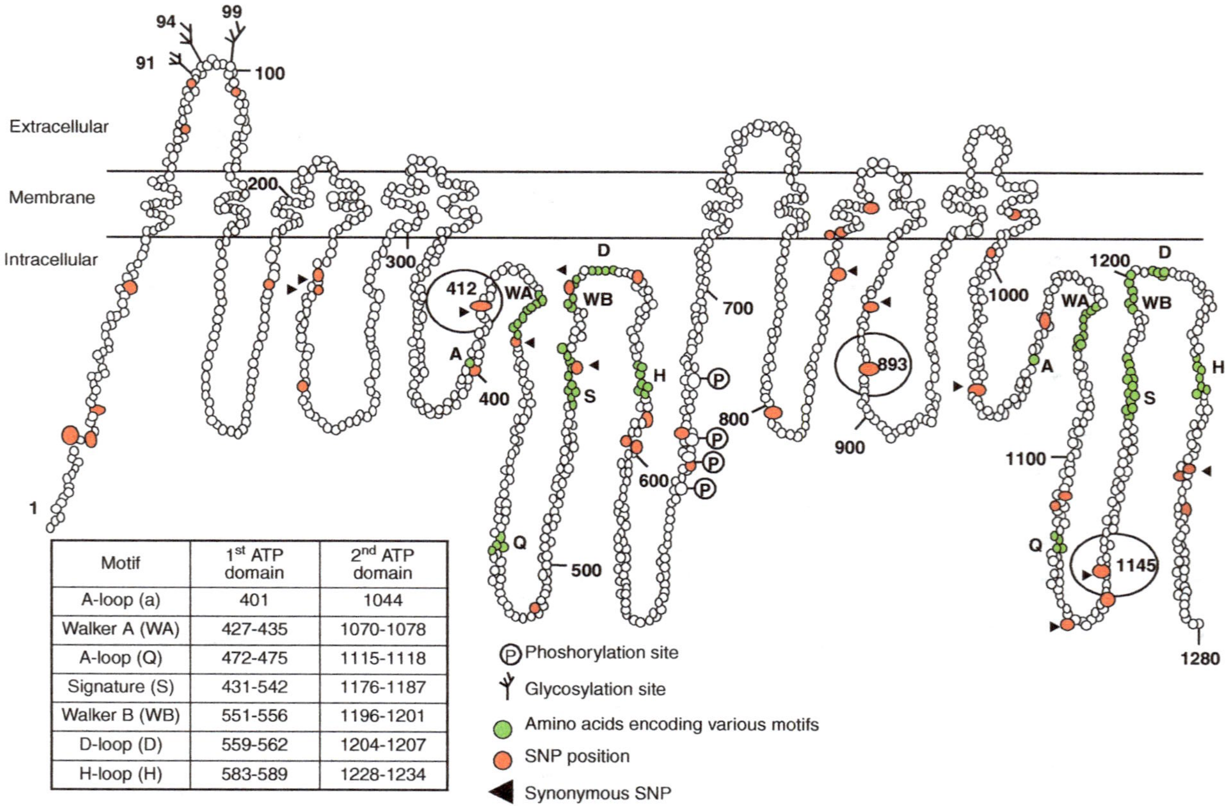

Motif	1st ATP domain	2nd ATP domain
A-loop (a)	401	1044
Walker A (WA)	427-435	1070-1078
A-loop (Q)	472-475	1115-1118
Signature (S)	431-542	1176-1187
Walker B (WB)	551-556	1196-1201
D-loop (D)	559-562	1204-1207
H-loop (H)	583-589	1228-1234

Figure 12.1 Hypothetical two-dimensional model of ABCB1 highlighting single-nucleotide polymorphsims and amino acid exchanges. Red circles represent the amino acids affected by SNPs. Amino acid residues affected by synonymous SNPs are marked with triangles. The amino acids, affected by the three major ABCB1 SNPs often related to certain diseases or drug disposition (c.1236C>T, p.412G=G; c.2677G>T/A, p.893A>S/T; and c.3435C>T, p.1145I=I) are circled. Phosphorylation and glycosylation sites are also shown. Adapted from Fung and Gottesman (2009)

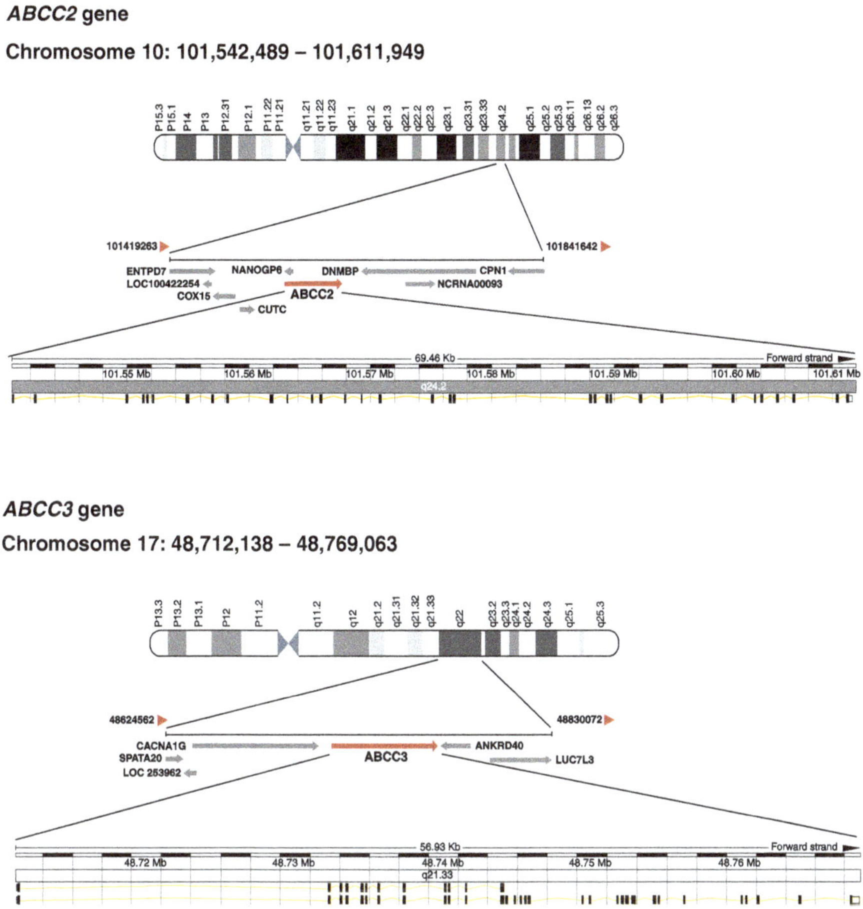

Figure 15.1 Genomic location of the human *ABCC2* gene on chromosome 10q24.2 and of the *ABCC3* gene on chromosome 17q22. For *ABCC2*, only one transcript has been identified (NM_000392.3). For *ABCC3*, two transcript variants exist: a longer variant (NM_003786.3, lower track) encoding full-length MRP3 and a shorter variant (NM_001144070.1, upper track) lacking multiple 3′ exons. The resulting isoform is much shorter and has a different C-terminus, as compared to full-length MRP3. Data and diagrams are from Ensembl (Homo sapiens version 61.37f) and NCBI (Genomic Reference Consortium build GRCh37.p2).

ABCC2 protein

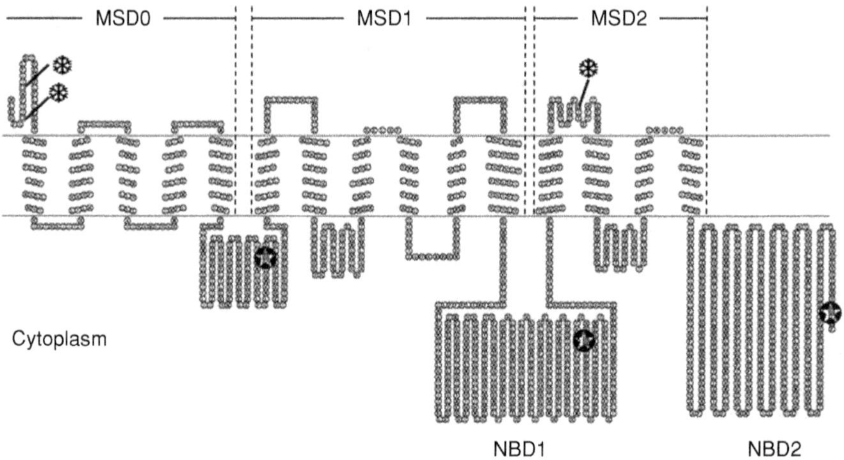

ABCC3 protein

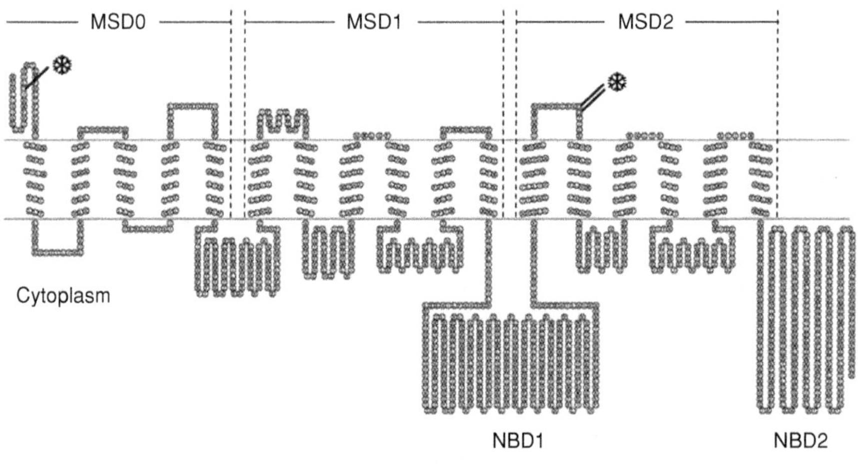

Figure 15.2 Topology models of the human ABCC2 and ABCC3 proteins. Model prediction was performed with Phobius (http://phobius.sbc.su.se) with the constraints that both nucleotide-binding domains are cytoplasmic. Topology models were drawn using TOPO2 (http://www.sacs.ucsf.edu/TOPO-run/wtopo.pl). MSD, membrane-spanning domain; ✶, potential *N*-glycosylation site; ✪, phosphorylation site.

Endogenous compounds and metabolites

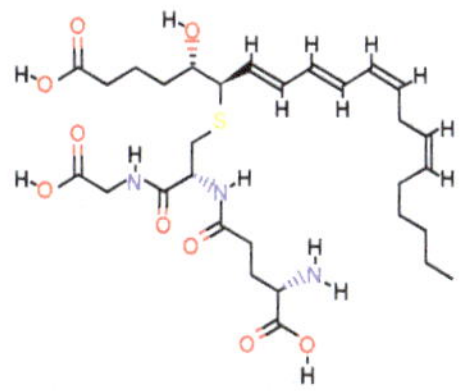

Leukotriene C_4

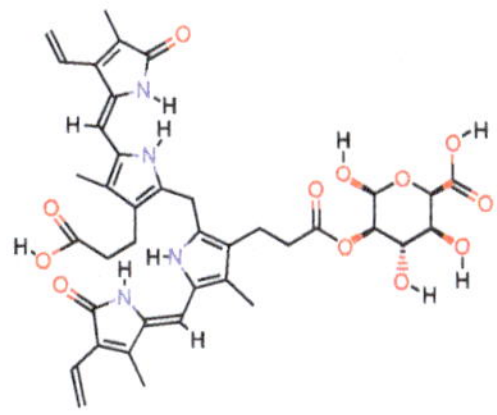

Bilirubin, monoglucuronoside

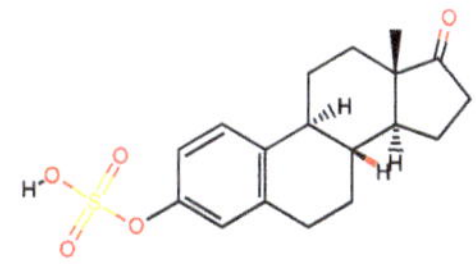

Estrone sulfate

Drugs

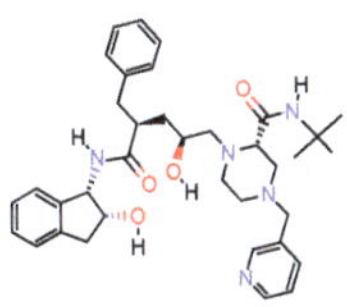

Indinavir

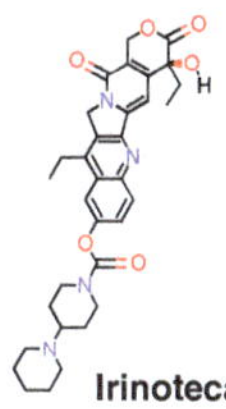

Irinotecan

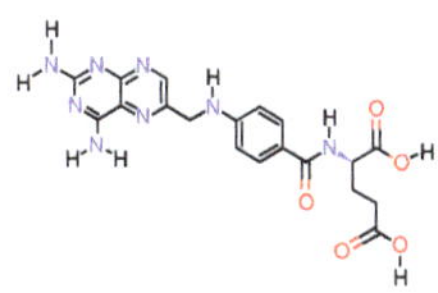

Methotrexate

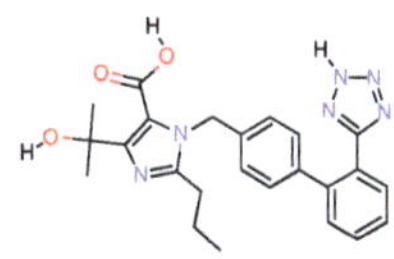

Olmesartan

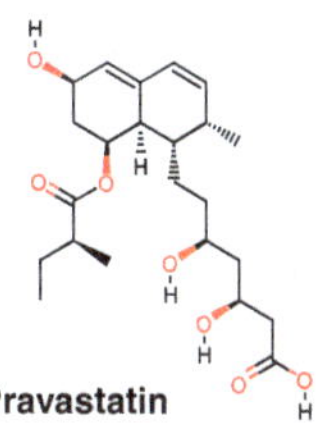

Pravastatin

Rifampin

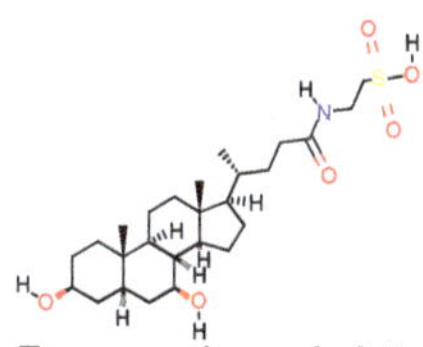

Tauroursodeoxycholate

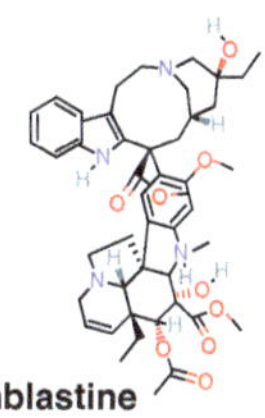

Vinblastine

Toxins

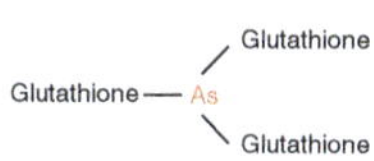

Arsenic triglutathione

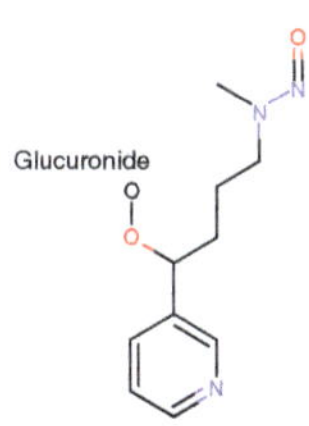

NNAL-*O*-glucuronide

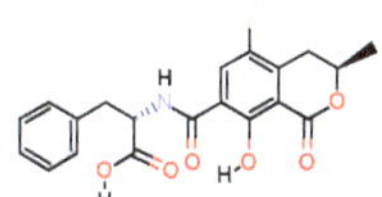

Ochratoxin A

Figure 15.3 Molecular structures of selected ABCC2 substrates. Structures are from the publicly available PubChem Compound database (http://www.ncbi.nlm.nih.gov/pccompound). NNAL, 4-(methylnitrosoamino)-1-(3-pyridyl)-1-butanol.

INDEX

Aac DNA polymerase, 113
Abacavir, 16, 50, 408
ABCB1 (P-glycoprotein), 9, 19, 47, 53–55, 88, 271–286
ABCB11 (BSEP), 92, 297–303
ABCC1 (MRP1), 57
ABCC2 (MRP2), 19, 26, 57, 58, 89, 53, 345–364, 372
ABCC3 (MRP3), 302, 345–370
ABCC2 and ABCC3 drug substrates, 348–350
ABCC4 (MRP4), 91, 119, 302, 365–380, 412
ABCC11 (MRP8), 387–389
ABCC12 (MRP9), 387
ABCG2 (BCRP), 18, 25, 48, 56, 57, 372, 404
Adefovir, 370
ADME, 13, 14, 15, 28, 30, 31, 39
Acquired intrahepatic cholestasis, 299
Acute myeloid leukemia, 357, 369
Alachidonate metabolite transport, 375
Allele-specific hybridization, 111
Allele-specific PCR, 111
Amphipathic molecules, 296
AmpliChip, 111
AmpliChip CYP450, 407
Anaplastic lymphoma kinase (ALK), 406
Anticoagulation, 41, 50
Antiviral drugs, 351, 356, 370
Apical membrane, 367
Apocrine gland, 390
Aquaporine, 295
Array-based GWAS, 404
Arylhydrocarbon receptor (AhR), 297, 312
ATP-binding cassette (ABC) transporter, 2, 8, 53, 54, 57, 110, 345–370
Axillary osmidrosis, 390
Azathioprine (AZA), 118
Azidothymidine (AZT), 370, 378
AZT triphosphate, 380

Basolateral membrane, 352, 209, 212, 367
Benign recurrent intrahepatic cholestasis, 299
Benzylpenicillin, 146, 147, 161
Bile salt, 142–145, 149, 159, 160, 375
Bile flow, 29
Bile formation, 295, 345
Bile salt, 295
Bilirubin, 26, 145, 147, 158, 161
Bilirubin glucuronoside, 345, 351
Bioavailability, 54
Biomarker, 16, 30, 65
β-lactam, 223
Blockbuster, 413
Blood–brain barrier, 56, 272, 280, 284, 312, 367
Bosentan, 300
BRAF, 406
Breast cancer risk, 393
Breast milk, 313
BRIC, 299
Brush-border membrane, 209–212, 215
BSEP (ABCB11), 92, 296
BSEP deficiency syndrome, 299

cAMP, 373, 374
Camptothecine, 323–325
Canalicular export system, 298
Canalicular membrane, 296, 351
Cancer chemotherapy, 372
Carbamazepine, 50, 63
Carnitine, 171, 193–195, 197–200
Carnitine transporter 1 (OCTN1), 171, 193–195, 197–199

Pharmacogenomics of Human Drug Transporters: Clinical Impacts, First Edition.
Edited by Toshihisa Ishikawa, Richard B. Kim and Jörg König

Carnitine transporter 2 (OCTN2), 171, 193–195, 197–200
CCAAT box, 312
CCD camera, 121
CD34+, 369
Cefazolin, 371
Cefmetazole, 371
Celecoxib, 46
Cell cycle control, 302
Cephalosporins, 373
Cetuximab, 50, 62
cGMP, 374
Chemical fragmentation code (CFC), 410
Chemotherapy, 24, 395
Cholestasis, 300, 375
Cholesterol, 24, 295
Cholyglycylamide-fluorescein, 298
Cisplatin, 213, 215, 216, 349
Clopidogrel, 16, 43, 50, 110, 408
CNT (SLC28A), 243, 244
Coding and noncoding regions, 73, 76
Combination chemotherapy, 331
Companion diagnostics, 406
Conjugated hyperbilirubinemia, 26, 345, 353
Constitutive androstane receptor (CAR), 297
CPA, 405
CpG island, 312
Crizotinib, 406
Crohn's disease, 271, 370
Cyclic nucleotide export, 373, 374
Cytochrome P450, 13, 40, 44
CYP2A6, 45
CYP2B6, 45
CYP2C8, 46
CYP2C9, 16, 40, 41–43, 50, 401
CYP2C9*2, 41, 42, 44, 46
CYP2C9*3, 41, 42, 44, 46
CYP2C19, 16, 40, 44, 46
CYP2D6, 16, 40, 51, 52, 401
CYP3A4, 55
CYP3A5, 47, 55
Cost-effective, 402

dbSNP, 109
Dehydroepiandrosterone 3-sulfate, 387
Dense granules, 369
Detoxification, 345, 350, 351
Diagnosis, 110
Dideoxynucleotide triphosphate (ddNTP), 112
Diflomotecan, 325
Digoxin, 19, 47, 55, 271, 272, 273, 275, 277, 278
Disease mechanism, 26
Draft guidance on clinical pharmacogenomics, 40
Drug development, 13, 14, 30
Drug disposition, 18, 27
Drug efficacy, 24, 31, 53
Drug-induced cholestasis, 300
Drug-induced toxicity, 24, 134
Drug interaction, 23, 39, 40, 50, 51 53, 54, 56, 59, 60, 65
Drug-metabolizing enzymes, 13, 40, 109
Drug resistance, 345, 349
Drug response, 54, 65, 133
Dubin-Johnson syndrome, 26, 345, 353–355

Earwax, 390
Edaravone, 373
Efficacy, 24, 64
EGFR, 62
Enalapril, 224
ENT (SLC29A), 243, 247
Enzymatic cleavage, 111, 112
Epilepsy, 272, 277, 280, 281, 282, 352
ERAD, 411, 392
Erlotinib, 328–329
Estradiol-17β-glucuronide, 300, 372
Ethnic difference, 411
Excretion of anionic conjugates, 350, 351, 357
Exposure-response, 65
Extensive metabolizer (EM), 40

Familial intrahepatic cholestasis, 298
Farnesoid X receptor (FXR), 296
FDA, 15, 40–42, 50, 51, 54, 56, 58, 59, 62, 64
FDA adverse event reporting system, 41
FDA clinical pharmacogenomic guidance, 64
FDA critical path, 29, 410
Flavopiridol, 330–331
5-Fluorouracil, 396
Function, 171, 175, 178, 183–193, 197–200
Functional genomic studies, 87
Functional validation, 323, 410

FXR (NR1H4), 296, 297, 302
FXR*1b allele, 297

Ganciclovir, 370, 372
Gefitinib, 56, 327–328
Gene expression, 125, 128, 129
Genetic polymorphisms, 14, 39, 216, 375
Genetic variants, 226, 233, 353–356
Genetic variation, 127, 129, 131
1000 Genome project consortium, 401, 402
Genome-wide association studies (GWAS), 6, 13, 17, 21, 24, 53, 110, 401, 403
Genotype to phenotype studies, 27, 74, 99–100, 353, 356
Glibenclamide, 300
GLUT9 (SLC2A9), 404
Gout, 313, 404

HAD, 405
Haplotype, 319, 404
HapMap, 109
Hematopoietic cells, 369
Hepatocyte, 25, 295, 351
Hepatocyte nuclear factor (HNF), 297
Hepatotoxicity, 301, 357
HER1/HER2 inhibitors, 301
Herpes simplex virus (HSV), 372
Heterogenity, 39
HGVbase, 109
HIV patients, 380
HLA-A*3101, 64
HLA-B*1501, 50
HLA-B*1502, 16, 63, 64
HLA-B*5701, 16, 50
HMG-CoA reductase inhibitor, 24, 146, 150, 157, 160
Homeobox genes, 302
HSV-TK modified cells, 372
α-HTBZ, 51, 52
β-HTBZ, 51, 52
Huntington's disease, 51
Hyperbilirubinemia, 26

Imatinib, 327
Inflammatory bowel disease (IBD), 117, 119, 379
Inflammatory cholestasis, 299
Interindividual variability, 28, 133, 134
Interferon-γ, 300
Interleukin-1β, 300
Intermediate metabolizer (IM), 40
International transporter consortium (ITC), 4, 410
Intestine, 312, 351
Intracellular trafficking, 235
Intronic SNP, 379
Investigational new drug (IND), 406
Invader assay, 111, 112
Irinotecan, 50, 325–326, 356, 408
Isothermal DNA amplification, 113, 114
Ivermectin, 271, 275

JSNP, 109

Kidney, 209–212, 215
Knockout mice, 5, 212, 225, 347, 352, 353
KRAS, 50, 63

Labeling, 40, 42, 50, 53, 54, 57, 60, 62, 63, 65
Lab-on-chip, 406
Lactation, 313
LAMP, 405
Leucovorin, 371
Leukemia, 272, 285, 286
Liver, 24, 141–143, 149, 159–161, 209–211, 215, 218, 219, 312, 345
LTB_4, 375
LTC_4, 345
Lymphoblastic leukemia, 372

MALDI-TOF MS, 112
Mammary gland, 313
Mass-based detection, 112
MATE1 (SLC47A1), 57, 61, 62, 94, 95, 209–219
MATE2-K (SLC47A2), 61, 62, 209–214, 216–219
MDR1 (ABCB1), 53, 54, 271
6-Mercaptopurine (6-MP), 50, 372
Metformin, 20, 61, 62, 99, 174, 178–192, 212–215, 218, 219
Methotrexate, 330, 356, 371
Microarray, 111
Mismatch-binding protein (MutS), 115, 116
Molecular dynamic (MD) simulation, 411
Molecular structure, 346–349
MRP2 (ABCC2), 19, 89, 345–370
MRP3 (ABCC3), 345–370
MRP4 (ABCC4), 91, 119, 365, 412

MRP4 (ABCC4) exon 1a, 365
MRP4 (ABCC4) exon 1b, 365
MRP8 (ABCC11), 387–389
MRP9 (ABCC12), 387
Multidrug and toxic compound extrusion, 94, 95, 209, 212
Multidrug resistance, 313, 349
Myopathy, 24, 156, 158, 404

NASBA, 405
National Human Genome Research Institute, 53
NCBI, 109
Nephrotoxicity, 370
Next generation sequencing (NGS), 402
9-nitrocamptothecin (9-NC), 326–327
NSAID, 375
NTCP (SLC10A1), 296
Nonsense-mediated decay, 365
Non-small cell lung cancer, 406
Nonsynonymous SNP, 378
Novel organic cation transporter, 93, 94
Nucleoside anticancer drugs, 248
Nucleoside antiviral drugs, 252
Nucleoside monophosphate, 365
Nucleoside transporters, 97, 98, 99, 118, 243
Nucleoside transporter expression, 254
Nucleoside transporter genetic variability, 256
Nucleotide analogs, 365

OAT1 (SLC22A6), 53, 57, 129–131
OAT2 (SLC22A7), 53, 57
OAT3 (SLC22A8), 129–132
OATP1A2 (SLCO1A2), 96, 125
OATP1B1 (SLCO1B1), 9, 18, 53, 57, 59, 296, 353, 404
OATP1B1*5, 23, 146, 150–153, 157–159, 161
OATP1B3 (SLCO1B3), 53, 296, 353
OATP2B1 (SLCO2B1), 296
OCT1 (SLC22A1), 9, 17, 20, 53, 57, 61, 62, 92, 93, 172–176, 178, 182–192, 209
OCT2 (SLC22A2), 20, 53, 61, 62, 92, 93, 172, 174–178, 180, 182–184, 186–193, 212
OCT3 (SLC22A3), 92, 93, 172, 174–178, 181, 183–184, 187–188, 190, 192–193
OCTN1 (SLC22A4), 93, 94, 171, 193–195, 197–199
OCTN2 (SLC22A5), 93, 94, 171, 193–195, 197–200
Oligopeptide transporter, 223
Olmesartan, 371
Organic anion, 142, 145
Organic anion transporter 1 (OAT1), 128, 212, 213
Organic anion transporter 3 (OAT3), 129, 212, 213
Organic anion-transporting peptide, 125, 404
Organic cation transporters, 53, 61, 92, 93, 212
Oseltamivir, 224, 368
OSTα/OSTβ, 302

Pain, 284
Panitumumab, 50, 62
PCR-RFLP, 111
Personalized medicine, 401, 402, 413
PEPT1 (SLC15A1), 223
PEPT2 (SLC15A2), 229
PGE_2, 375
P-glycoprotein (ABCB1), 9, 57, 88, 271, 275
PDZ-binding motif, 235
PFIC2, 298, 299, 375
PFIC3, 375
PGRN project, 409
Pharmacogenetics, 40, 42, 65
Pharmacogenomics, 39, 109
Pharamcokinetics, 39, 109, 128, 132, 183–185, 188–193, 209, 214, 215, 218, 219, 401
Pharmacodynamics, 39, 109, 209, 219, 40!
Phenotype ofABCC2 genetic variants, 353–356
Phenotype of ABCC3 genetic variants, 356, 357
Phosphoinositide-3-kinase, 296
9-(2-Phosphonyl methoxyethyl)adenine (PMEA), 370
Physiologically-based pharamacokinetics (PBPK), 65
PKI166, 301
Platelet, 369
Plavix, 44

PMEA-derived drugs, 372, 378
PMT project, 110, 409
Point-of-care, 110, 413
Polymorphism, 39–42, 52–54, 57–62, 65, 171, 173, 175, 178, 182, 184–193, 197–200
Pooling-of-blood, 404
Poor metabolizer (PM), 40, 43
Positron emission tomography (PET), 234
PPAR, 312
Pravastatin, 146, 148, 151–158, 160, 350
Pregnancy, 313
Pregnane X receptor (PXR), 297
Primary biliary cirrhosis, 299
Probenecid-sensitive, 370
Progesterone sulfate, 300
Prognosis, 39
Proof-of-concept, 409
Prostanoid transport, 375
Protein kinase A/CREB, 375
Protein stability, 109
Prostate cancer, 367
Proton (H+) gradient, 209, 212, 226

Q141K variant (ABCG2), 404
QSAR, 56, 410
QTc prolongation, 51

RAM, 405
RAS/MAP kinase pathway, 62
RCA, 405
Reactive oxygen species, 301
Regulation, 15, 30, 352
Regulatory, 39–41, 64, 65
Renal excretion, 215, 351
Retinoid X receptor (RXR), 296
RFLP, 405
Riboflavin, 313
Rifamycin, 300, 351
Rifampicin, 300
Ritonavir, 313
Rosuvastatin, 332
RPA, 405
RXR (NR2B1), 296, 300

SEARCH collaboration group, 403
S-enantiomer, 41
SHP-1 (NR0B2), 296
Simvastatin, 24, 404
Single nucleotide polymorphism (SNP), 3, 21, 387
SJS-TEN, 63
SLC superfamily, 141
SLC transporter, 53, 54, 57, 59
SLC28 genes, 243, 244
SLC28A (CNT), 118, 243, 244
SLC29A (ENT), 118, 243, 247
SLC29 genes, 243, 247
SLCO1B1 (OATP1B1), 9, 18, 48, 49, 60, 296, 297, 403, 404
SLCO1B3 (OATP1B3), 296
SLCO2B1 (OATP2B1), 296, 297
SLC2A9 (GLUT9), 404
SLC10A1 (NTCP), 296
SLC15A1 (PEPT1), 223
SLC15A2 (PEPT2), 229
SLC22A1 (OCT1), 9, 20, 61, 92, 93, 172–176, 178, 182–192
SLC22A4 (OCTN1), 193–195, 198
SLC22A5 (OCTN2), 193, 195, 196, 198, 199
SLC47A (MATE), 61, 209, 210, 212, 214, 216–219
SLC47A1 (MATE1), 57, 61, 62, 94, 95, 209, 210, 212, 214, 216–219
SLC47A2 (MATE2-K), 61, 62, 209, 210, 212, 214, 216–219
Small-heterodimer partner 1, 296
SMART, 405
SmartAmp method, 112, 405
Sodium-taurocholate cotransporting polypeptide, 296
Solute carrier (SLC) superfamily, 2, 7, 110
Species difference, 211
Statins, 24, 56, 331, 332
Steven-Johnson syndrome (SJS), 63
Strand-displacing DNA polymerase, 114
Sulfasalazine, 28, 315, 339
Sunitinib, 320, 329
SYBR Green I, 116
Synonymous SNP, 380

Tamoxifen, 110
TaqMan method, 112
Tariquidar, 313
Taurocholate, 296
Taurolithocholate, 298
TCN monophosphate, 370

Tenofovir, 370
Tetrabenazine, 50, 51
6-Thioguanine (6-TG), 372
6-Thioguanine nucleotide (6-TGN), 117
Thiopurines, 372
Thiopurine methyl transferase (TPMT), 50, 117, 40!, 411
Thymidine kinase (TK), 372
TMA, 405
TOPFRAG program, 410
Topotecan, 325
Toxicity target, 408
Transporter, 27, 28, 39, 40, 53, 54, 57, 59, 61, 62, 65
Transporter substrates, 126, 130
Tricyclic 7-deazapurine (TCN), 370
TRITON-TIMI 38, 44
Troglitazone, 301
Tubular secretion, 209, 211, 213–215, 218, 219
Tyrosine kinase inhibitors, 333

UGT1A1, 16, 26, 50
Uric acid, 313, 404
Ursodeoxycholate, 300

Vesicular monoamine tarnsporter 2 (VMAT), 16, 50
Vitamin B, 313
VKORC1, 41–43, 50, 401

Warfarin, 16, 41, 50, 110, 408

Xenazine, 52, 55

Zidovudine, 370

www.ingramcontent.com/pod-product-compliance
Lightning Source LLC
LaVergne TN
LVHW011828200926
848484LV00002B/338